P9-ELP-936

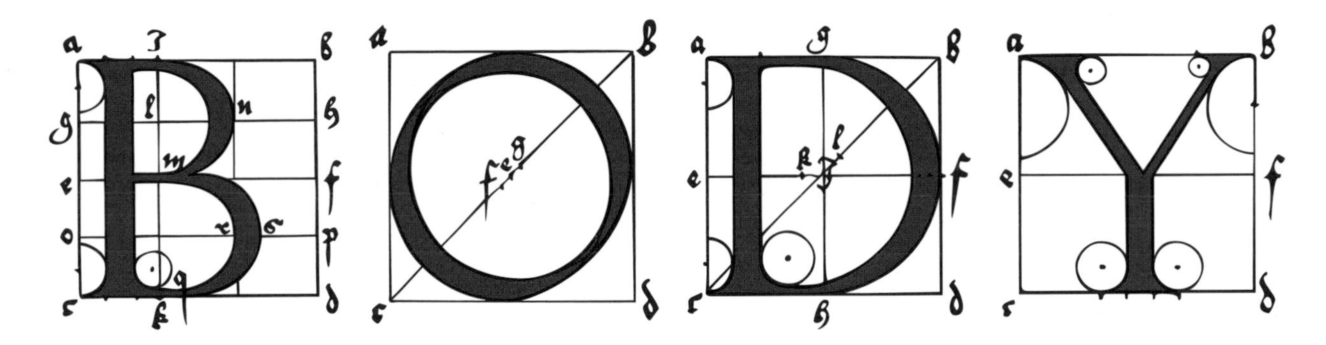

THE COMPLETE HUMAN

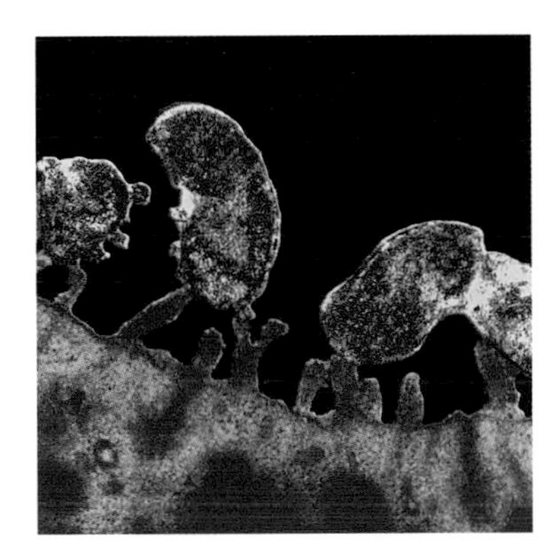

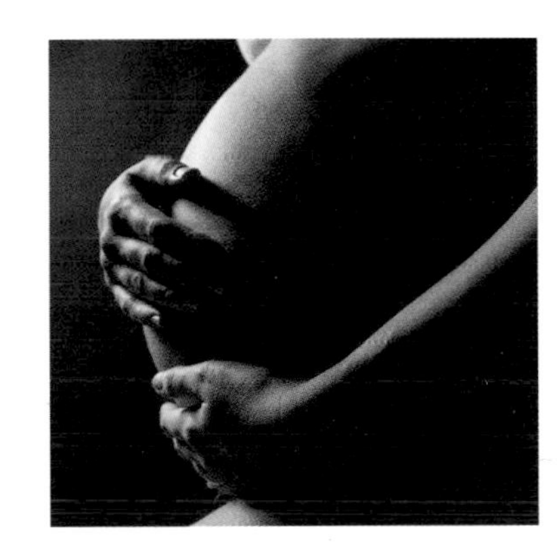

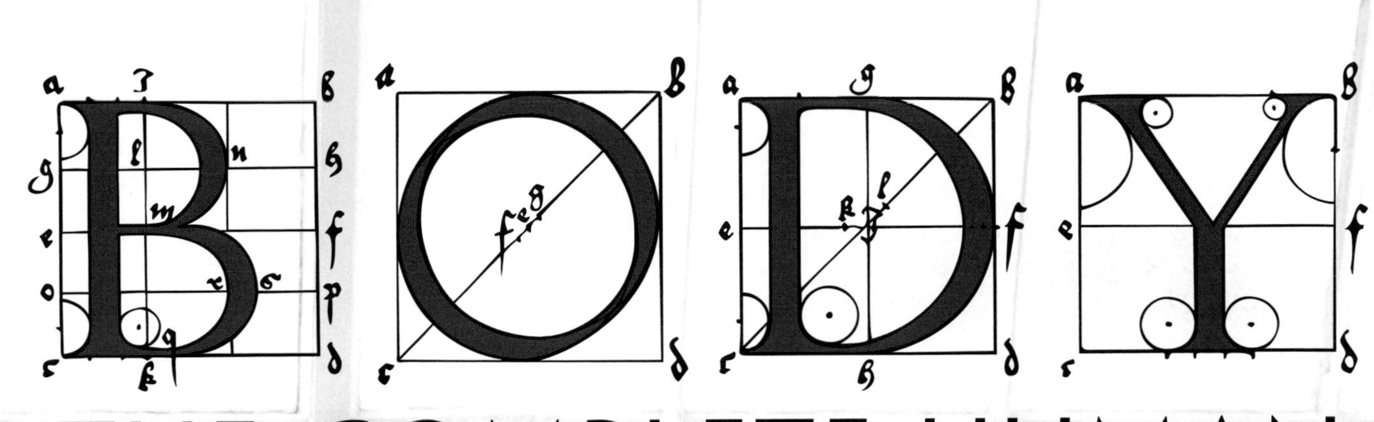

THE COMPLETE HUMAN

HOW IT GROWS, HOW IT WORKS, AND HOW TO KEEP IT HEALTHY AND STRONG

NATIONAL GEOGRAPHIC

WASHINGTON, D.C.

CONTENTS

LIBRARY OF CONGRESS CATALOGING-IN-PUBLICATION DATA

Body : the complete human / foreword by Richard Restak ; text by Patricia Daniels ... [et al.].
p. cm.
Includes index.
ISBN 978-1-4262-0128-8 (hc) -- ISBN 978-1-4262-0185-1 (deluxe)
1. Human physiology--Popular works. 2. Body, Human--Popular works. I. Daniels, Patricia, 1955-
QP38.B597 2007
612--dc22

2007026161

ISBN 978-1-4262-0128-8; deluxe edition 978-1-4262-0185-1
Printed in the United States of America

Founded in 1888, the National Geographic Society is one of the largest nonprofit scientific and educational organizations in the world. It reaches more than 285 million people worldwide each month through its official journal, NATIONAL GEOGRAPHIC, and its four other magazines; the National Geographic Channel; television documentaries; radio programs; films; books; videos and DVDs; maps; and interactive media. National Geographic has funded more than 8,000 scientific research projects and supports an education program combating geographic illiteracy.

For more information, please call 1-800-NGS LINE (647-5463) or write to the following address:
National Geographic Society
1145 17th Street N.W.
Washington, D.C. 20036-4688 U.S.A.
Visit us online at www.nationalgeographic.com/books

For information about special discounts for bulk purchases, please contact National Geographic Books Special Sales: ngspecsales@ngs.org. For rights or permissions inquiries, please contact National Geographic Books Subsidiary Rights: ngbookrights@ngs.org

The medical illustrations in this book were created by Scientific Publishing Ltd., a research and development publishing company specializing in high-quality medical and scientific illustrations and educational products.

The information published in this book is provided for informational purposes only and is not a substitute for professional medical advice. It is not intended to replace the services of a physician, to diagnose a medical and/or health condition, or to recommend a course of treatment. If a reader suspects that he or she has a problem, a healthcare provider should immediately be consulted.

BOARD OF ADVISERS

WILLIAM R. AYERS, M.D.
Professor of Medicine, Emeritus
Georgetown University
School of Medicine
Washington, D.C.

ROBERT M. BRIGGS, M.D.
Chairman, Department of Plastic and
Reconstructive Surgery (retired)
St. Barnabus Medical Center
Livingston, New Jersey

ELIZABETH BROWNELL, M.D.
Family Practice Physician
Waukesha Memorial Hospital, Division
of Family and Emergency Medicine
Waukesha, Wisconsin

TED M. BURNS, M.D.
Associate Professor of Clinical Neurology
University of Virginia,
School of Medicine
Charlottesville, Virginia

ROBERT A. CHASE. M.D.
Emile Holman Professor of Surgery
and Professor of Anatomy, Emeritus
Stanford University, School of Medicine
Stanford, California

D. ROBERT DUFOUR, M.D.
Consultant Pathologist and
Attending Physician, Veterans Affairs
Medical Center, Liver Clinic
Washington, D.C.

DAVID E. FLEISCHER, M.D.
Professor of Medicine
Mayo Clinic College of Medicine
Scottsdale, Arizona

JEAN L. FOURCROY, M.D., PH.D., M.P.H.
Medical Officer, Center for Devices
and Radiological Health (retired)
United States Food and
Drug Administration
Washington, D.C.

DONALD S. KARCHER, M.D.
Professor of Pathology; Acting Chair,
Department of Pathology; and Chief,
Chemistry and Flow Cytometry
George Washington University
Medical Center
Washington, D.C.

JOHN E. MORLEY, M.B., B.CH.
Dammert Professor of Gerontology
and Director, Division of Geriatric
Medicine and Endocrinology
Saint Louis University Medical Center
St. Louis, Missouri

POLLY E. PARSONS, M.D.
Professor of Medicine, Chief of
Pulmonary and Critical Care Medicine, and
Chair of the Department of Medicine
University of Vermont, College of Medicine
Burlington, Vermont

ROBERT S. ZOHLMAN, M.D.
Clinical Professor of Medicine (retired)
University of California, San Francisco
San Francisco, California

AUTHORS

RICHARD RESTAK, M.D.
Foreword

STEFAN BECHTEL
Epilogue

PATRICIA DANIELS
Chapters One, Five, Six, Nine, and Thirteen

SUSAN TYLER HITCHCOCK
Chapter Four

TRISHA GURA, PH.D.
Chapters Ten and Eleven

LISA STEIN
Chapters Two and Three

JOHN THOMPSON
Chapters Seven, Eight, and Twelve

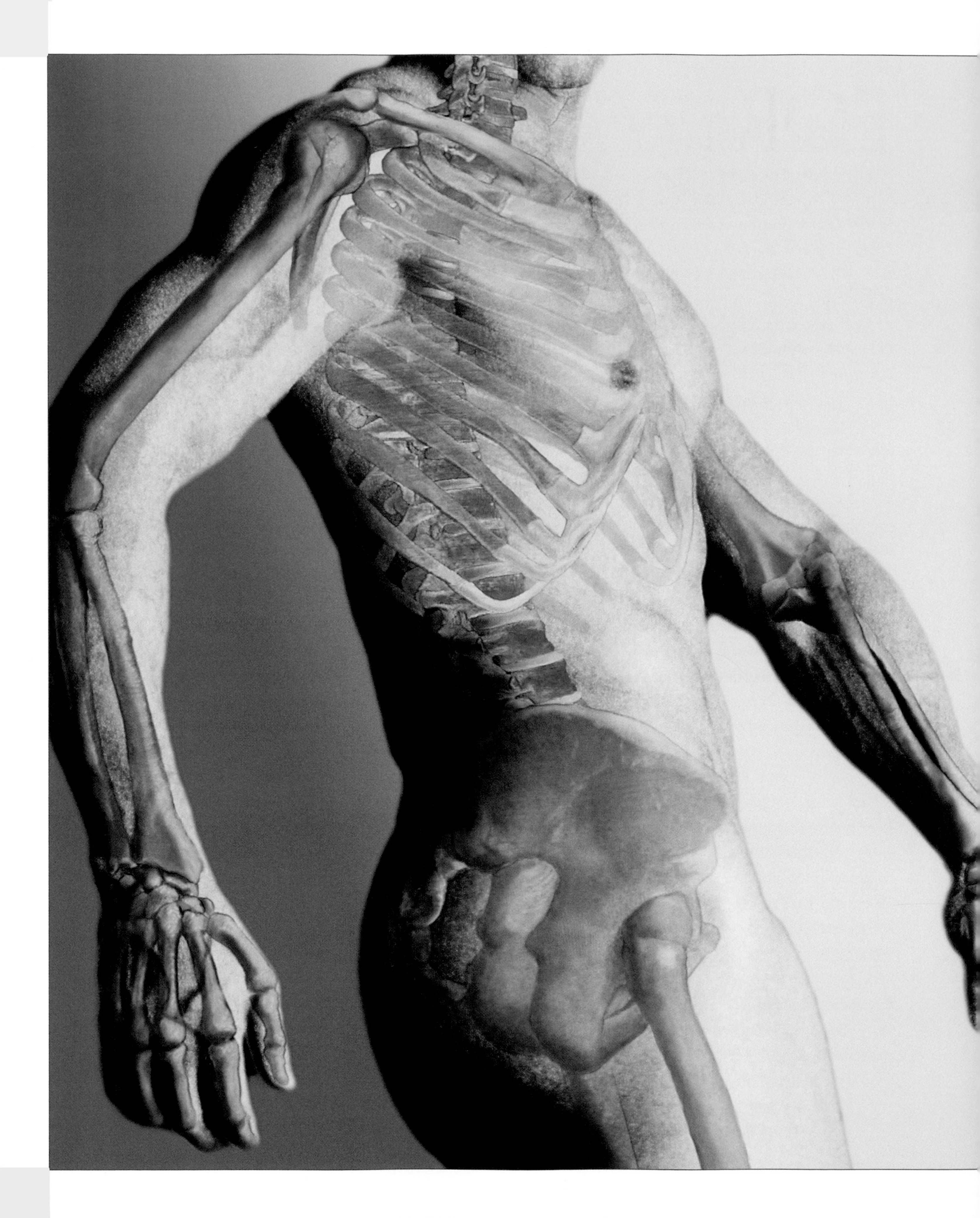

FOREWORD BY RICHARD RESTAK, M.D.

Body: The Complete Human successfully meets three challenging tasks. First, it provides new scientific information about the human body—much of it unknown until very recently—in ways that can be understood by the general public. Second, it illustrates how this knowledge relates to everyday life. Finally, it frankly acknowledges that our knowledge about the body is far from complete and will continue to evolve.

Body meets these challenges through the ingenious interweaving of the main text—a model of clarity, precision, and readability—with relevant sidebars and factboxes that provide both historical perspective and practical examples. As a result readers are rewarded with that rarest of publishing creations: a book that integrates science and humanism.

For example, the chapter detailing the kidney's extraordinary daily filtering capacity (more than 300 times the body's blood volume) contains an illuminating sidebar explaining the use of urine testing as a means of detecting "doping" in athletes. In the discussion about the skin, the largest organ of the body, we discover why there are no known ways of avoiding age-related skin changes and that given the inevitable we should perhaps take "a kinder view of wrinkles, accepting them as well-earned character lines, each with a story to tell, rather than as reminders of passing time." (See what I mean about humanism?)

When reading about blood we learn that our evolving knowledge about the body sometimes requires backward as well as forward vision: Medicinal bloodletting via the bloodsucking bite of the three-jawed segmented leech, which fell out of favor as a treatment in the 19th century, has now been resurrected and approved by the FDA for use in skin grafts.

The chapter on bones documents how minute electric currents and ultrasound waves can reduce healing time in arm and shinbone fractures by 35–45 percent. A survey of the bones of the skull segues gracefully into why most blows to the head do not result in concussion and why personality cannot be assessed, as claimed by 19th-century phrenologists, by palpating the skull in search of telltale "bumps." And for fans of murder mysteries comes this tidbit about the hyoid bone, a tiny U-shaped structure at the base of the tongue: "In forensic medicine the hyoid is examined to detect whether the deceased has been murdered—it sometimes snaps when a person is strangled or choked."

The section on the spine provides the clearest explanation I've ever encountered outside of a medical textbook about discs in the neck and back, how injuries occur, and what treatment options are available. And thanks to the entry on the knee joint, sports fans will understand why this powerful joint is injured so frequently by athletes and what can be done to repair such potentially career-ending injuries.

Thanks to its breadth of coverage, clarity of writing and marvelous integration of information from sources as varied as Renaissance art (Did you know Leonard da Vinci spent years studying the human body, dissecting more than 30 cadavers and producing hundreds of anatomical drawings?) to gymnastics (women's hips have a wider range of movement than men's—resulting in greater flexibility for splits and other maneuvers), *Body* will prove accessible and enjoyable to a broad spectrum of readers.

But discover this book's excellence for yourself by opening it to any chapter that interests you. I'm confident that after reading a page or two you'll agree that a lot of hard work and dedication has gone into the preparation of *Body*, a book whose overall excellence is likely to remain unsurpassed for a very long time to come.

Recreated from a scanned human data, this visualization shows how the skeleton supports and shapes the body.

HOW TO USE THIS BOOK

Each of the 13 chapters found in *Body: The Complete Human* is overflowing with information. A comprehensive, fact-filled narrative describes major components and systems, while complementary reference elements pepper every page with facts and figures about the body, descriptions of ailments and maladies, the stories behind medical breakthroughs, and the pioneering scientists and physicians who increased our understanding of how the body grows and functions.

❶ FACTBOX: Features fascinating facts and figures about the body
❷ SUBSECTION: Divides a chapter into subsections on the major components and duties of each body system
❸ DIAGRAM: Reveals the inner workings and hardworking anatomy of the body's systems and organs
❹ BIOGRAPHY SIDEBAR: Takes an in-depth look at the pioneering men and women who shaped scientific beliefs and medical practices
❺ CHAPTER/SUBSECTION DIRECTORY: Finds your place in a chapter. Located on right-hand page for easy reference
❻ WEBLINK: Links to the National Geographic Human Body website—nationalgeographic.com/humanbody—to connect with more interactive features on systems (see page 400 for more information)
❼ CROSS-REFERENCES: Makes valuable connections to related information in other chapters throughout the book
❽ BREAKTHROUGH SIDEBAR: Chronicles amazing discoveries that changed how the human body is understood
❾ CHAPTER GLOSSARY: Defines key terms found within each chapter
❿ KEEPING HEALTHY SIDEBAR: Documents the smart practices and tactics to keep everything running smoothly
⓫ TABLE: Organizes key information about the body and its systems into an easy-to-access format
⓬ WHAT CAN GO WRONG SIDEBAR: Explains what happens to the body when an injury or sickness occurs
⓭ HISTORY SIDEBAR: Tells the stories behind practices and beliefs about the body from throughout history
⓮ COMPARATIVE IMAGES: Aligns related pictures for quick, easy comparison

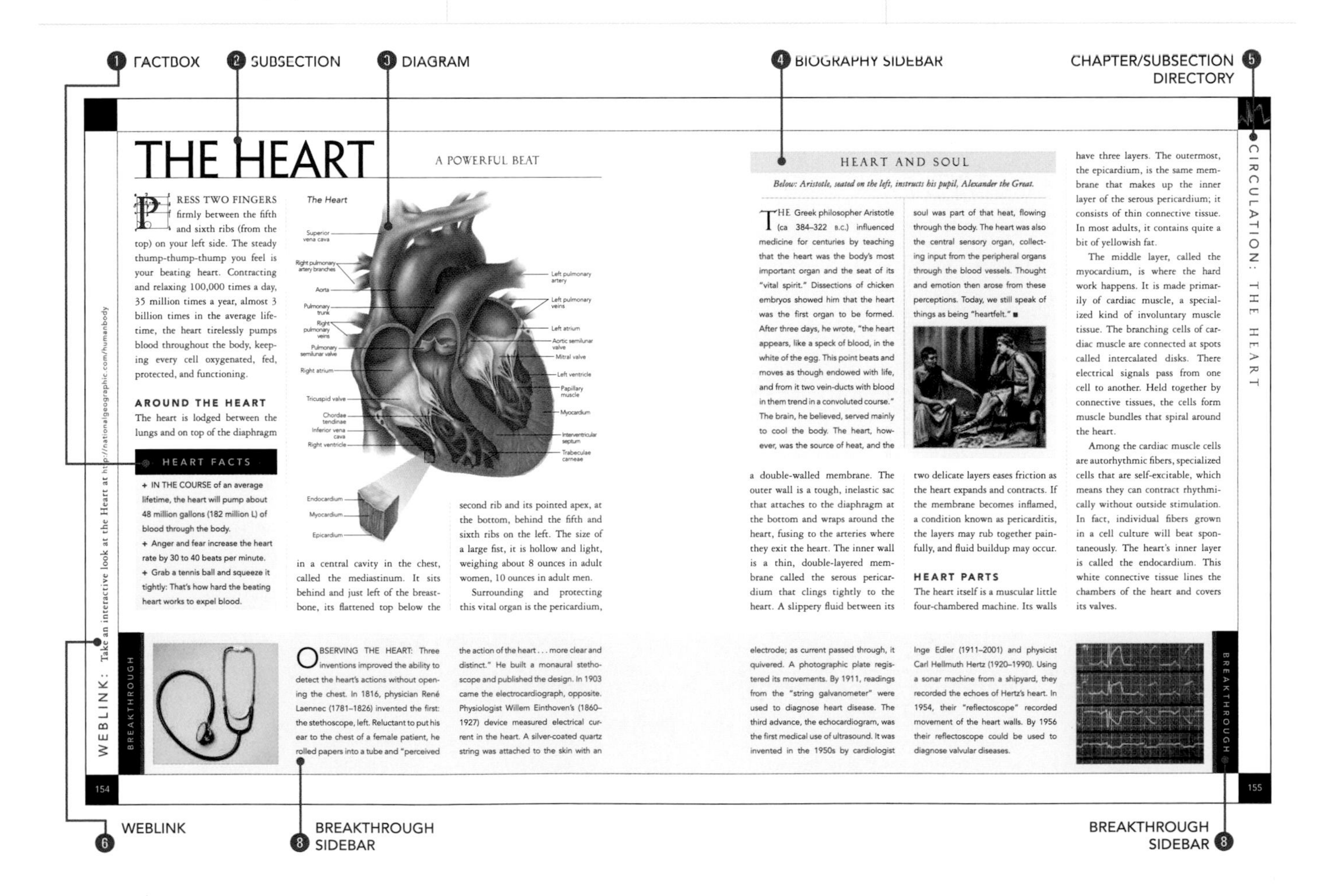

THE HEART

A POWERFUL BEAT

PRESS TWO FINGERS firmly between the fifth and sixth ribs (from the top) on your left side. The steady thump-thump-thump you feel is your beating heart. Contracting and relaxing 100,000 times a day, 35 million times a year, almost 3 billion times in the average lifetime, the heart tirelessly pumps blood throughout the body, keeping every cell oxygenated, fed, protected, and functioning.

AROUND THE HEART

The heart is lodged between the lungs and on top of the diaphragm in a central cavity in the chest, called the mediastinum. It sits behind and just left of the breastbone, its flattened top below the second rib and its pointed apex, at the bottom, behind the fifth and sixth ribs on the left. The size of a large fist, it is hollow and light, weighing about 8 ounces in adult women, 10 ounces in adult men.

Surrounding and protecting this vital organ is the pericardium, a double-walled membrane. The outer wall is a tough, inelastic sac that attaches to the diaphragm at the bottom and wraps around the heart, fusing to the arteries where they exit the heart. The inner wall is a thin, double-layered membrane called the serous pericardium that clings tightly to the heart. A slippery fluid between its two delicate layers eases friction as the heart expands and contracts. If the membrane becomes inflamed, a condition known as pericarditis, the layers may rub together painfully, and fluid buildup may occur.

HEART PARTS

The heart itself is a muscular little four-chambered machine. Its walls have three layers. The outermost, the epicardium, is the same membrane that makes up the inner layer of the serous pericardium; it consists of thin connective tissue. In most adults, it contains quite a bit of yellowish fat.

The middle layer, called the myocardium, is where the hard work happens. It is made primarily of cardiac muscle, a specialized kind of involuntary muscle tissue. The branching cells of cardiac muscle are connected at spots called intercalated disks. There electrical signals pass from one cell to another. Held together by connective tissues, the cells form muscle bundles that spiral around the heart.

Among the cardiac muscle cells are autorhythmic fibers, specialized cells that are self-excitable, which means they can contract rhythmically without outside stimulation. In fact, individual fibers grown in a cell culture will beat spontaneously. The heart's inner layer is called the endocardium. This white connective tissue lines the chambers of the heart and covers its valves.

HEART FACTS

+ IN THE COURSE of an average lifetime, the heart will pump about 48 million gallons (182 million L) of blood through the body.
+ Anger and fear increase the heart rate by 30 to 40 beats per minute.
+ Grab a tennis ball and squeeze it tightly: That's how hard the beating heart works to expel blood.

HEART AND SOUL

Below: Aristotle, seated on the left, instructs his pupil, Alexander the Great.

THE Greek philosopher Aristotle (ca 384–322 B.C.) influenced medicine for centuries by teaching that the heart was the body's most important organ and the seat of its "vital spirit." Dissections of chicken embryos showed him that the heart was the first organ to be formed. After three days, he wrote, "the heart appears, like a speck of blood, in the white of the egg. This point beats and moves as though endowed with life, and from it two vein-ducts with blood in them trend in a convoluted course." The brain, he believed, served mainly to cool the body. The heart, however, was the source of heat, and the soul was part of that heat, flowing through the body. The heart was also the central sensory organ, collecting input from the peripheral organs through the blood vessels. Thought and emotion then arose from these perceptions. Today, we still speak of things as being "heartfelt." ■

BREAKTHROUGH

OBSERVING THE HEART: Three inventions improved the ability to detect the heart's actions without opening the chest. In 1816, physician René Laennec (1781–1826) invented the first: the stethoscope, left. Reluctant to put his ear to the chest of a female patient, he rolled papers into a tube and "perceived the action of the heart . . . more clear and distinct." He built a monaural stethoscope and published the design. In 1903 came the electrocardiograph, opposite. Physiologist Willem Einthoven's (1860–1927) device measured electrical current in the heart. A silver-coated quartz string was attached to the skin with an electrode; as current passed through, it quivered. A photographic plate registered its movements. By 1911, readings from the "string galvanometer" were used to diagnose heart disease. The third advance, the echocardiogram, was the first medical use of ultrasound. It was invented in the 1950s by cardiologist Inge Edler (1911–2001) and physicist Carl Hellmuth Hertz (1920–1990). Using a sonar machine from a shipyard, they recorded the echoes of Hertz's heart. In 1954, their "reflectoscope" recorded movement of the heart walls. By 1956 their reflectoscope could be used to diagnose valvular diseases.

BREAKTHROUGH

WEBLINK: Take an interactive look at the Heart at http://nationalgeographic.com/humanbody

CIRCULATION: THE HEART

154

155

9 CHAPTER GLOSSARY

10 KEEPING HEALTHY SIDEBAR

11 TABLE

5 CHAPTER/SUBSECTION DIRECTORY

CIRCULATION: THE SYSTEM

CHAPTER GLOSSARY

ANATOMOSIS. An end-to-end joining of blood vessels (or lymphatic vessels or nerves).
ANGIOPLASTY. A technique for opening space within a clogged artery usually by temporarily inflating a balloon inside the artery.
ARTERIOLE. A tiny, muscular artery that carries blood to capillaries.
ARTERIOSCLEROSIS. A group of diseases characterized by stiff, thickened arterial walls.
ARTERY. Blood vessel that carries blood away from the heart.
ATHEROSCLEROSIS. A disease characterized by the buildup of plaque on arterial walls.
ATRIOVENTRICULAR (AV) NODE. A specialized mass of electrically conducting cells located between the heart's atria and ventricles.
ATRIOVENTRICULAR VALVES. Valves between the atria and the ventricles.
ATRIUM (pl. atria). Either one of the heart's two upper chambers.
AUTORHYTHMIC FIBERS. Muscle cells in the heart that produce electrical signals without outside stimulus.
CAPILLARY. A microscopic blood vessel that connects arterioles to venules; the site for interchange between blood and tissues.
CARDIAC CYCLE. The events of one complete heartbeat.
CARDIAC OUTPUT. Blood pumped by a ventricle in one minute.
DIASTOLE. The relaxation phase of the cardiac cycle.
ECHOCARDIOGRAM. An image of the heart's structure and function produced by ultrasound.
ELECTROCARDIOGRAM. A recording of the heart's electrical activity.
EMBOLISM. A blood clot that forms in a blood vessel and travels to another part of the body.
ENDOCARDIUM. The smooth, innermost layer of the heart wall.
ENDOTHELIUM. The inner lining of many body structures, including the heart and blood vessels.
EPICARDIUM. The thin membrane covering the heart.
ERYTHROCYTE. Red blood cell.
FIBRILLATION. Rapid, uncoordinated contractions of heart fibers.
FIBRIN. An insoluble protein formed during blood clotting.
FIBRINOGEN. A clotting factor in plasma; converted to fibrin.
HEMOGLOBIN. A substance in red blood cells that transports oxygen.
HYPERTENSION. High blood pressure.
LEUKOCYTE. White blood cell.
MITRAL VALVE. The valve between the left atrium and left ventricle.
MYOCARDIAL INFARCTION. A heart attack consisting of damage to or death of an area of heart muscle due to inadequate blood supply.
MYOCARDIUM. Muscular wall of the heart.
PERICARDIUM. Double-layered membrane that encloses the heart.
PLAQUE. Deposits of accumulated substances, such as cholesterol, fats,

KEEPING HEALTHY

Compression socks can prevent thrombosis.

BEDRIDDEN PATIENTS AND airplane passengers on long trips may share an unfortunate side-effect of immobility: deep-vein thrombosis. The condition is caused by blood flowing slowly or pooling in the deep veins in the lower legs or thighs of an unmoving person. The blood may not be able to wash away clotting factors, and a blood clot (thrombus) can result, blocking circulation. The limb becomes tender, swollen, hot, and red. The clot may break away and travel to the brain—causing a stroke—or to the lungs, heart, or other areas. Doctors may prescribe anticoagulants to the bedridden; people prone to the condition can wear compression stockings to prevent pooling. Plane or car passengers on a long trip should take periodic breaks to stand up and walk about, lest they suffer "economy-class syndrome." Another option is to do leg stretches to keep blood moving. ■

VEINS

After releasing their oxygen, nutrients, and other chemicals and picking up wastes and other cargo from the tissues, the capillaries merge to form small veins called venules, which drain into larger veins that carry the deoxygenated blood back to the heart.

Blood flowing through veins is at a relatively low pressure and veins are accordingly thinner-walled and less muscular than arteries. Some veins, particularly in the arms and legs, contain valves that prevent the blood from flowing backward as it is pumped upward toward the heart. Skeletal muscles in the legs help to pump the blood toward the heart as well by constricting around the veins and milking the blood upward through the valves. Leaky valves, sometimes damaged by stress, can lead to swollen varicose veins that hold pools of backflowing blood.

PRESSURE & PULSE

Blood pressure varies greatly from place to place within the circulatory system. During a physical, blood pressure is measured in the brachial artery in the upper left arm, with the arm held level with the heart. The first reading an examiner takes after relaxing a blood pressure cuff (which corresponds to the first sound heard through the stethoscope) is the systolic pressure, the pressure that the blood exerts on the big arteries as it leaves the contracting ventricles; the reading at which the sound suddenly becomes faint is the diastolic pressure, the pressure in the arteries as the ventricles relax. (See "What Can Go Wrong," below.)

Blood pumping through the elastic arteries creates a visible wave with each heartbeat—the pulse. The pulse can be felt in any artery close to the body's surface (see chart above) by pressing lightly on the skin. The pulse beats at the same rate as the heart, about 70 to 80 times a minute at rest.

WHERE TO FEEL THE PULSE

ARTERY	LOCATION
Superficial temporal artery	Temple, next to eye socket
Facial artery	Lower jaw on a line with the corners of the mouth
Common carotid artery	Neck, next to larynx
Brachial artery	Upper arm, underside of biceps
Femoral artery	Upper, inner thigh
Popliteal artery	Behind the knee
Radial artery	Bottom half of wrist
Dorsal artery of foot	On top of the instep

WHAT CAN GO WRONG

HEMOPHILIA is a rare, inherited disorder in which people are born without a necessary clotting factor in their blood. Hemophiliacs may bleed spontaneously, or extensively after minor traumas; they may also suffer painful, damaging bleeding in the joints, such as in the knees, elbows, or ankles. The two main types of the disorder are hemophilia A, or classical hemophilia, in which clotting factor VIII is missing (accounting for more than 80 percent of cases) and hemophilia B, linked to a missing factor IX. Both types are sex-linked conditions: They occur almost exclusively in males because the genetic mutation that accounts for the clotting factor is found on the X chromosome. (Women are protected because they almost always inherit a matching X chromosome without the mutation.) Women and men may be asymptomatic carriers of these kinds of hemophilia.

WHAT CAN GO WRONG SIDEBAR 12

2 SUBSECTION

13 HISTORY SIDEBAR

14 COMPARATIVE IMAGES

5 CHAPTER/SUBSECTION DIRECTORY

PROTECTION: IMMUNE SYSTEM

IMMUNE SYSTEM

KEEPING THE BODY HEALTHY

UNLIKE SOME OF THE body's other systems (such as the digestive and circulatory systems), the immune system is a functional system that operates throughout the human body, its work carried out by the body's trillions of immune cells and specialized molecules.

The first line of defense is the skin. Our body's covering of tissue protects us from all kinds of dangers large and small. One obvious invasion of this first defense is a bump, bruise, burn, or cut; nerves in the skin register pain and tell us to avoid the rock or the hot stove that caused the problem. A bump resulting in a bruise means that blood vessels have broken beneath the skin, and the escaped blood has formed little clots called hematomas. A cut allows microorganisms to breach the first wall—the skin—and trigger internal defenses, which we will examine in greater detail later in the chapter.

Injuries are generally the least of the immune system's problems. Armies of germs—viruses, bacteria, parasites, and fungi—make daily sorties against our body, whether we are injured or not. Through the mouth, nose, eyes, ears, genitals, and skin pores, these saboteurs infiltrate the body. Since skin can't seal out every possible opening (or else we could not breathe or eat), the first line of defense has

SEE ALSO: Chapter Two, "The Skin," PAGE 46; Chapter Five, "The Blood," PAGE 132

IMMUNOLOGY

A plague epidemic took the lives of more than 100,000 in 1720 Marseille, France.

IT HAS LONG BEEN known that the body has natural defenses and that it can, to some degree, heal itself when infected. Immunology, the science of how these defenses work on a cellular level, has developed only within the past one hundred years. It was known during the Revolutionary War that soldiers who injected themselves with a dose of smallpox had a better chance of not contracting the disease. Then in 1796 British physician Edward Jenner took the risk factor out of vaccination by creating one of the first successful vaccines for smallpox (see "Discovering Immunization," page 339). He did not know that smallpox was caused by a virus or the mechanisms by which the body's immune system could be primed by a small amount of it.

Not until the mid-1800s did the understanding that microorganisms caused infection and disease become accepted. Discoveries then began piling up. French chemist Louis Pasteur developed several ground-breaking vaccines for farm animals, including ones against rabies and anthrax in cattle. By the end of the century, scientists had discovered phagocytes, antibodies, and the fact that vaccines work by stimulating the production of antibodies.

During the first half of the 20th century, researchers began unlocking the mysteries of how antibodies respond to specific antigens. In the mid-1960s scientists described the workings of B cells and T cells, both adaptive immune cells.

Much research continues into the present time, with progress being made recently in understanding the genetic controls of the immune system and how each of the immune system's millions of different antibodies are produced. ■

a backup—the mucosal lining. Pathogens entering the gastrointestinal tract, genitourinary tract, and lungs are generally stopped by the sticky, acid-secreting mucous membranes lining the organs.

A second line of defense comes into play against those persistent invaders not blocked by the skin and mucosal lining. This second line is composed of cells, which include a broad category called phagocytes, or engulfing cells; basically their job is to eat the invaders. Among the phagocytes are white blood cells called neutrophils and eosinophils, which help initiate inflammation. Another broad category of generalized cells, the natural killer cells, are unleashed immediately against viruses, cancer cells, and other foreign cells. In addition to these immune cells, numerous chemical compounds respond to infection and injury, move in to destroy pathogens, and begin repairing tissue. While these cells and chemicals are at work, the body often makes another generalized response to foreign invasion—it heats up, or produces a fever.

The body's third line of defense—the adaptive defense system—is often slower to react, because the body must determine what kind of attack is underway and organize this final, more specific response. When the first two lines of defense fail, specialized warriors then go into action against the most perfidious of pathogens, whether they be microbes from the

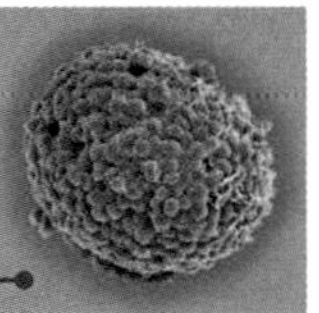

MAST CELL

This kind of white blood cell is activated by allergic reactions and physical injuries.

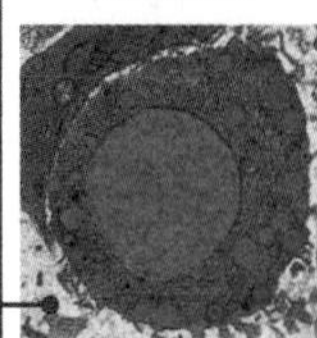

BASOPHIL

Basophils are the smallest and least common white blood cell at play in the immune system.

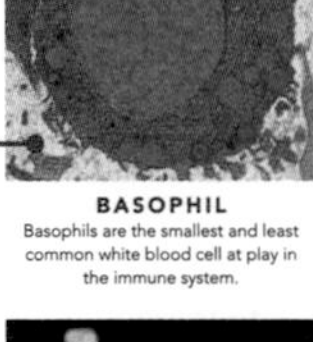

ANTIBODIES

These specialized proteins combat specific microorganisms.

outside or internal cells in revolt such as cancer cells. These elite fighting units are not just lying around in wait; instead, they are trained on the job—that is, they are created in response to a virus, for instance, that the body has not seen before. The adaptive defense system is thus an improvised response to a specific invader.

Chief among the immune cells of the adaptive defense system are white blood cells called lymphocytes that come from bone marrow and thymus and circulate through the body. The body also produces antigen-presenting cells that swallow pathogens or chemical constituents of the pathogens called antigens and cause lymphocytes to start reacting against the invaders. Two remarkable qualities of the adaptive defense system are that it works systemically—that is, once activated in one part of the body, it functions throughout; and it memorizes the antigens so that the next time they come along the body hits back quicker and harder.

REACTIONS

The nonspecific, or innate (natural), reaction requires no previous exposure to the antigen being encountered. Something potentially harmful enters the body, and its defenses go up. The nonspecific immune system can kill or limit the spread of pathogens, but it does not increase its efficacy against any specific invader—it doesn't learn to recognize one particular antigen

7 CROSS-REFERENCE

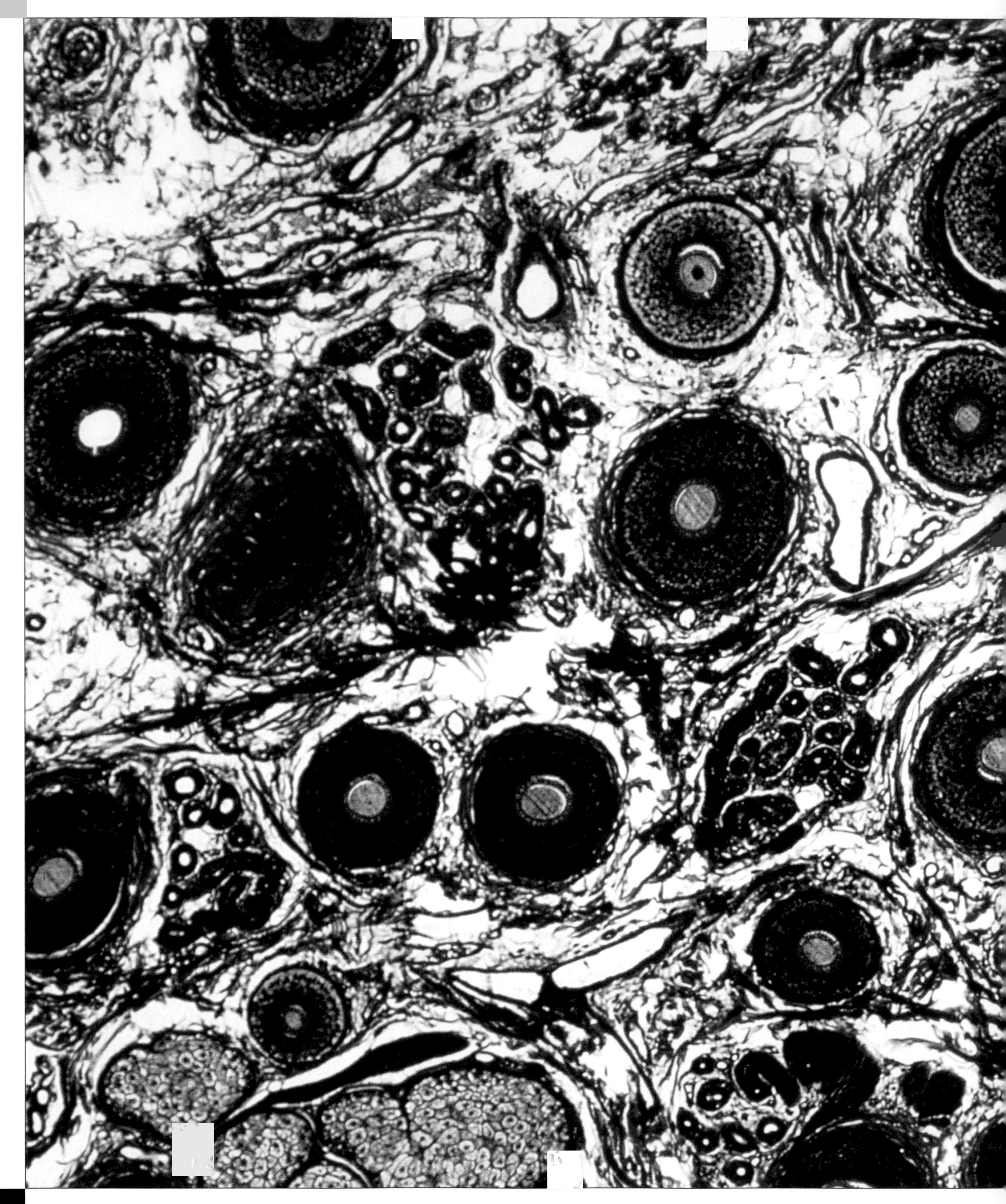

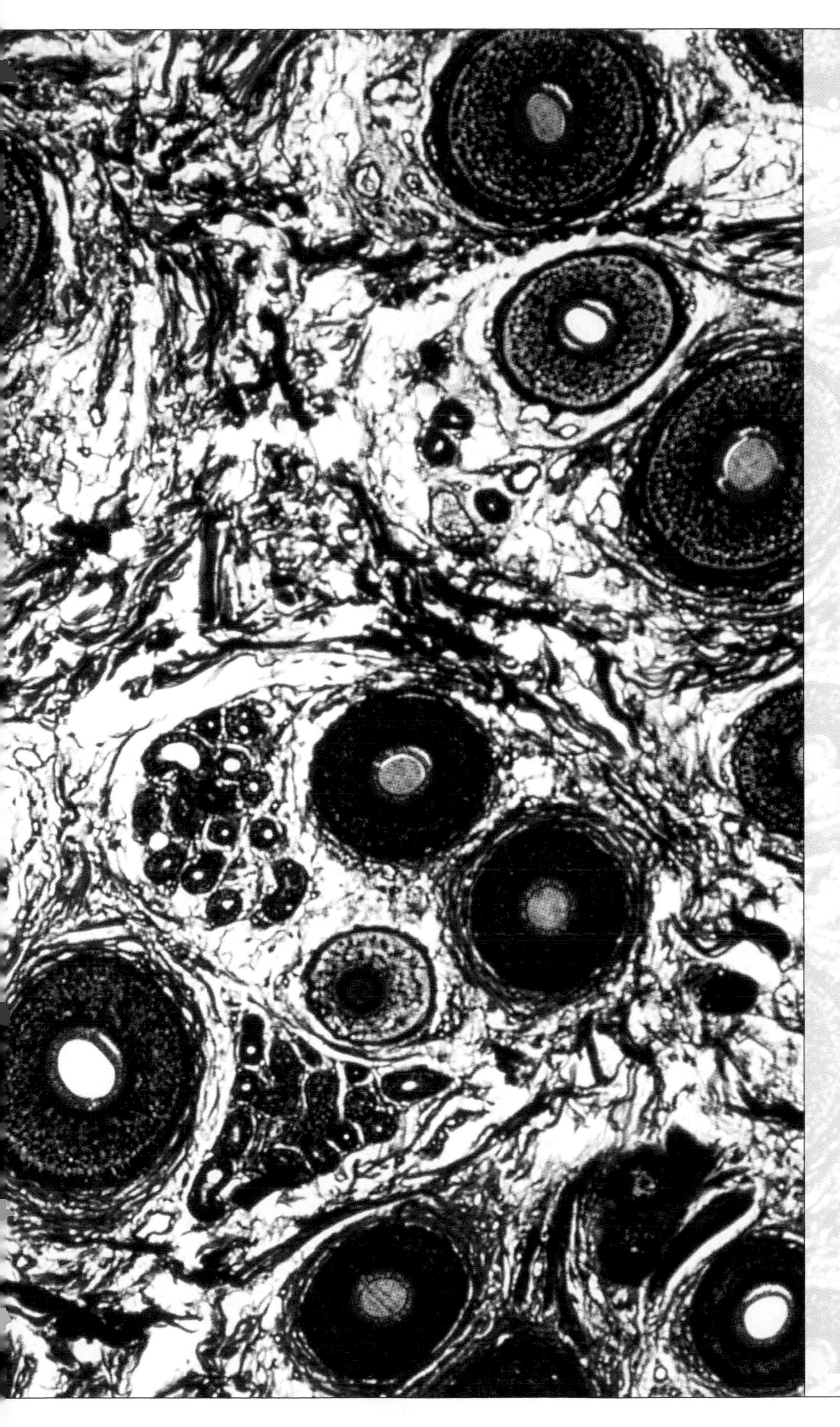

CHAPTER ONE

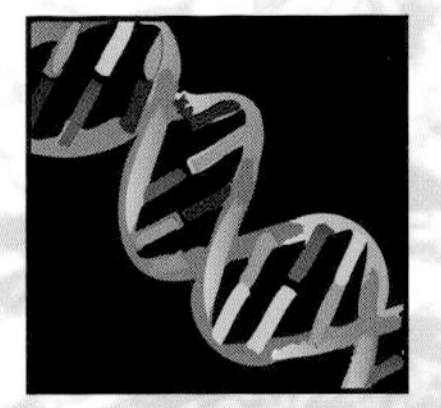

BEGINNINGS

TINY BUT MIGHTY, CELLS ARE the building blocks of the human body, the fundamental components of tissues, organs, and systems. Every human body starts with the meeting of two cells—a sperm cell and an egg cell—and develops from that union. Almost immediately, hundreds of other kinds of cells arise, each suited to a particular task and yet all made up of the same basic parts. Dividing, specializing, proliferating, and dying, the cells follow instructions encoded into their DNA to become the magnificently coordinated multicellular colony that is the human organism.

Skin cells enclose hair follicles (dark circles with pink centers), sweat glands (small dark circles with white centers), blood vessels, and nerves.

THE CELL

LIFE IN MINIATURE

JUST AS AN ATOM IS the basic unit of physics, so the cell is the basic unit of biology. All living things are made of cells; some organisms are no more than a single cell, while others contain trillions in hundreds of specialized forms. The human body is one of these complex, multicellular organisms. To understand its functions—from a sneeze to the birth of a baby—one must understand its cells.

Human cells have numerous shapes and tasks but share fundamental ingredients and structures. All consist mainly of carbon, hydrogen, oxygen, and nitrogen. Every cell has an outer boundary, the plasma membrane. Inside, the cell is divided into two main regions, the cytoplasm and the nucleus. The cytoplasm contains fluid and a variety of organelles, "little organs," that perform cell metabolism. The cytoplasm surrounds the nucleus, the largest organelle, which is protected by its own membrane. The nucleus is the control center and location of chromosomes, the genetic material.

The adult body contains 50 to 100 trillion cells in about 200 specialized forms. They range in size from the sperm, a few micrometers wide, to the egg cell, about 100 micrometers, the size of the period that ends this sentence. (Including the axon, the part of the nerve cell

ROBERT HOOKE

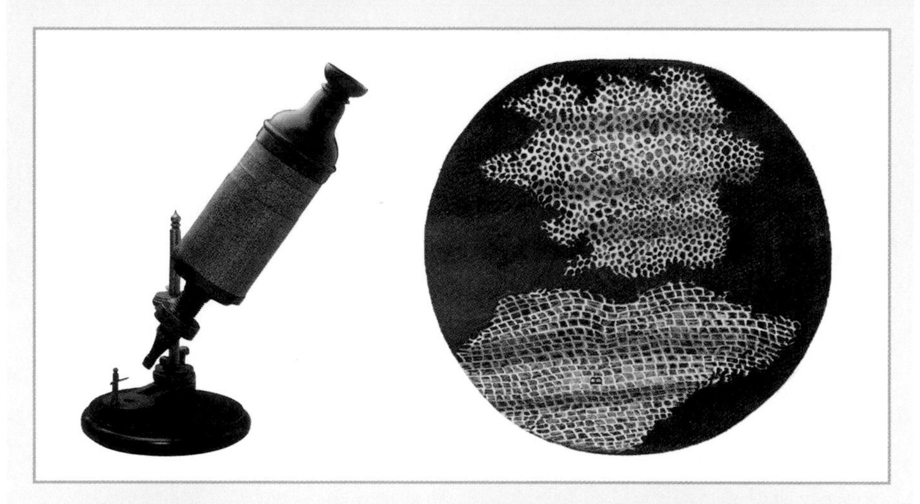

Hooke's hand-crafted microscope and his drawing of cork cells (color-enhanced)

ROBERT HOOKE (1635–1703) was a kind of scientist rarely seen in the moden era of specialization. The 17th-century British polymath made contributions to chemistry, astronomy, physics, paleontology, architecture, and engineering. Biologists remember him best as the man who discovered cells.

Hooke attended Oxford and by 1662 had become curator of experiments to the recently founded Royal Society. Among his achievements, he anticipated Isaac Newton in describing the inverse-square rule of a planet's attraction to the sun, oversaw the rebuilding of London after the Great Fire of 1666, and formulated an equation describing elasticity that is now known as Hooke's law. Hooke built one of the best early compound microscopes and used it to study insects, sponges, plants, and other small items. In *Micrographia* (1665), he published his observations of his microscopic subjects. Of a thin slice of cork, he wrote, "I could exceeding plainly perceive it to be all perforated and porous, much like a Honey-comb, but that the pores of it were not regular . . . these pores, or cells, . . . were indeed the first *microscopical* pores I ever saw." What Hooke saw were the walls of cork cells. Hooke was the first person to use the term *cell,* meaning "a small room," for these basic components of life. He didn't understand them as such, believing the structures to be cross-sections of channels through the plant, but his work would pave the way for the 19th-century pioneers of cell theory, Theodor Schwann and Matthias Schleiden (see "Cell Theory," page 17). ■

that conducts impulses, the nerve cells that run from the spine to the toes take the prize, at more than three feet long.) Cells of similar types usually group together as tissues and organs, and their distinctive shapes and internal workings are tailored to their uses. For instance, smooth muscle cells have a long, spindle shape and durable microfilaments that allow them to contract; cells of the intestines are covered with microvilli, tiny fingerlike extensions that increase intestinal surface area; and cells of the eye possess photoreceptors that can detect light.

Each kind of cell has its own life cycle. All are "born" from the division of preexisting cells; most will grow, divide, and eventually die, to be replaced by new cells. In some tissues, such as those of the skin, this cycle runs almost continuously. In others, such as those of the liver, cells reproduce slowly unless damaged; then their division speeds up, regenerating the tissue. In yet other tissues, such as those of the heart, mature cells cannot divide, and damage to the organ is repaired with scar tissue. The processes that control cell birth, death, and specialization are still not well understood, and solving these mysteries is a major goal of biologists today.

CELL PARTS

Although cells come in a veritable zoo of forms and functions, almost all have a similar basic anatomy. All, for instance, are contained by a plasma membrane—not a skin, but an oily fluid whose molecules are constantly shifting about on the cell's surface. This membrane quickly seals itself around a puncture—a handy trait for scientists who want to inject or remove something from a cell.

Consisting of a double layer of lipids—organic molecules of carbon, hydrogen, and oxygen—interspersed with proteins, the membrane is selectively permeable. Some molecules, like oxygen or alcohol, diffuse directly through them. Many others—glucose, for example, or amino acids—travel through protein channels or are transported by protein carriers.

Proteins in the membrane also serve as receptors that bind to particular molecules (such as insulin) as identity markers that distinguish familiar from foreign cells, and as linking molecules that hold cells together. Most cells have similar lipids in their membranes, but proteins vary dramatically among cell types.

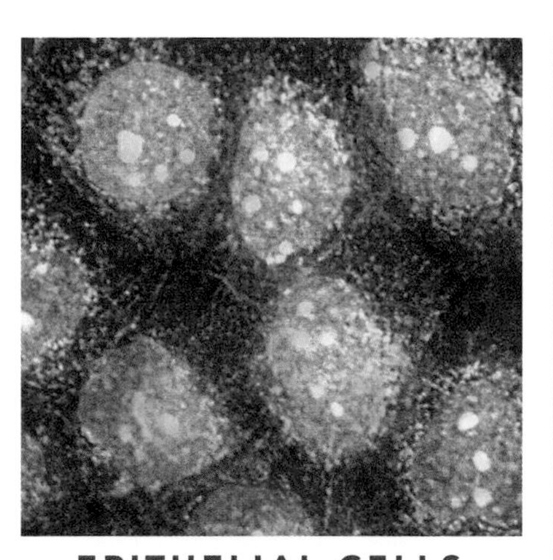

EPITHELIAL CELLS
A color-enhanced micrograph of cells from the pancreas reveals DNA (blue) and mitochondria (green).

NERVE CELLS
These cells form the brain's support system, feeding the neurons that transmit messages.

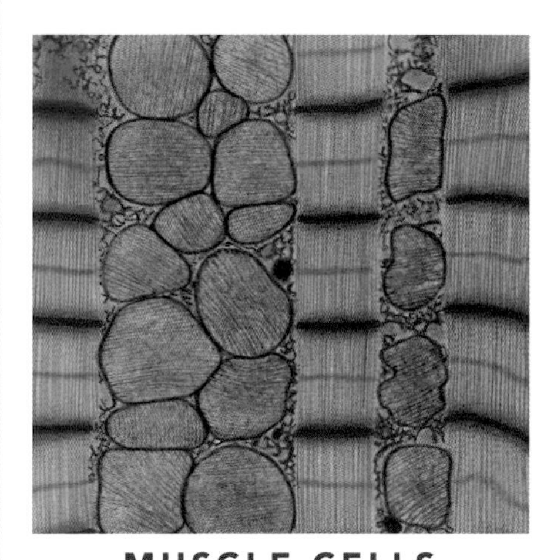

MUSCLE CELLS
A cross-section of heart muscle cells shows large mitochondria sandwiched between muscle fibers.

+ MICROSCOPIC +

+ MOST CELLS and cell structures are measured in micrometers (μm, also called microns) or nanometers (nm).
+ 1 μm = one-millionth of a meter
+ 1 nm = one-billionth of a meter
+ Typical animal cells measure 10–30 μm long. Bacterium are 2–8 μm long.
+ A human DNA molecule is about 2 nm wide but can measure several centimeters long.

Organelles in a typical cell (seen in cross-section here), such as the nucleus and the oval mitochondria, work together in a range of specialized jobs.

Beyond the membrane lies the cytoplasm, which includes all the structures of the cell, excluding the nucleus, and the fluid that contains them. The fluid, called cytosol, is made largely of water and has nutrients, wastes, and other chemicals dissolved or suspended in it. Packed into the cytosol are the organelles, specialized structures that carry out specific jobs within the cell in a smooth-running process that would be the envy of any industrial manager.

The organelles include the cytoskeleton, a network of tubules and filaments that supports the cell and gives it its shape. Microtubules surrounded by plasma membrane are at the core of the cilia (hairlike extensions) and flagella (whiplike tails) that can be found on some cells. Another organelle, the centrosome, contains tubelike centrioles that help to organize the stages of cell division.

Extending from the cell's nuclear membrane is a network of folded membranes called the endoplasmic reticulum (ER). The "rough" part of the ER is studded with little, granular ribosomes, which make proteins to be used by or secreted from the cell. (More ribosomes are scattered throughout the cytoplasm.) The smooth ER, lacking ribosomes, makes fatty acids and steroids, detoxifies drugs, releases glucose into the blood, and performs other tasks depending on the kind of cell it occupies.

The mitochondria are the cell's power stations. Anywhere from a few to thousands of these large, wiggling, oblong organelles dot the cytoplasm, with the largest number found in the most active cells, such as in the kidney or liver. In the process of aerobic respiration, mitochondria convert oxygen and nutrients into adenosine triphosphate (ATP). This important molecule releases energy that the cell uses for its own activities, including transporting substances around the cell, muscular contractions, and the movement of chromosomes during cell division.

Interestingly, mitochondria contain their own separate DNA and RNA; when the cell needs more energy, they can even reproduce. These organelles may once have been separate organisms, perhaps

CAMILLO GOLGI (1843–1926), above, was an Italian physician who won the 1906 Nobel Prize for the identification of nerve cells. Structures bearing his name include

+ Golgi cells: a type of nerve cell.
+ Golgi tendon organ: a sensor that detects changes in muscle tension.
+ Golgi complex: a cell organelle concerned with modifying and packaging proteins.

bacteria, that inserted themselves into ancient plant and animal cells and were incorporated into their structure over the course of evolution. Also intriguing to geneticists is that mitochondria appear to be inherited almost exclusively through the maternal line, undergoing less change over time than nuclear DNA: a useful trait in the study of evolution (see "Genetic Histories," page 28).

The Golgi complex (also called the Golgi apparatus) is an organelle made of flattened, disklike sacs. It modifies and packages proteins made in the endoplasmic reticulum before passing them on to other parts of the cell; some, for instance, may be incorporated into the cell membrane, while others may leave the cell altogether.

Organelles can destroy as well as produce; this job falls largely to the lysosomes, spherical bodies containing digestive enzymes. Lysosomes digest invaders "eaten" by the cells, such as bacteria and viruses; they also break down worn-out organelles and recycle their components. Similar, tiny organelles called proteasomes destroy damaged or unneeded proteins, while peroxisomes contain enzymes that neutralize free radicals.

Largest of the organelles, the nucleus is both the cell's boss and its archivist. It contains a complete copy of the body's genetic material and coordinates the cell's activities, including growth, protein production, and cell division.

CELL THEORY

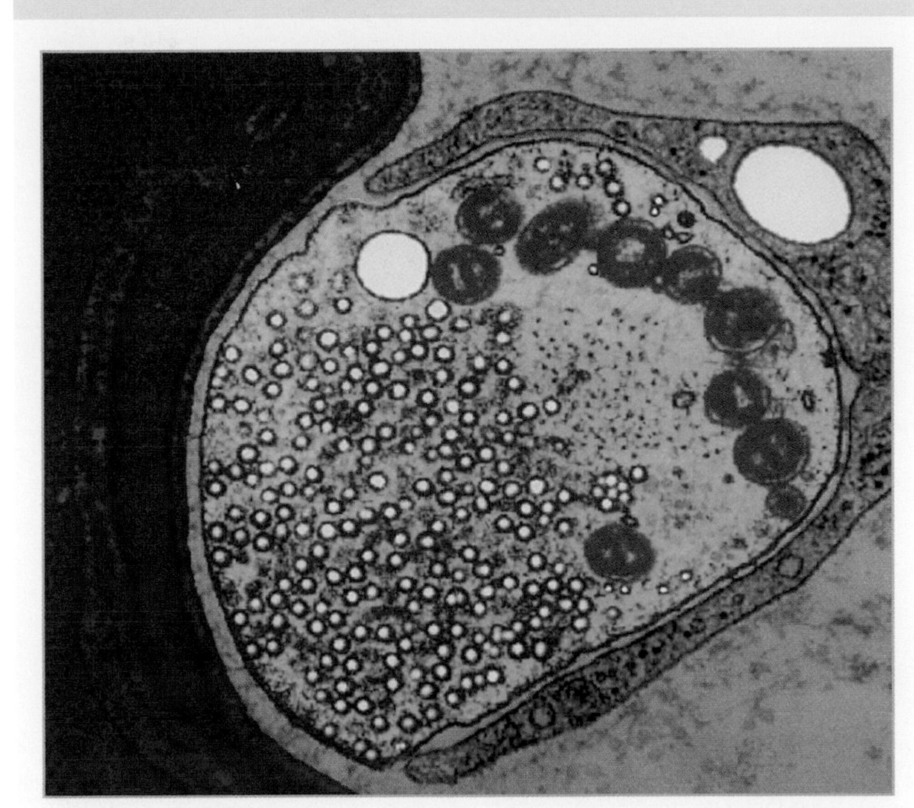

Named for a cell theory pioneer, a Schwann cell (green) supports a nerve cell (blue).

ALTHOUGH ROBERT HOOKE saw and named cells in the 17th century, neither the English scientist nor his contemporaries understood that cells themselves were living units. It was not until the 1830s, when lenses and microscopes had improved enough to reveal structures only a micrometer apart, that scientists really got a good look at cells and began to understand more.

Using such an instrument, in 1831 another Englishman, botanist Robert Brown, discovered that every plant cell he observed contained a little kernel, a "nucleus."

Just a few years later, in 1838, the German botanist Matthias Jakob Schleiden came to the conclusion that all plants were made of cells and that every young plant arose from a single cell. A year later, his colleague, physiologist Theodor Schwann, reached the same conclusion about animal cells.

In 1839, Schwann published his observations in a book that stated two basic tenets of cell theory: First, all organisms consist of one or more cells. Second, the cell is the basic unit of structure for all organisms.

In 1858, German pathologist Rudolf Virchow followed up on Brown's work with his book *Cellular Pathology,* which added a third tenet: All cells arise only from preexisting cells.Put together, these three rules form the basis of modern cell theory. ■

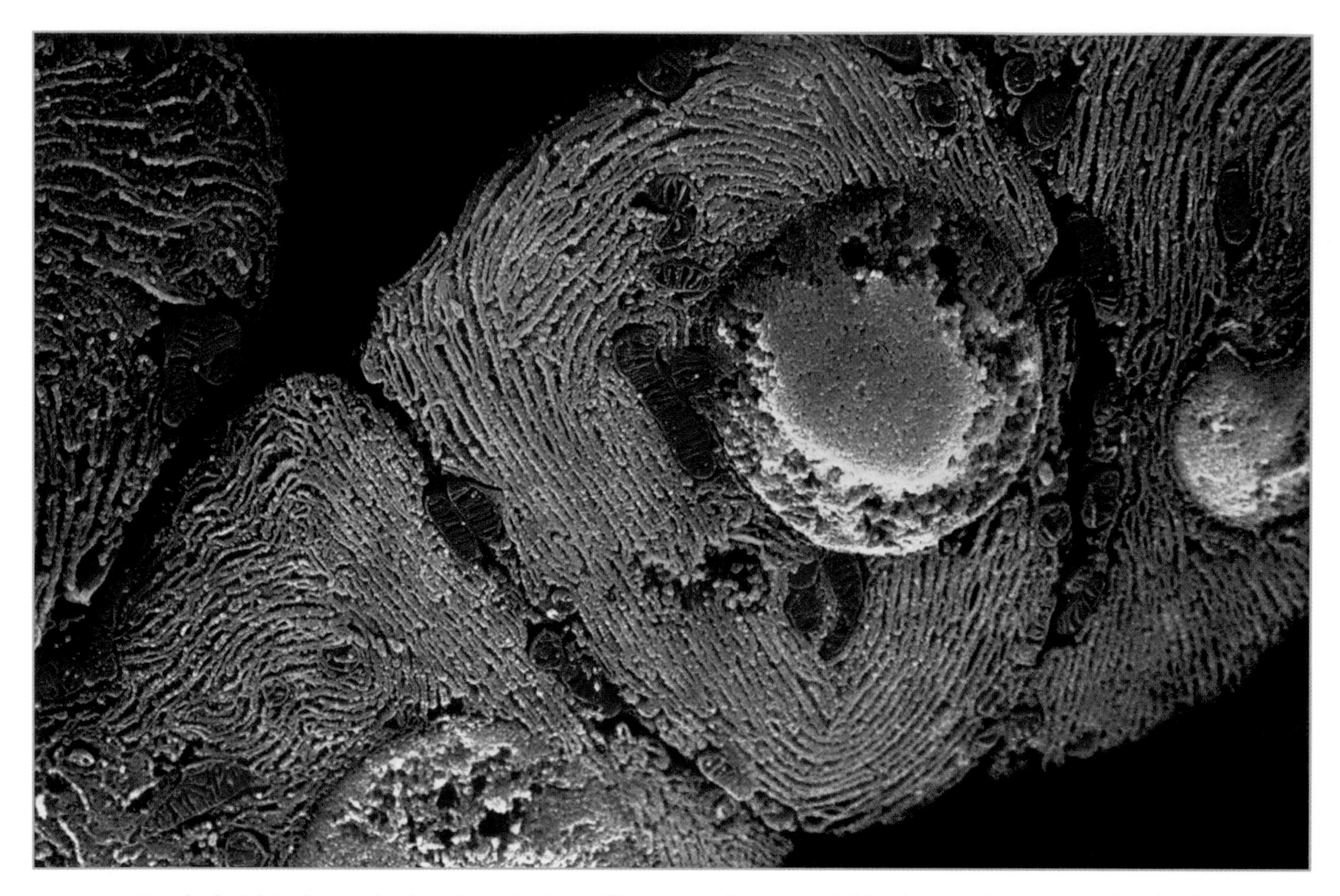

Nestled within the endoplasmic reticulum of human cells are nuclei (yellow) and mitochondria (red).

Almost all human cells have a nucleus, typically a spherical or oblong body about five micrometers wide. A few kinds of large cells, such as those in skeletal muscles, have more than one. Only mature red blood cells lack a nucleus; therefore, they can't reproduce or repair themselves and die within a few months.

The nucleus is enclosed by a fluid-filled double membrane known as the nuclear envelope. Minute pores penetrate both layers of the envelope to connect the inside of the nucleus to the rest of the cell; these portals allow proteins from the cytoplasm to enter the nucleus and RNA to leave it. Clinging to the outer membrane of the nucleus is the rough endoplasmic reticulum, pocked with little ribosomes. Like the cell itself, the nucleus is filled with a gel-like fluid, here called nucleoplasm. The fluid holds one or more little spherical bodies, nucleoli, that look like the nucleus's nucleus. These membraneless clumps of protein, DNA, and RNA assemble and then export the proteins that form the basic units of ribosomes.

Weaving its way through the nucleoplasm is the cell's motherlode: its DNA (see "Inheritance," page 24). In a mature, nondividing cell, this takes the form of chromatin. Chromatin is made of both DNA and little balls of proteins called histones. Each long strand of DNA coils tightly around the histones like string around sticky beads, so that the approximately six feet of DNA in the human nucleus fits compactly into its microscopic space. Some of this chromatin is genetically active, working constantly to dictate the cell's protein production; other portions are inactive and so are more condensed.

The coiled DNA strands of the chromatin are usually compacted by a factor of 40 (compared with the same strand stretched out straight). Then, as the cell prepares to divide, the chromatin coils in on itself far more, by a factor of 10,000, forming the dense little bundles of information known as chromosomes (see "Inheritance," page 24).

A CELL'S LIFE

A human cell's life cycle has two main portions: interphase, the period from formation to division; and the mitotic phase, when it reproduces by dividing into two identical daughter cells.

THE DOUBLE HELIX

IN 1944, Canadian biologist Oswald Avery showed that DNA carried the genetic information within bacteria—but just how it did it, no one knew. In 1953, American James Watson and Englishman Francis Crick, above, figured out the answer. They saw that the DNA molecule was a double helix composed of two sugar-phosphate strands linked by four repeating bases. The discovery began a revolution in human biology and medicine.

ROSALIND FRANKLIN

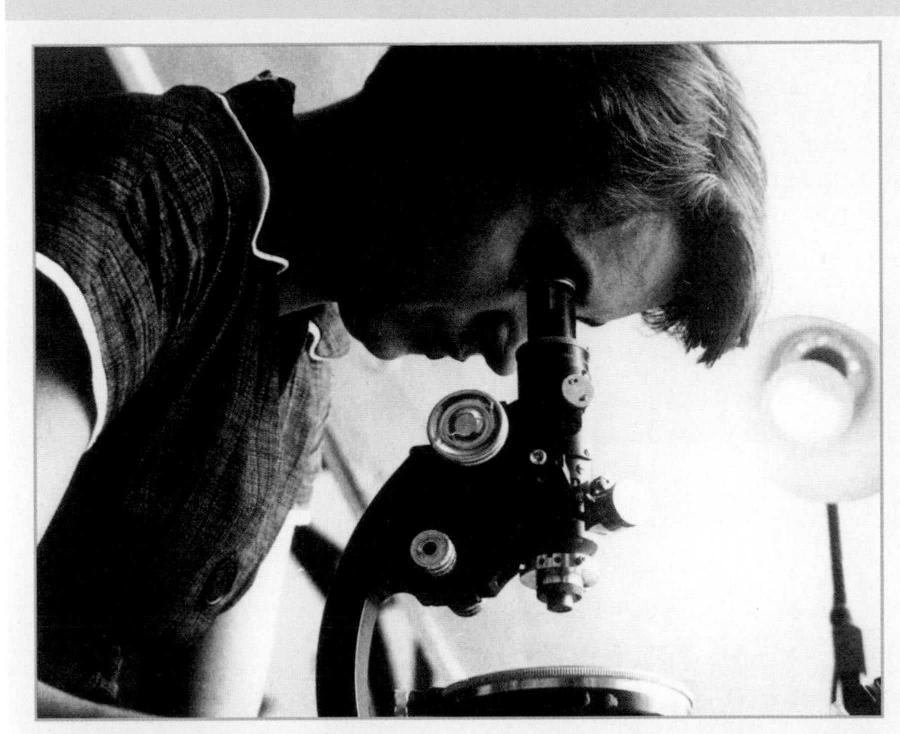

Chemist Rosalind Franklin's images of DNA provided key clues to its structure.

THE INSPIRED WORK of Watson and Crick in deciphering DNA owed much to a lesser-known scientist: Rosalind Franklin. Born in London in 1920, Franklin earned a doctorate in physical chemistry from Cambridge in 1945. After mastering the techniques of X-ray diffraction, in which X-rays are used to create images of crystallized solids, she began work as a researcher at a laboratory in King's College, London. She and colleague Maurice Wilkins were assigned to separate projects in the study of DNA.

Franklin improved X-ray diffraction methods in her study of the molecule, fine-tuning the X-ray beam and arranging the DNA fibers to better reveal their structure. Without her permission, Wilkins showed some of her work to Watson and Crick. "The instant I saw the picture," Watson wrote, "my mouth fell open . . . the black cross of reflections which dominated the picture could arise only from a helical structure . . . mere inspection of the X-ray picture gave several of the vital helical parameters." After viewing Franklin's images, Watson and Crick published their theory of DNA structure in 1953.

Wilkins and Franklin published their studies of DNA structure, but Franklin's contributions were eclipsed by the acclaim given to Watson and Crick. In 1962 the Nobel Prize was awarded to Watson, Crick, and Wilkins. Franklin had died of ovarian cancer in 1958 and was not posthumously eligible for the prize; many believe she was deserving of it. ■

Interphase is the time of cell growth and productive activity. During the first part of interphase, after a cell is "born" from the division of a parent cell, the new cell gets busy making proteins, carbohydrates, lipids, and the like, as well as organelles. These materials accumulate in the cytoplasm, and the cell gets bigger, its membrane expanding. As this stage comes to an end, the centrioles near the cell's nucleus begin to replicate in preparation for cell division.

The DNA within the cell's nucleus replicates during the second stage of interphase. First, the coiled threads of DNA in the chromatin begin to unwind from the histones. Then an enzyme unzips the two strands of the DNA molecule (see "Inheritance," page 24). The bases that form the teeth of the DNA zipper—adenine connecting to thymine, and guanine connecting to cytosine—pull apart as the enzyme breaks the hydrogen bonds that held them together. Free-floating DNA nucleotides in the nucleoplasm latch on to the separate DNA strands, forming an exact duplicate of the missing strand, to create two identical DNA molecules. The new molecules coil around newly made histones and condense to form a pair of identical chromatids, joined by a buttonlike centromere.

This middle part of interphase lasts about six to eight hours in a typical mammalian cell. When it is over, the cell is committed to reproduction. It enters the final, short period of interphase, continuing to grow and make proteins for perhaps four to six hours until cell division begins.

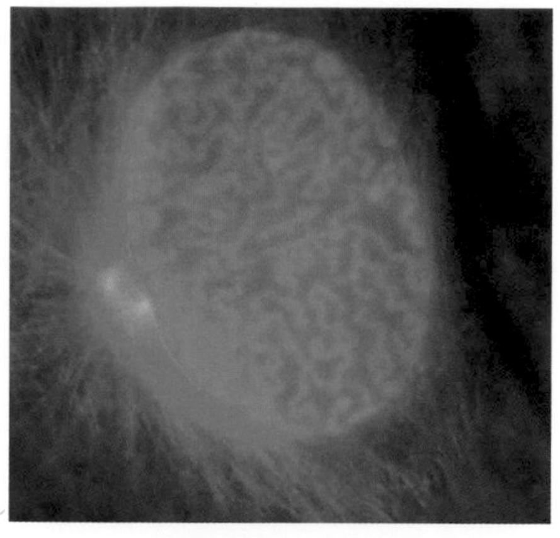

PROPHASE
A cell undergoing mitosis enters prophase; microtubules appear in green, chromosomes in blue.

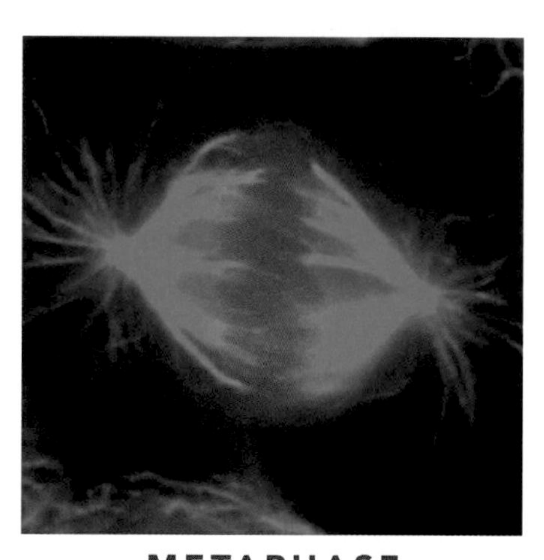

METAPHASE
During metaphase, chromosomes (blue) line up along the mitotic spindle (green).

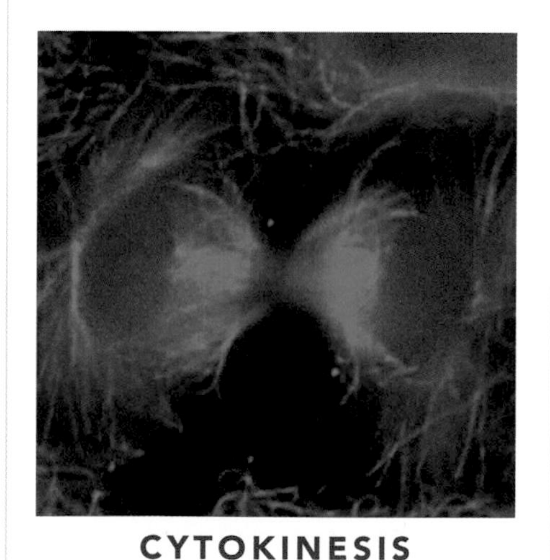

CYTOKINESIS
Microfilaments cinch a cell in half during the stage in which a cell splits in two.

When a cell divides, its nucleus splits first, in the process of mitosis, followed by the cytoplasm, in cytokinesis. (Reproductive cells—the sperm and egg—reproduce by meiosis; see "Meiosis," opposite.) The amazing dance of the chromosomes that takes place during mitosis is clearly visible through the microscope; biologists divide it into four steps known as prophase, metaphase, anaphase, and telophase.

In prophase, the chromatids condense into tightly coiled packages, each one made of two identical DNA molecules joined by a centromere. Around them, in the nucleus, the nucleoli disappear. The tubular centrioles in the nucleus, which had replicated themselves in interphase, separate into two identical pairs. From them grows a set of tiny tubules called the mitotic spindle, which stretches from one end of the cell to the other.

During the next step, metaphase, the paired chromatids line up neatly along the midpoint of the mitotic spindle. Then, during anaphase, the centromeres that held the pair of identical chromatids together split; the chromatids, now called chromosomes, are pulled apart and quickly dragged by their centromeres along the mitotic spindle's tubules toward opposite poles of the cell.

SEE ALSO: Chapter Eleven, "Female System," PAGE 300, Chapter Twelve, "Innate Defenses," PAGE 322

Once the two groups of chromosomes have reached their destinations, telophase begins. The chromosomes uncoil and become, once again, loose threads of chromatin. A new nuclear membrane forms around the chromatin, nucleoli reappear, and the mitotic spindle disappears.

Throughout the earlier phases of mitosis, the cell remains intact. During late anaphase, cytokinesis begins. An indentation forms in the middle of the membrane, where microfilaments tighten around the cell like a belt. Soon the microfilaments pinch the cell in half, creating two new cells with identical sets of chromosomes in their new nuclei. Despite its amazing mechanics, mitosis in mammalian cells typically takes less than an hour; the entire cycle from cell formation to division might last 20 to 24 hours, although some cells last a lifetime.

What prompts a cell to divide? Biologists don't know for sure. Cell size might be one trigger: When a cell reaches a certain ratio of surface area to volume—when it gets fat—chemical signals might prompt division. Normal cells also stop dividing when they begin to touch each other. Scientists are eager to decipher the signals that turn cell division on and off, since cancer cells notoriously ignore these normal controls.

CELL DEATH

Scientists would like to uncover the secrets governing cell division, but they may be even more eager to find out what causes cell aging and death. Most human cells eventually stop dividing and all of them die. Biologist Leonard Hayflick, working in the 1960s, found that cultured human cells divide about 50 times, on average, before they no longer reproduce and eventually perish: This upper boundary to a cell's division is now known as the "Hayflick limit."

Cell deaths can be planned or unplanned. Planned, or programmed, cell death is called apoptosis. Apoptosis prunes away unneeded tissues in the developing body, such as the webbing between fingers and toes in a fetus or unused neurons in the growing brain. Outside the womb, the process also kills cells that have been invaded by certain viruses. When the body determines that cells must go—a mechanism that is not fully understood—enzymes called capases destroy the cell's DNA and internal structures. The cell shrinks, detaches from its neighbors, and is promptly gobbled up by other cells.

Unplanned cell death—necrosis—comes about as a result of disease or injury. Cells that lose their oxygen supply, for example, or are frozen or burned may swell and burst. The immune response this triggers can cause inflammation (swelling and redness) in the surrounding tissues.

MEOISIS

SPERM AND EGG cells divide by meiosis. First, the 23 chromosomes replicate. Then, partnered chromosomes exchange genetic material; the cell divides into two cells with 23 chromosomes and two chromatids. Division occurs again, producing four daughter cells with 23 chromosomes and one chromatid. Each one is distinct from the parent cell and each other.

WHAT CAN GO WRONG

CANCER is a class of diseases marked by the uncontrolled growth of cells. Cancerous (malignant) cells grow into abnormal tissues called tumors, masses that may spread (metastasize) to other parts of the body. Almost any kind of cell can become cancerous, although cancers vary widely in their severity, with lung cancer, right, and prostate cancer being among the most deadly. Cells become cancerous as a result of mutations, typically damage to the genes that control cell division and growth. Hereditary factors can play a part, while exposure to carcinogens, like radiation or cigarette smoke, can also cause mutations.

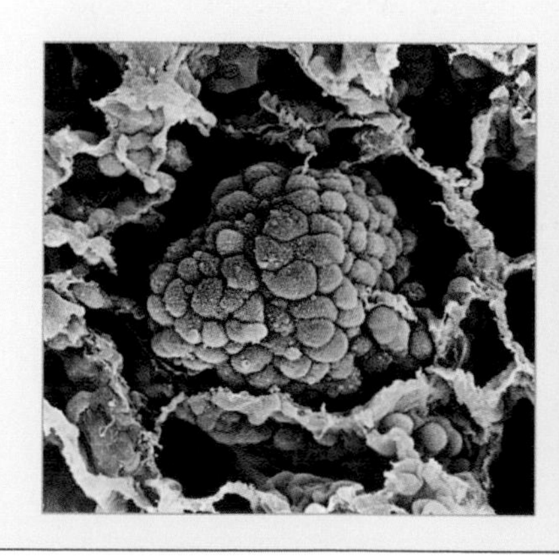

Cell aging is still a mystery. Simple damage over time may contribute to it as environmental toxins accumulate in the body. Free radicals may also play a part. These are molecules that are missing a single electron; when they enter a cell, they pull electrons from the cell's molecules, damaging them. Some substances, such as vitamins C and E, are antioxidants that may help prevent this damage.

Gradual genetic damage may promote aging, as well. Every chromosome is capped by a telomere, a long string of bases that carries no genetic information. Each time a cell reproduces, a piece of the telomere cap is snipped off. The erosion and loss of the telomeres may signal the cell to stop dividing.

This theory picked up support when, in the 1980s, biologists discovered an enzyme that protects telomeres. Dubbed telomerase, it can repair and extend the telomere string. Telomerase is virtually absent in most adult cells, but it is found in germ cells (the cells that produce sperm and eggs) and in cancer cells, which reproduce without aging.

Most likely, many factors working together cause cell aging and death, and those factors may vary from one cell type to another. Learning more about them could lead to cures for diseases ranging from cancer to Alzheimer's disease.

+ CELLULAR LIVES +

HUMAN CELLS vary widely in their life spans. Some examples include

+ Granulocyte (type of white blood cell): 13–20 days
+ Keratinocyte (type of skin cell): 14–28 days
+ Red blood cell: 120 days
+ Brain neuron: human lifetime
+ Heart muscle cell: human lifetime

A cancer patient prepares to enter a PET (Positron Emission Tomography) machine.

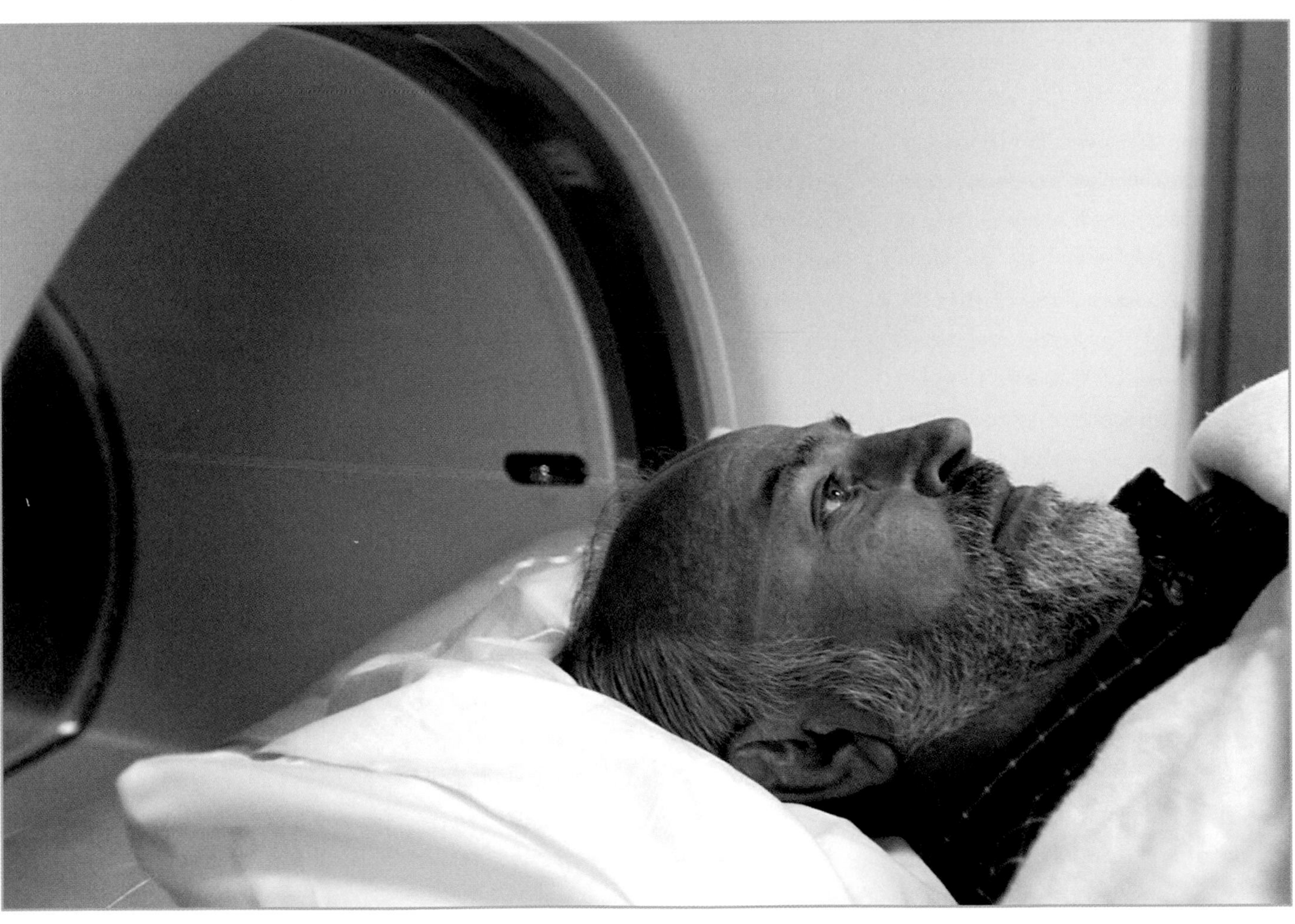

SEE ALSO: Chapter Thirteen, "Building Blocks," PAGE 352

CHAPTER GLOSSARY

ALLELE. A variant form of the same gene on a matching chromosome; different alleles lead to variations in inherited characteristics, such as eye color.

APOPTOSIS. Planned cell death.

BLASTOCYST. A hollow ball of cells that will develop into an embryo. It consists of a trophoblast (outer cells), blastocele (inner cavity), and inner cell mass.

CENTRIOLE. Two tubular structures, arranged at right angles, within a cell's centrosome.

CENTROSOME. A region of the cell containing centrioles that forms the spindle during cell division.

CHROMATID. One of a pair of replicated chromosomes.

CHROMATIN. The threadlike genetic material, made of DNA and histone proteins, that is present in the nuclei of nondividing cells.

CHROMOSOME. The structure found in the nucleus of a cell, made of DNA and contains genes In humans, chromosomes come in 23 pairs, making 46 chromosomes altogether in all cells but gametes.

CYTOPLASM. The contents of a cell, except for the nucleus.

CYTOSKELETON. A network of fibers in the cytoplasm that helps maintain a cell's structure.

CYTOSOL. The fluid portion of cytoplasm that holds the organelles.

DNA. Deoxyribonucleic acid; a nucleic acid molecule that makes up the 46 human chromosomes.

EMBRYO. The human organism from early stages of development to the end of the eighth week after fertilization occurs.

ENDOPLASMIC RETICULUM. A network of channels and flattened sacs running through a cell's cytoplasm; it serves to store, transport, and package molecules.

FETUS. Term for the developing human organism from nine weeks until birth.

GAMETE. Male or female reproductive cell; the sperm or egg. Contains 23 chromosomes.

GENE. The basic unit of heredity; a piece of DNA that codes for a specific protein or segment of protein. Found in chromosomes.

GENOME. All of the genes of an entire organism.

GOLGI COMPLEX. A cell organelle that modifies and delivers proteins and lipids within the cell and to the plasma membrane.

HETEROZYGOUS. Having different alleles that control the same trait on matching chromosomes.

HOMOZYGOUS. Having the same allele that controls the same trait on matching chromosomes.

LYSOSOME. A cell organelle containing powerful digestive enzymes.

MEIOSIS. A type of cell division that occurs only in egg and sperm cells; it involves two nuclear divisions and produces four daughter cells with half the number of chromosomes as the body's other cells.

MITOCHONDRION (pl. mitochondria). A double-membraned cell organelle that produces energy for the cell.

MITOSIS. A type of cell division in which each daughter cell receives the same amount of DNA as the parent cell and is genetically identical to the parent.

MUTATION. Any change in a gene that alters its sequence of bases.

NECROSIS. Cell death resulting from disease or injury.

NUCLEOLUS (pl. nucleoli). A spherical body within a cell nucleus that is the site for production of ribosomal units.

NUCLEOTIDE. A compound consisting of a nitrogenous base, a phosphate group, and a sugar; DNA and RNA are chains of nucleotides.

ORGANELLE. A structure within a cell that performs a specific job.

PEROXISOME. A cell organelle containing enzymes that utilize molecular oxygen to oxidize organic compounds.

PLASMA MEMBRANE. An outer membrane that separates a cell's contents from the outside environment.

PROTEASOME. A tiny cell organelle that destroys unwanted proteins.

RIBOSOME. A cell organelle that is the site of protein synthesis.

RNA. Ribonucleic acid; transmits instructions from DNA to the cytoplasm, where proteins are made, and does other metabolic tasks.

ZYGOTE. The fertilized egg.

INHERITANCE

PASSING ON TRAITS TO THE NEXT GENERATION

TWO KEY MOLECULES within the cell's nucleus, DNA and RNA, hold the key to the body's central mysteries: its birth and death, its diseases, its very identity.

DNA is deoxyribonucleic acid, an immensely long organic molecule. It's built like a ladder: The sides are chains of sugars called deoxyribose that alternate with phosphate groups. Four bases attached to the sugars pair up, each pair forming a rung: Adenine links to thymine, and guanine to cytosine. The DNA molecule twists around itself in the spiral called the double helix. Within the nucleus of almost every cell are 46 of these spiraling molecules in 23 matched pairs—46 human chromosomes.

The basic unit of each chromosome is a nucleotide formed of a sugar, a phosphate group, and a base: a piece of the side of the ladder attached to half a rung. Genes are simply groups of nucleotides, and the order of the nucleotides in the gene—for instance, ATTGCCA—dictates the gene's purpose.

Human DNA contains nearly 3.2 billion nucleotides; the average gene consists of 3,000 nucleotides, but some genes have more than one million. An organism's complete set of DNA is called its genome and contains some 30,000 to 40,000 genes. A large portion of DNA seems to consist of repetitive strands of nucleotides. These might be "extinct" genes, traits discarded in the course of evolution, or they might be strands whose meaning has not yet been deciphered.

TRANSCRIPTION

What do genes do, exactly? These molecular snippets loom large in the popular imagination, depicted as dictators of everything from criminal behavior to cold sores. Yet a gene is nothing more than a set of instructions for making a particular protein in a cell. Its molecular code—the sequence of A, T, C, G nucleotides that gives it its identity—spells out the order in which amino acids (the building blocks of proteins) must be strung together.

The gene doesn't even transmit the instructions. That job falls to ribonucleic acid, or RNA, and consists of two steps: transcription, when information from the gene is copied onto a strand of RNA, and translation, when the information in the RNA is used to make the protein in the cell's ribosomes.

During transcription, the DNA helix pulls apart. One of its strands serves as the template for a new strand of RNA, whose complementary bases match those on the DNA—except that RNA contains uracil (U) instead of thymine. A

An unwinding strand of DNA picks up nucleotides to form two new, matching strands of DNA.

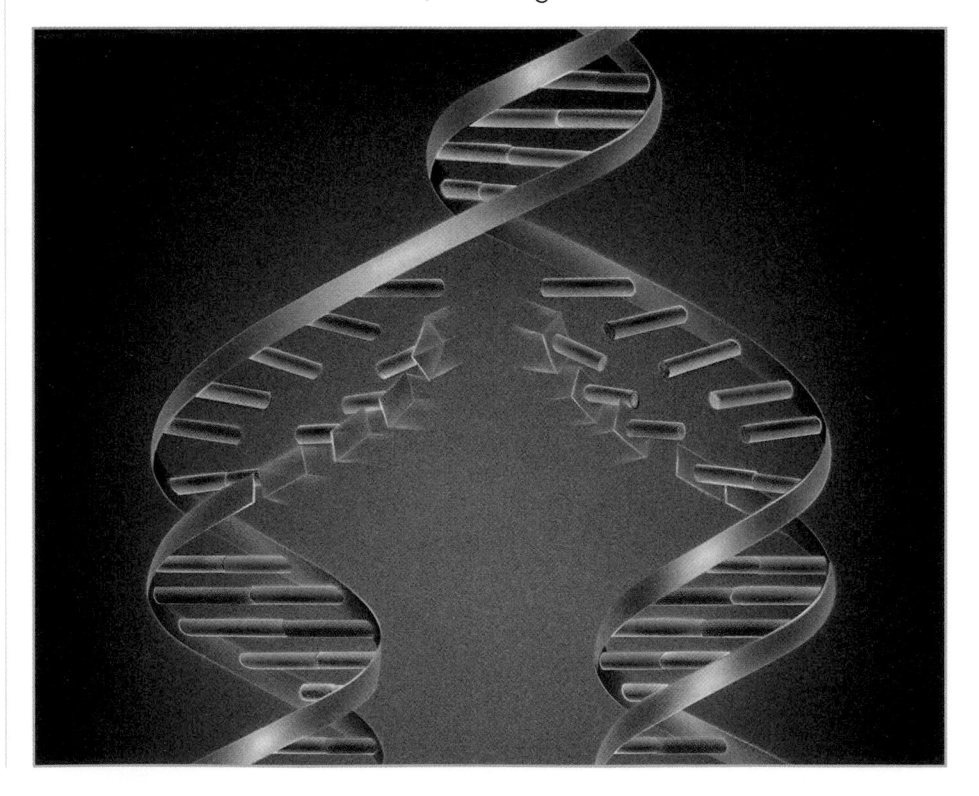

SEE ALSO: Chapter Eleven, "Conception," PAGE 306

strand of DNA with TTAACC would therefore produce the RNA sequence AAUUGG. Three kinds of RNA are made in this way: messenger RNA (mRNA), which carries the protein instructions out of the nucleus to the ribosomes; ribosomal RNA (rRNA), which helps the ribosomes build the proteins; and transfer RNA (tRNA), which brings amino acids, dissolved in the cytosol, to the ribosomes.

So all that genes do is direct a cell's protein production. Still, this is an extraordinarily powerful job. Proteins form the basic material of the human body, accounting for up to 30 percent of the mass of each cell. They include enzymes that carry out a cell's chemical reactions; they make up the hemoglobin in the blood, the collagen in the skin, hormones that regulate growth, antibodies that fight disease, and innumerable other vital parts and functions. In controlling proteins, genes control the growth and workings of every cell in the body. They also dictate the growth, specialization, and development of every cell in a human embryo.

Virtually all cells contain an identical, complete set of the genetic code in their nuclei. Why, then, does a kidney cell look and act differently than a brain cell? It's because different genes are active in different cells. In any given cell, only a small percentage of the genes are at work, most performing functions that are basic to any cell. Just a few are activated for the tasks that identify that particular cell or tissue. Thus, all cells contain the genes for making insulin, for example, but only in the pancreas are those genes active.

All kinds of signals can activate genes, and this is where the environment and even emotions interact with the subtlest workings of the body. Molecules within the cell may switch on genes, but genes are also turned on or off by hormones or other chemical signals that enter the cells from outside. Fear or stress, for instance, can signal genes to produce proteins in response and to stimulate or inhibit the activity of neurons in the brain. The interplay of genes and environment is one of the more exciting new areas of biology.

SIMILAR GENES

+ THE HUMAN GENOME is 99.9 percent identical throughout the world.
+ The genes of humans and chimpanzees differ by about 1.2 percent.
+ Humans and fruit flies share about 60 percent of their genes.
+ Cro-Magnon DNA has similarities to Europeans' but not Neandertals', suggesting Neandertals did not contribute to the European gene pool.

HEREDITY

Genes are inherited. Because they are produced by meiosis, the gametes (or sperm cells and egg cells) are the only cells to contain just 23 chromosomes (see "Meoisis," page 21). When a sperm merges with an egg, their chromosomes combine to form the usual 46—half from the mother, half from the father—that will be duplicated again and again as the embryo grows and develops.

These inherited chromosomes are not, however, exact copies of those in the parents' cells. Unlike cell division elsewhere in the body, every time sperm cells and egg cells divide, their genes cross over and are randomly reassigned

BREAKTHROUGH

DECODING THE GENOME: In 1990, the Human Genome Project began work on an ambitious program: mapping the entire human genome by identifying the sequence of all three billion base pairs on our chromosomes. Research centers in the United States, Europe, and Asia collaborated, completing the work in April 2003, ahead of schedule. As well as identifying the order of every nucleotide in human DNA, the project pinpointed the location of the genome's 30,000 to 40,000 genes and created maps to link inherited traits to certain genes. Today, the entire sequence is now available online, for free.

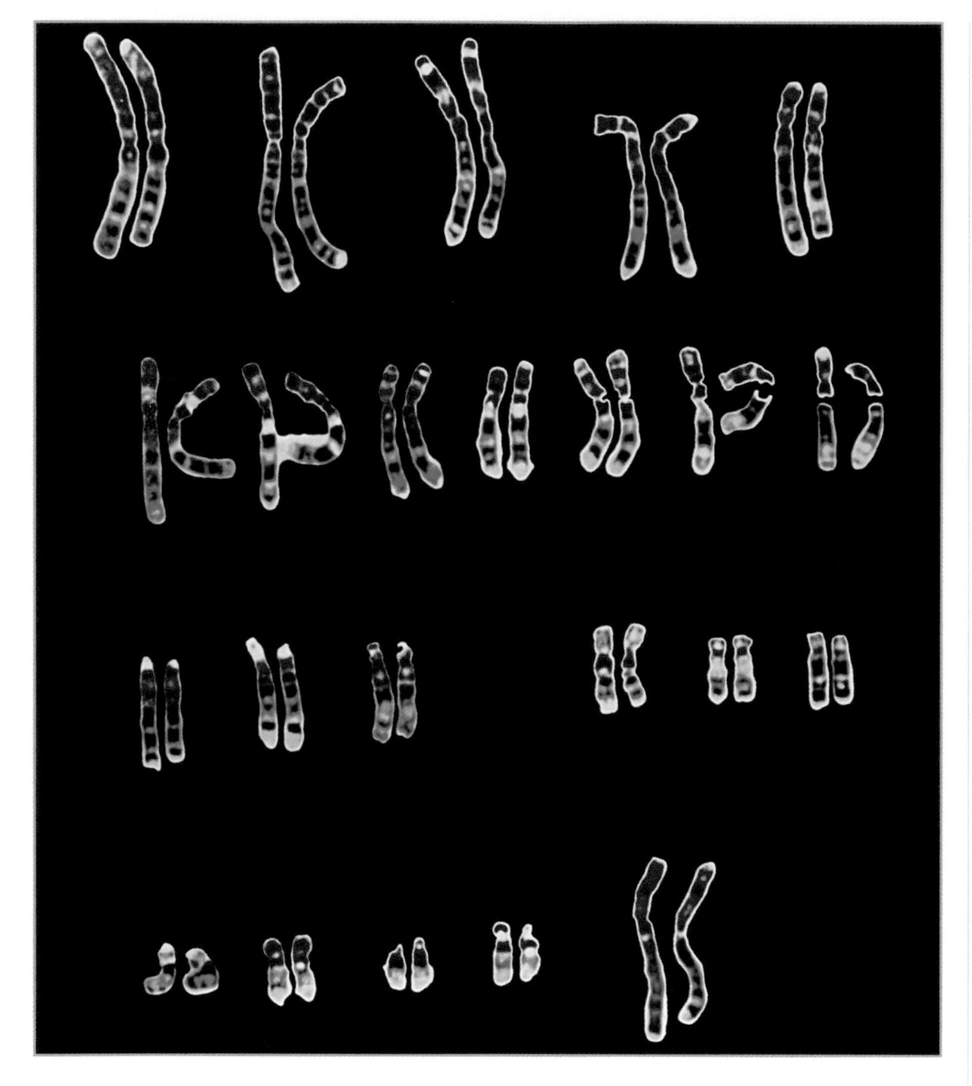

Twenty-three pairs of matching chromosomes end with two X chromosomes, showing that this set belongs to a woman.

between matching chromosomes. As a result of this process, every sperm cell and egg cell is slightly different genetically.

The genes on each of the linked chromosomes in every cell code for the same proteins and the same traits. This redundancy provides critical backup for the cell: If one gene is damaged, the matching gene on the linked chromosome will usually pick up the slack. The matched genes—called alleles—aren't always identical, however. If they do code for the same form of a trait, they are called homozygous. If they code for a different version, they are heterozygous.

Sometimes, one heterozygous allele suppresses the expression of its partner. Such an allele is called dominant, while its partner is called recessive. Geneticists typically depict this pairing with capital and lowercase letters. For example, brown-eyed man #1 may have inherited dominant genes for brown eyes (B) from both of his parents, resulting in a BB genotype. Brown-eyed man #2 may have inherited one dominant gene for brown eyes and one recessive gene for blue eyes, giving him a genotype of Bb.

The mix-and-match game of inherited traits begins when sperm meets egg. Let's say brown-eyed man #1 (BB) marries blue-eyed woman #1 (bb). Each of their children will have a Bb pair of genes, and will therefore have brown eyes.

Now let's say brown-eyed man #2 with a recessive blue-eyed gene (Bb) marries a similar brown-eyed woman (Bb). The odds are 3 out of 4 that a child will have brown eyes (BB or Bb) and one in four that it will have blue (bb). Note that these are probabilities only. It is possible for the second couple to have four blue-eyed children.

It's usually not that simple. Few traits are expressed by one gene: Almost all result from several genes working in tandem. Some traits—such as ABO blood types—are expressed by multiple allele forms, in which two can be co-dominant. Alleles are incompletely dominant, such as those that can cause sickle-cell anemia (see "What Can Go Wrong," page 30).

Nor are dominant traits necessarily more common, or desirable, than recessive ones. Huntington's disease and polydactyly (extra fingers or toes) is caused by dominant genes; normal vision and normal cartilage formation are found on recessive genes.

KEEPING HEALTHY

ALMOST EVERYONE can identify an ailment that "runs in the family": heart disease, perhaps, or asthma. It is true that many diseases have a genetic component, with family members sharing an increased risk of developing the same illness.

Compiling a family medical history can give a clearer understanding of your own health risks and help you and your doctor plan your own health care, including lifestyle changes, further medical tests, or even genetic testing.

A family medical history should include parents and siblings, and preferably grandparents, aunts, uncles, and half-siblings. Try (tactfully) to assemble the following information for each person:

+ Birth date
+ Death date (if applicable)
+ Major medical conditions, such as diabetes, coronary artery disease, cancer, stroke, depression, or Alzheimer's disease
+ Other health issues, like allergies, asthma, arthritis, migraines
+ Miscarriages or stillbirths, and causes (if known)
+ Dominant racial and ethnic background (some diseases, such as sickle-cell anemia, are linked to race or ethnicity)
+ Lifestyle factors and choices, such as smoking, diet, exercise, and alcohol consumption
+ Cause of death (if applicable) ■

Families share more than hair color or winning smiles: Health conditions ranging from allergies to heart disease can also be inherited.

In the game of heredity, two chromosomes stand out from the rest: the X and Y sex chromosomes. The Y chromosome contains the genes that determine maleness; every man's sex chromosomes are an X and Y set, while every woman's are XX.

The X chromosome is much larger and richer in genetic information, containing about 1,300 genes versus several hundred on the Y. This means that most characteristics on the X chromosome will be expressed in men, because they are not countered by matching genes on the Y. Traits that are linked to genes on the X or Y chromosomes are called sex-linked; muscular dystrophy and hemophilia are examples of X-linked traits (or more precisely, damage to genes on the X chromosome).

INHERITING ILLNESS

Genetic diseases can be linked to a single damaged gene, to entire misplaced chromosomes, or to many genes working together.

DNA PROFILES

ONLY IDENTICAL TWINS have identical DNA, so most people can be identified by their DNA. First, researchers cut chromosomes into pieces of different lengths and sort the fragments by size. They look for 13 specific DNA pieces using a complementary DNA set as a probe. If all 13 areas match in two samples, then they come from the same person.

GENETIC HISTORIES

EVERY HUMAN ALIVE today—from you and your neighbor to Inuit fishermen and Indonesian schoolgirls—is descended, DNA studies suggest, from hunter-gatherers who migrated from Africa roughly 60,000 years ago. They made their way past the shores of the Red Sea and into western Asia. Among the first modern humans to leave humanity's ancestral homeland, they then spread out to Australia, and East Asia, followed by Europe and the Americas.

RESEARCH

The story of their epic journey is written in our genes. Genetic markers in DNA present a picture of our ancestors and their migrations, allowing us to trace the broad pathways of ancient human exploration. Using techniques that have become available only in the past 20 years, some geneticists are now mapping ancient human history.

One of the leading inquiries into human migration is the Genographic Project, sponsored by National Geographic and IBM, with funding for field research provided by the Waitt Family Foundation. The project is led by population geneticist Spencer Wells, a National Geographic explorer-in-residence. At ten research centers around the world—in sub-Saharan Africa, North Africa/Middle East, India, East/Southeast Asia, Australia/Pacific Ocean, North Eurasia, Central/Western Europe, North America, and South America—geneticists are collecting blood samples and cheek swabs from living people in order to study their DNA. At an 11th center, in Australia, another researcher studies DNA from ancient fossils. Researchers hope that comparing samples will reveal new insight into how people populated the world.

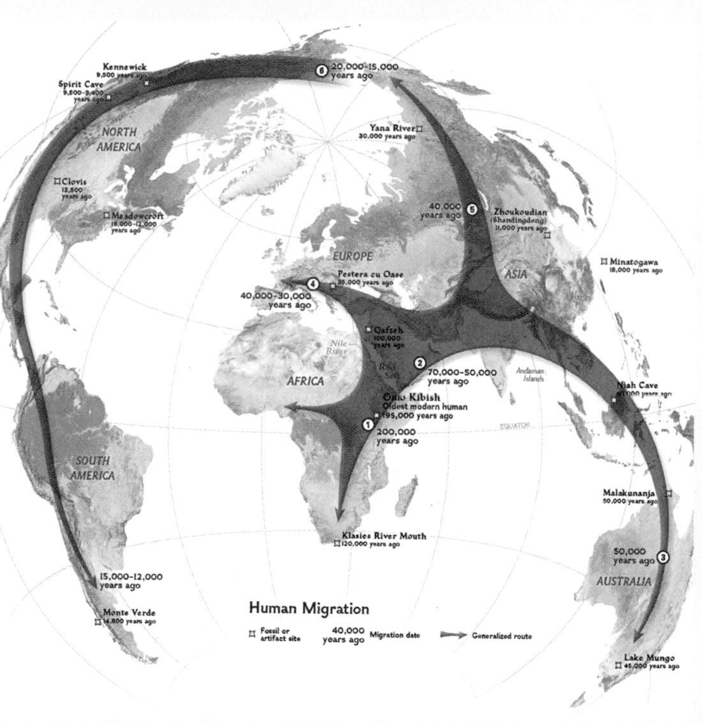

DNA studies have been used to map human migration out of Africa into the rest of the world.

MARKERS

The genetic markers that make this possible are found in two places: the Y chromosome and the DNA of mitochondria, the energy-producing organelles found in each cell (see "The Cell," page 14). Although virtually all of our DNA is recombined with every generation, these two parts of the genome remain for the most part unshuffled from parent to child.

Only men possess a Y chromosome (their sex chromosomes are XY, whereas women's are XX), so that chromosome is passed down virtually unchanged from father to son. Mitochondrial DNA, on the other hand, is passed down only through the maternal line, from mothers to children of both genders, because the mitochondria in a man's sperm cells are concentrated in the sperm's tail, which drops off before fertilization of the egg. Therefore this DNA, too, escapes reshuffling.

Very rarely, but apparently at a steady rate over time, a harmless mutation will occur in the DNA of the Y chromosome or the mitochondria. This mutation, called a genetic marker, will then be passed down through the generations;

all people with the same genetic marker share the same ancestor.

By comparing shared and distinct markers within populations, geneticists can determine when certain groups of people split off from others. They can even trace the markers back through time to the most recent pair of ancestors common to living humans, dubbed "Y-chromosome Adam" and "Mitochondrial Eve," two Africans who lived about 60,000 years ago and 150,000 years ago, respectively.

FINDINGS

Though much remains to be learned, the broad outlines of human migration have now been drawn. Modern humans arose about 200,000 years ago in Africa (a finding reinforced by fossil evidence). These hunter-gatherers spread across the continent until, about 70,000 to 50,000 years ago, one group migrated into the Arabian Peninsula.

Some of them then moved rapidly across Asia and into Australia some 50,000 years ago. Because of an ice age, sea levels were remarkably low at the time, so that just 62 miles of open water separated the Australian landmass from Asia; the newcomers may have navigated by sight from island to island. Others traveled into Central Asia (about 40,000 to 50,000 years ago) and some into Europe (about 40,000 years ago).

One question with regard to Europe that the Genographic Project's findings could possibly answer is the contribution, if any, of the Neandertals to the genetic code of modern humans. European descendants of prehumans who left Africa around 500,000 years ago, Neandertals mysteriously died out about 30,000 years ago.

This Kazakh girl's ancestors may have reached Mongolia 50,000 years ago.

One theory holds that the newcomers from Africa gradually pushed the Neandertals aside. Some scientists, however, citing fossil evidence, have recently suggested that there may have been interbreeding between the two groups, or even that commingling caused the Neandertals to disappear, as they were absorbed into the other group. Although mitochondrial DNA analysis shows that the two groups were distinct species, their mating is not an impossibility (as a horse might mate with a mule). One problem with this theory is that unlike the Neandertals, most human populations today have poor genetic defenses against cold weather. As yet, no trace of Neandertal genes has been found in the genetic code of modern humans.

The exact arrival date of humans in the Americas is in dispute, but DNA evidence suggests that sometime between 15,000 and 20,000 years ago several waves of immigrants from Asia crossed from Siberia to Alaska. It is estimated that they then spread into South America, 12,000 to 15,000 years ago.

Today, the Genographic Project is working to collect DNA from around the globe, particularly from isolated, indigenous populations before they are assimilated into the broader genetic pool. Not all groups welcome or cooperate with the project or others like it; some are concerned that the results may overturn traditional ideas of cultural origin and beliefs about race and identity. The project's researchers hope to collect 100,000 DNA samples within five years from around the world.

Those linked to a single gene—say, hemophilia—are the easiest to track down, but many disorders are probably caused by the complex and difficult-to-trace interaction of genes and environment.

Normal genes do not code for disease. When we say that someone has the gene for muscular dystrophy or breast cancer, what we mean is that the person has a mutated version of a particular gene, and therefore the protein encoded by the gene is changed or missing in that person's cells.

A mutation is a permanent change in the structure of a gene. Mutations may be inherited, passed down from parent to child in every cell; they can also be triggered in selected cells by a host of environmental factors, including malnutrition and exposure to solar radiation. Possible mutations include a swapped base in the genetic code—adenine in the place of cytosine, say—or missing or repeated stretches of DNA.

Many mutations appear meaningless genetically. Some produce harmless variations in a population: different hair color or blood types, for example. The mutations of most concern are the destroyers. Several thousand diseases result from genetic damage. In some, the location of the mutation determines the severity of the disorder. Hundreds of different mutations can affect the genes that control mucus production, for example. Some have no effect; some result in a mild case of cystic fibrosis and others in a severe one.

Inherited genetic disorders typically follow the standard rules of inheritance. For example, if one parent has a dominant mutated gene and the other parent's gene is normal, each child has a 50-50 chance of inheriting the condition. People who have one recessive mutated gene are called carriers. Carriers do not usually have the disease, but each child of two carriers has a one-in-four chance of having the disease, a similar chance of being healthy, and a two-in-four chance of being a carrier himself.

In recent years, geneticists have begun to pinpoint the location of some genetic diseases. The neurological disorder Huntington's disease, for instance, is caused by excessive repetitions of the nucleotide triplet CAG on chromosome 4. Most people average about 18 of these triplets in a gene that codes for the protein huntingtin. In people with Huntington's disease, the triplets keep repeating 40, 50, or more times, resulting in an abnormal protein that negatively affects the brain. Although symptoms of the disease typically manifest around middle age, in some cases the strings of triplets increase as they are inherited, leading to an earlier onset of the disease in succeeding generations.

Because scientists can identify the malformed gene, they can

COLOR BLINDNESS

RED-GREEN COLOR blindness is an X-linked trait. Women who carry the mutated gene on one X chromosome rarely have the disorder because they usually have a normal gene on the other X chromosome to offset it. A man who inherits the mutated gene from his mother will find the trait expressed, because he has no matching normal gene on the Y chromosome.

SEE ALSO: Chapter Five, "The Blood," PAGE 132

WHAT CAN GO WRONG

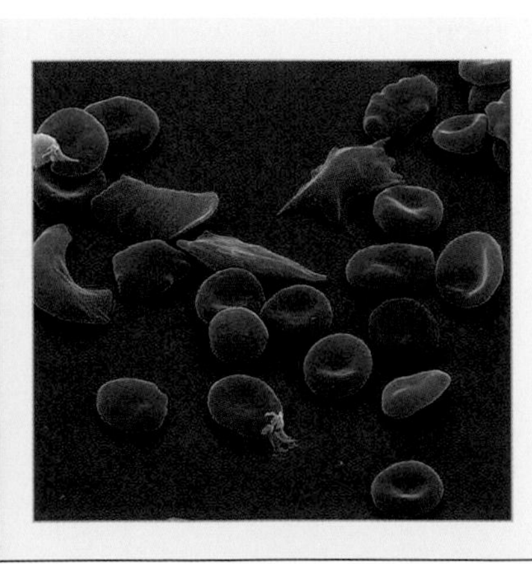

SICKLE-CELL ANEMIA is one of the most common and best understood genetic diseases. Mainly occurring in people of African, Hispanic, Mediterranean, and Middle Eastern heritage, it affects about 1 in 500 African Americans. People with this condition have an abnormal form of hemoglobin, the oxygen-carrying molecule; this variant is stiff and spiky, causing the red blood cells to take on a crescent, or sickle, shape, as they release oxygen to the body's tissues. When oxygen demand is high, the oddly shaped cells, left, can clog blood vessels, blocking bloodflow to tissues and organs—a painful and dangerous

A young woman with Down syndrome most likely has an extra copy of chromosome 21.

tell people whose parents have Huntington's disease whether they, too, will develop it. The disease currently has no cure, and the decision to be tested can be agonizing.

Some of the most prevalent genetic disorders are caused not by one mutated gene, but by the replication of all or part of an entire chromosome. An extra copy of chromosome 21, created in error during meiosis of a parent's sperm or egg cell, causes most cases of Down syndrome. Found in 1 out of every 800 to 1,000 babies, Down syndrome brings with it a host of medical problems including mental retardation, poor muscle tone, and heart and digestive defects.

Other trisomy ("three body") disorders include trisomy 18 (marked by three copies of chromosome 18) and trisomy 13 (three copies of 13), both linked to mental retardation and other problems. One or more extra X chromosome in boys (giving them XXY sex chromosomes) causes Klinefelter syndrome, a disorder whose onset (typically during puberty) leads to abnormally low levels of testosterone and development of some female characteristics.

Genetic disorders may also arise when a single cell in an early embryo mutates. As the embryo divides and grows, some cells will contain the mutation and others will not, a condition called mosaicism.

Rarely, two unrelated chromosomes will break and exchange chromosomes during cell division, a process called translocation. When this occurs, for instance, a piece of chromosome 21 might switch places with a piece of chromosome 14. Outwardly, the person carrying these swapped genes will have no symptoms, but when he or she produces sperm or egg cells, some of them will have the extra chromosome 21 material. The unsuspecting carrier could then parent a child with Down syndrome.

crisis. Sickle cells die more rapidly than bone marrow can replace them, resulting in anemia. The disorder is caused by a mutation to a single base in one gene on chromosome 11. Because the trait is recessive, it must be inherited from both parents to be expressed. If only one parent has the gene, red blood cells will be normal, right. Interestingly, the disease is strongly correlated with malarial areas in Africa; malaria parasites have difficulty surviving in blood containing sickle cells, so the disorder actually confers some protection from malaria. This positive effect may account for the persistence of an otherwise harmful condition.

Genetic disorders range widely in their severity. Some mutations are fatal from the outset; perhaps 50 percent of fertilized eggs fail to develop normally for this reason and are silently miscarried in the first days of pregnancy. Other traits, like male-pattern baldness, are not life-threatening.

Many common diseases—such as cancers and heart disease—probably involve a combination of genetic and environmental factors. Scientists are beginning to pinpoint the location of genes connected with some diseases, but mutations on these genes typically indicate only a tendency to develop a disease in certain circumstances. Jane Doe, a nonsmoker with cancer-related mutations on chromosomes 3 and 11, may never develop the disease; her identical twin, Joan, a smoker, might not be so lucky.

GENE THERAPY

Few areas of genetic research hold more promise than gene therapy. With the mapping of the human genome and advances in the manipulation of genetic material come the possibility that doctors could cure a host of genetic diseases using genes themselves.

This could be done in one of three ways: Scientists could replace a damaged gene with a healthy copy, introduce new genes into the body to fight disease, or deactivate or repair a misbehaving gene.

New genes usually won't incorporate into a cell's genetic material if just injected into the cell. They need to be delivered by a carrier, called a vector. Most researchers use a deactivated virus, such as an adenovirus (cold virus), that will infect the target cell without causing illness. The researcher splices the gene into the DNA of the vector and injects the vector into the

THE STORIES IN OUR GENES

Researchers have identified disorder-causing genes on each of the 23 chromosomes. Most carry a few defective genes with no signs of disease, and many genes only contribute to susceptibility. Lifestyle choices and environmental factors can raise or lower disease risk.

CHROMOSOME	DISORDER	CHROMOSOME	DISORDER
1	Breast cancer, Alzheimer's disease	13	Retinoblastoma, pancreatic cancer
2	Early-onset glaucoma, congenital hypothyroidism	14	Goiter, leukemia / T-cell lymphoma
3	Small-cell lung cancer, susceptibility to Asperger's syndrome	15	Tay-Sachs disease, colorectal cancer
4	Huntington's disease, polycystic kidney disease	16	Endometrial carcinoma, Crohn's disease
5	Allergic rhinitis, diabetes mellitus	17	Dementia, ovarian cancer
6	Estrogen resistance, epilepsy	18	Leukemia / B-cell lymphoma, Paget's disease of the bone
7	Cystic fibrosis, growth-hormone-deficient dwarfism	19	Myotonic dystrophy, cystinuria
8	Prostate cancer, myopia	20	Fatal familial insomnia, Creutzfeldt-Jakob disease
9	Fructose intolerance, dilated cardiomyopathy	21	Autoimmune polyglandular disease, amyotropic lateral sclerosis
10	Congenital cataracts, leukemia	22	Ewing's sarcoma, giant-cell fibroblastoma
11	Sickle-cell anemia, albinism	X	Color blindness, hemophilia
12	Lung cancer, morbid obesity	Y	Turner's syndrome, gonadal dysgenesis

Above: Blue light helps a Mennonite girl with Crigler-Najjar syndrome, a rare genetic disease, rid her body of toxic bilirubin.

affected tissue. There, it infects the cell and incorporates the healthy gene into the cell's DNA. The cell and its descendants then produce functional proteins.

Gene therapy is still being researched and is available only in experimental trials. Since the first test in 1990 there have been hundreds of trials with limited success and some tragic failures. In 1999, an 18-year-old patient with a liver disorder died when the treatment resulted in massive organ failure. In 2002, two French children who had been successfully treated for immune system disorders developed a leukemia-like condition.

Right now, gene therapy is not long-lasting and must be repeated periodically in the patients, but this can stimulate a harmful immune response. Disabled viruses make useful vectors, but they could possibly regain their ability to cause illness once in the patient.

Ethical questions also surround the procedure: What will happen if a parent decides that short stature, below-average I.Q., or impulsive behavior is a disorder to be cured? Who will decide? What should the guidelines be?

On the other hand, gene therapy's potential is undeniable, especially for single-gene diseases. The same experiment that resulted in two cases of leukemia also successfully treated several other children. With more study, today's risky business could be tomorrow's spectacular cure.

FIRST GENE THERAPY

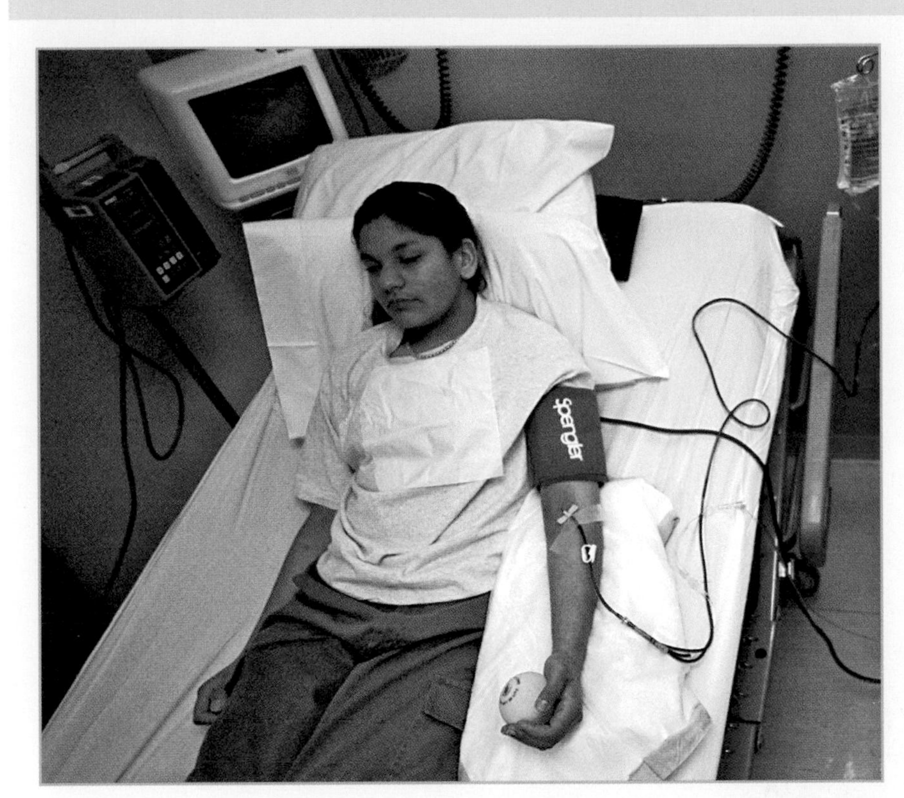

Ashanthi De Silva in 1999 receiving follow-up care for her immune disorder.

A LITTLE GIRL from the Cleveland suburbs made medical history September 14, 1990, when she became the first person to receive gene therapy. Four-year-old Ashanthi De Silva suffered from adenosine deaminase (ADA) deficiency. People with the rare disorder have defects in the genes that code for ADA, an enzyme needed for the normal development of T cells, key elements of the immune system. Without a healthy number of these white blood cells, the body is virtually unable to fight off infection. De Silva had been desperately ill since birth. She had to live in a cloistered, sterile environment, and her chances of surviving to adulthood were small.

After four years of struggle, De Silva's parents took her to the National Institutes of Health, where gene therapy was in its vigorous infancy. There, doctors removed some T cells from her blood, cultured them in the lab, inserted healthy ADA genes into the cells, and infused the cells back into her bloodstream.

The genetically altered cells took hold to some extent, and De Silva's body eventually produced about 25 percent of the necessary ADA. The new genes seem to falter after a while, so the therapy has to be repeated frequently. De Silva is healthier, although not completely cured. She has become an active adult and an example of gene therapy's potential. ■

DEVELOPMENT

SINGLE CELLS TO HUMAN BEINGS

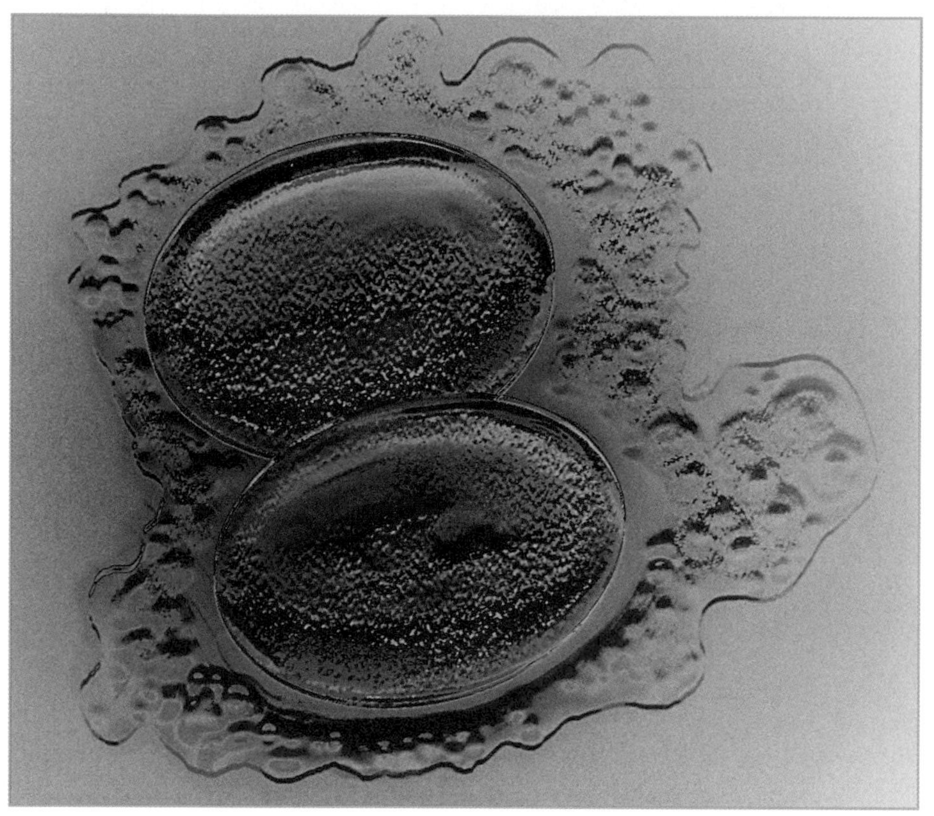

A two-cell zygote is the first product of fertilization. From these identical cells will come all the structures of the human body.

TRANSFORMING a single fertilized cell into a human infant is one of the most extraordinary processes in biology. Even today, with our understanding of cell biology advancing on a daily basis, many of the complex mechanisms that govern embryonic growth and human development remain enigmatic, but promising subjects for further research and study.

For the human egg, the literal journey toward fertilization begins when the female's egg pops out of an ovary and drifts into the nearby fallopian tube. At this point, the egg is midway through the process of cell division by meiosis; it carries two sets of 23 chromosomes, one set in the main body of the egg (called a secondary oocyte) and the other encapsulated in a little polar body, which is later discarded. As it moves down the fallopian tube, propelled by hairlike cilia in the tube's lining, the egg is chaperoned by a cloud of encircling smaller cells called the corona radiata.

Meanwhile, some 250 million sperm are furiously swimming up the reproductive tract. Consisting of little more than heads, filled with 23 chromosomes, and tails (or flagella) Packed with mitochondria to produce energy, the sperm die by the millions as they traverse the uterus and enter the fallopian tube where the egg is slowly making its progress down the tube. Only a few hundred sperm will survive to meet the egg. The first to arrive won't even get to fertilize it. Instead, as they begin to batter the outside of the egg, they release enzymes that disperse the corona radiata cells surrounding it and gradually expose its surface.

Finally, one sperm penetrates the egg's outer covering, the zona pellucida, and the egg internalizes it. As soon as this event happens, calcium ions flood the cell and its charge changes from negative to

TWINS

+ ABOUT 12 OUT OF every 1,000 births in the United States result in twins, although multiple births are increasing due to fertility treatments.
+ Fraternal, or dizygotic, twins, are produced when two eggs, released simultaneously in the mother, are fertilized by different sperm. These twins will be genetically distinct, no more alike than any other siblings.
+ Identical, or monozygotic, twins result when a single fertilized egg splits into two embryos, almost always within eight days of fertilization. Monozygotic twins are genetically identical and always the same gender.

SEE ALSO: Chapter Eleven, "Conception," PAGE 306

positive, repelling all other sperm. The egg completes its process of meiosis and ejects the polar body. The nuclei of the sperm and egg, each containing 23 chromosomes, then merge.

The egg cell is now officially fertilized. Almost immediately, its chromosomes duplicate and the cell divides by mitosis, creating two identical daughter cells. This tiny entity, still packed within the confines of the egg's tough membrane, is called a zygote.

Proceeding slowly down the fallopian tube, the zygote divides again, into four cells, and then eight—each new cell just half the size of its predecessor because the zygote is still packed within the egg's hardened shell. About three days after fertilization, the zygote is a lumpy little ball of 16 or more cells, all identical. At this point, owing to its appearance, it is known as a morula, from the Latin word for "mulberry."

After four or five days, the morula finally arrives at its destination, the uterus. There, it begins to change. Its outer shell breaks down, allowing fluids from the uterus to enter the ball of cells. The cells, still duplicating, reorganize into a hollow sphere with a lump of cells, called the inner cell mass, at one end. The tiny mass—still no bigger than the original egg—is now called a blastocyst. The inner cell mass will eventually develop into the embryo, while the other cells, called the trophoblast, will become

CONJOINED TWINS

Conjoined twins Chang and Eng Bunker in a formal portrait, circa 1870.

CONJOINED TWINS (formerly called Siamese twins) are identical twins whose bodies are physically joined at birth. This unusual condition, occurring in an estimated 1 in every 100,000 live births, begins some time after the eighth day following fertilization, when a zygote starts to divide into two separate cell masses but, for unknown reasons, does not separate completely. The twins continue to develop, but share organs or body parts, most typically at the chest, abdomen, or back. Most—again, for unknown reasons—are female.

Although the first successful operation to separate conjoined twins took place as early as 1689, until the 1960s most conjoined twins remained attached for life. Such was the case with the two men who made "Siamese" twins famous, Chang and Eng Bunker. Born in Siam (now Thailand) in 1811, the twins were joined at the waist and soon began to be exhibited as curiosities. Both twins married and fathered several children before they died within a day of one another in 1874.

Fortunately, modern society no longer relegates conjoined twins to sideshows, and modern medicine has developed surgical techniques that can separate some, but not all, conjoined twins. The surgery can be highly risky and may result in the death of the weaker twin; the decision to go forward is a difficult one for doctors and families alike. ■

KEEPING HEALTHY

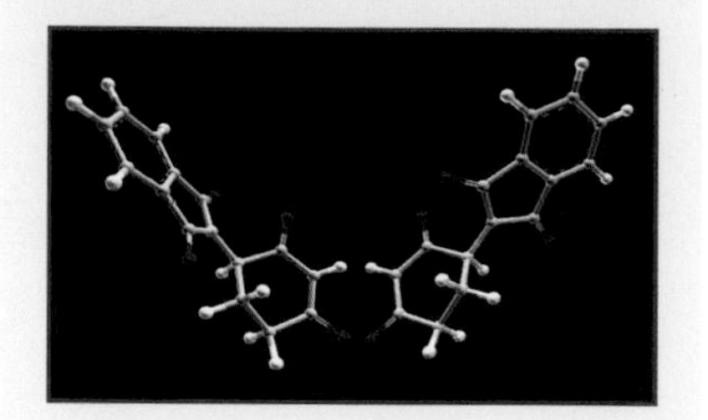

Molecular models of Thalidomide

FROM 1957 to 1961, pregnant women across Europe, and some in the United States, took a new drug called Thalidomide to ease morning sickness and sleeplessness. Almost 10,000 children born of those pregnancies had birth defects, many in the form of shortened or missing limbs. Too late, doctors realized that Thalidomide was a teratogen—an agent that causes developmental damage to an embryo or fetus.

Although the placenta serves as a filter between the maternal and fetal bloodstreams, as a rule whatever the mother ingests, the fetus also takes in. Many substances that barely affect the mother can hurt the fetus. Teratogens include alcohol, which can cause the irreversible brain damage and malformations of fetal alcohol syndrome; environmental chemicals such as mercury or herbicides; radiation; infectious agents such as rubella or toxoplasma; some prescription and over-the-counter medications; and illegal drugs. The most delicate period usually is during the first trimester of pregnancy, when the embryo is growing rapidly. ■

the fetal portion of the placenta. (The inner cell mass is also one possible source of stem cells; see "Powerful Dividers," page 38.)

IMPLANTATION

About six or seven days after fertilization, the blastocyst "implants" as it attaches to the wall of the uterus. The inner cell mass is closest to the wall. Producing enzymes that eat away at the uterine tissue, it extends its outer layer and embeds itself in the uterine lining. Its outer cells also begin to produce the hormone human chorionic gonadotropin (hCG), which helps to prepare the uterine lining and prompts the empty egg follicle back in the ovary, which has turned into a little gland called the corpus luteum, to continue its hormonal secretions. For the next few months, the corpus luteum will secrete the hormones estrogen and progesterone, which sustain the embryo as it grows.

A week after fertilization, the blastocyst is still no larger than a speck of dust. In its second week of development it will begin the spectacular process of differentiation and growth that will transform it into an embryo, then a fetus, and eventually an infant.

The process of implantation takes about a week. During this period, structures begin to form that will protect and nourish the developing embryo and, as it is called at a later stage, the fetus: the placenta, the amniotic sac, and the umbilical cord. The outermost cells in the blastocyst multiply and extend fingerlike projections called chorionic villi into the uterus, where they develop blood vessels. These blood vessels will eventually

ORGANOGENESIS

In an embryo, three layers of cells give rise to the body's different structures.

	ECTODERM	MESODERM	ENDODERM
STRUCTURES	All nervous tissue	Kidneys, gonads, reproductive ducts	Epithelium of digestive tract
	Epidermis, hair, sweat glands, nails	Skeletal, smooth, and cardiac muscle	Thyroid, parathyroid, and thymus glands
	Melanocytes	Cartilage, bone, connective tissue	Liver and pancreas
	Cornea and lens of eye	Blood, bone marrow, lymphoid tissues	Epithelium of respiratory tract, auditory tube, tonsils
	Tooth enamel	Synovial membranes of joints	Epithelium of urethra and bladder

SEE ALSO: Chapter Eleven, "Pregnancy," PAGE 312

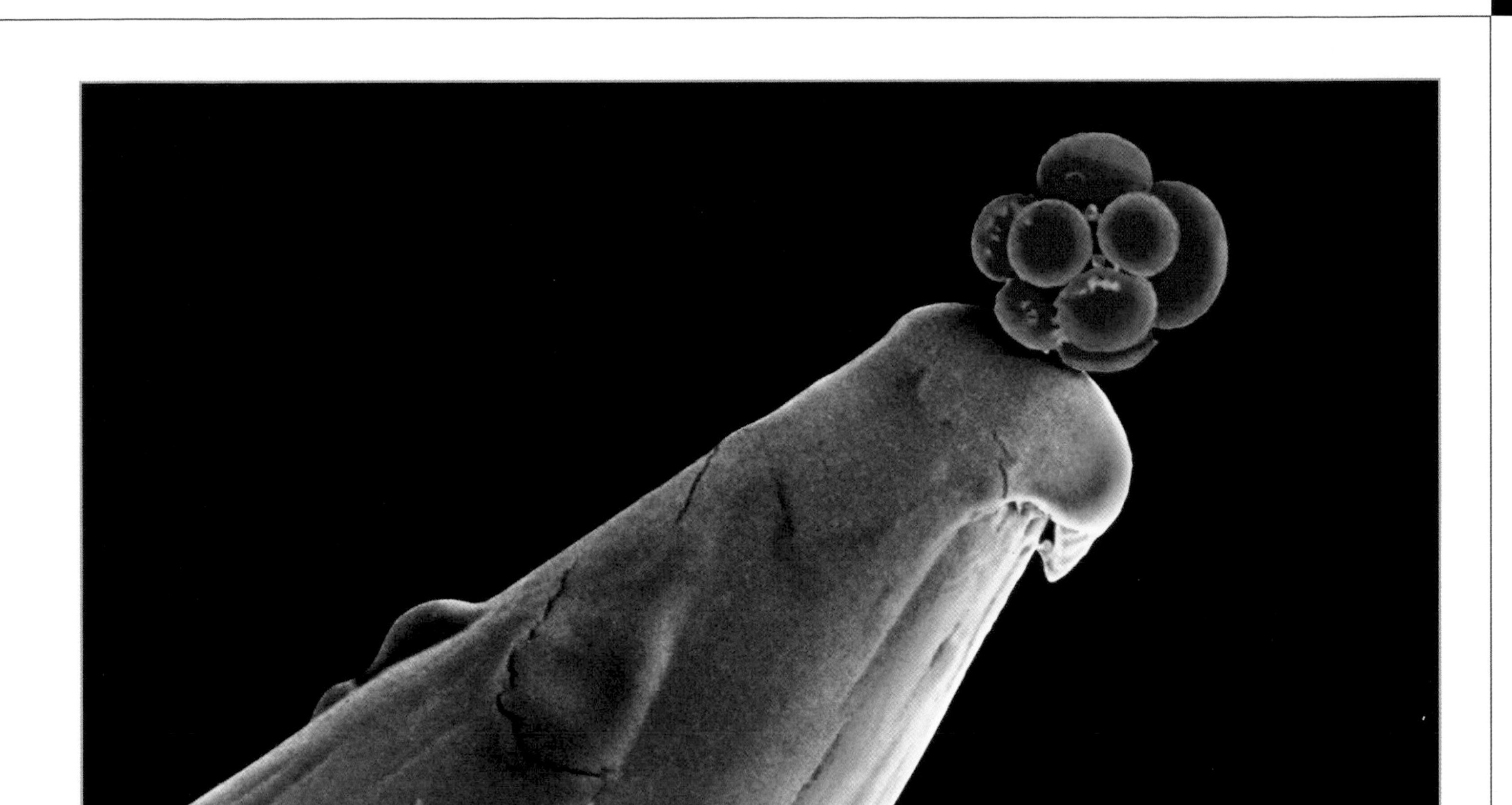

Perched on the head of a pin, this ten-cell embryo—known as a morula—is about three days old.

connect the embryo to the mother's blood supply through the umbilical cord. Cells of the uterus around the chorionic villi form a layer called the decidua basalis. Together, the villi and the decidua basalis will become the placenta.

Meanwhile, the cells of the inner cell mass are shuffling about, reorganizing themselves into two small, hollow balls pressed together inside the blastocyst. One of the balls will become the amnion, the sac that will enclose the growing embryo. The other ball will develop into a yolk sac (yes, even human eggs have yolks). This yolk sac is the site of red blood cell production in the early weeks of pregnancy; later it will be incorporated into the growing embryo, forming part of the gut, blood vessels, and gonads (ovaries or testes).

The most significant action takes place where the yolk sac and amnion are pressed together, forming a double layer of cells called the embryonic disk. During the third week after fertilization, in a process called gastrulation, the two layers of the embryonic disk become three. In response to some unknown signal, a raised ridge with a groove down the middle—the so-called primitive streak—forms on the embryonic disk.

Cells from opposite sides of the disk swim toward the streak in the middle. Other cells then go into motion. Surface cells scoot over to the primitive streak and dive down into the groove, emerging underneath to form a bottom layer of cells called the endoderm. Other cells follow, forming a mesoderm

+ FETAL AGE +

WHEN A DOCTOR tells a woman that she is 12 weeks pregnant, he means that her fetus is 10 weeks old. Confused? Physicians typically date a pregnancy from the last menstrual period, which is usually two weeks before conception; this is the gestational age of the baby. The fetal age is the actual age of the fetus, dating from fertilization. When a woman reaches full term, her pregnancy is 40 weeks along, but her fetus is 38 weeks old.

POWERFUL DIVIDERS

AT ALMOST EVERY STAGE in human life, in almost every part of the human body, cells are locked into a specific shape and function. Muscle cells contract. Nerve cells transmit impulses through their long axons. Red blood cells tote oxygen away from the lungs.

Only stem cells are different. These are the blank slates of human biology, generic cells with the potential to develop into a great variety of others. This potential holds tremendous promise: It means that, in theory, stem cells could be used to repair damaged tissues, as well as to test new drugs and to research birth defects. Among the conditions most often mentioned as candidates for stem cell therapies are the biggest killers and cripplers in the medical lexicon: diabetes, stroke, heart disease, Parkinson's and Alzheimer's diseases, rheumatoid arthritis, and spinal cord injury.

An 11-year-old with sickle-cell anemia awaits a bone-marrow transplant in a sterile environment.

CELL TRAITS

Regardless of where they are found, all stem cells share three key characteristics. First, they are unspecialized, without any tissue-specific structures. Second, they can divide and replicate themselves for long periods, unlike most cells with their limited life spans. Third, they have the ability to differentiate into specific cells, such as blood, muscle, or brain cells, in response to inner or outer signals.

Stem cells are found in a range of locations, and some seem to have more potential than others. Embryonic stem cells come from the inner cell mass of a blastocyst—technically a pre-embryonic stage of development (see "Development," page 34)—which, at five days after fertilization, contains more than a hundred identical cells. Those found in laboratory cultures are derived from blastocysts created in an in vitro fertilization clinic; some blastocysts have not been chosen for implantation and are scheduled to be discarded, so they may be provided for research with the consent of the donors. With the ability to transform into any kind of cell in the body, except those that develop a fetus, embryonic stem cells are called *pluripotent*.

Adult stem cells are scattered in minute quantities in some adult tissues, including the bone marrow, skin, brain, liver, skeletal muscle, peripheral blood, blood vessels, and cornea. They remain dormant in those tissues until activated to repair an injury.

With the right stimulus, adult stem cells can develop into a limited range of cells related to the tissue in which they are found: neural stem cells, for instance, can become glial cells or neurons. The adult stem cells' potential has been put to use since the 1960s in bone marrow transplants, because stem cells in transplanted bone marrow lodge in the recipient's bone marrow and begin to produce new blood cells. Some research indicates that adult stem cells might be able to differentiate even more widely if given the right stimulus.

A source of hematopoietic (those capable of forming blood and immune cells) stem cells is the blood from the placenta and umbilical cord, tissues normally discarded after delivery. For two decades, cord blood transplants have been done to treat blood and immune disorders typically in children, because the number of stem cells in cord blood is too low to meet adult needs. Stem cells are also found in amniotic fluid; little is known about these cells yet, but like embryonic cells, they may be pluripotent.

PROS & CONS

Embryonic stem cells and adult stem cells have different advantages and disadvantages. Embryonic stem cells can be maintained indefinitely in laboratory cultures, growing into millions of identical cells; each genetically identical set of embryonic stem cells is called a stem cell line.

Embryonic cells are also pluripotent. Transplanted embryonic stem cells or tissues made from them, however, might be rejected by the recipient, just as some transplanted organs are. Adult stem cells are harder to find and very difficult to culture, but if they could be produced in large quantities, doctors might be able to transplant a patient's own cells or tissues made from them, avoiding rejection.

Stem cell research is still in its early stages and faces serious challenges. Adult stem cells have not been successfully cultured in such a way that they can replicate rapidly enough to be useful. To culture embryonic stem cells, researchers move the inner cell mass into a laboratory dish coated with a "feeder layer" of embryonic cells, typically those of a mouse. With luck, the cells begin to divide without differentiating (or picking up mouse viruses). When they crowd the culture dish, they are spread among several dishes and cultured again, until they become a cell line of several million cells. These can then be frozen and distributed to other laboratories.

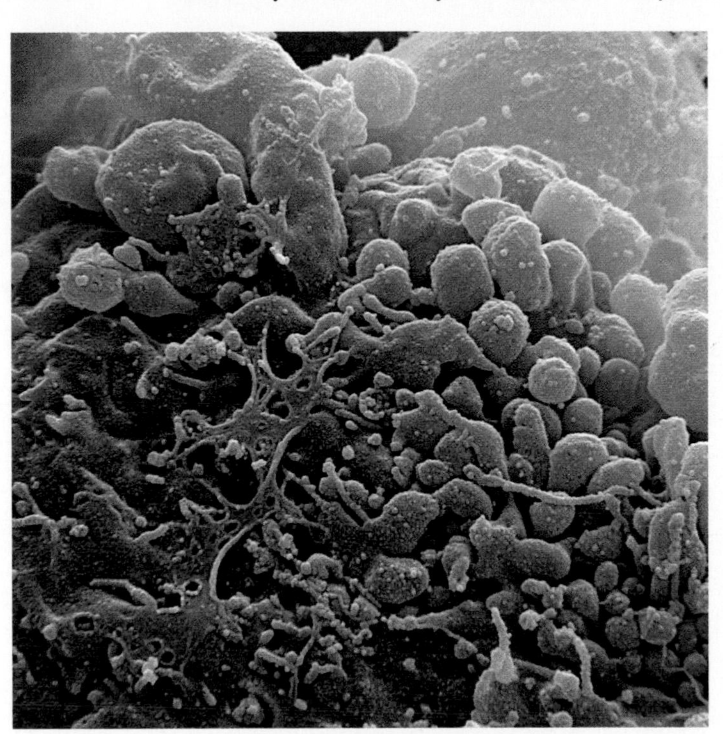

High magnification reveals the busy surface of a human embryonic stem cell.

To be viable for human therapies, however, the cells must be coaxed into differentiating into the desired type; certain genes must be turned on or off—and scientists still don't understand exactly what triggers this process. In some experiments, differentiating cells have grown into tumors or veered from the desired path and become different kinds of tissue—bone, say, instead of bladder. To be useful in human therapies, cells also have to remain healthy, to be able to be transplanted without rejection, and to survive for the lifetime of the recipient. Research continues as scientists work to meet these conditions.

ETHICS

Nor are all the challenges medical. Ethical questions surround stem cell research. Some people believe that destroying human embryos, even at such an early stage, is tantamount to murder. The possibility of creating new human tissues in the lab also raises the eerie specter of body farms or cloning mills. If a way can be found to culture adult or amniotic stem cells, then the thorny issue of embryo use may be sidestepped. Regardless of the controversies around it, it is unlikely that scientists will abandon such a hugely promising field.

layer on top of the endoderm. The cells that remain on top are called the ectoderm.

From these three layers will arise all the structures of the human body in a process called organogenesis (see "Organogenesis," page 36). Exactly how this process is controlled—how the cells know where to go, and how they decide to turn on and off the genes that regulate their development, directing them to specialize into different tissues—is still one of the key mysteries of biology.

The three-layered, ridged disk is now an embryo, and it develops at a furious pace. Some of the mesodermal cells collect into a lengthwise supporting rod called the notochord. The notochord releases chemicals that cause the cells of the ectoderm to thicken into the neural plate.

The plate then curls up at the edges until it forms an enclosed pipe, the neural tube. The top end of this tube will become the brain, and the rest the spinal cord. (If this process goes awry and the tube does not close completely, the baby may be born with a birth defect such as spina bifida or anencephaly.) Cells nearby move about the embryo to become, eventually, such diverse features as sensory nerves, bones of the face, and connective tissue.

The heart and blood vessels also develop early. In the third week after fertilization of the egg, blood vessels form and link up throughout the embryo. A primitive heart tube arises from the mesoderm, bends into an S shape, and starts to beat. It connects to the blood vessels to create a rudimentary cardiovascular system.

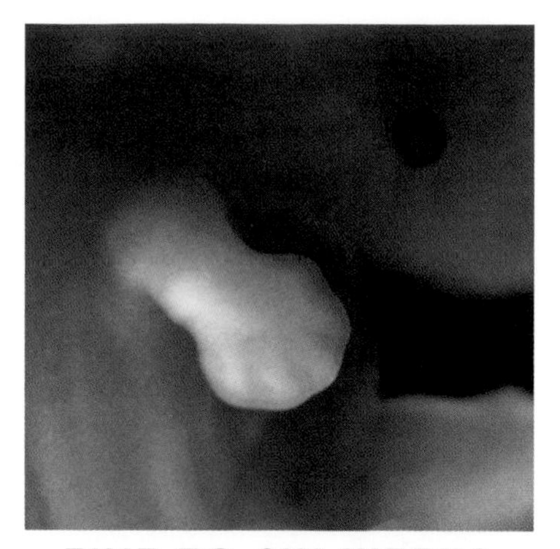

FIVE TO SIX WEEKS
A few weeks into its development, an embryo's upper limb buds begin to look like arms and hands.

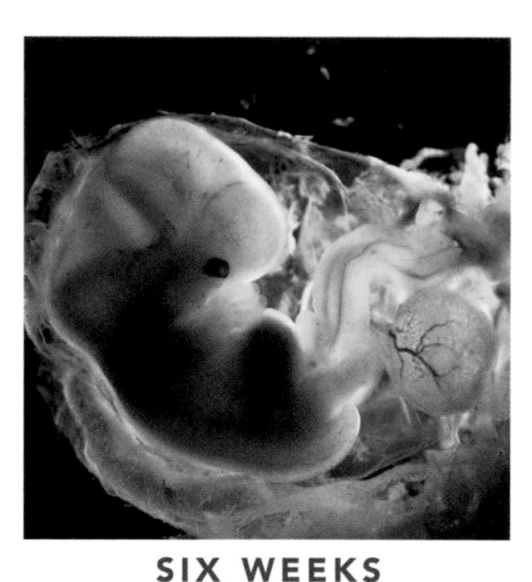

SIX WEEKS
The heart and liver show up as dark masses within the 1/2-inch-long embryo.

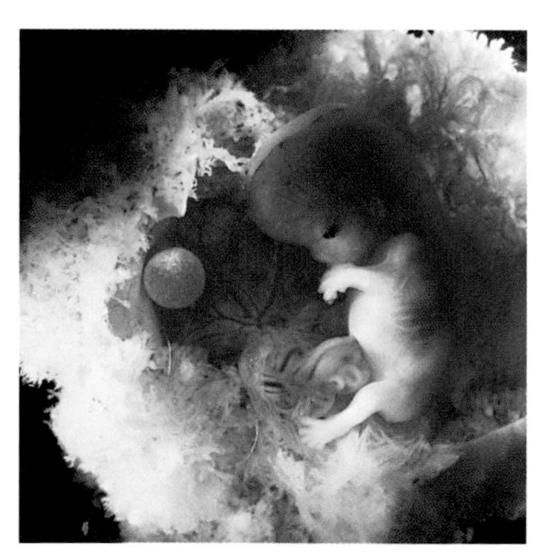

EIGHT WEEKS
Though only the size of a walnut, the fetus now has all its organs; its brain is growing rapidly.

During the fourth week, the embryo triples its size (although it is still well under an inch in length). The formerly flat embryonic disk curls into an irregular cylinder with a recognizable head and tail. Part of the yolk sac now merges with the embryo's midsection to form a primitive gut. Eyes and ears begin to appear on the head, while upper and lower limb buds form, eventually to be transformed into arms and legs.

From the fifth through the eighth week after fertilization, almost all remaining structures and organs, such as the bones, liver, and lungs, begin to appear. The brain grows rapidly, and with it the head. Arms and legs lengthen, and fingers and toes appear; at first the hands and feet are webbed, but soon the webbing disappears. Muscles begin to twitch weakly. At the end of the eighth week, the embryo is about 1.2 inches long and weighs approximately 0.03–0.06 ounces (2 grams). It is ready to move into the next stage of its life, as a fetus.

FETAL DEVELOPMENT

At nine weeks, the foundations of all of a fetus's major organ systems have been laid down. From that time until 38 weeks after fertilization, its body begins a campaign of

SEE ALSO: Chapter Eleven, "Pregnancy," PAGE 312

growth, development, and consolidation. From week to week, and from month to month, the fetus comes closer to being able to survive outside the womb.

From 9 to 12 weeks, the tiny body elongates to about 3.5 inches from the top of its head to the end of its rump, but the head still makes up about half its total length. The face is formed and the eyelids are developed, but closed. The prominent liver takes over as the primary producer of red blood cells until late in pregnancy. The brain and spinal cord grow as well—the brain alone is adding up to 250,000 neurons a minute—and the fetus's sex can be determined from the genitals.

During weeks 13 to 16, more muscle and bones develop; the fetus can move vigorously. The kidneys and other gastrointestinal organs become more mature. The fetus's skin is almost transparent, and a fine hair, called lanugo, appears on its head. Growing rapidly, the fetus reaches a length (head to rump) of 5.5 to 7 inches.

From weeks 17 to 20 after fertilization, more muscle development and overall growth mean that the mother can now feel the fetus move. Fine lanugo hair covers its entire body. During the next month, in weeks 21 to 24, fine details appear: Footprints and fingerprints form, as well as eyebrows and eyelashes; the eyes are well developed, and air sacs (alveoli) take shape in the lungs. The fetus starts to gain weight quickly.

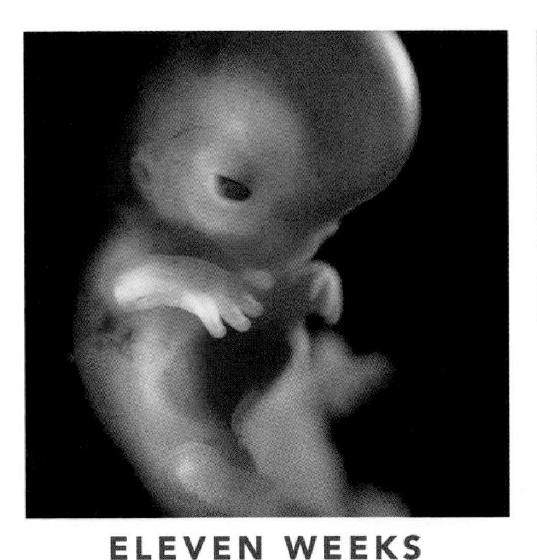

ELEVEN WEEKS
Fingers appear as the cells between them die in the programmed cell death known as apoptosis.

SIXTEEN WEEKS
Visible in a translucent hand are the beginnings of a fetus's bones.

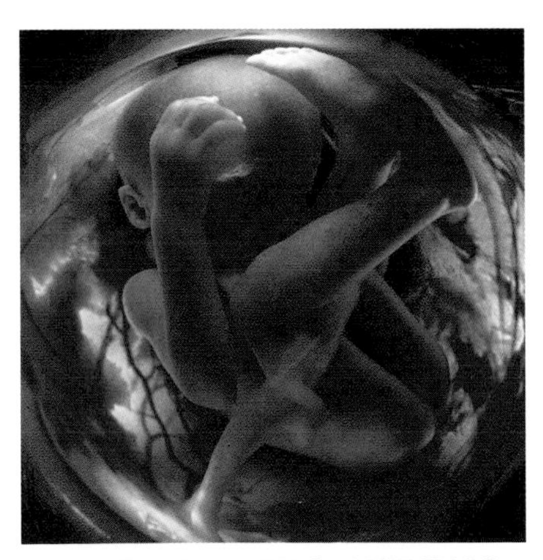

TWENTY-TWO WEEKS
This ten-inch-long girl looks well-developed, but she is not yet ready for life outside the womb.

In the seventh month (weeks 25 to 28), the brain grows rapidly and the nervous system matures enough to be able to control some body functions. The bone marrow takes over production of red blood cells, a role it will maintain after birth and through adulthood. By the end of this period, a baby born prematurely might be able to survive, but only with a high risk of complications.

From weeks 29 to 36, the fetus grows from about 15 to 19 inches head to toe and gains approximately 4 pounds. The bones are fully developed, eyes are open, and fingernails and toenails appear. Fat is laid down under the skin. Babies born toward the end of this time have a good chance of survival.

Beginning sometime from 36 to 38 weeks, thicker hair starts to grow on the head. The fetus is considered full term now and ready for birth, reaching an average size of 19 to 21 inches and weighing between 6 and 10 pounds.

Throughout this period of growth, of course, the fetus has been immersed in amniotic fluid. It does not breathe, nor does it eat in any conventional sense. It survives during the long months of development because of remarkable adaptations of its cardiovascular system. The blood that flows through a fetus's veins and arteries picks up all its oxygen and nutrients from the mother's blood as it is filtered through the placenta.

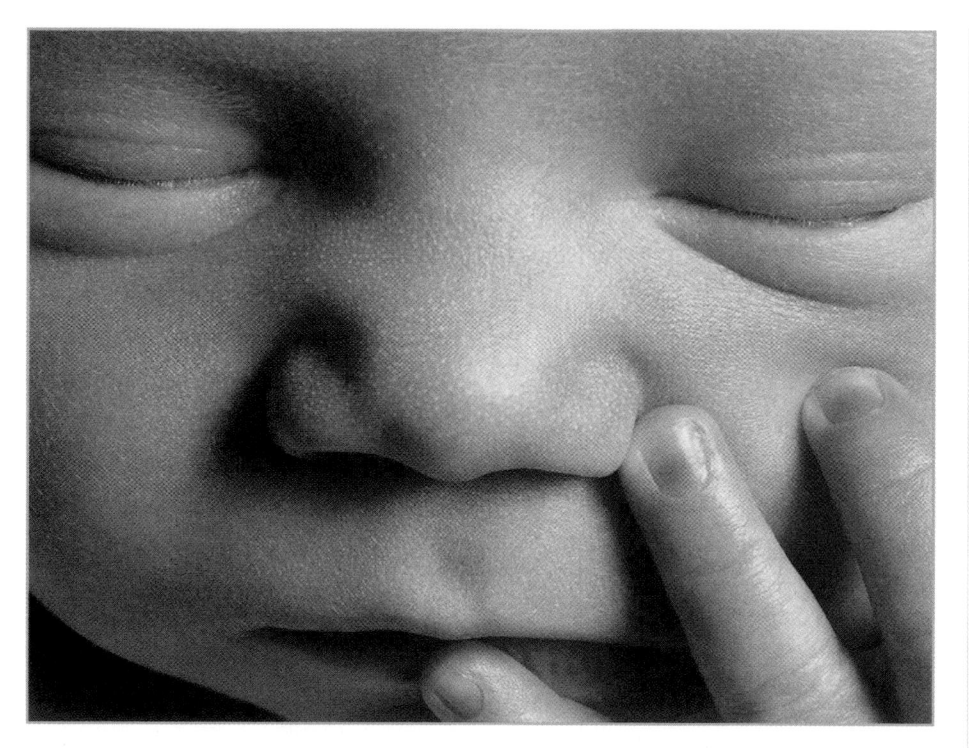

A newborn baby boy in a rare quiet moment

Having already traveled in the mother's body, the maternal blood is somewhat depleted of oxygen. To make up for this, fetal blood carries a special form of hemoglobin that is better than adult hemoglobin at binding to oxygen (and which, incidentally, never carries sickle-cell anemia).

The fetus's blood, returning from the placenta, enters its body through the umbilical vein. It circulates through the heart and body but largely bypasses the lungs and liver, using three special blood vessels as shunts around these organs. Just enough blood flows through the lungs and liver to keep those organs alive and growing.

Eventually, the deoxygenated blood, now laden with wastes, leaves the fetus's body via two umbilical arteries. All of these shunts will shut themselves off neatly at birth, when the fetus emerges from the mother's body as a newborn baby.

BIRTH

When the fetus is full term, hormones in the mother's bloodstream interact with hormones released by the fetus itself to initiate labor. The mother's uterus begins to contract, pushing the baby toward the dilating cervix.

Squashed and stressed by the contractions, the baby's body responds by releasing epinephrine and norepinephrine, hormones that begin to clear the baby's lungs of fluid and pump a greater volume of blood to the heart and brain.

As it is exiting the mother's body, the baby no longer receives oxygen from the mother's placenta. Rising levels of carbon dioxide in its blood signal respiratory centers in its brain to jump-start the process of breathing. Outside the womb, the baby takes a breath,

APGAR SCORING SYSTEM

	0	1	2
HEART RATE	Absent	Less than 100 beats per minute	More than 100 beats per minute
RESPIRATION	Absent	Slow; irregular weak cry	Good; strong cry
MUSCLE TONE	Limp	Some flexing of arms and legs	Active motion
REFLEX	Absent	Grimace	Grimace and cough or sneeze
COLOR	Blue or pale	Body pink; hands and feet blue	Completely pink

SEE ALSO: Chapter Eleven, "Pregnancy," PAGE 312

filling its lungs with air for the very first time.

With the lungs working, the baby's cardiovascular system must also kick in fully to transport the oxygen. The foramen ovale—an opening between the atria of the fetal heart—closes, sending blood to the lungs. Blood vessels that bypassed the lungs and liver also close off, as do the umbilical arteries.

After a baby is born, its systems still continue to grow and develop outside the womb. This is particularly true of its nervous system and brain. Nerve fibers throughout the baby's body are still acquiring the myelin sheaths that ease the transmission of nerve impulses. "Lower" brain functions, such as breathing, circulation, and swallowing, are working, but "higher" functions, such as sophisticated thinking and coordination, are still developing as the brain continues to add cells and connections.

In the first three months, a baby's synapses—the connections necessary for learning to occur—multiply twentyfold. A young infant's brain development is marked by great advances in motor control, social awareness, and language. At birth, a baby cannot hold up her head, see clearly past about 12 inches, or communicate except through cries or coos. By one year, she may well be able to walk, play games, and speak a few words. She is a recognizable member of the human family—all from the union of two single cells.

VIRGINIA APGAR

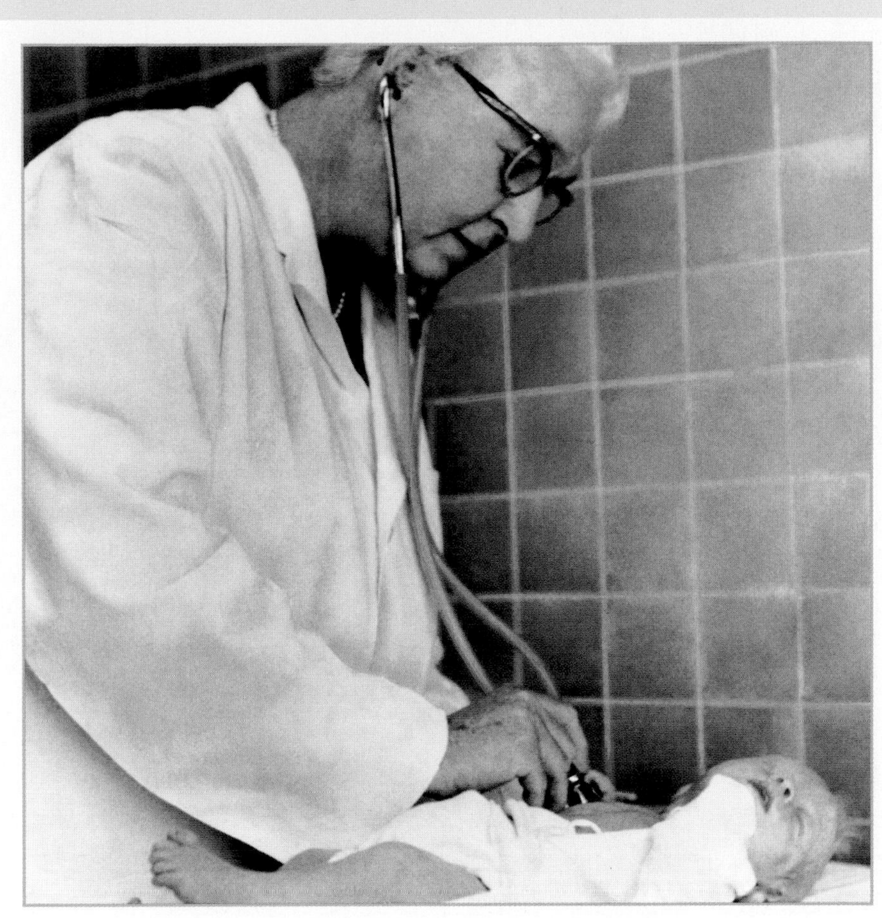

Dr. Virginia Apgar assesses the health of an infant in 1966.

THE WOMAN WHO devised the widely used method for evaluating the health of newborns—the Apgar score—was a medical pioneer in many ways. Virginia Apgar (1909–1974) graduated from the Columbia University College of Physicians and Surgeons in 1933. Discouraged from entering the male-ruled field of surgery, she turned to anesthesiology, becoming the first female board-certified anesthesiologist and the head of anesthesia at Columbia. At 50, she earned a master's in public health and devoted herself to preventing birth defects.

An interest in the effects of anesthesia during childbirth led her in 1952 to create a system to assess a newborn health. At the time, no standard practice existed. Apgar's easily performed tests awarded 0, 1, or 2 points for each of five criteria: heart rate, respiration, muscle tone, reflexes, and color. A nurse or intern tests the baby at one minute after birth and then again at five minutes. Healthy newborns typically score 8 or 9; not yet fully oxygenated, they usually still have bluish hands or feet. The test (especially at five minutes) has proved effective in predicting an infant's survival. ■

CHAPTER TWO

BODY ARMOR

WHAT'S ON THE OUTSIDE OF the body is often overlooked, until there's a problem: a rash itches, a blemish blooms, or a wrinkle emerges. But the skin—also known as the integument, from the Latin word *integumentum*, meaning covering—is far more than a wrapper for the body. Together, with the hair and nails, the skin forms the integumentary system, one of body's strongest defenses. This system performs a specific set of essential functions to protect, heal, and regulate the body.

A healthy head of hair can protect the body from getting too much sun.

THE SKIN

WORKING PARTS OF THE LARGEST ORGAN IN THE BODY

BEAUTY MAY BE ONLY skin deep, but how far down does the skin go? The skin is made up of a series of layers, each with its own tasks and goals.

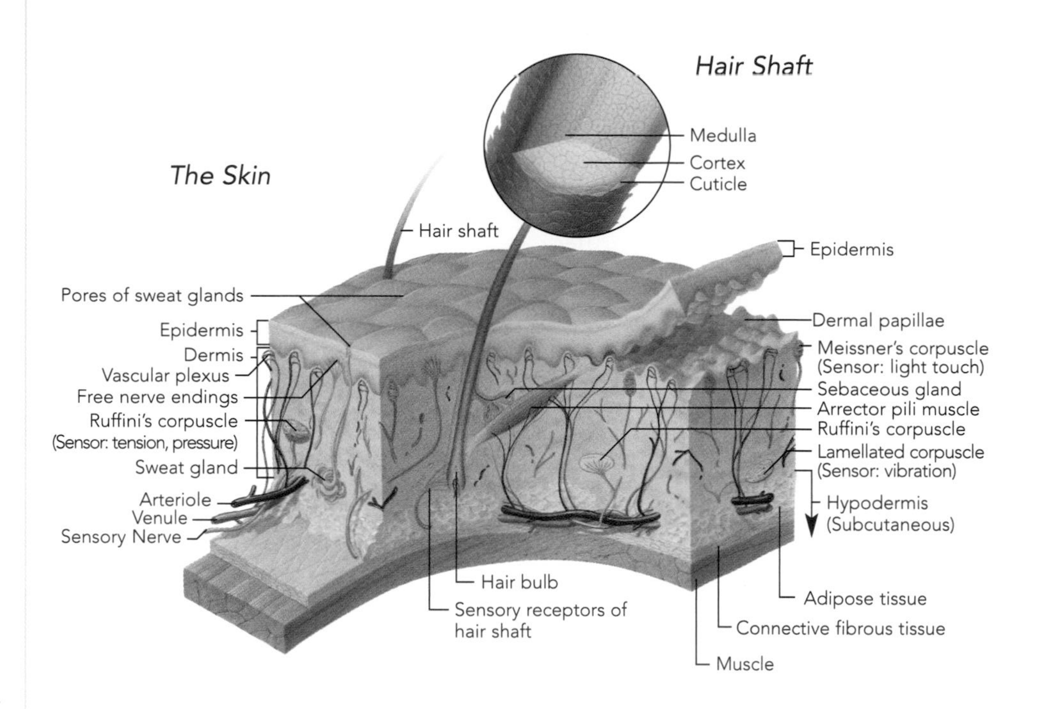

LAYER BY LAYER

In contrast to the body's other organs, the skin is nearly flat with varying thicknesses. The thinnest skin (from 0.004 to 0.006 inch) is found on the eyelids, while the thickest (about 0.18 inches) is on the palms of the hands and the heels of the feet. That narrow expanse contains two kinds of tissue. The moisture-proof epidermis is the protective outermost layer, which can be seen and touched. Below it, the thicker dermis provides support and gives the skin both strength and suppleness.

The epidermis is multi-tiered. Its outer surface, the stratum corneum, or horny layer, is a sheet of dead skin cells that serves as a barrier against the outside world. These squamous (scale-like) cells originate in the epidermis's corrugated lower region—the stratum basale, or basal layer—where new basal cells are generated through mitosis. Neighboring Langerhans' cells scout for disease-causing intruders and other foreign substances. The epidermis also contains melanocytes, cells that make melanin, a radiation-absorbing pigment, which gives skin its color and protects it from the sun's ultraviolet (UV) rays. Occasional Merkel cells, found in the lowest regions of the epidermis, function as sensory receptors for touch.

SKIN STATS

+ THE SKIN of an average adult, if laid flat, would cover about 18 to 22 square feet.
+ The skin accounts for as much as 15 percent of body weight.
+ Some 30,000 to 40,000 dead skin cells are sloughed off every minute, or about 40 pounds (18 kilograms) in an average lifetime.
+ Fresh skin cells continuously replace shed cells, resulting in a new epidermis about once a month.

As the basal cells move up through the epidermis, they fill with keratin—a fibrous, waterproof protein that also makes up nails and hair—and bind together in tightly packed sheets. Cut off from the blood supply of the lower epidermis, the keratinized cells, called keratinocytes, die and rise to the skin's surface, where they make up the visible, outermost layer of skin. The body sheds millions of old skin cells daily, making way for a fresh group of cells.

GOING DEEPER

Just beneath the epidermis is the dermis, a tough sheath of connective tissue that makes up the bulk of the skin. The dermis is often referred to as the "true skin"

WEBLINK: Take an interactive look at the Skin at http://nationalgeographic.com/humanbody

because, unlike the continually renewing epidermis, it stays the same throughout life. The dermis contains blood and lymph vessels, nerve endings, hair follicles, sweat and sebaceous glands, and the proteins collagen and elastin.

Where the dermis and epidermis meet are bumpy features, called dermal papillae, that project up into the epidermis. These contain different structures; some papillae contain blood vessels and others contain nerve endings and touch receptors. Blood vessels nourish skin cells by bringing them oxygen and nutrients and by removing waste produced by cell metabolism. Sensory receptors in the dermis detect pressure, vibration, and light touch; bare nerve endings sense painful stimuli such as irritating chemicals and extreme heat and cold. Collagen in the dermis gives skin its flexible strength; elastin enables the skin to return to its original shape when tugged and twisted.

Beneath the dermis lies the subcutaneous tissue—a separate region, not technically part of the skin, that anchors it to underlying structures. Generally much thicker than the dermis, the subcutaneous tissue—also known as the hypodermis (a region "below the dermis")—varies in thickness depending on the person, because this is where fat cells are located. Fat is deposited into the cells of adipose tissue, which makes up most of the subcutaneous. Despite its poor reputation, fat plays an important role in the body: It holds in heat, helps to cushion organs and tissues from injury, and serves as a fuel reserve when the body uses more energy than it consumes.

The subcutaneous tissue also serves as a bridge between the skin and the rest of the body. It hooks onto underlying muscles and, in spots where there are no muscles, such as the knuckles, attaches directly to bones. While skin seems to fit tightly, it's actually loosely hinged to allow the body to move freely.

SKIN TYPES

The skin is divided into two major types, based on function and structure: thick skin, which covers the soles of the feet and palms of the hands, and thin skin, which covers the rest of the body.

Thick skin isn't all the same thickness. It ranges from 0.024 inches (0.6 mm) to 0.18 inches (4.5 mm). Thick skin contains more epidermal cells than thin skin and features five tiers of epidermis (from the inside out): basal layer, or stratum germinativum (where cells are made); stratum spinosum (spiny, interlocked cells that support the skin); stratum granulosum (thin middle layer where keratinization takes place and cells begin to die); stratum lucidum (a translucent sliver of flattened, dying cells); and the stratum corneum (the outermost layer). Thick skin contains many more sweat glands than thin skin, but it does not have sebaceous glands, hair follicles, or arrector pili (the smooth muscles attached

CELLS OF THE EPIDERMIS

CELLS	FUNCTION
Keratinocytes	Produce keratin, the protein that endows skin with its protective properties. The most numerous cells in the epidermis
Melanocytes	Spider-shaped cells that produce melanin, the pigment that gives skin its color. Found in the lower level of the epidermis
Langerhans' cells	Protective cells that work with the immune system to fight invaders. Found throughout the epidermis
Merkel cells	Sensory receptors for touch. Found in the lowest levels of the epidermis

Above: Magnification makes the skin on the hands resemble a criss-crossed road map.

STRETCH MARKS

SKIN'S ELASTICITY is a marvelous ability, but it is possible to stretch it too far and too fast. Growth spurts, pregnancy, and other rapid fluctuations in weight can all damage the skin and tear the dermis. The streaks left behind, commonly called "stretch marks," are scars created by the dermal tearing. At first, stretch marks may appear reddish or purple, but they eventually lighten and may disappear over time.

to hair follicles that make the hairs lie flat or stand erect).

In thin skin—with an epidermis from 0.004 inches (0.10 mm) to 0.006 inches (0.15 mm) thick—the epidermal layers are less dense, and there is no stratum lucidum. Thin skin is softer and more flexible than thick skin, has fewer sweat glands, and contains hair follicles, sebaceous glands, and arrector pili.

The sensitivity of thick and thin skin varies, depending on the number and type of sensory receptors for touch, pressure, heat, cold, and pain. In general, nerve endings are more densely clustered in thick skin. Fingertips, for example, have many receptors and are very sensitive to touch. The thin skin of the upper arm, on the other hand, is less sensitive, because it has very few touch receptors.

The thin-skinned lips are an exception; they have many nerve endings, which makes them extremely sensitive to touch and temperature. The skin of the lips is composed of a very thin epidermis and dermis; there are no sweat glands or sebaceous glands, which leaves them susceptible to chapping. The lips also have no hair follicles or arrector pili.

The lips contain many dermal papillae, which nourish cells in the basal layer. The proximity of these capillaries to the skin's surface gives the lips their color. In lighter-skinned people, lips may appear pinker because the cells contain less pigment (which also makes them vulnerable to damage from the sun's UV rays). Darker-skinned individuals may have some melanin in their cells, making the pink color less prominent and the skin less vulnerable to UV damage.

BARRIER

Acting like armor, the skin is an effective defense system, keeping invading bacterium from penetrating it (unless it is broken, as through a bite, cut, or puncture). Because it is water resistant yet also a powerful excreter of wastes and sweat, it helps to keep body fluids in balance and regulate temperature. The skin cushions the organs and tissue against injury and, by detecting touch, pain, pressure, and degrees of temperature, tells the body about the world.

To protect the human "fortress" the skin employs multiple defenses. Most would-be invaders never get past the acid mantle of sweat and sebum that coats the skin and contains antibacterial and antifungal substances, including human defensin, which literally punches holes in bacteria. The ongoing action of shedding the outermost cells deters other microbial invaders from the skin's surface. Normal flora do colonize the skin and attack any invaders that manage to land. Because these "home" bacteria have already occupied most of the available colonization sites, their sheer

WHAT CAN GO WRONG

ACNE, left, develops when sebum, dead skin cells, and bacteria block hair follicles, the tubular openings in which hair grows. The blockages, or comedones, prevent drainage of the sebum through the pores. If a pore is completely blocked (closed comedo), a bump called a whitehead forms. When the blockage occurs at the skin's surface, melanin from the trapped skin cells reacts with the air, causing the exposed portion of the plug to darken, forming a blackhead (or open comedo). Bacteria that normally live on the skin (especially *Propionibacterium acnes*) multiply in the blocked follicle, where

SEE ALSO: Chapter Twelve, "Innate Defenses," PAGE 322

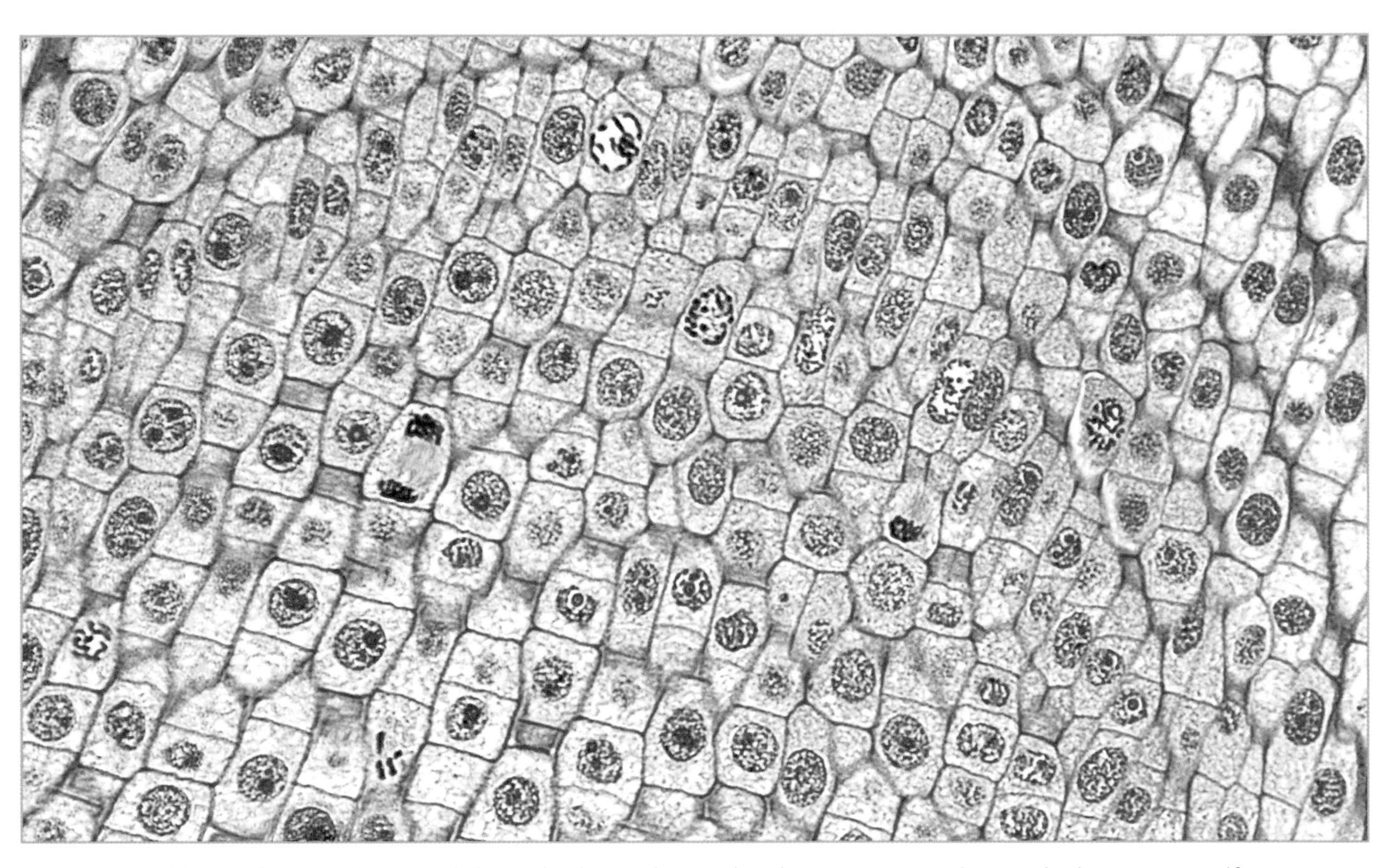

Billions of skin cells are tightly packed together in the skin's outermost layer, which renews itself every 25 to 45 days. Millions of skin cells are sloughed off daily to make way for new ones.

numbers deter newcomers. They also produce proteins called bacteriocins that kill specific species and make other substances, including fatty acids and peroxides, that inhibit nonspecific invaders. In addition to these defenses, the epidermis's multilayered sheets of tightly packed keratinocytes form a dense barricade against microbes trying to penetrate the interior. When microorganisms gain entry through damaged skin, another army of cells rushes to the body's defense (see "Healing" page 54).

The layers-thick armor of keratinized cells, together with sebum on the skin's surface, also forms a moisture barrier, regulating the flow of fluids in and out of the body and helping to maintain the proper chemical balance. Some substances can penetrate the skin in limited amounts. Among them are: water (which is absorbed by the epidermis, as when fingers plump up after a long bath); gases such as oxygen

WHAT CAN GO WRONG

they feed on the sebum and release waste material that irritates the skin. White blood cells defend against the infection, resulting in mild to marked inflammation (swelling, redness, heat, and pain) and forming bumps called pimples. More severe forms of inflammatory acne are characterized by surface lesions that may be solid (nodules) or pus-filled (cysts and abscesses). Acne most often results from higher levels of hormones, particularly androgens (such as testosterone), which are produced during puberty. They trigger the sebaceous glands of the face, neck, back, shoulders, and chest to pump out excess sebum. Acne is often associated with teenagers and young adults (about 85 percent of the population between ages 12 and 24 have at least a mild case), but 12 percent of women and 3 percent of men will suffer from chronic acne until about age 44.

and carbon dioxide; fat-soluble vitamins A, D, E, and K, which are not repelled by the skin's oily coating but dissolve into it, and from there into the epidermis and dermis; and certain drugs such as steroids.

PIGMENTATION

Excessive exposure to the sun's UV rays can be harmful to the body, but some sunlight exposure is necessary to help the body produce vitamin D, essential to healthy bones. To protect the body from the harmful effects of too much sun, the skin produces melanin, a radiation-absorbing pigment that also determines the wide variety of colors of the human skin (and hair). Melanocytes in the basal layer of the epidermis produce melanin in tiny cell structures called melanosomes. There, the enzyme tyrosinase interacts with the amino acid tyrosine to create pigment. When melanosomes are packed with melanin, the melanocytes pass them to the keratinocytes, the cells in the upper layers of the epidermis.

Skin color depends primarily on the concentration of melanosomes and the amount of melanin they contain. Fair skin has much less melanin, fewer melanosomes, and lower levels of eumelanin than darker skin. The quantity and kind of melanin a person has is genetically determined, but other factors play a role in skin color, including UV exposure (tanning is the result of the skin's producing more and bigger melanin granules after being stimulated by UV rays), stress, hormones, and some medications.

Increased melanin production can result in temporary or permanent hyperpigmentation, such as brown splotches called liver spots (or age spots) caused by years of cumulative sun exposure. In melasma, a typically temporary condition, dark spots appear on the cheeks, forehead, and upper lip, triggered by hormonal changes from menopause or pregnancy (as well as by oral contraceptives and hormone replacement therapy).

Another common change in skin color—freckling—is an accumulation of pigment in the bottom layer of the epidermis. Freckles first develop in childhood from repeated exposure to the sun. Freckling may be a protective mechanism, since it is most common in those with fairer complexions. Freckles are not associated with skin cancers or other illnesses.

Students from a Washington, D.C., school illustrate the wide range of variation in human pigmentation.

Decreased melanin production may cause hypopigmentation, or loss of skin color. Scars from cuts or burns, for instance, commonly lack pigment. In albinism, a rare, inherited condition, some or all melanocytes are missing tyrosinase, the enzyme needed to make melanin. People with albinism do not have pigment in the skin, hair, and eyes; they may have extreme sensitivity to the sun. A condition called vitiligo creates white patches of skin. It is believed to be an autoimmune disorder in which the immune system attacks some melanocytes.

Changes in skin tone are common and usually harmless, but sometimes they signal problems elsewhere in the body. A bluish purple cast to the skin and nail beds may indicate cyanosis (poorly oxygenated blood), which can occur during heart failure and severe respiratory disorders. Transient reddening of the face and neck, ears, and upper chest is associated with embarrassment or anxiety (blushing), but it may indicate an underlying condition. Hypertension can cause flushing, as can allergies to some food additives and medications.

Pale skin can indicate distress as well. There are numerous causes of pallor, including stress and illness. Unusual paleness may indicate blood supply problems. Blood may be diverted away from the skin when blood pressure is low or, if there is an infection someplace else in the body, to vital organs. The most common cause of paleness is iron-deficiency anemia (especially in young women), resulting in a decrease in red blood cells.

Other disorders can cause the skin to look yellow or bronze.

BIRTHMARKS

NEWBORN SKIN may have distinctive red patches called hemangiomas (clusters of blood vessels). Some, like strawberry marks (red, raised spots) or stork bites (flat, pink to red, above), may fade with time. Others, like port-wine stains (flat, maroon, or dark red) or cavernous hemangiomas (bluish-red marks that extend into the dermis) often persist and must be removed surgically.

+ SWEAT +

EVERY DAY, your 2.6 millon sweat glands release about 2 cups (400–500 ml) of sweat. The body has two types of sweat gland. Active from birth, eccrine glands are found all over the body and secrete an odorless liquid composed mostly of water. Developing at puberty, apocrine glands are concentrated under the arms and in the genital area. Stimulated by stress or sexual arousal, they produce thicker, milkier secretions that develop a strong smell after interacting with bacteria on the skin.

A yellow cast usually indicates liver disease in which bilirubin (yellow bile pigment, a component of bile secreted by liver cells) builds up in the blood and is deposited in body tissues. In infants, jaundice can be a normal, transient condition. A bronze appearance to the skin is a sign of Addison's disease, a disorder of the adrenal glands.

REGULATOR

The skin plays a major role in maintaining a healthy body temperature, which typically hovers around 98.6° Fahrenheit (F) or 37° Celsius (C). Slight variations are typical; normal body temperature can fluctuate a degree in either direction. Extreme changes, however, can be deadly, as they create an unlivable environment for organs and tissues.

When the body is too cold, blood vessels in the dermis constrict, decreasing the flow of warm blood to the skin and conserving it around the central organs. To cool down, blood vessels dilate, bringing warm blood closer to the skin's surface, which is why exercise may make the face redden. At the same time, the hypothalamus, the brain's heat regulator, signals the

A highly magnified sweat gland (in purple) opens on the skin's surface. Skin cells flake off around the pore's opening.

SEE ALSO: Chapter Eight, "The Kidney," PAGE 212

eccrine sweat glands to produce more sweat—a fluid containing mostly water with small amounts of ammonia and urea, sugar, and salts. The hotter the body becomes, the more it perspires. In contact with the air, the sweat evaporates, cooling the skin's surface. (It's also common to sweat when nervous, because sweat glands respond to over-stimulated muscles and over-stimulated nerves.)

After the water in sweat evaporates, it leaves behind the salts sodium chloride and potassium. The rate of evaporation depends on the humidity of the surrounding air. In muggy weather, the air is already saturated with water vapor, evaporation slows, and the body takes longer to cool down. To keep everything running smoothly, replenishment is essential. The body needs fluids to replace both water and electrolytes (mineral salts, such as sodium and potassium) to avoid dehydration.

Sweat glands also aid the body in eliminating waste. The skin is sometimes called the "third" kidney because of this role (although minor by comparison). Sweat glands excrete waste products such as urea, a by-product of protein metabolism. If urea in the cells is not disposed of regularly, headaches, nausea, and, in extreme cases, death can result. Sweat also helps to flush out toxic metals absorbed from pollution and to eliminate lactic acid, which causes stiff muscles and contributes to fatigue.

FINGERPRINT RECOGNITION

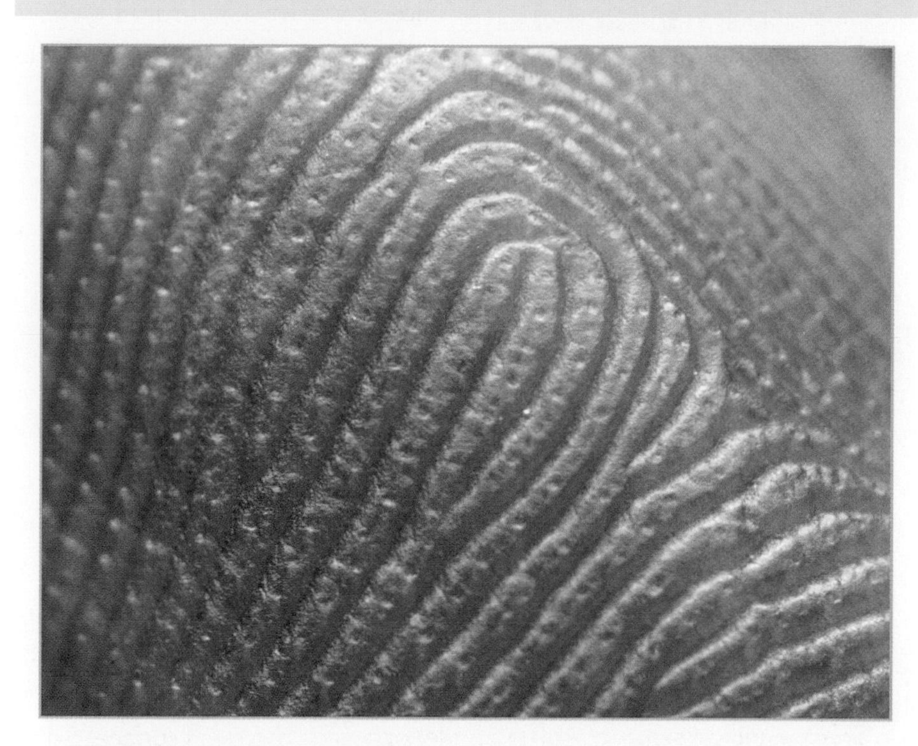

Minute ridges in the epidermis form unique patterns that make fingerprints.

THE DISTINCTIVE whorls, loops, and arches of fingerprints have become crucial identification markers. No two are alike and they never change. The first to recognize their uniqueness was British physician Henry Faulds, who began studying "skin furrows" in the 1870s. In 1880, he published an article that said fingerprints were constant and unalterable. He described how to take fingerprints from a thin layer of ink spread over a smooth board or sheet of tin. Faulds suggested that fingerprints might "lead to the scientific identification of criminals," a suggestion rejected by the London police in 1886. Prompted by Faulds's findings, British scientist Francis Galton (1822–1911) assembled the first extensive collection of fingerprints for studies on heredity. In 1892, his book, *Fingerprints,* presented the first classification system and encouraged its use in forensics. Galton established what Faulds had suspected: Fingerprints remain the same for life, and no two people share them. He later opened a bureau to register civilians using their fingerprints. Sir Edward Richard Henry (1850–1931), a senior British police official, simplified Galton's classification system. In 1901, he established England's first fingerprint bureau, the United Kingdom Fingerprint Branch at New Scotland Yard. Henry's system serves as the basis for modern fingerprint classification still in use today. In 1902, the New York Civil Service Commission became the first U.S. organization systematically to use fingerprints to identify applicants. ■

HEALING

REPAIRING THE SKIN, PROTECTING THE BODY

How a Cut Heals

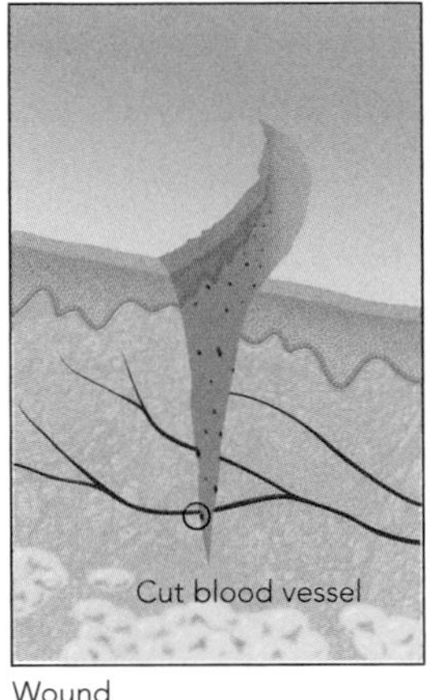

Wound

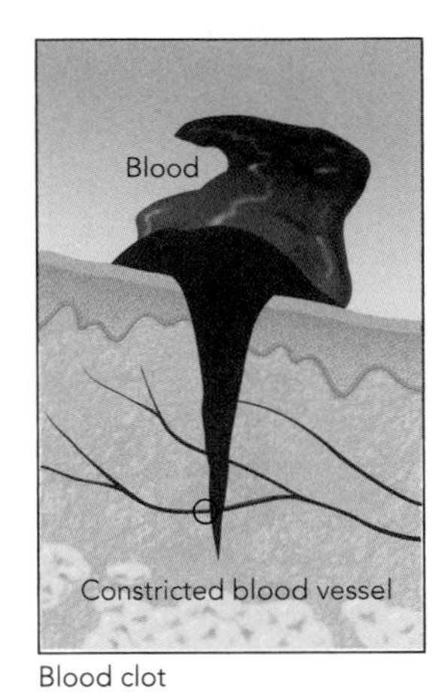

Blood clot

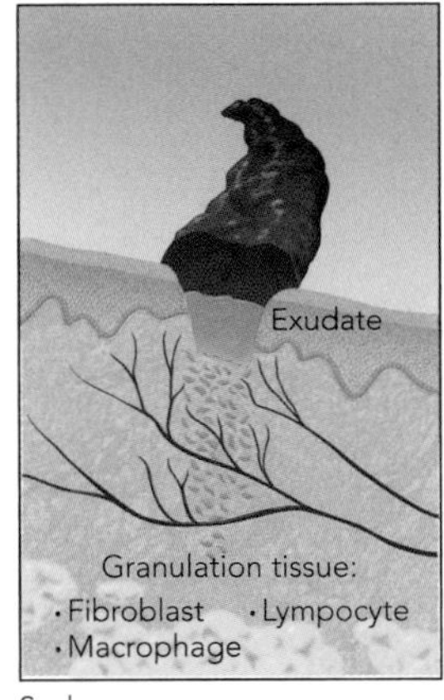

Scab

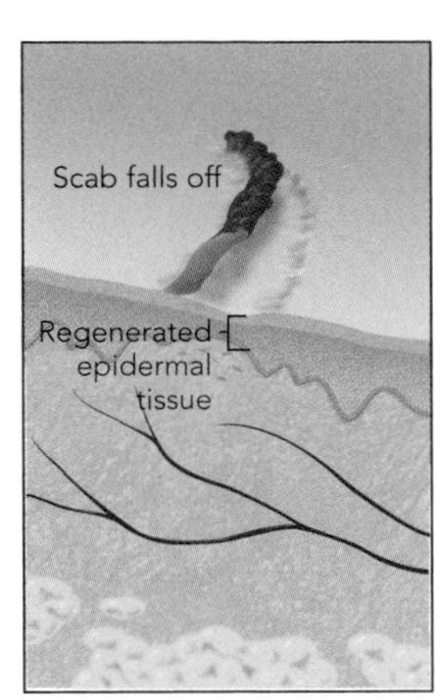

Scar

IN THE COURSE OF a lifetime the skin suffers countless injuries, from nicks and scrapes and more serious wounds, to irritations, infections, and infestations, to burns, small and large. Although the body's internal systems do most of the work of repairing the skin, treatment ranging from minor first-aid to a doctor's care may be needed, depending on the problem, to help ease discomfort, minimize damage, and speed the healing process.

WOUND HEALING

When superficially injured, the epidermis speeds up cell replacement. The new cells surround the wound and migrate in sheets across it. In a few days the damaged area is completely resurfaced, and the abrasion is gone. Wounds that penetrate the dermis and the subcutaneous tissue, where destroyed cells do not regenerate, are another matter. Because multiple tissue layers must be mended, the healing process is more complex, involving a series of cellular and vascular activities that occur in three phases: inflammatory, proliferative, and remodeling.

In the inflammatory phase, blood from severed blood vessels floods the wound, washing away microbes and debris. To control the bleeding (a process called hemostasis), the broken vessel quickly constricts to slow blood flow. Platelets, a component of blood, line the wound and clump together to patch punctures in blood-vessel walls. The platelets interact with blood proteins (called clotting factors) to make fibrin, a protein whose strands fan out from the platelets, creating a mesh that traps red blood cells and other platelets to form a blood clot. The clot stops further bleeding, holds the edges of the wound loosely together, and effectively walls in the injured area; this prevents bacteria, debris, and other harmful substances from spreading to surrounding healthy tissue. The part of the clot exposed to air hardens, forming a reddish brown scab.

Meanwhile, the damaged tissue releases inflammatory chemicals that signal the body to respond to the injury. Local blood vessels dilate, increasing the blood supply to the area, and become more permeable, allowing fluid and infection-fighting white blood cells (neutrophils and monocytes) to pass into the injury site within minutes. The neutrophils and monocytes devour invading bacteria and debris.

After several hours, the number of neutrophils (which are short-lived) declines. Macrophages now take over. These are monocytes that

SEE ALSO: Chapter Five, "The Blood," PAGE 132, Chapter Twelve, "Innate Defenses," PAGE 322

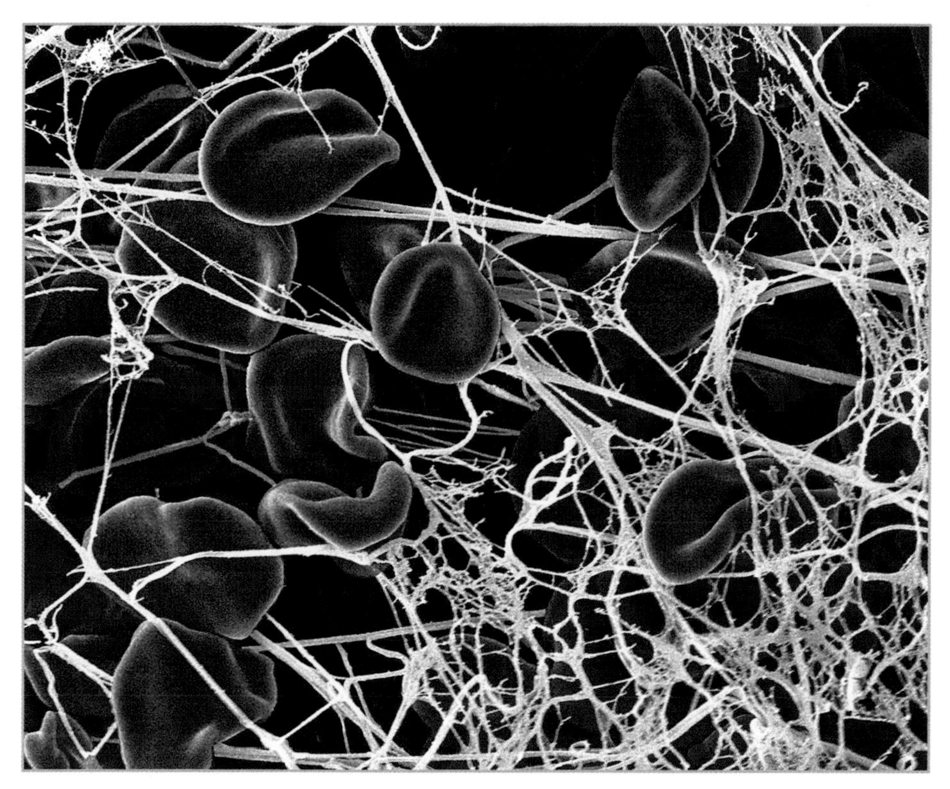

When skin is cut, blood cells spring into action to begin the healing process. Produced by platelets, strands of fibrin bind blood cells to form a scab.

BANDAGES

+ AS EARLY as 2000 B.C., Sumerian wound treatments included salves made of dust, plants, mud, milk, wine, beer, oil, and flour.
+ There are records of sutures, made of flax, hemp, and hair, being used 2,000 years ago in India.
+ Ancient Egyptians invented an adhesive bandage by lining linen bandages with honey—now known to kill bacteria and combat swelling.
+ The ancient Greek physician Galen (A.D. ca 129–199) recommended bathing wounds in seawater and dressing them in honey and wine. He wrapped ulcerated lesions in figs, later found to contain the enzyme papain, which can speed healing.

have grown larger and produced granules, which contain enzymes to digest bacteria and other invading cells. The macrophages engulf and consume remaining invaders and debris, dead and damaged cells, and old neutrophils. Swelling occurs around the wound, caused by the buildup of fluid in the tissues. The area becomes red and warm, from the increased blood supply, and pain results from inflammatory substances that stimulate the nerves. Inflammation, a sign of healing, usually lasts 24 to 48 hours but can last up to two weeks.

REBUILDING

The proliferative phase of healing is characterized by robust growth underneath the scab. Damaged blood vessels regenerate. In the dermis, cells called fibroblasts produce growth factors and collagen fibers to create new tissue in the injured areas. Some of the fibroblasts have contractile properties that enable them to pull the separated margins of the wounds together.

During this stage, delicate pink granulation tissue fills the wound beneath the scab, taking the place of the clot. Macrophages eat the original blood clot and destroy remaining dead and dying cells. The granulation tissue creates a new capillary bed. (The tissue is actually named for the capillaries, which jut out from its surface, giving it a granular appearance.) These capillaries are fragile and bleed easily (as demonstrated when someone picks a scab). Epithelial cells proliferate and then migrate over the granulation tissue.

During the remodeling phase, the epidermis returns to its normal thickness, and eventually the scab falls off. Damaged tissue has been replaced, and there's underlying scar tissue. The scar (a permanent fibrous patch) may be visible or invisible, depending on the severity of the wound. Raised scars within the borders of the original wound are known as hypertrophic scars; those that extend beyond the boundaries into normal tissue are called keloids. Scars give extra strength to the skin in deeply wounded areas but may lack flexibility and elasticity. Collagen fibers in scar tissue are more tightly packed than in normal skin, and there are fewer blood vessels, which may cause scars to be lighter in color than the surrounding skin.

Keeping the wound covered with a bandage can speed healing. Sutures (external or buried), butterfly bandages, and wound closure tapes can also quicken the pace by bringing divided edges closer, decreasing the distance that cells have to travel to bridge the wound. Debridement, the surgical removal of dead or diseased tissue in wounds, also speeds healing and lowers the risk of infection.

IRRITATION

Skin can be irritated by many things: bacteria and viruses, mites and mosquitoes, poison ivy and poison oak, and many other potential causes. Genetic or other underlying disorders may also reveal themselves through skin irritation. The most common symptom is pruritus, or itching, which can be triggered by a primary skin condition, allergies, or systemic diseases including liver disease, kidney failure, diabetes, and cancer. Sometimes the skin itches when there are no visible lesions. Generalized itching at night can be a symptom of Hodgkin's disease, a lymphoma.

At times, rashes accompany itching. They may come in a variety of shapes and sizes, depending on the trigger, and may make the skin warm, tender, lumpy, cracked, and dry. The pattern and evolution

FERDINAND RITTER VON HEBRA

Hebra, left, documented many skin conditions, such as psoriasis, right.

DERMATOLOGY owes a debt to Ferdinand Ritter von Hebra (1816–1880), considered by many to be the father of modern dermatology. Hebra was the first to classify skin diseases by structural alterations (the foundation of present-day classification) and to recognize that parasites, fungi, and other local irritants could cause them. Born in Brno, in Austrian Moravia, Hebra graduated from the medical school at the University of Vienna in 1841. At the General Hospital of Vienna, he ran the ward popularly known as the "scratch department," reserved for patients with skin problems. Hebra suspected that most of the patients were suffering from the same condition (scabies, a microscopic infestation of parasitic mites) and began observing them closely after the recommended therapies—elixirs taken internally—failed. He may have read earlier physicians' accounts of women relieving the itch by picking mites off the skin with sewing needles. In 1844 Hebra became the first to prove that the "itch" mite caused the condition and to advocate local treatment. Hebra recognized that underlying disease also triggered many skin conditions, such as hives, or urticaria. He reported his findings in his groundbreaking *Textbook of Skin Diseases*, written with dermatologist Moritz Kohn Kaposi. Hebra and Kaposi also published the *Atlas of Dermatology*, which illustrated skin conditions. The books were long considered the bibles of dermatology. Within a few years Hebra had turned his department into a world center for dermatology. He was awarded the noble title Ritter von ("knight of"). In 1879 the Viennese Society of Physicians elected Hebra president, testament to the respect that he brought to the field of dermatology. ■

SEE ALSO: Chapter Twelve, "Innate Defenses," PAGE 322

of a rash is key to pinpointing its source. The red, slightly bumpy rash present with the measles virus, for example, almost always starts on the head (after a fever) and spreads down the body. While most rashes are minor, some may be signs of serious diseases, such as Lyme disease, typically caused by deer ticks infected with *Borrelia burgdorferi*, a bacterium contracted from rodents. The rash begins as a tiny red spot at the tick's site and expands over days or weeks. If the rash looks like a bull's-eye with a clear center, it should be examined by a physician.

Dermatitis and eczema are interchangeable terms for several conditions that cause skin inflammation. They have different causes but similar symptoms: patches of skin with raised red lumps and/or blisters. Contact dermatitis occurs after the skin comes in contact with a substance that either irritates it or causes an allergic reaction. The skin itches and may even blister, depending on length of exposure to the substance. Common irritants include laundry detergents, metals from jewelry or clothes fasteners, some rubber products, some cosmetics, plants such as poison ivy, and some medications.

Urticaria, or hives, can be triggered by many different factors, including stress, allergies, and reactions to drugs or food additives (such as monosodium glutamate). Some people break out in hives after perspiring, after being in the sun, or in extremely cold temperatures. After exposure to one of these factors, the skin responds by producing the chemical histamine, which sets in motion a chain of immune system reactions that result in skin inflammation. Outbreaks usually begin with itching followed by a rash that consists of pale bumps (wheals) surrounded by a clearly defined area of redness. In some cases hives may indicate a serious allergic reaction and might require prompt medical attention.

BURNS

Burns occur when extreme heat damages tissue. Fire is a common cause, but many other things—like sunlight, friction, steam, hot liquids, electricity, radiation, and chemicals—can burn the skin. A burn can range from mild to life-threatening, depending on the depth and the amount of affected surface area. Superficial burns respond to simple hygiene and are usually gone in a few days without any blistering or scarring. Critical burns can melt away the skin's layers and structures in a moment and, in horrific instances, go even deeper to damage muscle and bone. At the surface, severely damaged skin can no longer carry out crucial functions, leaving burn victims vulnerable to disease, environmental assaults, extremes of temperature, dehydration, and death.

The three types of burns are classified according to the damage they cause. First-degree burns, like a mild sunburn, are the least serious because they only damage the skin's top layer. Affected skin turns red and may sting, but the burn usually heals by itself in three to six days. Peeling or flaking may occur, but scarring is unlikely.

A second-degree burn penetrates the epidermis and parts of the dermis. Pain, swelling, redness, and blisters result. Severe sunburns with blisters are second-degree burns. Blisters (lymph fluid that collects between the epidermis and dermis after burning or friction, causing the layers to separate) are a sign of deep damage: The more that burned skin blisters, the deeper the burn has gone into the dermis. If there is no infection,

WHY WE ITCH

AN ITCH activates the neural pathways of the movement-controlling part of the brain, which signals for a scratch. Scratching may create a slight pain to override the itch (pain impulses travel more quickly). Things like poison ivy (above), allergies, bug bites, and dryness or heat can trigger itches, but the cause of itching absent other symptoms remains unclear.

PROTECTING THE SKIN

SPENDING HOURS IN THE sun can be hazardous to your health: Chronic exposure to ultraviolet (UV) radiation is the leading risk factor for developing skin cancer, an abnormal growth of cells that can turn deadly.

WAVELENGTHS

Changes in skin color, like tans and sunburns, indicate skin cell damage. They're caused by UV radiation and can increase the chances of developing cancerous growths. Ultraviolet radiation is broken into three wavelengths: UVA, UVB, and UVC. Nonthreatening UVC rays are completely absorbed by Earth's atmosphere and pose little threat to the skin, unlike UVA rays, which are not absorbed by the ozone layer and can penetrate the dermis and damage collagen and elastin fibers. UVB is partly absorbed by the ozone layer; it penetrates the skin less deeply but causes at least 75 percent of UV damage, including premature aging, burning, and skin cancer. UVB damages DNA—even if a sunburn does not occur—creating mutations that can cause cancer in epidermal skin cells.

The Food and Drug Administration (FDA) determines the risk of sun damage on a scale from 1 to 6 for different skin types. People with fair skin tend to burn easily, because they produce less radiation-absorbing melanin; those with higher-number skin types have darker skin; they can still burn but not as easily. The same people who are most likely to burn are also most vulnerable to skin cancer.

This 84-year-old Australian lifeguard has had over 600 melanomas removed. Every sticker represents a former lesion.

DIFFERENT CANCERS

There are three main forms of skin cancer—all on the increase. Basal cell carcinoma, the most common, is curable in 99 percent of cases because of its slow growth rate and tendency not to spread, or metastasize. It occurs in cells in the lowest layer of the epidermis. Lesions are usually shiny, round, reddish growths with raised edges but can vary greatly. Basal cell carcinomas often result from cumulative exposure to UV rays and develop on exposed areas such as the face, hands, and forearms.

Squamous cell carcinoma originates in keratinocytes of the epidermis, in the layer just above the basal layer. It may grow and spread rapidly, but it can be cured if diagnosed early and removed surgically or by radiation therapy. (The mortality rate is 1 in 100 diagnosed cases.) This kind of skin cancer generally begins as a flat, red, scaly area that may become raised and bumpy as it grows and develop into an open sore. Any suspicious lesions should be examined by a doctor.

The most dangerous form of skin cancer is melanoma, a malignancy of melanocytes, the pigment-producing cells. Melanoma occurs in the skin and can appear in the eye. Compared to other skin cancers, melanoma is more unpredictable, spreads more rapidly, and is more resistant to chemotherapy. Although melanoma occurs in the skin, some suggest that its behavior is closer to that of a sarcoma, a cancer of deeper, connective tissues (such as tendons and cartilage). Continuing research into the exact nature of melanoma could bring about earlier detection methods and greater cure rates.

A melanoma may be a flat brown patch with uneven edges and small black spots; a raised brown patch with black, blue, red, or white spots; or a black or gray lump. It may also develop in a mole. Normal moles are overgrowths of melanocytes, but most are noncancerous, or benign. A new mole or a mole that becomes bigger, changes color or shape, bleeds, itches, hurts, or becomes inflamed should be examined by a doctor.

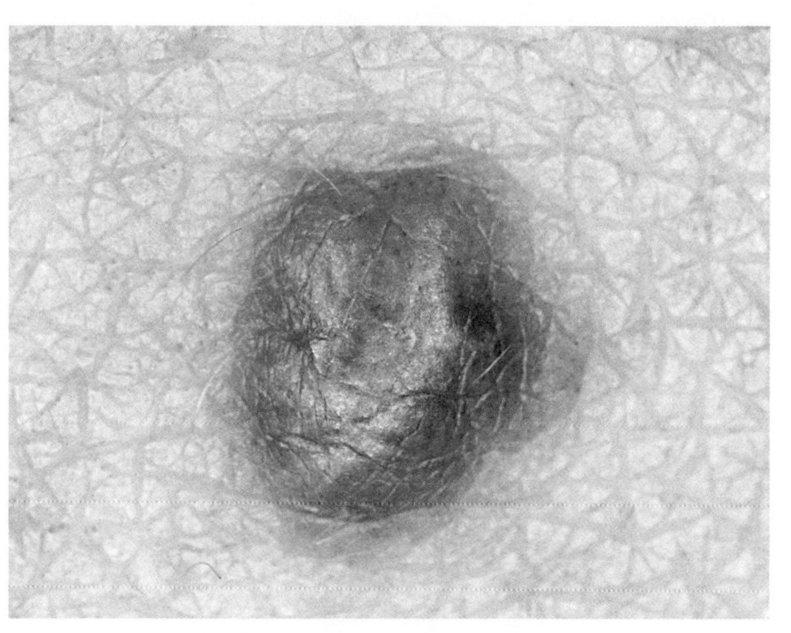

Melanoma, an unpredictable and fast-spreading form of cancer, may be fatal if not treated early.

Melanoma lesions that have penetrated the skin more than 0.03 inches (about 0.80 mm) are likely to have spread through lymphatic and blood vessels to other body areas, sometimes bypassing adjacent lymph nodes. Melanoma accounts for about 3 percent of skin cancers but the large majority of skin cancer deaths. Twenty percent of people with malignant melanoma die within five years of being diagnosed, so early detection is key to survival.

Researchers recently discovered that moles, which are harmless clumps of pigment cells, start the same way as malignant tumors—with a mutated gene that allows cells to divide and reproduce profusely. In moles, the process suddenly grinds to a halt, and the mole becomes a benign tumor whose cells likely won't ever divide again. Scientists are optimistic that this observation could lead to a new, highly effective way of stopping malignant growths.

PREVENTION

Skin cancer is largely preventable by taking some simple precautions, among them wearing sunscreen and protective clothing. Sunscreen protects the skin by absorbing and/or reflecting UVA and UVB rays. The FDA requires all sunscreens to carry a sun protection factor (SPF) label that indicates the amount of protection provided. Sunscreens with an SPF of at least 15 are recommended; an SPF of 15 protects the skin from 93 percent of UVB radiation; an SPF of 30 protects against about 97 percent.

Although SPF ratings apply mainly to UVB rays, many sunscreens now include ingredients that protect the skin from some UVA rays. To get the full effect of a sunscreen, it should be applied 20 minutes before going outside. Between 60 and 80 percent of UV radiation passes through clouds, so sunscreen is needed on cloudy days, too. The FDA also requires that "water resistant" sunscreens be able to maintain their SPF for 40 minutes after submersion.

The American Cancer Society recommends limiting time in the sun when its the strongest (from 10 a.m. to 4 p.m.), staying in the shade as much as possible, wearing sunglasses with 99 to 100 percent UVA and UVB protection, and wearing a wide-brim hat. Snow, water, and sand increase sun exposure by reflecting incoming UV rays, making it especially important for boaters, beachcombers, and skiers to cover up with sunscreen and protective clothing to avoid damaging their skin.

CHANGES

FROM CHILDHOOD TO MATURITY

DURING A PERSON'S lifetime, the skin undergoes many changes as the body grows taller and larger. Some are structural and part of the body's natural maturing process, with stages through which everyone progresses. Others, often associated with aging, result from cumulative damage to the skin caused primarily by sun exposure, other environmental factors, and lifestyle choices (like smoking or diet). These aspects can vary greatly from one person to another. In these instances, heredity plays an important part too.

INFANCY

Babies have a very thin outer layer of skin coupled with a much thicker layer of subcutaneous fat. The combination creates that enviable "baby soft" skin. The extra-thick fat layer serves not only as a cushion, however, but also as a barrier to keep toxins from reaching the rest of the body through the thin keratin layer, which is more permeable than in adult skin.

Newborns may exhibit various temporary skin conditions. At birth, the skin is reddish, and for a few hours the fingers and toes may appear bluish from poor blood circulation. Some babies have bruises and other marks from the birth process that will soon disappear. In the first week after birth, a mild rash often develops where clothing rubs the skin, and there may be dryness and peeling at wrists and ankles. Many normal newborns, particularly premature babies and those that are breastfed, develop jaundice, a yellowing of the skin and the whites of the eyes caused by a buildup of bilirubin in the blood (see "The Skin," page 46). As the spleen disposes of old red blood cells, it breaks down their hemoglobin, creating bilirubin, a yellow

Elastin and lubricating oils make young skin pliable. As skin ages, these substances decrease, which causes wrinkling.

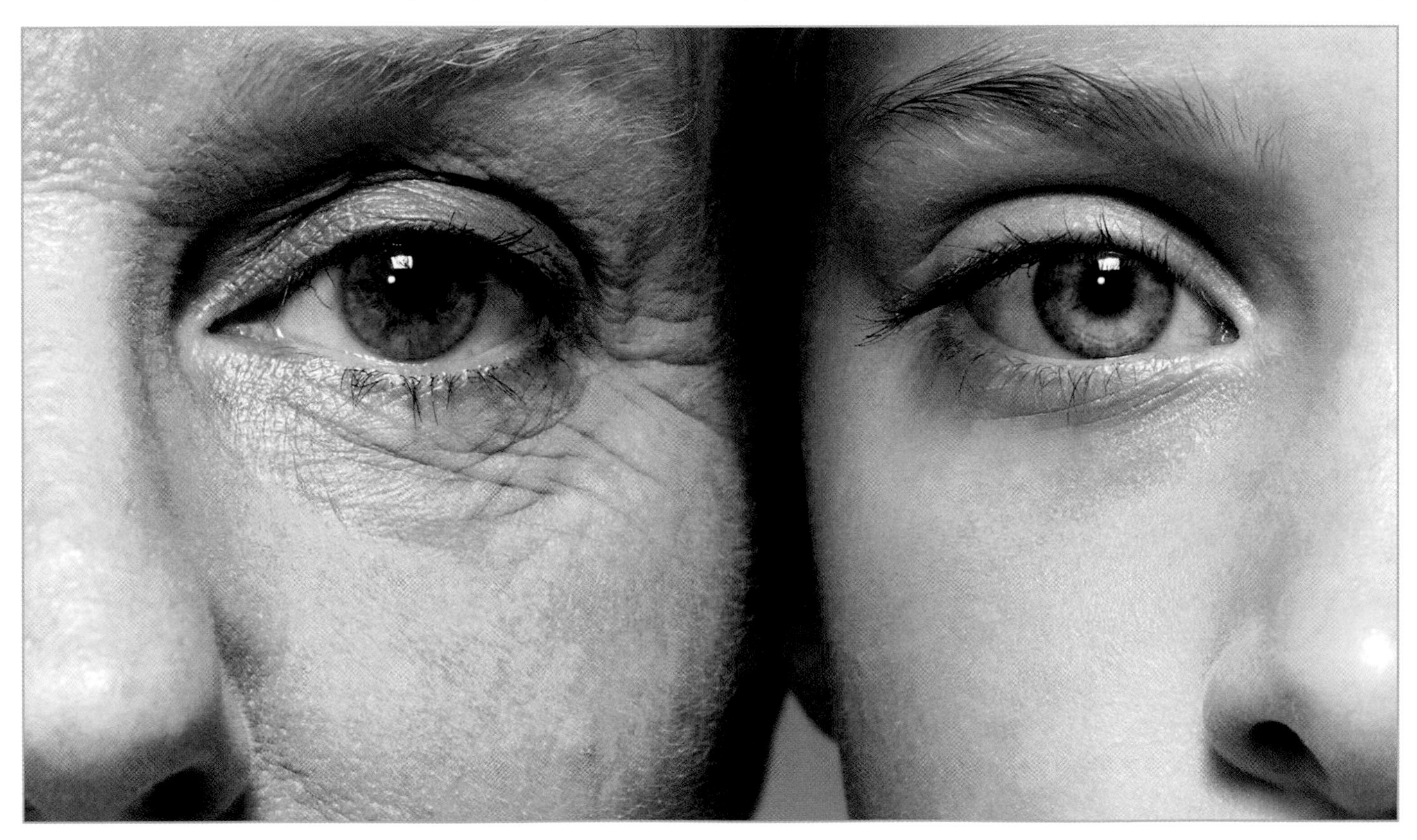

SEE ALSO: Chapter Thirteen, "Building Blocks," PAGE 352

pigment, which is transported to the liver for eventual excretion. The condition usually clears up within a few weeks but should be monitored carefully by a doctor.

PUBERTY

As children approach puberty, their skin begins to undergo significant changes. A surge in hormones stimulates the sebaceous glands, which have been relatively inactive throughout childhood, to secrete more sebum. The eccrine glands produce more sweat, and the apocrine sweat glands become active, secreting a thicker, milky perspiration. Acne and other adolescent skin concerns usually clear up by early adulthood as sebaceous and sweat glands calm down (see "What Can Go Wrong," page 48). During this stage, the skin is soft, supple, and strong, thanks to a bountiful combination of collagen, elastin, subcutaneous tissue, and lubricating oils.

MATURITY

As the body ages, both the dermis and the epidermis thin. The skin becomes drier; collagen and elastin fibers in the dermis split, weaken, and decrease in number. The skin starts to sag and fold over on itself, forming wrinkles. As a person nears 40, creases may appear around the eyes, nose, and mouth, but the pronounced effects of aging are not usually seen until the late 40s or early 50s. Generally, women show signs of aging before men do because their skin tends to be thinner and dryer. In the elderly, the combined loss of subcutaneous fat, collagen, and elastin makes the skin look almost transparent and blood vessels more apparent.

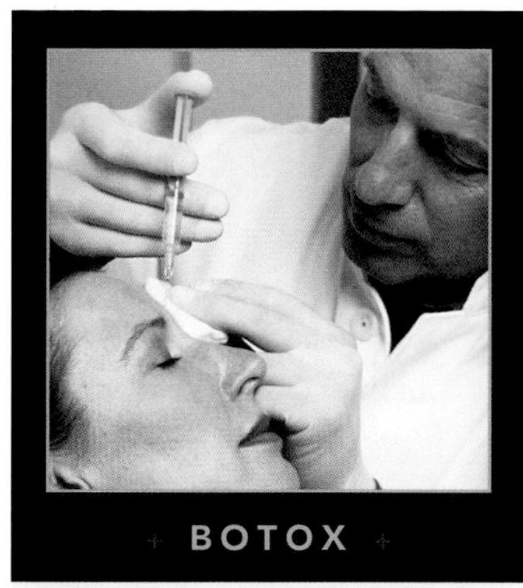

BOTOX

AMERICANS SPEND more than $1.4 billion annually on injections of Botox, the trade name of a toxin produced by botulism-causing bacterium. Botox was first used to treat muscular disorders of the eyes. A Canadian ophthalmologist noticed it also softened the wrinkles around her patients' eyes, which led to its now popular cosmetic applications.

With advancing age, the number of Langerhans' and other disease-fighting skin cells decreases and cell turnover in the epidermis slows, so wounds begin to take longer to heal. These changes, combined with diminished circulatory capacity, also make the elderly more vulnerable to infection.

Meanwhile, the sebaceous glands become smaller, causing dry and broken skin. The number and output of sweat glands declines, leaving the elderly more vulnerable to heat stroke; at the same time, the loss of insulating subcutaneous fat makes them more sensitive to the cold.

Older skin produces fewer melanocytes, too, and they tend to cluster in areas exposed to the sun, causing darker brown blotches that appear on the face, hands, and chest, commonly called called liver or age spots. Lighter-skinned individuals, who have less melanin to start, usually show these age-related changes more rapidly.

Ultraviolet (UV) light markedly accelerates aging by activating enzymes that break down collagen and elastin; the sun's UV rays also make skin rough and can overstimulate pigment cells, causing unsightly age spots and other skin conditions, including skin cancers (see "Protecting the Skin," page 60). Smoking also can damage the skin, causing changes that mimic the effects of aging. Smoking causes narrowing of blood vessels in the outermost layers of the skin, which impairs blood flow, depleting the skin of oxygen and important nutrients such as vitamin A. Smoking also appears to speed the decline of collagen and elastin fibers, creating thick, leathery skin.

We can stretch and lift and fill in the creases, but there are no known ways to avoid true age-related changes to the skin. Given the inevitable, one could try taking a kinder view of wrinkles, accepting them as well-earned character lines, each with a story to tell, rather than as reminders of passing time.

THE HAIR

THE LONG AND THE SHORT OF IT

SOME FIVE MILLION hairs grow on the human body, most obviously on the head, although the typical scalp contains only about 100,000 hairs. Body hair, found everywhere except the lips, palms, soles of the feet, nipples, and external genitalia, accounts for the rest. The hair plays an important role in protection. It protects the scalp from the sun's ultraviolet rays and prevents heat loss. Eyelashes and eyebrows shield the eyes from sun, foreign particles, and perspiration. Hair in the nostrils and ears helps keep out dust and other irritants.

A thicket of eyelash hairs (colored blue in this scanning electron micrograph) emerge from follicles in the underlying layer of skin.

HAIR PARTS

Each hair shaft—the visible part of the hair—is just a flexible strand of dead tissue. Hair grows in follicles, tube-like structures that originate in the dermis of the skin. In the base of the follicle is the hair bulb, a region of cell creation. A piece of dermal tissue, or papilla, packed with capillaries (tiny blood vessels) projects into the bottom of the bulb, providing nutrients to the undifferentiated cells, which are pushed upward as more cells are created. In the upper part of the bulb, the cells develop into distinct types and group themselves into concentric layers. The hair cells wither, die, and are compressed into a hard, dense shaft growing out of the skin. As new cells push the older sections of hair up, the overall shaft grows longer. Like the outermost layer of the skin, the hair shaft is made up almost entirely of keratin proteins. In the follicle, each hair bulb is surrounded by a hair root plexus, a network of sensory nerve endings that are sensitive to light touch and send out impulses if a hair is disturbed. That's why it hurts when hair is plucked or pulled and tickles when a bug alights. Haircuts are painless, however, because there are no sensory receptors in the hair shafts.

TYPES AND GROWTH

There are three types of human hair that occur during different stages of life. In the womb, extremely fine, nonpigmented hair called lanugo covers nearly the entire fetus by the fifth month of pregnancy and sheds before birth, except on the scalp, eyebrows, and eyelashes. Short, downy "peach fuzz" called vellus replaces lanugo hair a few months after birth. Vellus hair grows in most places on the human body in both sexes. Terminal hair is fully developed hair that is generally longer, coarser, thicker, and darker than vellus hair, including eyelashes, eyebrows, scalp, armpit and pubic hair, and, in men, hair on the face, chest, abdomen, legs, and arms.

Some 95 percent of the hair on the body of an average man is terminal and the rest vellus. In contrast, vellus hair accounts for about 65 percent of the hair on a woman's body; the rest is terminal. The reason for the difference is that during puberty, boys produce massive amounts of androgens (male hormones), stimulating the growth of terminal hair over much of the body. In girls, the ovaries and adrenal glands produce only small amounts of androgens, which promote hair growth in the armpits and pubic region.

Hair growth is cyclical, alternating between periods of continuous growth and rest. In the anagen portion of the cycle, growth is continuous. (As much as 90 percent of the hair on a person's head is in this phase.) Next, in the catagen (regressive) period, growth abruptly stops and the hair bulb shrivels; only about 1 percent of hair is in this stage, which lasts two to three weeks, at any time. There is no activity during the telogen (resting) phase, which lasts about three months. Then the follicle sheds the hair, and a new hair begins growing.

Growth rates vary with sex, age, and body location. Hair that can grow long, such as scalp hair, has a prolonged growth phase; shorter hair—eyelashes and eyebrows, for instance—has a relatively short growth period. Eyelashes grow for only one to six months, and eyebrows remain active for three to four months, which is why they don't become as long as hair on the head. Each scalp hair, for example, grows about half an inch (12.5 mm) per month for two to six years, then rests for a few months, sheds (about 50 to 100 head hairs normally fall out daily), and is replaced. If it isn't cut, scalp hair can grow up to five feet (152.4 cm).

+ GROWTH +

+ EACH HAIR follicle goes through 10–20 lifetime growth cycles. The duration of active growth varies significantly depending on the kind of hair.
+ Hair follicles on the scalp go through growth cycles that can last for 2 to 6 years.
+ Scalp hair grows, on average, about 0.01 inch (0.35 mm) per day.
+ The growth cycle for leg hair runs from 4 to 8 months.
+ Hair generally grows faster in women than in men, and faster on the top of the head than at the temples.

HAIR COLOR

Melanocytes in the base of the follicle produce pigment, melanin, that is passed to cells in the hair shaft. Hair color results from the ratio of eumelanin (black/brown melanin) to phaeomelanin (sometimes called red melanin). Black and brown hair contain a high amount of eumelanin; red and blond hair have more of the lighter pigment. Red hair is also colored by trichosiderin, an iron-containing pigment. Melanin varies in the hair of different parts of the body, which is why their colors may be different.

Melanocytes are genetically programmed to produce a certain amount of pigment at specific ages. Graying results from a progressive decline in the amount of tyrosinase, an enzyme needed to make melanin. Some people start graying as early as their teens; others, not until their 60s, but graying ordinarily begins in the 30s. The process is usually gradual; it may take more than ten years for a person to become completely gray. Gray hair retains some pigment but white hair has none.

WHAT CAN GO WRONG

EVEN THOUGH both men and women lose hair with age, baldness is most common in men. Less than a fifth of women experience hair thinning, but male-pattern baldness (right), or androgenetic alopecia, affects one quarter of men before the age of 30, about half over the age of 40, and two-thirds before 60. The condition is thought to be caused by a gene that changes the response of the hair follicles to the male hormone dihydrotestosterone (DHT). The condition can be inherited from either side of the family. There is no cure, but drugs have been found to stimulate hair growth in some people.

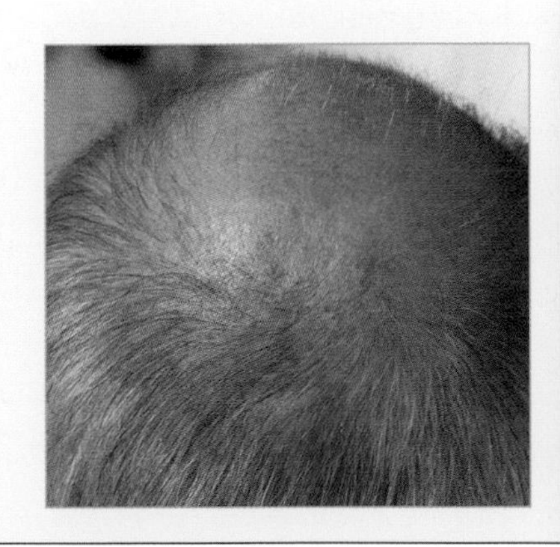

THE NAILS

USEFUL TOOLS ON THE FINGERS AND TOES

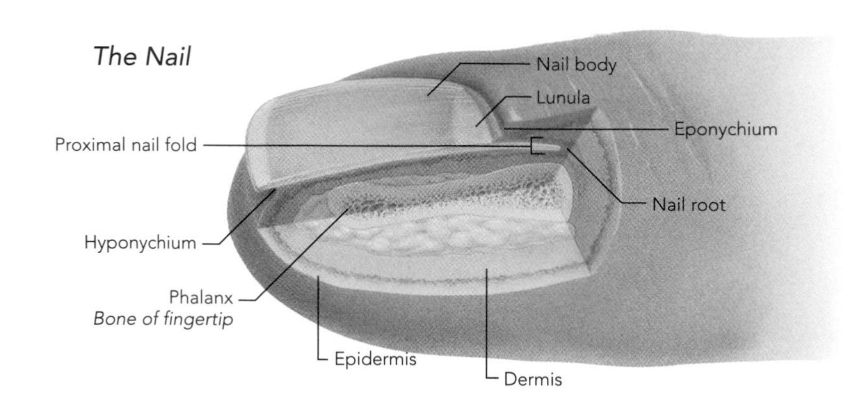

IN OUR EVERYDAY encounters with objects, the fingernails and toenails are often the first parts of the body to make contact. They absorb the force of impact while also acting as feelers, sending messages and conveying touch to the sensitive tissue in the tips of the fingers and toes. As a counterforce, they enhance that sense of touch. When we grasp an object, for instance, pressure on the fingertips causes resistance in the fingernails. The degree of resistance helps us gauge the object's texture and weight.

Fingernails are also practical tools. They make it easier to pick up small objects, unknot a shoelace, and peel an orange. Like the claws from which they have evolved, they can scratch an attacker or, more usually, an itch. The nails also give doctors a window into the body; changes in the nails' appearance can be an indicator of distress or disease in other systems.

NAIL PARTS

The nails are made up of several parts. The nail body is the most visible one; the hard part of the nail, it is a modification of the stratum corneum, or outer layer of epidermis. This translucent plate is composed of three tight layers of dead, keratinized cells that originate in a specialized epidermis called the nail matrix.

Nail growth begins in a part of the matrix called the nail root, an area of intense cell production. The lunula ("moon" in Latin)—the white crescent under the base of the nail plate, and most visible on the thumbnails— forms one end of the root; the other end passes below the thin flap of skin at the base of the nail plate, called the cuticle or eponychium.

After nail cells are produced in the root, they become heavily keratinized. The nail body grows out over the nail bed, which extends from the lunula to the nail's free edge—the part of the nails that are trimmed in a manicure. Parallel ridges in the underside of the nail plate interlock with ridges in the nail bed, forming a tight bond. Beneath the free edge a thick ridge of skin, called the hyponychium, binds the nail plate to the tip of the finger (or toe).

Folds in the skin overlap the base and sides of the nail plate, holding it in place and guiding the direction of its growth. The proximal nail fold, at the base of the nail plate, produces a thin, tough outer epidermis, the cuticle. This skin adheres to the nail's surface, which creates a waterproof seal. The cuticle is renewed as skin cells flake off. Many people also be trim it as part of their grooming.

Nails grow continuously and don't shed. The nail plate consists

NAIL FACTS

+ THE EMBRYO begins growing nails in the ninth week. Fingernail length is sometimes used to estimate whether the baby is pre-, full-, or post-term.
+ It takes 6 months to grow a complete fingernail, 18 months for a toenail.
+ Children's nails grow fastest on the middle finger, in the daytime, and in summer.
+ Typically, toenails are thicker than fingernails. Men's nails are thicker than women's.

SEE ALSO: Chapter Nine, "The Nerves," PAGE 250

Fingernails on a 5-month-old fetus are still developing. After the 36th week of a pregnancy, the fingernails will grow beyond the end of the fingertips.

of about 25 tightly packed layers of dead, keratinized cells organized into three distinct layers. The top layer is a few cells thick; the bottom layer, only one or two. A random, overlapping arrangement of keratin fibers in those layers—not unlike the hair shaft—gives the nail its flexibility, resiliency, and strength. In the center layer, however, which makes up about three-quarters of the nail plate, long keratin fibers are arranged parallel to the lunula. The combination allows the nails to tear easily across the width of the nail plate but not down its length, preventing the pain and infection that would result from tears that exposed the nail bed. This self-trimming mechanism would have kept early humans' nails neat and functional. Besides containing keratin, the nail plate includes trace elements (mainly iron, zinc, and calcium) and small amounts of lipids and water.

The normal nail has a smooth upper surface, but vertical ridges may appear with aging. The nail plate appears pink from dermal capillaries below, while clusters of new cells give the lunula its white color. Physicians can diagnose some autoimmune connective tissue conditions by examining the dermal capillaries, which form unique patterns that are altered by such disorders.

Damage to the cuticle, nail folds—hangnails, for instance—or hyponychium can allow germs to enter the nail bed and cause infection. Biting the cuticles and nails may introduce bacteria from the mouth into the numerous blood vessels alongside or underneath the nail. The fingertip bone, or phalanx, is more vulnerable to infection because there is no subcutaneous tissue between the nail and bone to intercept bacteria such as *staphylococci* or *streptococci*, which can spread rapidly.

ILLNESS INDICATORS

Many ailments not only affect nail growth, but can cause disfiguring changes in the nails, alerting doctors to the existence of the underlying condition. Brittle, concave, spoon-shaped, or ridged nails, for instance, may indicate iron-deficiency anemia. Nails separating from the nail bed could be a sign of a hyperthyroid disorder such as Graves' disease; black marks that look like splinters under the nail plate could indicate respiratory or heart disease; hard, curved, yellow nails could indicate serious bronchial problems or a fluid backup in the lymph glands.

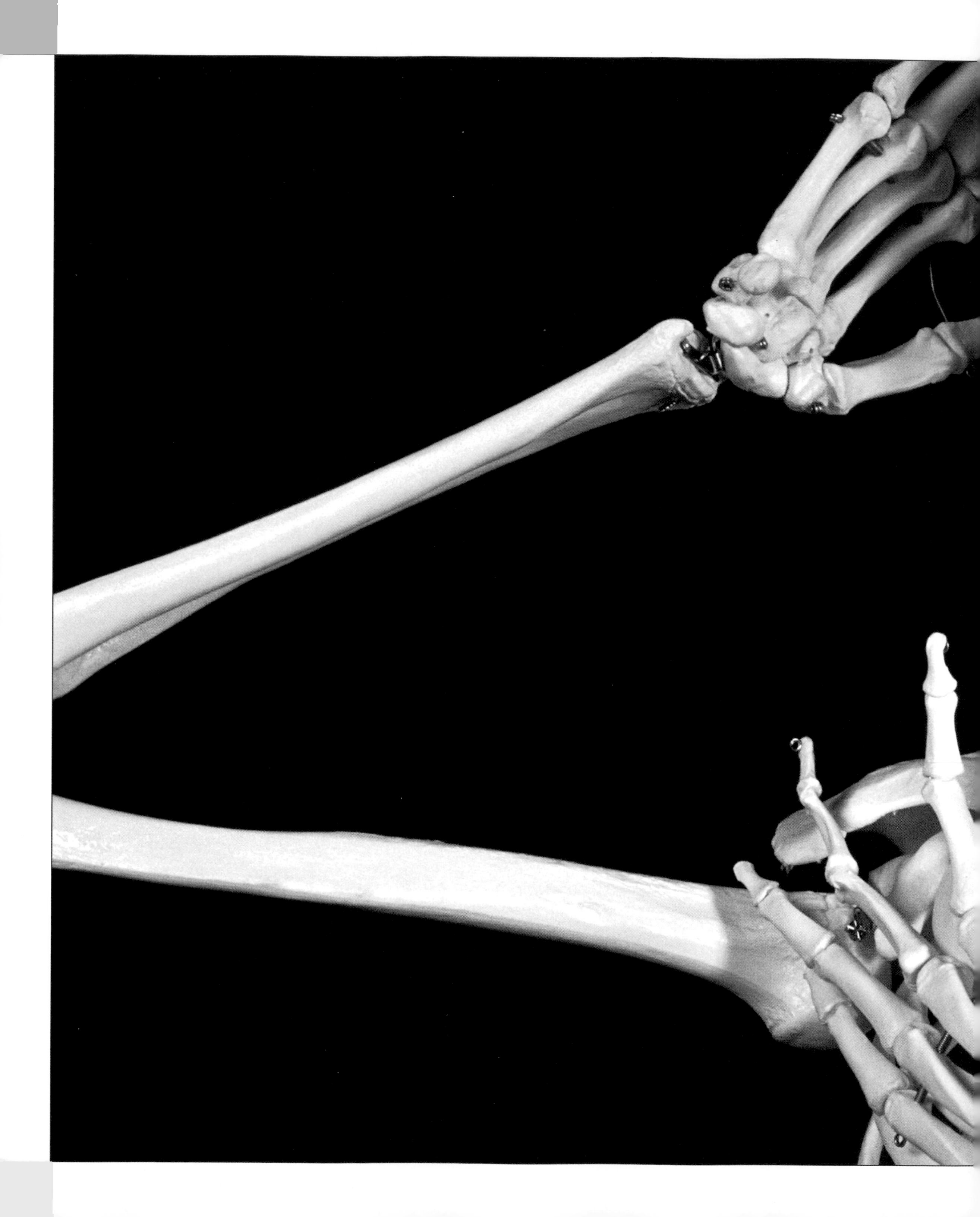

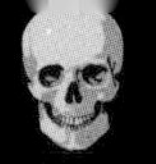

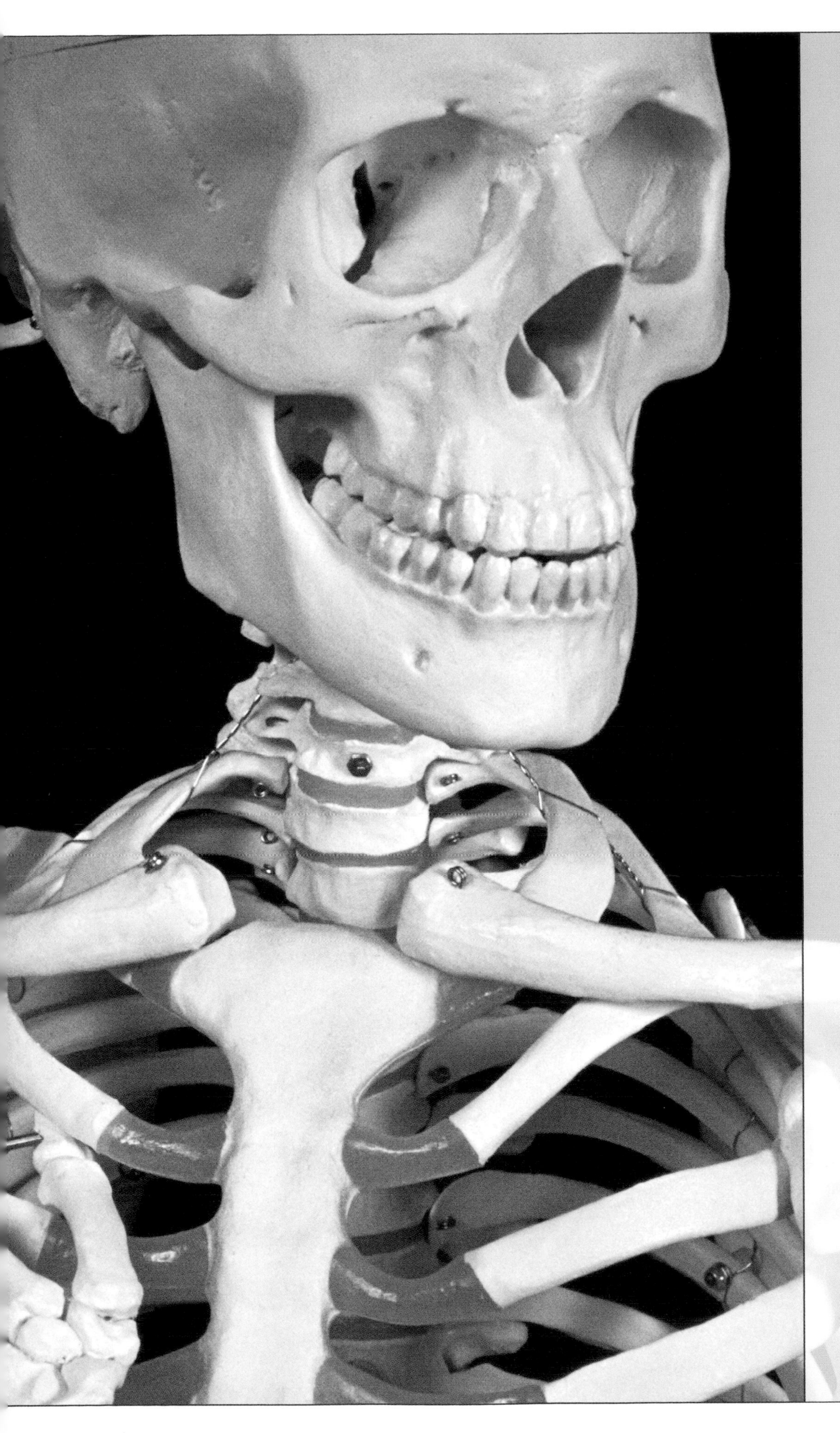

CHAPTER THREE

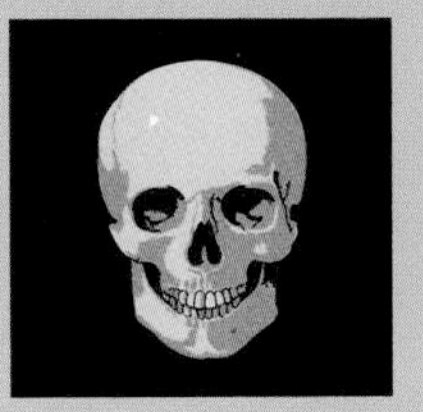

STRUCTURE

THE SKELETONS WE SEE AT Halloween may look frail, but the living skeleton is a powerful, elaborate structure composed of bones, joints, cartilage, and ligaments. It supports the body, gives it shape, and anchors the muscles that move it. Its bones not only protect internal organs, they also manufacture blood cells and store mineral salts such as calcium and phosphorus, needed for healthy bone and muscle function.

A grinning skull is made of more than 20 separate but fused bones.

THE SKELETON

SUPPORT AND PROTECTION

THE HUMAN SKELETON is made up of 206 bones of different shapes and sizes. The skull, backbone, and ribs form the axial, or central, part of the skeleton. Extending symmetrically on either side, the appendicular skeleton includes the bones of the arms and hands, legs and feet, and the pectoral and pelvic girdles, which attach the appendages to the axial skeleton.

FUNCTION

The skeleton's major tasks are to support the body, provide a framework for the muscles to enable movement, and shelter the internal organs. The 12 pairs of ribs that curve around the chest cavity from the spine (also called the backbone) protect the heart, lungs, and organs of the upper abdomen; the spine protects the delicate spinal cord; the pelvis protects the uterus and bladder; and the skull safeguards the brain. The brain may be a mighty thinker, but its soft tissue would be extremely vulnerable to injury were it not for the thin yet sturdy skull encasing it (see "The Skull," page 90).

The skeleton enables humans to stand upright and to resist the constant pull of gravity. The spine provides some flexibility (hence, our ability to bend, curl, sit, twist, and tumble), but it also enables humans to stand tall and, together with cervical muscles, to hold their hefty heads high (the brain and skull together weigh approximately 15 pounds).

Bones, although relatively light—the average adult skeleton weighs about 20 pounds (9 kg)—can bear an enormous amount of weight and withstand forces of movement with the strength of cast iron. They are said to be "twice as tough as granite" in resisting compression. When a person takes a leisurely stroll, for instance, each foot strikes the ground with a force of about three times that of his or her weight. Increase the tempo to a brisk walk, jog, or run, and the impact increases to five or six times the person's weight. This means that in someone who weighs, say, 150 pounds, his or her lower limbs are subjected to a walloping 450 to 900 pounds of force during normal physical activity.

DEVELOPMENT

When the skeleton first forms, it consists of rubbery cartilage. Bone-making cells called osteoblasts deposit mineral salts in this soft template, gradually converting most of it to bone. Ossification, or bone building, continues after a baby is born and lasts throughout life. For instance, the xiphoid process, a flap of cartilage just below the sternum (breastbone) often doesn't ossify, or harden, until a person reaches 40 years of age or even older.

Bones are living tissues that grow with the rest of the body and can repair themselves. They undergo continuous renewal, even in adulthood, during a process called remodeling. This dynamic process allows bones to develop proper proportions, to gain strength, and to mend when they break. If the remodeling process is interrupted, it can result in potentially debilitating conditions such as the bone-thinning disease osteoporosis (see "What Can Go Wrong," page 78).

Initially, bones are only partially ossified and their ends consist of cartilage. As the cartilage grows, the ossified areas also expand until the adult bone is complete. Most growth (in length) occurs in the rounded area at the end of the bone, called the epiphysis. A baby's skeleton consists of about 300 "soft" bones (composed mostly of cartilage) that eventually fuse to form the 206 bones in the adult skeleton.

During childhood, the cartilage grows and is slowly replaced by solid bone in a process that may continue until a person reaches the mid-20s. The cartilage almost completely disappears in an adult skeleton; it remains only in certain parts of the ear, nose, mouth, trachea and bronchi, the front of the ribs, and on the surface of the

SEE ALSO: Chapter Thirteen, "Building Blocks," PAGE 352

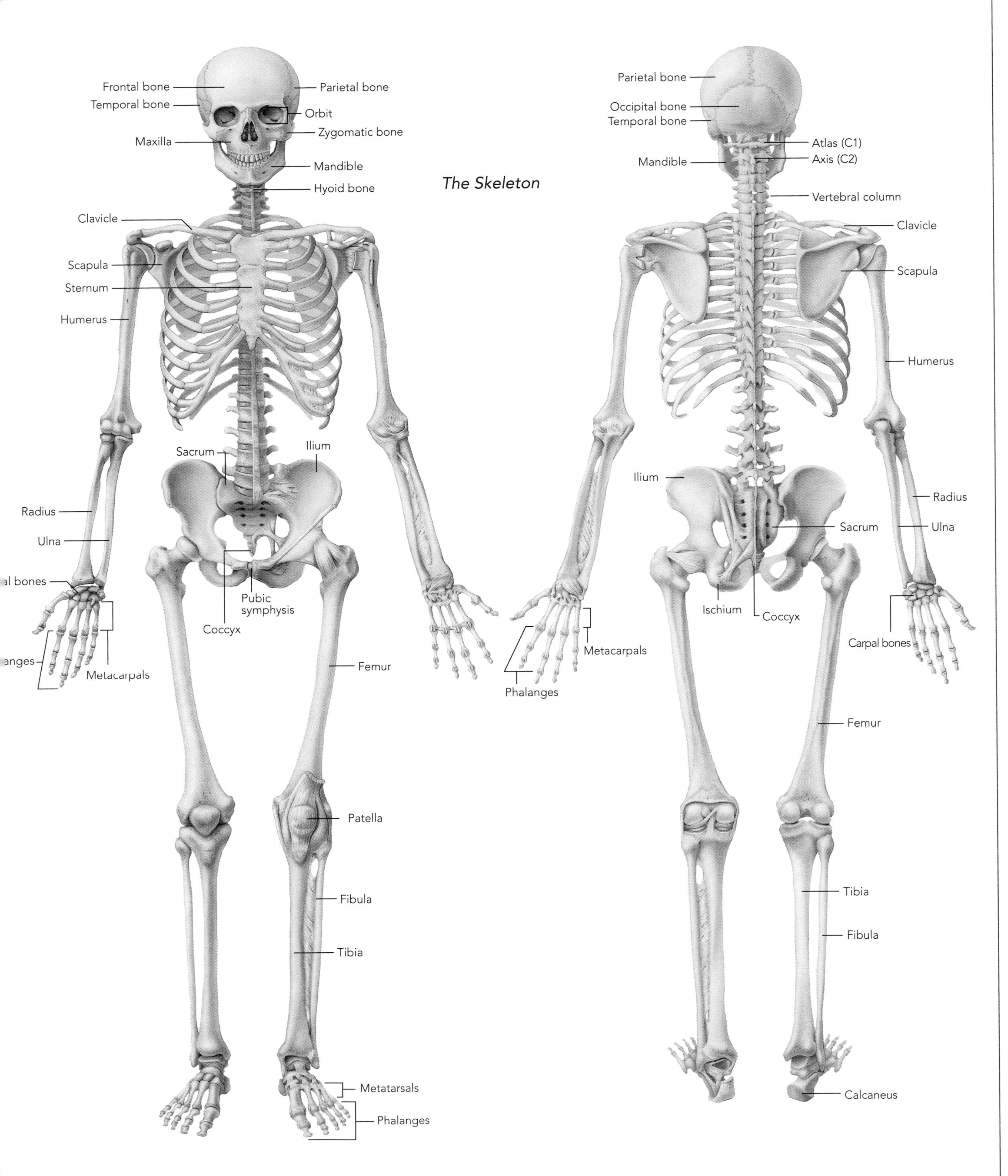
The Skeleton
Frontal bone
Temporal bone
Maxilla
Parietal bone
Orbit
Zygomatic bone
Mandible
Hyoid bone
Clavicle
Scapula
Sternum
Humerus
Sacrum
Ilium
Radius
Ulna
al bones
anges
Metacarpals
Pubic symphysis
Coccyx
Femur
Patella
Fibula
Tibia
Metatarsals
Phalanges
Parietal bone
Occipital bone
Temporal bone
Mandible
Atlas (C1)
Axis (C2)
Vertebral column
Clavicle
Scapula
Humerus
Ilium
Radius
Ulna
Sacrum
Ischium
Coccyx
Metacarpals
Phalanges
Carpal bones
Femur
Tibia
Fibula
Calcaneus

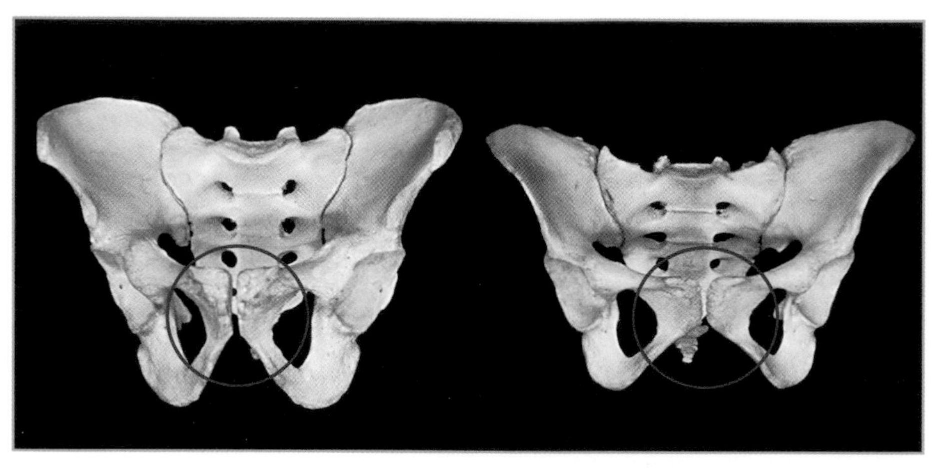

MALE PELVIS FEMALE PELVIS

Above: Differences between the male and female pelvis: A male's pelvic inlet (on the left) is much narrower than that of a female's (on the right). Below: The graceful skeleton supports a wide range of movements.

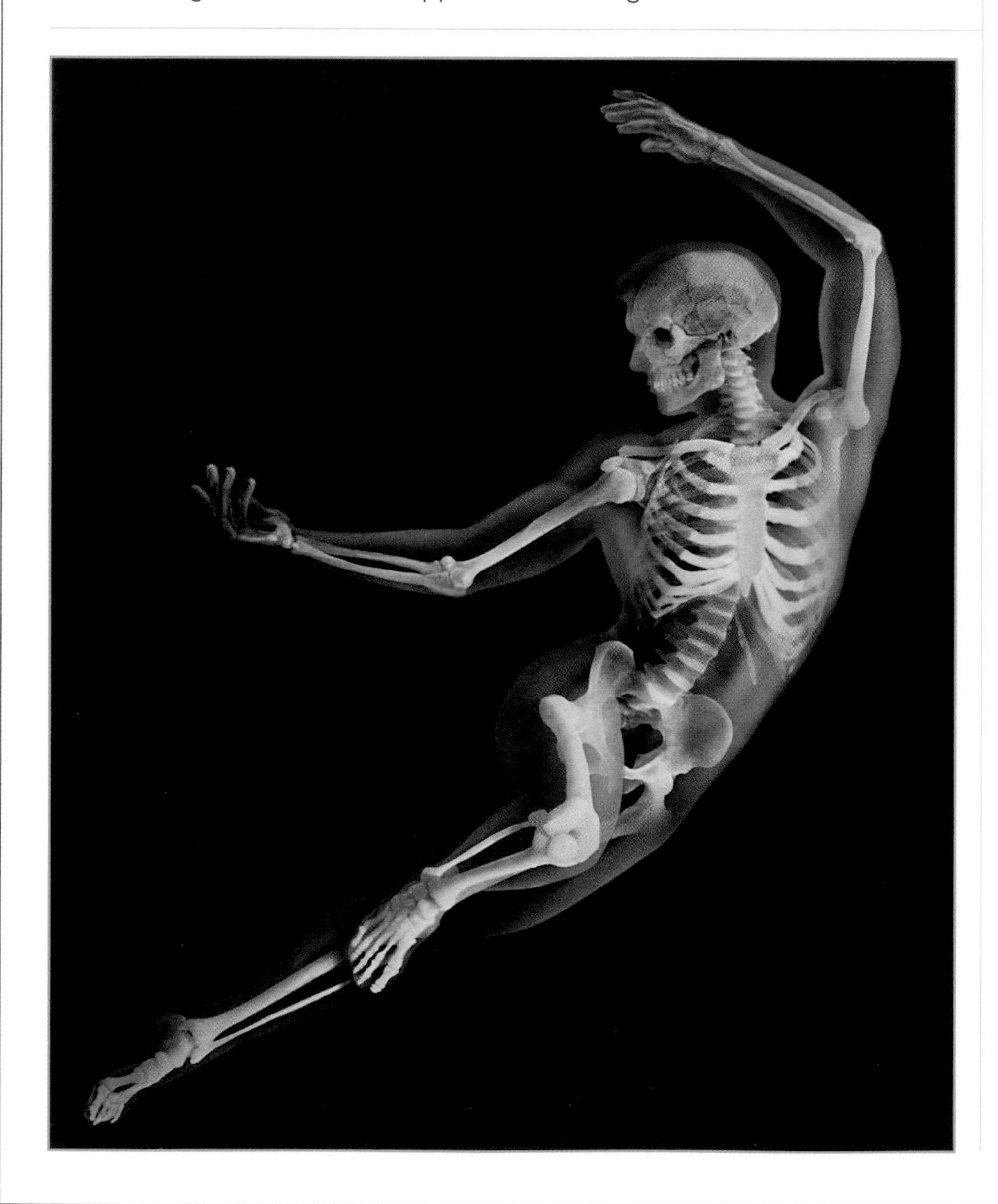

joints. Fetal cartilage serves as a model for the bones; mineral salts, such as calcium, are deposited onto the template, and the cartilage becomes calcified, or hard, a process called endochondral ossification. Most bones, called endochondral bones, are formed this way.

BONE GROWTH

Ossification starts in the middle of the bone in the medullary cavity. Osteoblasts continue to ossify the original cartilage cells. As a child ages, the osteoblasts that convert cartilage to bone become mature bone cells called osteocytes and retire from the bone-making process. During childhood and adolescence, the rounded ends of the long bones in the arms and legs are separated from the shaft (or diaphysis) by a cartilage growth plate, also called the epiphyseal plate.

As the cartilage cells divide and reproduce, the bone lengthens; when the cartilage cells in the growth plate stop dividing, the plate ossifies, leaving only a thin epiphyseal line. The bone stops growing in length, and adult height is reached, usually between the late teens and early 20s. Even though bone growth halts, however, mass may still be added or lost. The body manufactures normal bones by balancing the operations of bone-building osteoblasts and other bone-destroying cells called osteoclasts. At the same time that osteoblasts deposit protein and mineral salts to make new bone, osteoclasts remove old and aging bone cells. Cell material that's no longer needed is excreted as waste, but calcium is recycled through the bloodstream and used by osteoblasts to create new bone.

GENDER GAPS

There are several key differences between the male and female skeletons. In most individuals, the female skeleton is smaller and lighter than the male skeleton. The easiest differences to see are that women tend to have broader hips and their so-called true pelvis—the ring formed by the pubic bones, ischium, lower part of the ilium, and the sacrum—is wider and rounder. The sacrum is tilted back more in a woman, and the coccyx (tailbone) moves more easily, because the opening has to be wide and flexible enough for a baby to pass through during birth.

The female skull is, on average, 20 percent smaller than the male skull, and a woman's facial frame is generally less angular—her chin is less square, her jawbone is smaller and less pronounced, and her forehead tends to be more rounded and longer from top to bottom. In addition, the ocular orbits, or eye sockets (cavities that house and protect the eyeballs), generally appear higher and rounder in women than in men.

Women generally have narrower shoulders and shorter arms than men do, which is why they typically have less throwing leverage. In addition, their broader hips increase the angle between the pelvis and thighbone, making it dificult for women to raise their knees as high or to push off the ground with as much force as men can while, for example, running or jumping. The trade-off: The female body is generally more stable, because women's lower hips and slender shoulders create a lower center of gravity. Women also have a wider range of movement in their thighs than men do because the acetabulum, a hollow area in the hip that the head of the femur (thigh bone) fits into, is shallower and the right and left acetabulum are farther apart in women, providing greater flexibility.

+ BONE FACTS +

+ FROM HEAD to toe, there are 206 bones in the adult human body.
+ The femur, or thighbone, is the longest bone in the body. In an average adult, it measures 18 inches (46 cm) in length.
+ The shortest bones are the ossicles, "little bones" in the ear that are named for their shape. Smaller than the hammer (maleus) or anvil (incus), the stirrup (stapes) is just 0.1–0.13 inches (2.6–3.3 mm) long.

AXIAL SKELETON

The skeleton is divided into two parts: the axial and appendicular systems. The axial skeleton protects the major organs of the nervous,

A flexible female gymnast performs a split while in a handstand. Because of bone structure, women's hips have a wider range of movement than men's.

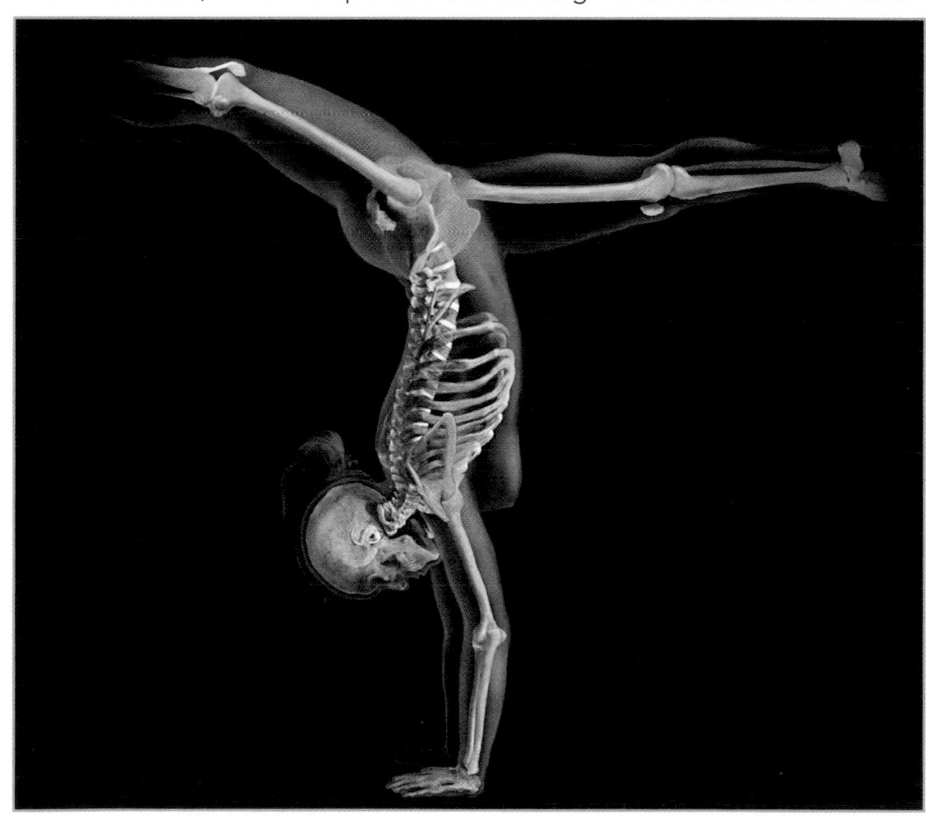

ANDREAS VESALIUS

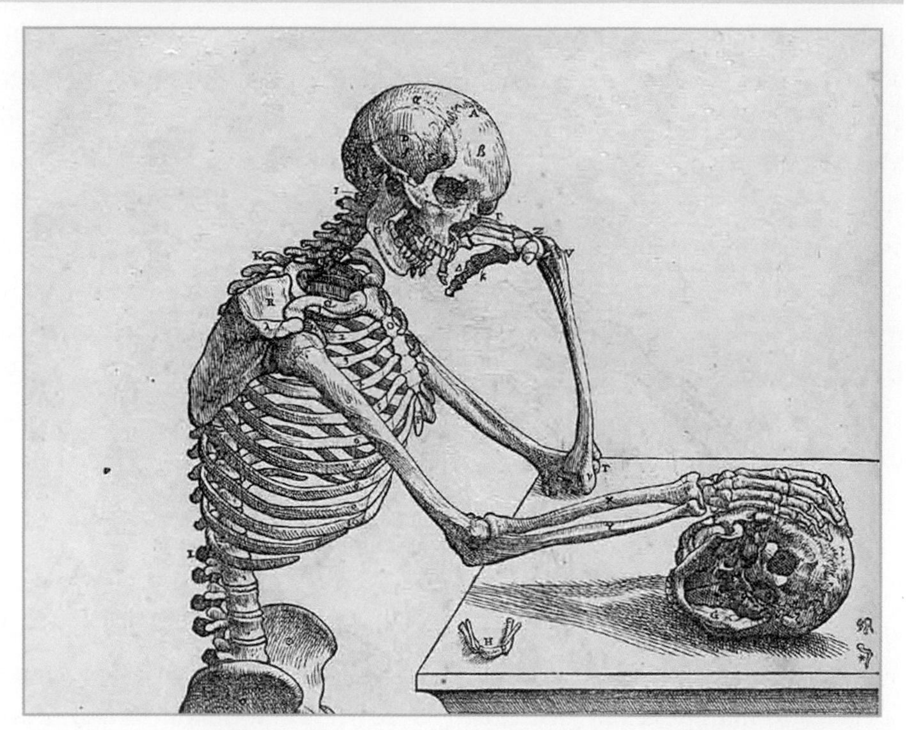

Vesalius captioned this image with the phrase, "Genius lives on, all else is mortal."

FOR MORE THAN a thousand years, knowledge of the human body was based on the often inaccurate teachings of the second-century Greek physician and surgeon Galen. Galen's influential work *On the Bones,* for instance, contained many errors about the skeleton, as Andreas Vesalius (1514–1564), a medical student from Brussels, learned when he began articulating skeletons, a practice he continued as a professor of anatomy at Padua.

In 1543 Vesalius published *De Humani Corporis Fabrica (On the Structure of the Human Body),* containing more than 200 detailed illustrations (some accurate enough to be used today). Based on his own dissections, it heralded a new era of medicine and anatomy founded in observation. The *Fabrica* rectified many of Galen's errors, including that the human sternum had seven parts, instead of the actual three.

In the first book, on the bones and joints, Vesalius advises students to learn by articulating a skeleton. The necessary tools are described and illustrated, with instructions on how to dismember a corpse, clean the bones, and reassemble them using copper wire and tubing. Students, he said, should also keep a second skull handy to study its internal structure, as the illustration above suggests. For purposes of comparison they would do well, he said, to have available the skeleton of an ape—an animal with a sternum made of seven bones, and the one, Vesalius realized, on which Galen had based his claim. ■

respiratory, and central circulatory systems. The term "axial" refers to an imaginary line, or axis, that cuts through the center of the body. Eighty bones make up this portion of the skeleton, including the vertebral column (or spine) and the bones of the skull, the ribs, and the sternum (breastbone).

The axial skeleton also includes the hyoid bone, a tiny U-shaped structure at the base of the tongue (just above the larynx, or voice box), which anchors the tongue and muscles used during swallowing. (In forensic medicine, the hyoid is examined to determine whether the deceased has been murdered—it sometimes snaps when a person is strangled or hanged, choking the victim and causing death.) The hyoid is the only bone in the body that does not articulate with another bone; it is supported by ligaments attached to the styloid processes of the temporal bones and multiple muscles.

APPENDICULAR

The appendicular skeleton consists of the 126 bones of the body's "hanging parts," or appendages. Working with the muscles, this group of bones can manipulate objects and move around itself. Its bones protect the organs of digestion, excretion, and reproduction. The appendicular skeleton is organized into the upper extremities (arms and hands), the lower extremities (legs and feet), and the bones that attach them to the axial skeleton. The arms and legs have the same number of bones (30 in each limb) and share the same fundamental structure: a single long bone (upper arm, thigh) connected by a joint (elbow, knee) to a pair of long bones (forearm, lower leg), connected by another joint (wrist, ankle) to a collection of tiny hand or foot bones.

In the upper arm, the bone is called the humerus; the large bone in the forearm is the ulna and the other bone is the radius. In the leg, the femur is the huge thighbone, the tibia is the large shinbone, and the fibula aligns alongside.

GIRDLES

The appendicular skeleton also includes two "girdles," the bones that connect the limbs to the axial skeleton. Whenever someone jumps, kicks a ball, swings a tennis racket, or writes a letter, he or she is using limbs and girdles (and, of course, muscles, tendons, and ligaments). The girdles are rings of bones ("girdle" means to encircle) that attach each limb pair to the axial skeleton. At each shoulder, the upper arm bone (specifically, the humerus) attaches to the pectoral girdle, consisting of the clavicle (collarbone) and scapula (shoulder blade). The thigh bone (femur) connects to the hooplike pelvic girdle. When you place your hands on your hips, they rest on the sloping upper edges of the pelvic girdle.

The pectoral girdle is the weaker, and has only one true joint between it and the axial skeleton, the sternoclavicular joint. It is also the more flexible of the two girdles, which gives the arm bones great range of motion. Coupled with the flexibility of the hand bones, it makes the upper limbs perfect for manipulating objects. However, the intricate connections at the shoulder joint allow more motion, but leave the shoulder susceptible to dislocation if wrenched or twisted.

The leg bones are firmly attached to the stationary pelvic girdle, which enables them to support the body's weight. Two curved hip bones (each composed of three fused bones, the ilium, ischium, and pubic bone) make up the pelvic girdle. The pelvic bones meet at the front of the girdle, while at the rear they are firmly attached to the sacrum at the sacroiliac joint. The pelvic girdle and sacrum together form the pelvis, a bowl-shaped structure that supports and protects lower digestive, reproductive, and urinary organs. The opening in the center of the pelvis, the pelvic inlet, is much larger in diameter in females than in males, to provide sufficient space for a baby's head to squeeze through during birth.

LIMBS

The bones in limbs each have to absorb and stand up to a great deal of pressure. Human leg bones (and hip, knee, and ankle joints) must be powerful enough to support the body in an upright position. To help meet this demand, the human thighbone has evolved into the strongest, largest bone in the body; both the femur and the muscles attached to it must be strong enough to withstand the intense stress that walking, running, jumping, and other movements place on them. (Walking upright liberated the upper limbs and hands of early human ancestors.)

BREAKTHROUGH

THE X-RAY: In 1895 German physicist Wilhelm Röntgen (1845–1923) discovered a form of radiation that could penetrate the body, the x-ray, right. X-rays are absorbed by the dense bones, which appear as shadows on film; they pass through softer tissues, which appear less distinctly. The x-ray allowed doctors to see inside the body without surgery, and it was quickly adopted as a diagnostic tool. X-rays are used to diagnose bone fractures and breaks, reveal dental cavities, and identify foreign objects. Many soft-tissue abnormalities can be detected with computed tomography (CT)—computer-enhanced x-rays.

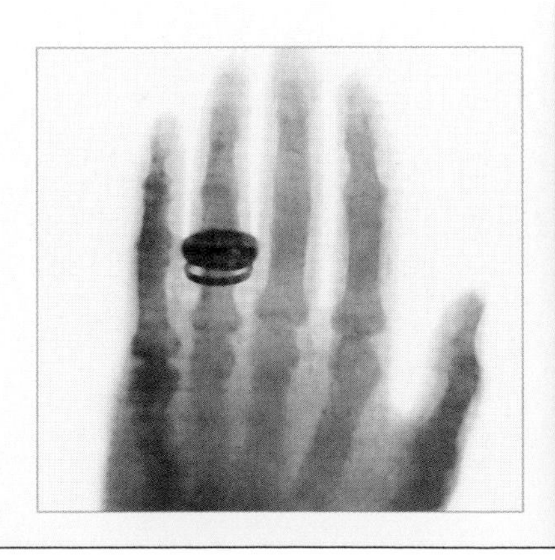

THE BONES THE MULTITASKERS OF THE BODY

BONES ARE STRONG and, because of their internal structure, surprisingly light. As living organs, they bustle with activity. Inside, blood cells are being created by bone marrow. Outside, competing cells continually build and break down bone in a process known as remodeling. Bone-making cells called osteoblasts build new bone by secreting a matrix of collagen, other proteins, and minerals such as calcium and phosphorus. At the same time, cells called osteoclasts absorb old bone cells, releasing their calcium into the blood, where it can be used again to make new bone. (Any material not re-used is excreted as waste.)

BONE GROWTH

During childhood, bone formation outpaces bone breakdown, enabling the bones to become longer, denser, and heavier. Osteoblasts continue to calcify the original fetal cartilage, and growth takes place throughout the bone. As a child ages, osteoblasts become trapped in the matrix and mature into bone-maintaining cells called osteocytes. Eventually, the only areas of growth are near the ends of certain bones in two small pads of remaining cartilage called growth plates, or epiphyseal plates. In each plate, the inner edge of cartilage is gradually ossified; at the same time, new cartilage grows at the outer edge, lengthening the bone. Bones stop growing when these plates fully ossify.

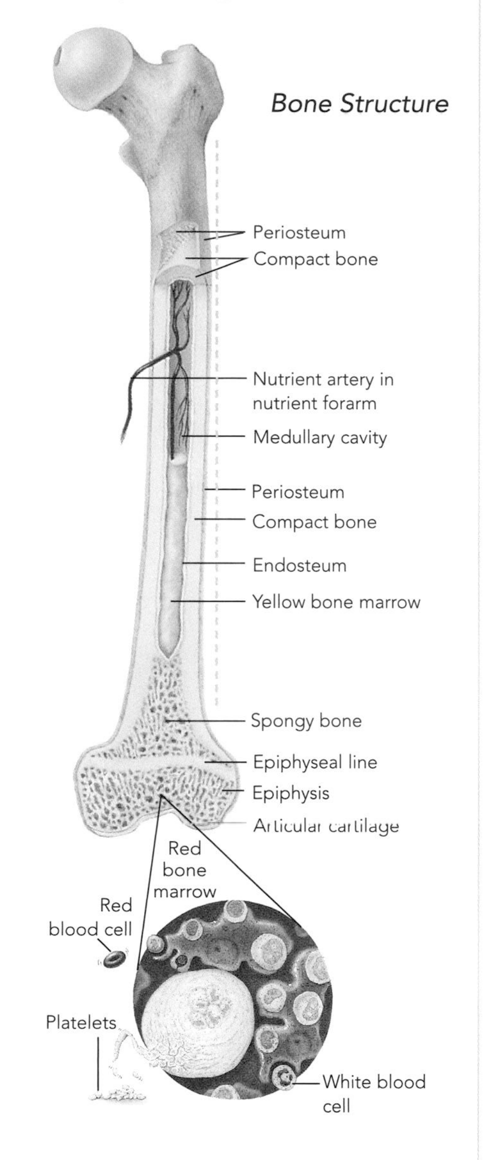

How long? How tall? Growth and maintenance of bone depends on nutrition and levels of certain hormones. Large amounts of calcium and phosphorus are needed for healthy bone growth, as are smaller amounts of fluoride, magnesium, iron, and manganese. Vitamin A helps keep osteoblasts functioning. Vitamin C supports the manufacture of collagen—the main bone protein—and the development of osteoblasts to osteocytes. Vitamins K and B12 also are needed for protein synthesis. Vitamin D promotes absorption of calcium and phosphorus from the intestines and is essential to bone mineralization, growth, and repair.

When and how much the bones of a child grow depends on levels of insulin-like growth factors (IGFs) produced by bone tissue and the liver. IGFs promote cell division at the epiphyseal plate and in the periosteum (outer bone membrane) and enhance synthesis of the proteins needed to build new bone. Production of IGF-1 in the liver is stimulated by human growth hormone (HGH) secreted by the pituitary gland.

Hormones from the thyroid gland (thyroxin) and from the pancreas (insulin) are also needed for normal bone growth. At puberty, sex hormones (estrogen and androgens) stimulate increased osteoblast activity and synthesis of bone matrix, leading to the sudden growth spurts that occur during the teenage years. Ultimately, the sex hormones shut down growth at the epiphyseal plates. At this point, bones stop growing longer. By the time a person reaches his or her mid-20s, the bone-building and bone-destroying processes begin to

SEE ALSO: Chapter Five, "The Blood," PAGE 132, Chapter Twelve, "Adaptation," PAGE 328

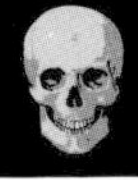

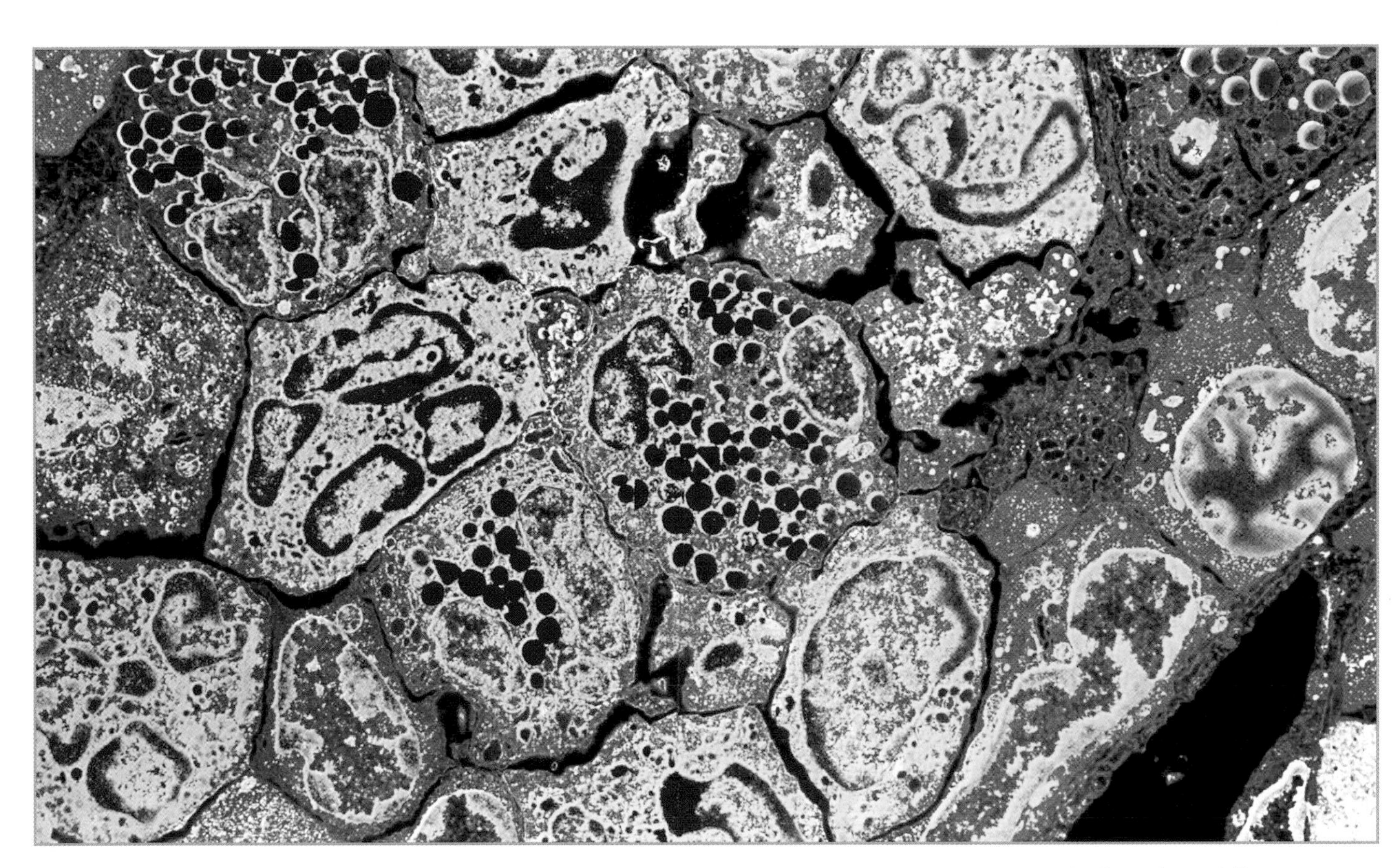

A colored electron micrograph of bone marrow shows the various precursive blood cells before they differentiate.

occur at about the same rate; by the mid-30s, bone loss starts to exceed bone formation.

BONE MARROW

Bones consist of two kinds of bone tissue: compact bone, which forms the hard outer layer; and spongy bone, a lacy tissue in the interior. Gel-like tissue called marrow is found in the gaps of some spongy bone, such as in the long bones of the arms and legs.

Red marrow is the body's blood cell factory, churning out more than 100 billion new blood cells every day. In babies, all bone marrow is red, but in adults, red marrow is found only in the ends of the long bones of the limbs and in the breastbone, spine, ribs, shoulder blades, pelvis, and skull. Yellow bone marrow serves mainly as a storage bin for fat tissue; it can be found in the medullary cavity in the shaft of long bones.

Blood cells begin as stem cells in red marrow that differentiate as they continuously divide, becoming red blood cells, white blood cells, or platelets. Red blood cells transport oxygen from the lungs to cells throughout the body and carry away carbon dioxide. Platelets, small pieces of blood cells, help blood to clot. The white blood cells produced in bone marrow play a key role in fighting disease and keeping the body's immune system fit. (White blood cells are also produced in the lymph nodes, spleen, and thymus gland.) This is why bone marrow transplants can help people with

BONE TYPES

BONES ARE typically classified into groups by shape.

+ Long bone: humerus, radius, ulna (arms); femur, fibula, tibia (legs); phalanges (fingers, toes); metacarpals (hands); metatarsals (feet)
+ Short bone: include the carpals (wrists) and tarsals (ankles)
+ Flat bone: include the scapulae, ribs, most skull bones, sternum
+ Irregular bone: vertebrae, mandible
+ Sesamoid bone: the patellae, or kneecaps

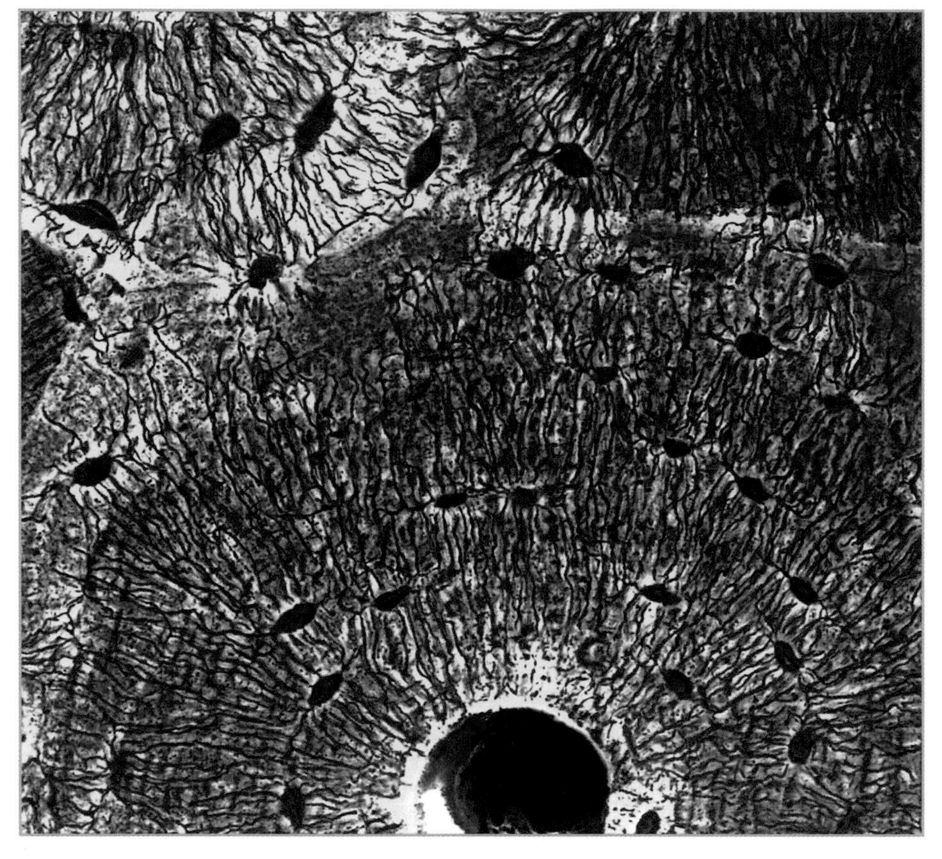

In this highly magnified cross section of compact bone, osteoctyes (dark spots) branch out to connect with other cells.

diseases or treatments that attack and weaken the immune system.

BONE STRUCTURE

Bones are so light that they account for only about 14 percent of a human adult's total body weight, and yet they are about as strong as reinforced concrete. Deposits of hard minerals such as calcium and phosphorus make the bones firm. The minerals form a complex called apatite salts. The salts stick to the flexible collagen fibers, forming a rigid casing. Bones are densest and toughest near the external surface, where they bear the most stress. Inside, spongy bone forms a lattice-like scaffolding that helps to distribute loads while minimizing bone weight.

The end of the bone, where it meets another bone, is coated with articular cartilage both to absorb shock and reduce friction at the joints. The rest is covered by the periosteum, a double-layered, white membrane of fibrous connective tissue that contains osteoblasts supplied by lymph vessels, nerves, and blood vessels—essential for bone growth, repair and nutrition. Immediately beneath this lining is a layer of densely packed compact bone, and, underneath that, is a layer of spongy bone (also called cancellous bone) .

Other than tooth enamel, compact bone is the toughest material in the body. It contains mature bone cells (osteocytes) that live in individual lacunae, or pockets of bone. Sheets of lacunae, arranged in concentric layers, or lamellae, form tubular networks called osteons, or Haversian systems; the Haversian canal at the core of the system serves as a conduit for blood vessels and nerves from the periosteum to supply the bone

WHAT CAN GO WRONG

OSTEOPOROSIS is a condition where bone loss outpaces bone regeneration, causing bones to weaken and lose mass. Osteoporosis occurs in both types of bone, but spongy bone, shown left at normal density, seems more vulnerable. As bone weakens and becomes more porous, shown opposite, fractures can result. They are most likely to occur in the hips, wrists, and vertebrae. The extent of osteoporosis (and of its precursor, osteopenia) can be determined by a simple test, a bone density scan. Osteoporosis strikes women more often than men. Starting at about age 30, bone begins to be lost

SEE ALSO: Chapter Thirteen, "Building Blocks," PAGE 352

tissue. Smaller canals, canaliculi, contain interstitial fluid that carries hormones, nutrients, and wastes between outlying osteocytes.

If the skeleton were made entirely of compact bone, it would be far too heavy to move. That's where spongy bone comes into play. Despite its name, this tissue is not soft. Rather, spongy bone looks somewhat like a honeycomb; it is a web of tiny spikes, called trabeculae, that is relatively light but can endure extraordinary stress, because of the way the spikes are angled and arranged. Spongy bone is thickest closest to the joints, which absorb the brunt of bumps and jolts.

Compact Bone

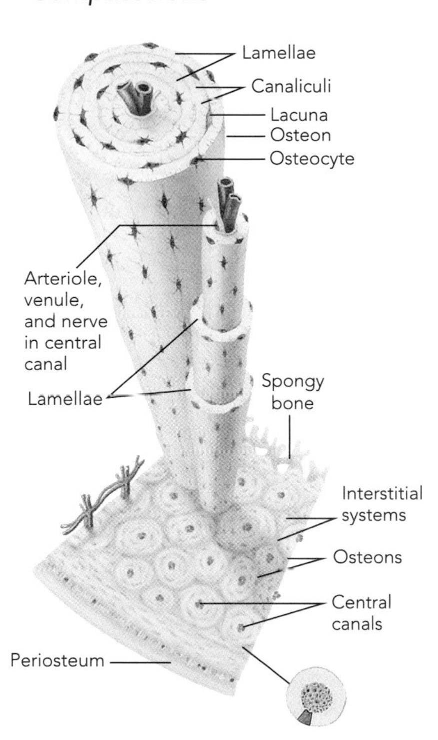

BONE STRUCTURE

Bones are typically classified by shape (see "Bone Types," page 77). Flat bones shield the brain and the organs in the chest cavity and pelvis. They are generally thin and composed of two nearly parallel plates of compact bone, providing generous space for muscles to attach. In flat bones, the spongy bone is called the diploe ("folded") and is sandwiched between the layers of compact bone.

Long bones are powerful, weight-bearing bones, chiefly those of the arms and legs, the clavicles, and those of the hands and feet. They have greater length than width and consist of a tubular shaft, or diaphysis, and a variable number of epiphyses, or ends. They are usually curved somewhat to distribute the stress of the body's weight at several points, helping to prevent fractures. The diaphysis consists mostly of compact bone tissue surrounding the medullary cavity, a central opening lined by tissue called endosteum. Spongy bone fills the epiphyses and, interrupted by the cartilaginous epiphyseal plate, extends a small distance into the shaft.

Short bones are small cubelike bones found in the wrists (carpals, other than the pisiform, which is sesamoid) and ankles (tarsals, except for the calcaneus, which is irregular). Range of motion is very limited between carpal and tarsal bones. Short bones consist of spongy bone surrounded by a thin layer of compact bone.

Irregular bones come in a variety of shapes and usually have projections that muscles, tendons, and ligaments can attach to. They vary in the amount of spongy and compact bone present. Such bones include the vertebrae of the spine and some facial bones.

Sesamoids are a type of short bone that develops in tendons where there is considerable friction and stress, such as in the palms and soles, to help reduce wear. They are usually sesame seed–shaped (hence their name). Sesamoids are not always completely

more quickly than the body makes it. Then, during and after menopause, levels of estrogen, a hormone that slows bone loss, decrease. Other risk factors include race (Asians and Caucasians are more susceptible than blacks), body type (thin, petite women are more likely to lose bone density as they age), and low calcium levels (low blood calcium prompts the body to drain calcium from bone, causing bone loss). To prevent osteoporosis, a healthy diet rich in calcium and vitamin D can help maintain bone density. Regular exercise—such as walking or jogging—can stave off bone loss, as can quitting smoking.

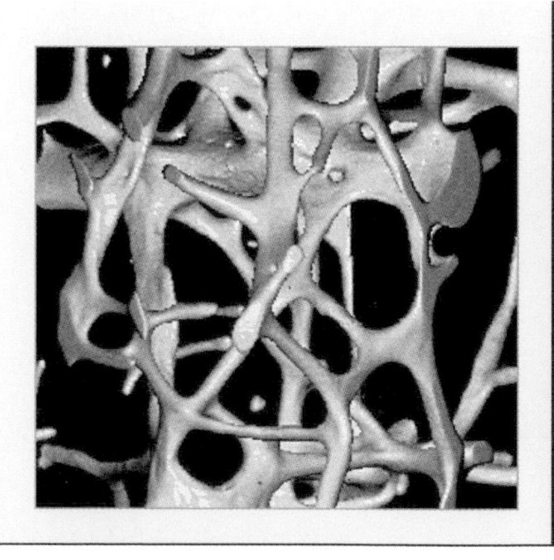

HOW ANCIENTS MENDED BROKEN BONES

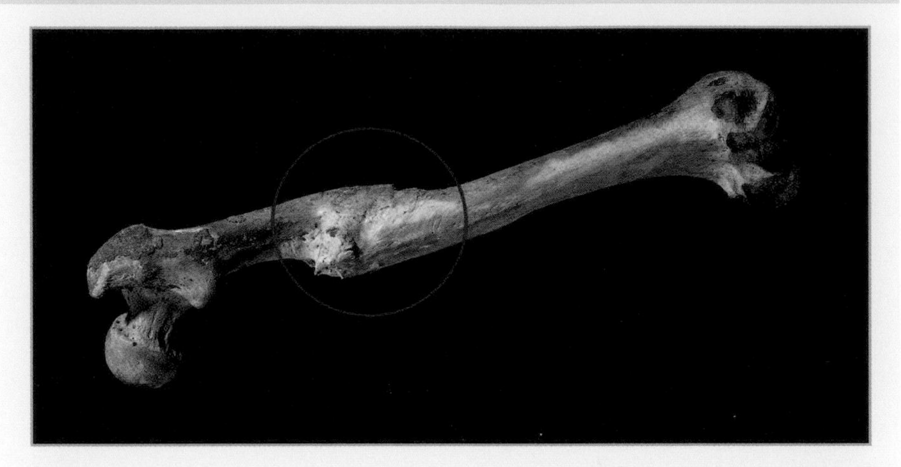

A bone with a healed fracture dates to Egypt's New Kingdom, ca 1539–1075 B.C.

THE ART OF bone-setting dates back thousands of years and arose in multiple cultures. There is written evidence that as long ago as 10,000 B.C. the Egyptians treated fractures by stretching affected limbs until the bones realigned and then setting the bones with splints made of tree bark wrapped in linen. They also used stiff bandages for support that probably employed mummification methods. Excavations at the pyramids have unearthed bones with simple and multiple fractures that appear to have healed in good alignment, indicating that they were set correctly and that traction might have been used.

Other cultures left traces of how they treated breaks. Ancient Hindus set fractures with bamboo splints. The Greek physician Hippocrates (ca 460–377 B.C.) described using wooden splints and prescribed exercise to prevent the muscles from atrophying while immobilized. The Greeks also used waxes and resins to create hard, supportive bandages, whereas the first-century Roman author Celsus explains in his medical treatise *De Medicina (On Medicine)* how to use splints and bandages stiffened with starch. Arab doctors stiffened bandages with lime from sea shells and albumen from egg whites. Later, the Greek Oribasius (ca A.D. 325–400), who published the works of his famous fellow physician Galen (ca A.D. 129–200), describes mechanical appliances being used to treat fractures, including screw traction and a multiple-pulley system.

In medieval Europe, blacksmiths often doubled as bonesetters, repositioning broken bones with their hands. They used casts made of egg white, flour, and animal fat to set them. Later bonesetters used rope-and-pulley systems to pull broken bones back into place. By the 16th century, the great French military surgeon Ambroise Paré (1510–1590) was making casts consisting of wax, cardboard, cloth, and parchment that hardened as they dried. ■

ossified and typically only a few millimeters wide. Notable exceptions are the kneecaps, large bones in the patellar ligament.

HEALING & REPAIR

When a bone breaks or fractures, blood vessels in the bone and periosteum, and perhaps in surrounding tissues, are torn and bleed. A hematoma (blood clot) forms at the fracture site. Soon bone cells deprived of nutrition die, and the tissue at the site becomes red, swollen, and painful.

Within days, several events lead to the formation of soft granulation tissue. Capillaries grow into the hematoma, and germ-fighting white blood cells called phagocytes race to the region and begin sweeping up debris to prevent infection. Meanwhile, fibroblasts and osteoblasts arrive from the nearby bones and begin reconstruction.

The fibroblasts produce collagen fibers that fill in the damaged areas. Some differentiate into cartilage-making cells called chondroblasts. The blood clot is replaced by fibrous tissue (made by osteoblasts), which forms a callus. Blood vessels grow through the callus, and osteoblasts start forming new bone tissue and collagen. The missing bone will be almost completely replaced by new bone within about two months. Osteoclasts complete the healing process by absorbing any remaining debris.

Most fractures in children and teenagers result from trauma that

SEE ALSO: Chapter Twelve, "Innate Defenses," PAGE 322

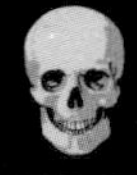

yanks or twists the bones (sports injuries, car crashes, and falls). In the elderly, more fractures occur as bones thin and weaken.

When the break is simple (doesn't puncture the skin), a physician will manipulate the broken edges back into place by hand and put a plaster or fiberglass cast on the area to immobilize it until the bone knits properly. Some injured bones don't need special assistance. A broken rib, for instance, is held to neighboring ribs by the surrounding chest muscles, and, though painful, will heal on its own in time.

More serious fractures—such as those in which the bone snaps in several places or even punctures the skin—may require surgery. In such cases, a surgeon may insert metal plates, rods, or screws to piece together and secure the broken pieces. The skin is sutured and a cast or splint is applied to protect the repaired area; sometimes traction is used to maintain proper position while the bone fuses. Fractures in small and medium bones typically heal in six to eight weeks. Healing can take longer in large, weight-bearing bones and in bones of the elderly (because of poorer circulation).

NEW TREATMENTS

Some medical advances may improve healing after a break. They may even reduce recovery time. Sending minute electrical currents through the break sites can speed healing. Another technique, daily exposure to low-intensity ultrasound waves can reduce healing time in arms and shinbones by 35 to 45 percent. The pulsations appear to hasten callus formation.

Free vascular fibular bone graft technique uses pieces of the fibula to replace missing or severely battered bone. In the past, extensive bone grafts (typically, pieces of hip bone) failed because blood could not get into the substitute and the graft died. This new technique consists of a bone graft with its blood vessels being attached to vessels at the recipient site.

Scientists are also researching bone substitutes. In the past, a paste of crushed bone from cadavers or synthetic material was sculpted into the shape of the desired bone or packed into spaces in broken bones, but both had drawbacks: As with any transplant, cadaver bone carries a risk of hepatitis or HIV infection, and foreign material may be rejected by a recipient's immune system. A new material made from coral has proved more successful than previous bone fillers. The derivative is heated to kill living coral cells and convert its mineral structure to hydroxyapatite, the salt in bone. The graft is then shaped, sterilized, and coated with a natural growth substance. Other types of artificial bone, including tricalciumphosphate (TCP), a biodegradable ceramic, are being explored, as are stem cell therapies.

BROKEN BONES

FRACTURE TYPE	DESCRIPTION
Simple	Bone breaks cleanly, does not penetrate skin
Compound	Broken ends of bone protrude through skin, soft tissue
Greenstick	Bone splits and bends, does not break, like a green twig. Seen only in children
Stress	Hairline fracture caused by extensive or repeated stress
Spiral	Ragged break caused by twisting motion. Most commonly seen in sports injuries
Compressor	A bone weakened by a degenerative condition (like osteoporosis) collapses
Depressed	Broken bone presses inward. Often seen in skull fractures
Impacted	Broken bone ends are forced into each other. Typically occurs when breaking a fall
Comminuted	Bone shatters into smaller pieces, above. Can be caused by high-impact trauma (like a car accident) or by a degenerative condition (such as osteoporosis)

CONNECTIONS
JOINING THE BODY TOGETHER

THE HUMAN BODY can perform an incredible array of movements, from typing to threading a needle, to pitching a curve ball, to tumbling, jogging, skiing, and literally jumping through hoops. These actions are possible because of movable joints, the connections between the bones of the skeleton.

There are more than 400 joints, also known as articulations, in the body. Muscles span joints, crossing from one bone to another. When these muscles contract, they tug on the bones, producing skeletal movement at the joint. Muscle contraction and movement are controlled by signals sent from the brain or ordered by reflex action.

Bones work like mechanical levers, and the joint between them is the fulcrum, or fixed point, on which they move. Bands of strong tissue called ligaments support the bones around a joint. Cords called tendons join muscles to the bones, which move when the muscles contract. Resilient tissue called cartilage cushions the bones' ends and helps joints function smoothly.

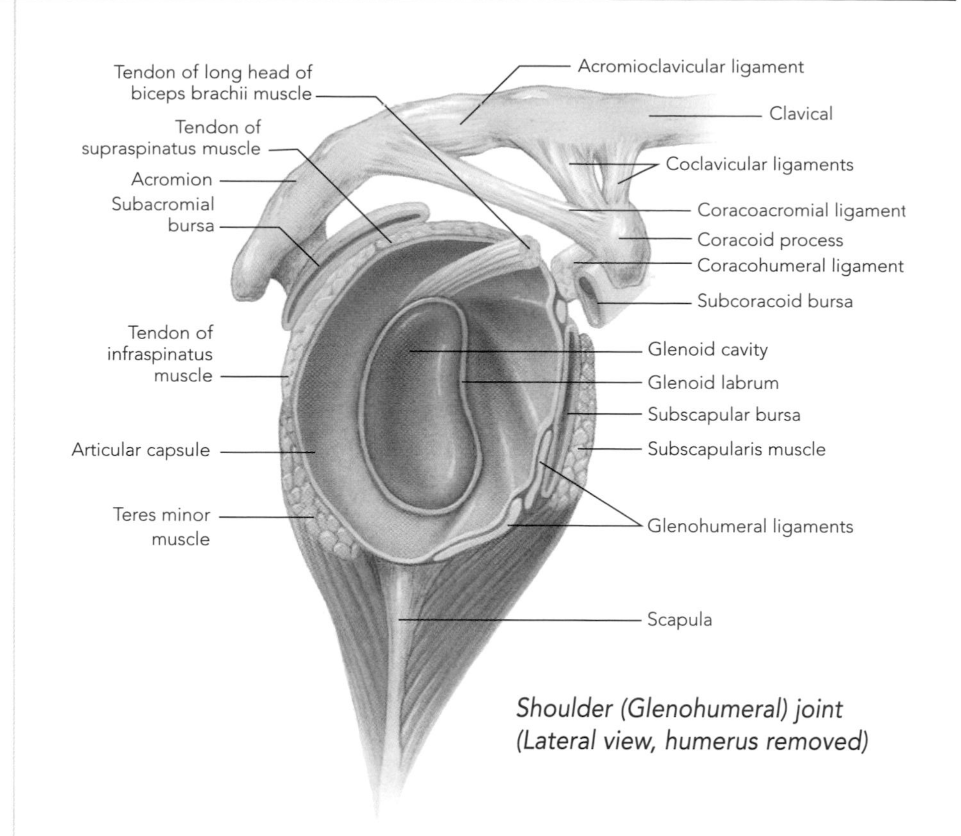

Shoulder (Glenohumeral) joint (Lateral view, humerus removed)

LIGAMENTS

Ligaments are bands of tough, fibrous collagen and stretchy elastin that connect bones to one another; they also surround and bind the joints to help strengthen and stabilize them, permitting movement only in certain directions. For instance, ligaments on either side of the fingers prevent side-to-side bending, and ligaments that stretch across the palm keep fingers from hyperextending.

Ligaments are the most important stabilizers in the complex knee joint. These powerful, tight cords connect the thighbone (femur) and shinbone (tibia), keeping the knee from wobbling. Unlike the hip, for example, which is securely anchored in a deep socket, the knee has virtually no built-in bony stability and would roll around like a ball were it not for ligaments, along with the upper and lower patellar tendons and hamstring muscles, to rope it in place. Collateral (side) ligaments—the medial collateral (MCL) and lateral collateral (LCL)—prevent the knee from wobbling too far to either side; cruciate (crossing) ligaments—the anterior cruciate (ACL) and posterior cruciate (PCL)—prevent the knee from slipping too far backward or forward. (Ligaments also support various organs, including the liver, bladder, and uterus, as well as the female breast.)

TENDONS

Tendons are rigid cords of tissue, containing mostly collagen, that secure muscles to bones and to other muscles. Usually they are

SEE ALSO: Chapter Four, "Movement," PAGE 116, Chapter Thirteen, "Building Blocks," PAGE 352 round or flat bundles, but a few, such as those in the abdominal wall, are flat sheets called aponeuroses. Tendons are covered by lubricated sheaths that allow them to glide easily without disturbing surrounding tissue. When a muscle contracts, it pulls on a tendon, which then tugs on the attached bone and—presto—the bone moves at the joint. The fingers, for instance, are moved mainly by muscles in the forearm that are connected to them by long tendons.

Tendons also help stabilize joints. In the knee, for example, the upper and lower patellar tendons help keep the patella (knee cap) from slipping out of place. The Achilles tendon, which attaches to the heel bone and connects the calf muscles to the foot, is probably the best-known tendon in the body. Easily damaged during overstrenuous exercise, it is named after the mythical Greek hero Achilles, who died when an arrow pierced his heel, his only vulnerable spot.

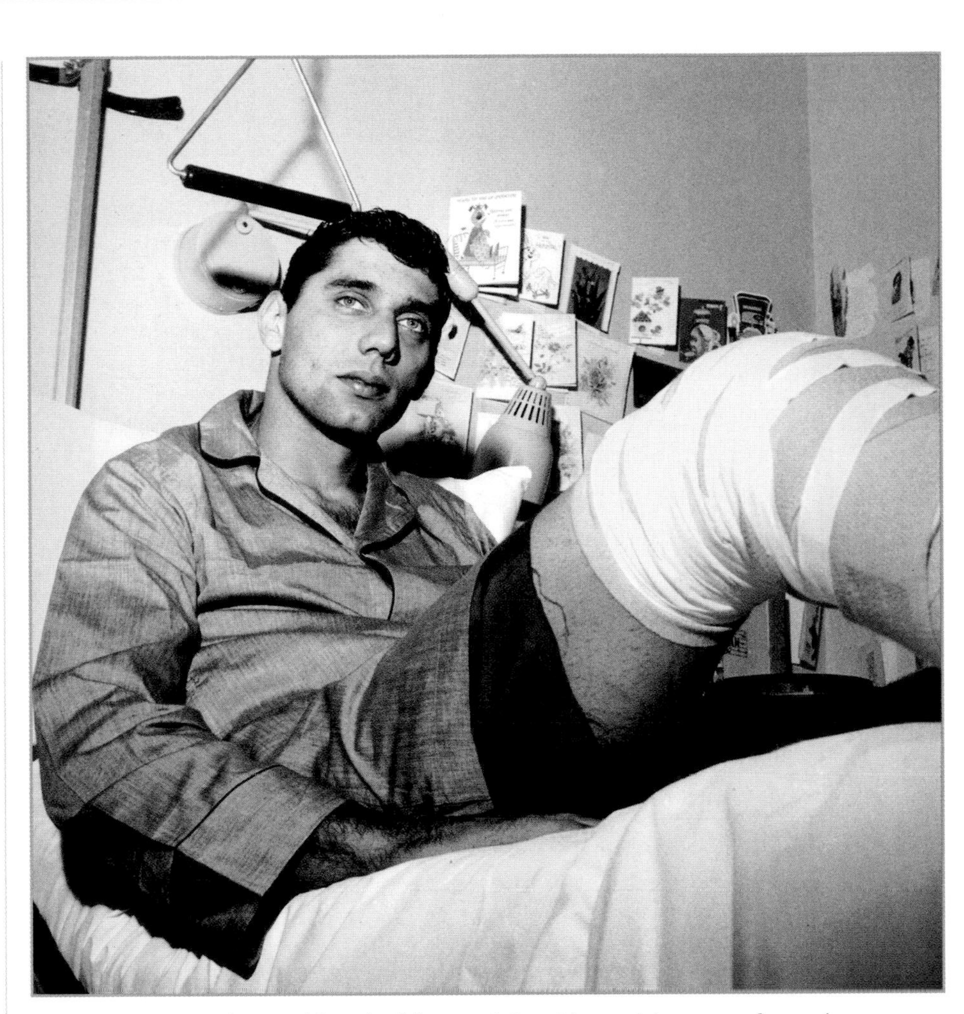

Knee injuries plagued football legend Joe Namath's career from the start. He suffered a serious knee injury during his senior year in college.

CARTILAGE

No joint in the body is under more stress and pressure than the knee, which is why it needs so much support. In addition to ligaments and tendons, it has thick pads of protective cartilage. Cartilage is a dense, rubbery connective tissue composed of collagen fibers, water, and a gel-like substance. It is key to preventing wear and tear on joints. A layer of cartilage covers the ends of joint bones, cushioning them and reducing friction during movement. Cartilage also supports and shapes various structures such as the ears, nose, and intervertebral disks.

There are three types of cartilage. Hyaline cartilage is the most widespread type found in the body. It covers the articular bone surfaces, where one or more bones meet at a joint; extends between the ribs and sternum, cushioning them against impact; and is the flexible tissue that forms the external nose. It is also the cartilage inside bones that is the foundation for bone growth, making up most of the embryonic skeleton. Fibrocartilage is a tough cartilage located in areas needing dense cushioning, such as between

WATER ON THE KNEE

AFTER A TRAUMA to the knee, synovial fluid builds up within the joint, causing pain, inflammation, and stiffness, a condition often called water on the knee. The fluid can be drained with a needle; an injection of corticosteroids can reduce inflammation and prevent new buildup. Football great Joe Namath, above, at times had his knees drained during a game.

A VULNERABLE JOINT

THE KNEE IS THE MOST complex joint in the body. An engineering marvel, it supports while allowing for a wide range of movement; but it is this design that also makes it vulnerable to a range of injuries. A knee injury can happen quickly, as in a hyperextension, or it may occur over a lifetime, as with osteoarthritis. Prevention and care are crucial to keeping the knees healthy.

HYPEREXTENSION

Hyperextension is an injury that occurs when the knee is extended beyond its normal fully straightened position. It is relatively common in sports that feature a lot of twisting and jumping, such as basketball, football, or gymnastics. It often results from an off-balance landing or when the upper and lower parts of the leg pull in opposite directions.

The injury may cause swelling and pain when the knee is extended. Treatment depends on the severity of the damage. A hyperextended knee may also lead to a partial or complete ligament tear, especially in the anterior cruciate ligament (ACL). A person should seek medical attention if an injury causes severe pain, swells dramatically, impairs mobility, or causes the knee to lock or give out.

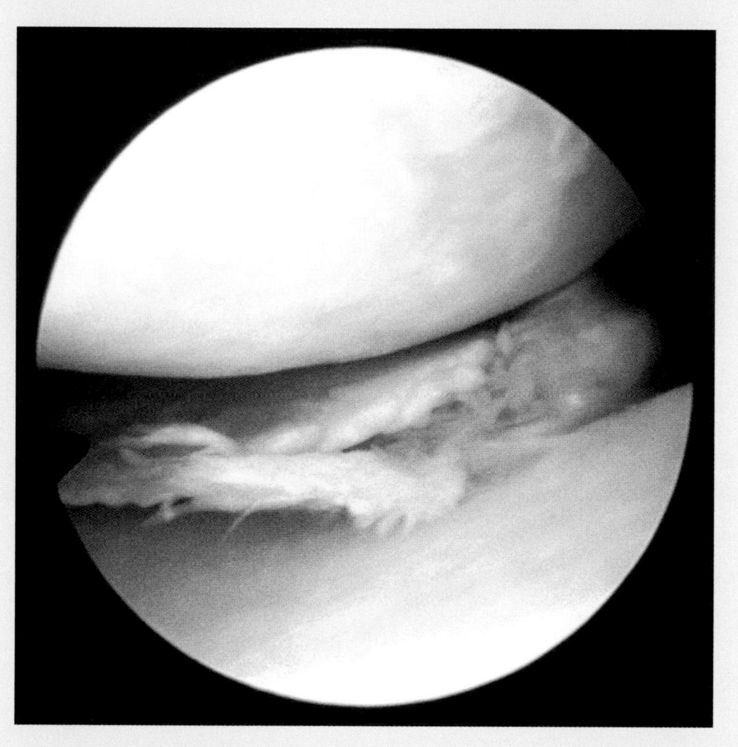

The ragged edges of the medial meniscus bears signs of degenerative damage to the knee.

LIGAMENT TEAR

Ligaments are strong bands of tissue that connect bones and support the knees and various organs. They can be bruised if they're suddenly twisted or pulled. Injuries range from minor stretches and strains to total tears. Commonly injured ligaments are the medial collateral ligament (MCL), which runs along the inner knee, and the anterior cruciate ligament (ACL) in the front center of the knee. The MCL usually heals on its own. If the ACL rips, it generally requires surgery; if it isn't repaired, it can result in permanent instability, meaning the knee might give out at any time.

About 70 percent of traumatic knee injuries involve the ACL. They frequently occur during sports in which the feet are planted while the body sharply shifts direction. An ACL tear can be a career-ender for many professional athletes, especially football and soccer players.

When a ligament is torn, the pain is swift and severe. If the knee swells, a physician will likely aspirate it—remove fluid with a needle. Blood in the fluid may indicate a torn ligament or a fractured patella. In mild cases, a knee brace can keep the ligament still enough to heal, but for many people surgery is the only option.

The ACL can be reconstructed by creating a new ligament from a tendon near the knee. The operation is usually performed using an arthroscope (a tiny, fiberoptic camera). In arthroscopic knee surgery, small incisions are made for inserting the arthroscope and an irrigator, which pumps sterile fluid into the joint to expand it. Other surgical instruments may be needed to cut, shave, or repair tissue. Recovery from ligament surgery is a long process: The patient may be on crutches for several weeks and require several months of physical therapy to improve the knee's range of motion and to regain muscle strength.

MENISCUS TEAR

The meniscii, wedges of protective cartilage sandwiched between the femur (thighbone) and tibia (shinbone), one on each side, may be torn if the knee is twisted too far while bearing weight—say, when a person abruptly turns to hit a tennis ball. Symptoms of a torn meniscus include swelling, pain (especially when straightening the knee), buckling, clicking, and locking.

Magnetic resonance imaging (MRI) is used to confirm the diagnosis. Traditional x-rays cannot detect mensicus injuries, but physicians may take them to rule out degenerative damage or any arthritic change. For years, surgeons treated such injuries by removing the meniscus. They discontinued that treatment after discovering that some patients developed arthritis or deformities later on as a result.

Today, physicians repair and preserve as much of the meniscus as possible with arthroscopic surgery, removing damaged sections and leaving the healthy portions in place. Recovery from such surgery typically takes about a month.

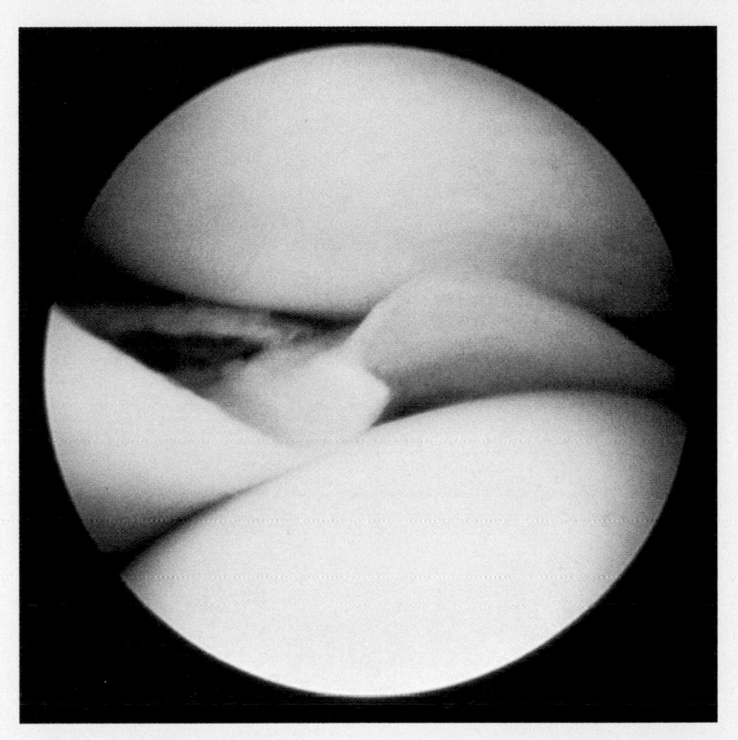

When cartilage in the knee is suddenly torn, the injury may appear crisper compared to cumulative damage.

OSTEOARTHRITIS

Osteoarthritis is the most common form of the 100-plus types of arthritis, affecting some 21 million Americans. It mostly occurs in people over 60 but can strike at any age. Osteoarthritis used to be considered a natural part of aging caused by normal wear and tear on cartilage. Osteoarthritic cartilage, however, is chemically different from cartilage with age-related changes. Many experts now believe that microscopic changes in cartilage structure and composition are often the precipitating factor. Other factors include heredity, obesity, or injury.

In healthy joints, cartilage covers the ends of bones, reducing friction and absorbing shock. Osteoarthritis causes that cartilage to erode. As a result, affected bones grind directly against one another, causing accelerated wear. As the disease advances, bone density decreases and osteophytes (bone spurs and cysts) may develop. This is because the body tries to repair the damage but manufactures bony protrusions instead of normal bone. Such growths irritate surrounding soft tissues, often causing swelling.

Osteoarthritis most often affects the joints of the knees, hips, lower spine, and neck. Men are more likely to develop osteoarthritis in their hips, knees, and spines; women suffer more in their hands and knees.

Symptoms may develop slowly over many years. The first sign may be a bit of pain and stiffness. Symptoms of arthritis in the knee include pain, swelling, stiffness, and a locking sensation, especially during attempts to straighten it. Treatment may include exercise to strengthen supportive muscles; physical therapy with heat to reduce stiffness; treatment to restore cells in cartilage with small defects; injections of a joint fluid compound into the joint; and, in severe cases, joint replacement surgery.

the vertebrae. It is also found in the knees where it forms two disks, each called meniscus, between the femur and tibia.

Participants in contact sports often injure this tissue (see "A Vulnerable Joint," page 84). The third type, elastic cartilage, is found in the external ears and the epiglottis. As its name suggests, this type of cartilage is both flexible and resilient, because of the abundance of elastic fibers it contains. There is little blood in the tissue of tendons and ligaments and none in cartilage. Consequently, healing in those tissues can be slow when compared to the rest of the body.

JOINT TYPES

Each joint is constructed differently, based on its function, and each has its own range of movement. Some joints move freely, some move very little, and some don't move at all. Some joints are big and strong enough to bear all of the body's weight, while others are so small (like those between the tiny ossicles, or ear bones) that they're barely visible.

The skeletal system consists of three types of joints: fixed (also called sutures), cartilaginous (symphysis), and synovial. Fixed joints are slim bands of fibrous tissue that cement one bone to the next and do not move. Among them are the jagged sutures that seal the platelike bones of the skull; a thin layer of connective tissue joins them.

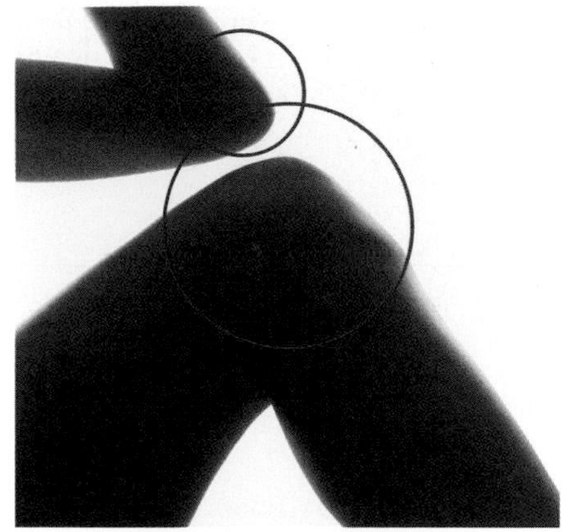

HINGE JOINT
In addition to the knees and elbows, hinge joints are found in the fingers and toes.

PIVOT JOINT
The neck enjoys a wide range of versatile motion because of this joint's structure.

ELLIPISODIAL JOINT
The bones in the wrist permit side-to-side and up-and-down movements.

A symphysis or cartilaginous joint occurs when two bones join and are held firmly together so that they function as one bone. Such joints are made up of tough cartilage plates that flex and permit limited motion. An example is the sacroiliac joint, where the sacrum and pelvis join. The disks between the vertebral bones in the spine are also cartilaginous joints, but they are thicker and allow more movement than the sacroiliac joint. In these joints, there is no so-called synovial space between the bones, which means that the adjoining bones have little room to move. In the spine, however, while each of the joints between neighboring vertebrae has limited flexibility, working altogether they allow a lot of movement.

Synovial joints are the most common and most mobile joints. They include the shoulders, elbows, wrists, fingers, hips, knees, ankles, and toes. The ends of bones in these loose and limber joints are wrapped in slick, resilient

BURSA

A BURSA is a fibrous, fluid-filled sac that acts like a cushion in places where tendons, ligaments, muscles, or skin pass over bone. They serve to reduce friction, especially around hardworking joints such as the shoulder, elbow, and knee. Overuse, inflammation, gout, or other factors can cause bursitis, a painful inflammation of the bursa.

SEE ALSO: Chapter Four, "Movement," PAGE 116

cartilage, which reduces friction when they slide over one another. Synovial joints have a lining of tissue, a synovial membrane. The joint is filled with synovial fluid that keeps the bones lubricated so that they glide as smoothly as the parts of a well-oiled machine.

MECHANICS

The way joints move depends on their shape. Some are hinged like doors, allowing limited back-and-forth movement. Others are constructed more loosely, allowing motion in many directions.

While everyone has the same number of joints, the range of motion may vary slightly from person to person. Some, for instance, are said to be double-jointed because they have unusually flexible fingers or other joints. Others, such as gymnasts and ballet dancers, are more limber. These differences, however, are largely attributable to differences in ligament length and suppleness.

The most common and movable types of joints are synovial joints. In most of these joints, the bones either hinge onto each other or rotate. There are six types of synovial joints. A pivot (or swivel) joint is one in which a cylinder-shaped projection on one bone snaps into a ring of another bone to allow movements like a wheel spinning on an axle. There's a pivot joint between the first (atlas) and second (axis) cervical vertebrae (neck bones) at the top of the spine that allows the head to turn left and right.

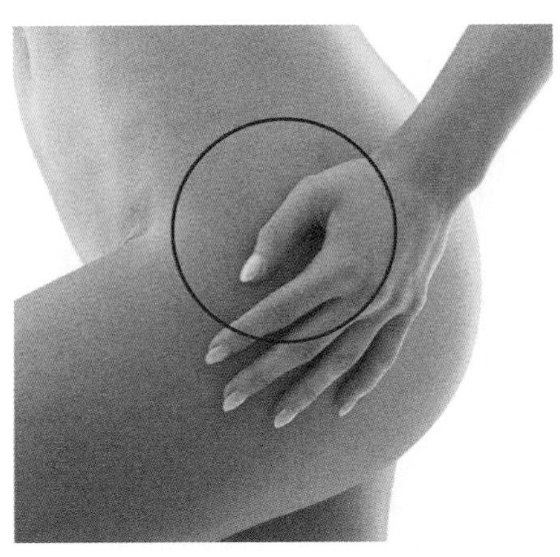

BALL & SOCKET JOINT
This construction is present in both the hips and the shoulders, which enjoy a wide range of motion.

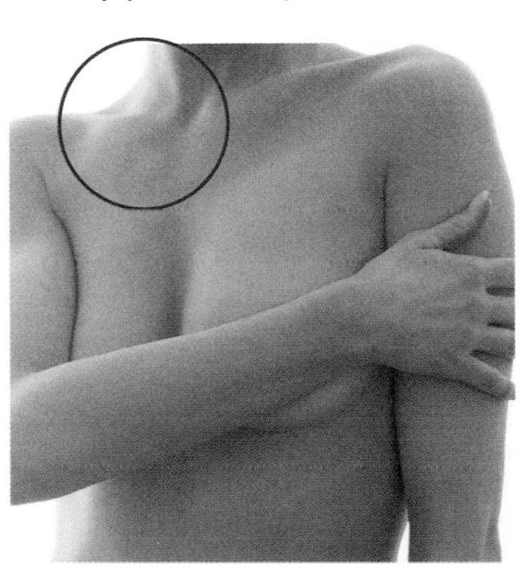

PLANAR JOINT
Bone surfaces in these joints tend to be smooth, allowing for seemless gliding motion.

SADDLE JOINT
The thumb can rotate and bend forward and back because of its connection at the wrist.

In ball-and-socket joints, such as the hip and shoulder, the rounded end of one bone sits in the cup-shaped socket of the other and can move in almost any direction. This kind of joint with loose ligaments allows a wider range of movement than any other in the body.

An ellipsoidal, or condyloid, joint is one in which the dome of one bone fits into the recessed, oval-shaped socket of another, allowing limited movement. The joints of the wrists between the radius and carpals are an example.

The bones of saddle joints interlock like a pair of curved saddles. This type of joint occurs between trapezium of the wrist and metacarpal of the thumb where it joins the hand. Its loose ligaments allow great mobility with considerable strength (as in thumb wrestling).

Planar joint bones have nearly flat surfaces where they meet. They are held together tightly, so that they have only limited side-to-side and back-and-forth gliding movement. This kind of joint can be found in the carpals of the hands, the tarsals of the feet, and between the clavicle and sternum.

Hinge joints have one bone that rotates inside a cylindrical hollow. These joints are located in the knees and elbows and between some of the bones in the fingers and toes. Hinge joints allow the moving bone to flex and extend.

The largest and most complex joint in the human body is the

CHAPTER GLOSSARY

APPENDICULAR SKELETON Group of bones in the skeleton that includes the limbs and their girdles.

ARTICULAR CARTILAGE. Occurs in joints where bone meets another bone. Cushions and reduces friction.

ARTICULATION. Joint or junction of two or more bones.

AXIAL SKELETON. Group of bones in the skeleton that includes the skull, spine (or vertebral column), and rib cage (or bony thorax).

BURSA. Sac filled with synovial fluid that is found close to joints between tendon and bone. Decreases friction during movement.

CARTILAGE. Smooth, rubbery connective tissue. Found in the skeleton, as well as other body structures.

COMPACT BONE. Hard, dense tissue that forms outer bone layer.

DIAPHYSIS. Shaft of a long bone.

EPIPHYSEAL PLATE. Cartilage separating diaphysis and epiphysis; area of bone growth.

EPIPHYSIS. Bone end containing spongy bone and red bone marrow.

FONTANELLES. "Soft spots" on an infant's skull where sutures have not yet closed.

FORAMEN. Central channel in bones through which blood vessels, nerves, and ligaments pass.

LIGAMENT. Band of fibrous tissue that connects bone to bone.

MEDULLARY CAVITY. Core of diaphysis; contains yellow marrow.

MENISCUS. Curved fibrous cartilage in some joints such as the knee.

OSTEOARTHRITIS. Form of arthritis generally associated with aging in which joints wear out.

OSTEOBLAST. Bone-forming cells.

OSTEOCLAST. Bone-destroying cells.

OSTEOPOROSIS. Common disorder where bone density is lost, making the bone more fragile.

PERIOSTEUM. Tough, fibrous connective tissue that covers and nourishes bones.

RED BONE MARROW. Spongy tissue in bones where blood-cell formation takes place.

RHEUMATOID ARTHRITIS. Autoimmune condition that can cause a variety of symptoms, including inflamed and swollen joints.

RICKETS. Bone disorder caused by calcium and vitamin D deficiencies in early childhood.

SPONGY BONE. Lattice-like porous bone that gives strength with minimal weight; found in ends and inner portions of long bones; contains red bone marrow. Also called cancellous bone.

SUTURE. A fused, immobile joint.

SYMPHYSIS. Where two bones join and are held firmly together so they function as one bone. Also known as a cartilaginous joint.

SYNOVIAL FLUID. Lubricating fluid present in the joint cavity.

SYNOVIAL JOINTS. Freely movable joints in the body. Filled with synovial fluid and linked with synovium.

TENDON. Band of tough, fibrous tissue attaching muscle to bone.

knee, which is designed for its own protection. It is completely surrounded by a joint capsule that is strong enough to hold the joint together but flexible enough to allow motion. The capsule is lined with synovial tissue, which secretes synovial fluid (clear in healthy joints, cloudy in diseased joints, and bloody in injured joints) that bathes the joint.

Pads of cartilage coat the ends of the thighbone (femur) and shinbone (tibia) that join to form the knee, protecting them from wear. Two menisci (cartilage pads) act as cushions between the two bones and help distribute body weight in the joint. Ligaments along the the sides and back of the knee reinforce the joint capsule, adding stability. The knee cap (patella) protects the front of the joint.

PROBLEMS

Synovial joints are tough but their mobility makes them easier to damage when compared to other parts of the skeleton. Sudden wrenches or twists may result in a sprain—a painful injury involving damage to muscles and ligaments. During strenuous exercise, the knee's internal ligaments can be torn, and pieces of cartilage may break away, jamming or locking the joint (see "A Vulnerable Joint," page 84). Injuries can also result in joints being displaced: Luxation is the full dislocation of a bone from its joint, and subluxation is a partial displacement.

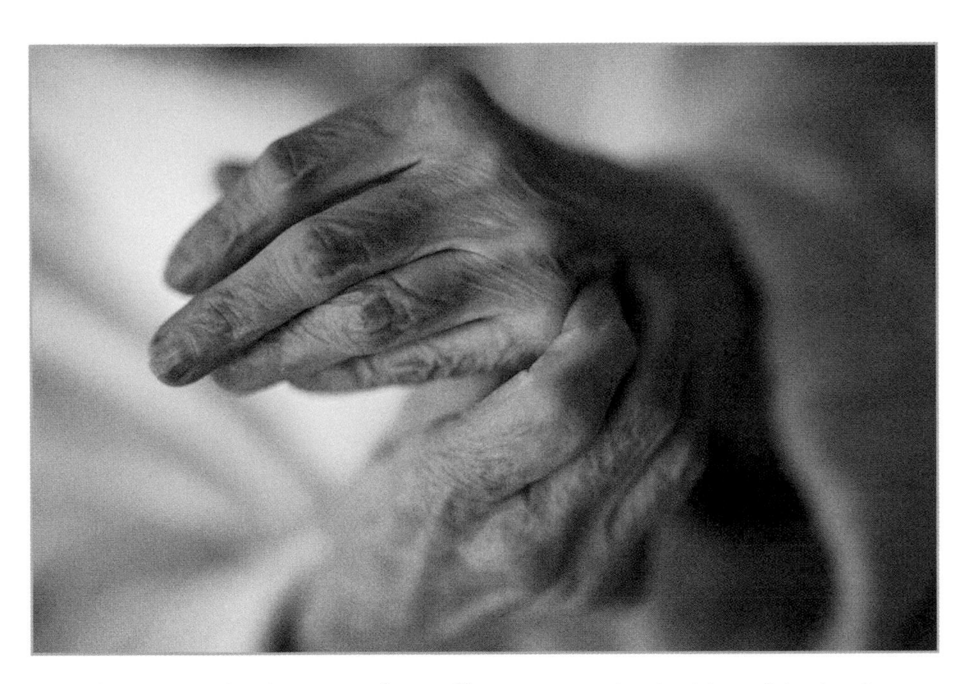

Rheumatoid arthritis tends to afflict joints on both sides of the body. If symptoms appear in the right hand, they are likely to occur in the left.

Synovitis is an inflammation of synovial membrane that results in swelling and pain. It can be caused by an injury, infection, or irritation produced by damaged cartilage.

A bunion, (a condition also known as hallux valgus), is an abnormal angulation of the joint at the base of the big toe. This misalignment puts pressure on the other toes as the big toe is forced sideways. Those who wear tight-fitting, pointed, or high-heeled shoes put themselves at risk for this condition, which can often only be treated surgically.

A group of degenerative diseases that inflame or damage the joints are collectively known as arthritis. The most common form, osteoarthritis, is most common among the aged and is attributed to normal wear and tear. Articular cartilage is broken down, and new tissue does not replace it. Bones begin to grind directly against one another, eventually wearing down the cartilage and forming bony spurs where the exposed bone tissue has thickened. A crunching noise, called crepitus, may be heard when the roughened surfaces rub against each other. Today, an estimated 21 million Americans live with the disease.

Another is an inflammatory, autoimmune disease, rheumatoid arthritis. In it, the immune system attacks the body's own tissues. The synovial membrane proliferates and triggers production of enzymes that break down cartilage, damaging the joint. Symptoms may range from stiffness and pain to restriction or loss of movement.

ARTHRITIS

ARTHRITIS IS a general term for more than 100 different types of inflammatory or degenerative diseases that damage the joints. To varying degrees, these forms share similar symptoms: pain, joint stiffness, and swelling. Today, arthitis is among the most common diseases in the world, but it may also be one of the oldest. Remains of skeletons dating to the Ice Age bear traces of damage to their joints that resembles osteoarthritis.

BREAKTHROUGH

TO RELIEVE JOINT pain and loss of movement caused by osteoarthritis, other diseases, or injury, surgeons can often replace the damaged joint with an artificial one. Hips and knees are the joints most commonly replaced. A prosthesis is joined to natural bone with an adhesive, or the surgeon may use a replacement joint made of porous material into which new bone cells migrate. Eventually, enough new bone is produced to firmly anchor the shaft. In joint replacements (as in a hip replacement, right), points of contact between prosthesis and bone are coated with a special plastic to reduce friction.

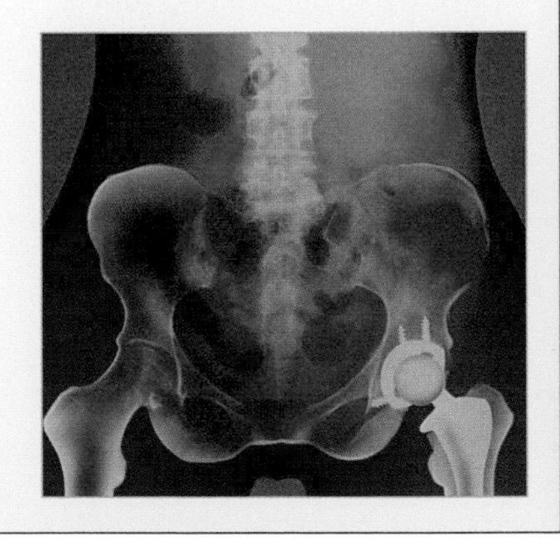

THE SKULL

THE MOST COMPLEX BONES IN THE BODY

THE MOST COMPLEX portion of the skeleton is the skull, a network of bones that gives shape to the head and face as it cradles and protects the brain. Given the nature of its important function, many assume the skull to be a heavy mass of solid bone. But, appearances are deceiving. The skull's structure is lighter, more complicated, and more specialized than it looks.

The skull looks at first glance as though it were made of one bone, but it's composed of 22 separate, mostly flat bones held together by five rigid joints (called sutures), which resemble jagged stitching. The only bone in the skull that is freely movable is the mandible, or lower jaw. The condyloid joints that connect it to the skull enable the multiple angular motions that allow humans to chew and talk (see "Connections," page 82).

There are two sets of bones in the skull: the cranial and facial. Eight cranial bones form the cranium, the upper dome, or cranial vault, that encases, supports, and shelters the brain and hearing apparatus; it also provides attachment sites for head and neck muscles. The 14 facial bones form the base for the face and jaw, provide openings for air and food, secure the teeth, and connect to facial muscles that enable the myriad expressions of which the human face is capable.

The Skull

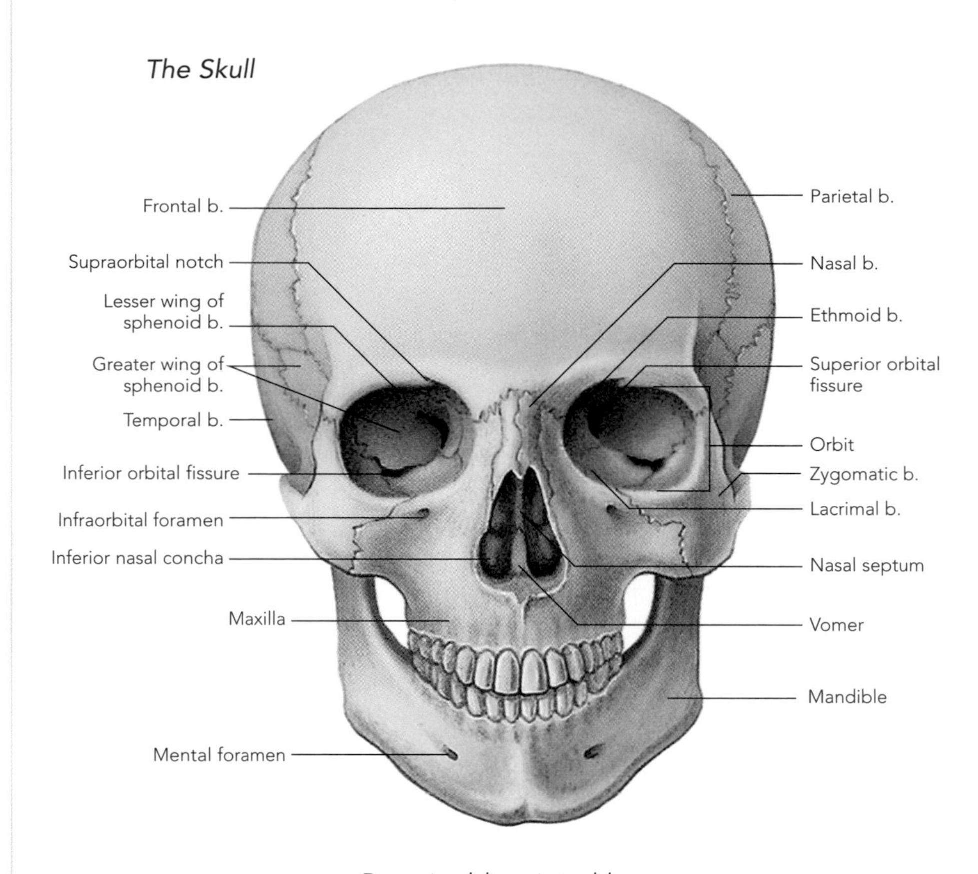

Bone is abbreviated *b.*

The skull has four large cavities—the cranial, which contains the brain; nasal (nose); and two orbits, or eye sockets, which house the eyeballs (cushioned by fatty tissue). There are also many smaller cavities, including those of the middle and inner ear.

Some of the bones of the skull, most notably surrounding the nasal cavity, contain sinuses. These hollow air chambers serve to lighten the load of the skull, but as allergy and cold sufferers know, they can make the head feel heavy and cumbersome when they fill with fluid. Sinuses also act somewhat like echo chambers, making the voice sound richer and more resonant. During speech or other vocalization, such as singing, the sound waves reverberate in the sinuses. The reverberation prolongs and intensifies the sound of the voice. By the same token, when a person has a cold and the sinuses are blocked, the voice can sound muffled and dull.

There are small holes in the skull to allow blood vessels and nerves to pass through: for example, the optic nerves to the eyes and the olfactory tracts to the nose. The largest opening is the foramen

SEE ALSO: Chapter Nine, "The Brain," PAGE 234

magnum in the base of the occipital bone at the back of the skull, through which the spinal cord passes to join the brain.

CRANIAL BONES

The brain's bony protector includes the two parietal bones, which form most of the roof and upper sides of the skull. These bones make up the bulk of the cranial vault, or calvarium, where the brain sits safely enclosed within.

The shell-shaped frontal bone forms the frontal squama (forehead), the upper margin of the eye sockets, and a portion of the nose, or nasal cavity. Temporal bones form the sides (just above the ears) and part of the base of the cranium. The butterfly-shaped sphenoid bone forms part of the base of the skull and part of the bottom and sides of the eye sockets. A sunken portion of the sphenoid bone called the sella turcica ("Turk's saddle") shelters the pituitary gland.

The delicate leaf-shaped ethmoid bone forms part of the nose, orbits (it comprises most of the area between the eyes), and the floor of the cranium. It contains several sections called plates, including the cribriform plate, which shapes the top of the nasal cavity; nerves from the olfactory (smell) receptors in the nasal cavity pass through tiny holes in the cribriform (pierced) plate to the brain. The perpendicular plate forms part of the nasal septum, which divides the nasal cavity into right and left nasal passages. The occipital bone forms the back wall of the skull and, with the sphenoid bone, the base of the cranium.

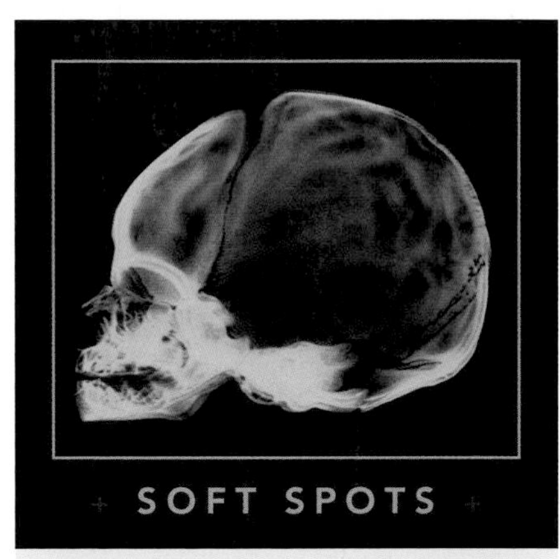

SOFT SPOTS

THE CRANIAL BONES of newborns, above, are not fused but connected by fibrous membranes called sutures. At six places where the bones abut each other, the sutures are expanded. These "soft spots"—two on top, two on each side—are called fontanelles. The fibrous sutures and fontanelles allow the bones to overlap during the birth process and to expand evenly as the brain grows. The fontanelles normally are replaced by bone when the infant is 6 months to 2 years old; sutures ossify by early adulthood.

FACIAL BONES

The remaining skull bones form the skeleton of the face; all but the mandible (lower jaw) and vomer come in pairs, giving the face its symmetrical structure. Zygomatic bones form the cheekbones and sides of the eye sockets. The nasal bones, two thin rectangles, form the bridge of the nose, one on either side; the lower, movable portion of the nose is supported by flexible cartilage. Inferior nasal conchae are the thin, scroll-like bones that form part of the interior lateral wall of the nose. Palatine bones form the floor of the nasal cavity and part of the hard palate (roof) of the mouth.

The slender vomer bone, along with the ethmoid bone and a section of cartilage, forms the the nasal septum. The nasal septum is the structure that divides the nasal cavity into right and left airways; it's also the part of the nose that can be easily pushed off track (called deviation) by a swift blow. More commonly, uneven growth rates between bone and cartilage can cause the septum to deviate. This often occurs during the growth spurt that takes place in the teen years. The deviation can block one of the nasal airways, resulting in mild to severe difficulty in breathing. It can also produce a bump in the bridge of the nose. Surgery can usually correct the problem, unblocking the air passage and restoring an even appearance to the nose.

Maxillary bones, or maxillae, form most of the upper jaw. The mandible, or lower jawbone, and maxillae both contain sockets for the teeth. The horseshoe-shaped mandible is the largest, strongest facial bone and the only movable bone of the skull. The mandible is attached to the temporal bones, one on each side of the skull, just in front of the ears, at the two temporomandibular joints.

The temporomandibular joint, or TMJ, is one of the most complex joints in the body. Supported by muscles, tendons, and ligaments, it can open and close like a hinge but also move from side to side and slide forward and backward. A disk of cartilage, similar to the meniscus disk in the knee, provides extra padding at this much-used joint. TMJ disorders can produce a range of symptoms including muscle pain, headache, and an inability to open the jaw wide. Causes vary and can include clenching and grinding of the teeth during sleep.

Shaped like fingernails, the lacrimal bones form the inside walls of the orbits (eye sockets) at the inner angle of the eyes. Ever wonder why your nose runs when you cry? It's because tears flow from the eyes to the nasal cavity through a passageway called the nasolacrimal canal, located between the lacrimal bones and the nose.

BUMPS & BREAKS

The domed structure of the skull is able to distribute loads, making the skull very strong while allowing it to be relatively light. The skull is thick and dense at the top, where there are no muscles to pad it, as there are on the sides and back, areas where the bone is thinner. It's nevertheless advisable to protect the skull in situations where it could receive a severe blow, as when playing sports such as football and ice hockey, auto racing, and bicycling.

THE PRACTICE OF PHRENOLOGY

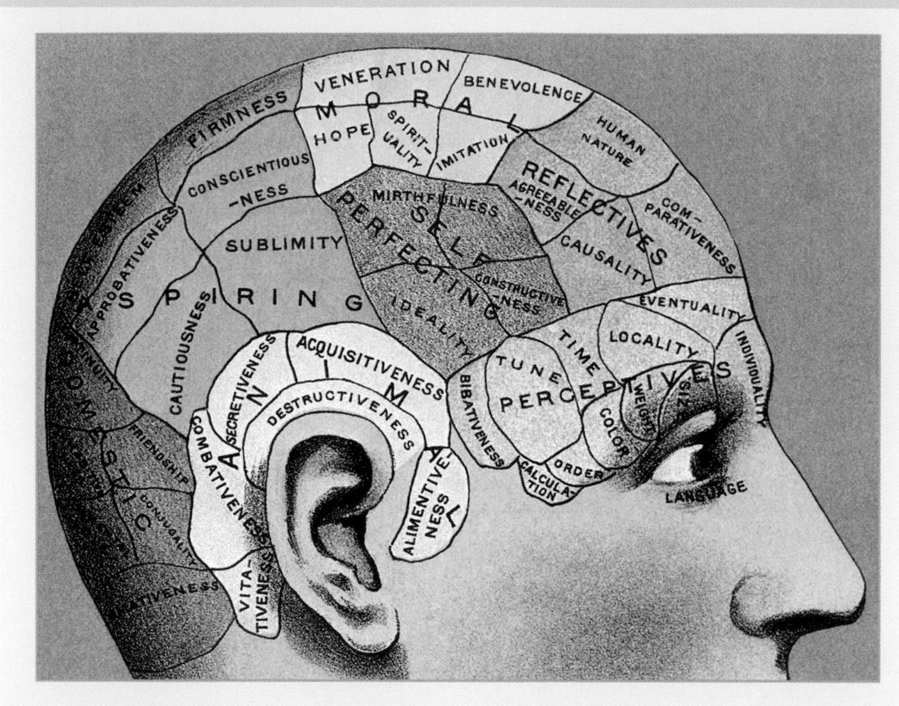

A phrenology diagram maps out the organs of thought and associated traits.

INTEREST IN PHRENOLOGY—a pseudoscience that held that personality could be assessed by the topography of the skull—has existed since the Austrian physician Franz Joseph Gall (1758–1828) first proposed the theory in Paris in the 1790s. Gall claimed that the brain contained 27 (later increased to 37) "organs" of thought, each linked to a particular trait, such as pride, benevolence, musical affinity, poetical talent, and even criminality. An organ corresponded to a bump or contour on the skull. Thus a large bump over the location of the pride organ, for instance, meant that the organ was large and much used; an even larger bump signaled vanity. A depression over an organ meant it was small and the characteristic weak. Gall also claimed that these formations could be manipulated to alter a particular trait.

Soon hosts of "bump doctors" around the globe were giving lectures and phrenological analyses. Phrenological societies were established and supporting literature was widely disseminated. Phrenology's popularity began to wane in mid-century but was revived in the 1860s by an American family, the Fowlers, who traveled the country promoting the theory and analyzing the skulls of their growing audiences.

In the 1920s, however, phrenology died out in the United States under increasing pressure from the scientific community. Perhaps its most enduring legacy are the porcelain phrenological busts now much sought after by collectors. ■

SEE ALSO: Chapter Nine, "The Brain," PAGE 234

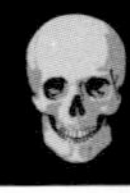

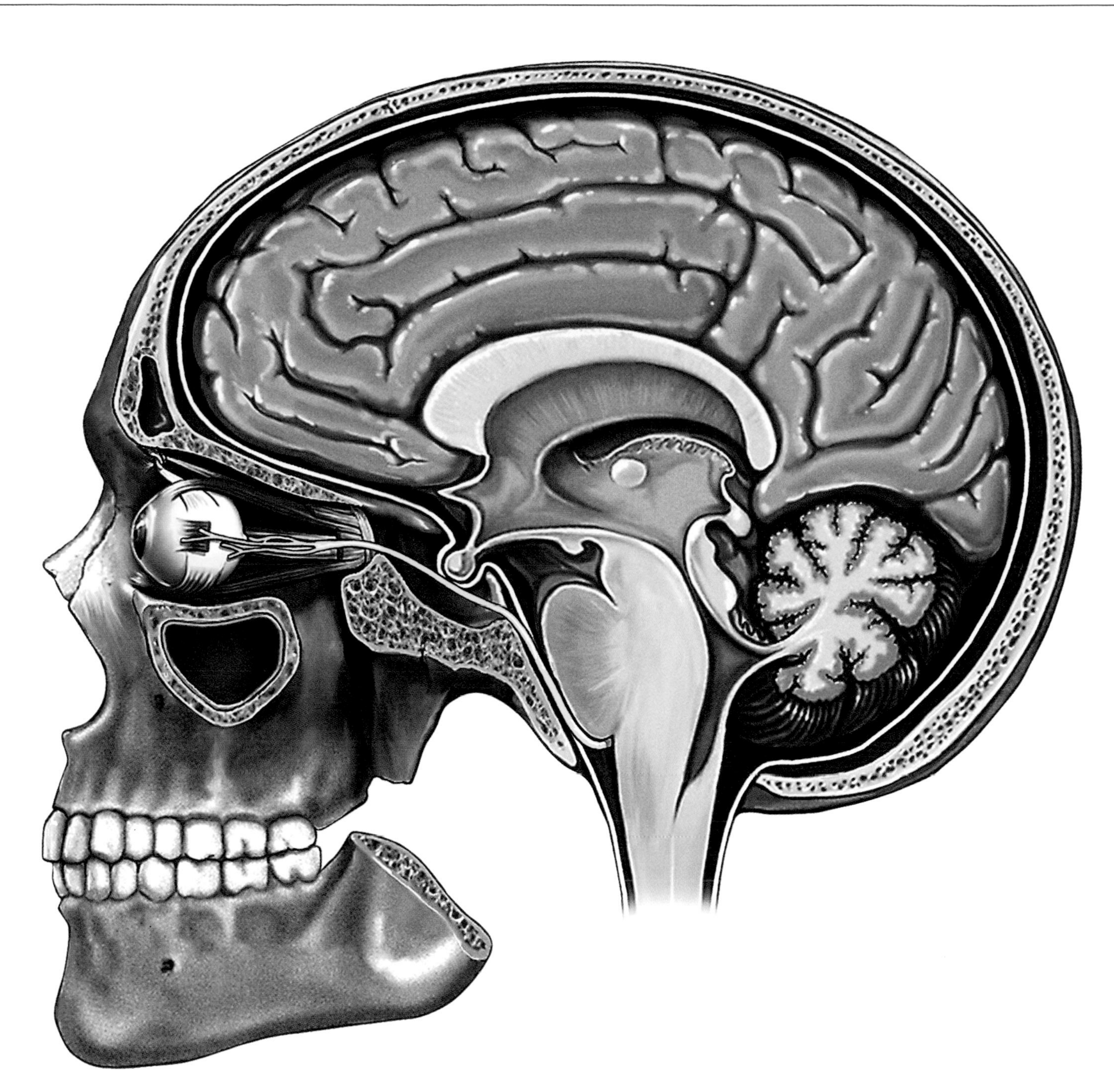

A cross-section of the skull reveals how the brain is perfectly housed and protected inside.

If a person gets a bump on the head, the skull is designed to absorb the impact, working with the cerebrospinal fluid that surrounds the brain inside. But if a person takes a severe blow, injuries to the brain, such as concussions, can occur—even if there is no skull fracture. Symptoms of a concussion can include dizzyness, vomiting, headaches, ringing in the ears, and blurred vision. If a concussion is suspected after a head injury, it's wise to be examined by a doctor.

When the skull fractures, surgery is not always necessary. If the fracture is depressed, however, with bone pressing inward, brain damage may occur. The bone may also tear the tissue layers covering the brain, allowing bacteria to enter and causing infection and brain damage. In these situations, surgery is necessary to clear away debris, make what repairs are possible, and restore fragments to their proper position.

THE TEETH

BUILT TO BREAK DOWN FOOD

A BRIGHT, HEALTHY smile starts with a set of pearly white teeth, the living, calcified structures embedded in the jawbones. The teeth play a vital role in digestion: breaking up food through chewing.

COMPOSITION

The teeth sit in the jawbones but are not composed of bone. The exterior of the visible portion, or crown, is enamel, the hardest tissue in the body. Beneath the enamel shell is a framework of bonelike dentin that makes up the bulk of the tooth. This hard, yellow tissue extends into the long root portion of the tooth, below the gumline. A chamber within dentin contains a soft pulp, rich in blood vessels, nerves, and nerve endings that detect pain, cold, and heat. Blood vessels course through the jawbone and into the pulp, supplying it with nutrients. The pulp transmits sensations such as temperature and pain to the roots of the tooth, embedded in a socket in the jawbone.

Cementum is a thin, bony layer of tissue covering the dentin in the root portion. It serves as an area of attachment for the tooth, which is connected to the jaw by the periodontal ligament, a two-sided membrane that lines the socket. Collagen fibers from the ligament attach to the cementum, on one side, and to the bone, on the other side, both anchoring and cushioning the tooth. The teeth are surrounded by soft smooth tissue called the gums.

The Teeth

Crown
Neck
Root
Enamel
Dentin
Odontoblast layer
Dental pulp cavity
Gum
Inserted periodontium
Protective periodontium
Cementum
Periodontium membrane
Root canal
Apical formen
Root apex
Mandible
Artery, vein, and nerves

A tooth consists of four layers

Enamel is the white, highly calcified outer layer. It is the hardest substance in the body and highly resistant to acids and other corrosive agents. Enamel is without feeling.

Dentin is a hard, yellow layer of tissue beneath the enamel that forms the bulk of the crown. It is softer than enamel and transmits sensations such as temperature and pain to the root.

Cementum is a thin, bony layer covering the root portion of the tooth. It is connected to the jaw bone by collagen fibers that pass through the periodontal membrane to hold the tooth in place.

Pulp is the soft tissue in the inner cavity of the tooth. It contains the nerve fibers and blood vessels and supplies nutrients to the tooth. Pulp extends into the jaw bone and is highly sensitive to pain and temperature.

TOOTH TYPE

Humans have two sets of teeth. Primary, or baby, teeth are the first to develop. These typically emerge between 6 and 24 months. There are 20 primary teeth (10 upper, 10 lower), which are then naturally shed, starting at about age 6, to make way for secondary, adult teeth. Healthy baby teeth play a key role in a child's ability to chew, to develop normal jaw structure and facial characteristics, and to speak clearly.

The adult jaw holds 32 secondary teeth arranged in two arches,

BY THE NUMBERS

- + THIRTY-TWO: a full set of adult teeth
- + Eight incisors for chopping
- + Four canines for tearing
- + Eight premolars for crushing
- + Twelve molars (including four wisdom teeth) for grinding

SEE ALSO: Chapter Four, "Movement," PAGE 116, Chapter Seven, "Ingestion," PAGE 184

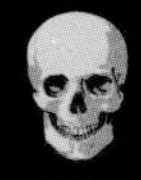

16 in the upper jawbone and 16 in the lower (see "By the Numbers," opposite). The incisors (four upper, four lower) are chisle-edged, center-front teeth used to cut food. Canines (cuspids) border the incisors, one on each side, and work with them to rip food, support the lips, and guide jaw alignment. Next to each canine are two premolars (bicuspids) with broad, pointed surfaces and rounded cusps for crushing and grinding. Six pairs of molars (three upper, three lower) follow the premolars and work with them. The four rearmost molars, one in each back corner of each jaw, are the wisdom teeth, the last to grow. Some people may never develop them.

DECAY

After a tooth erupts from the gum, the cells that make enamel begin to decay, which means that any breaks in enamel must be artificially repaired by a dentists. Tooth decay is caused mainly by plaque, a thin, sticky film of food debris, mucus, and bacteria. Plaque forms an acid that gradually erodes enamel. Once a cavity begins, erosion continues into the dentin and, if left untreated, spreads to the pulp and nerve, causing a toothache or, more seriously, gum disease.

Protective bacteria in the saliva helps the teeth by neutralizing acids. Brushing the teeth twice a day can also help control acid levels and remove food particles. Flossing is another way to remove plaque from between the teeth.

FALSE TEETH

The springs on George Washington's false teeth helped keep them in his mouth.

THE FIRST RECORDED sets of false teeth were left on the nightstand back in 700 B.C. Designed by the Etruscans, the dentures were crafted from ivory or bone and secured by gold bridgework. They continued to be used for centuries (up through the Roman Empire); only the wealthy could afford to buy them.

Later forms of dentures, dating from the 15th century or possibly earlier, were carved from bone or ivory. In some specimens, individual teeth might be scavenged from corpses or purchased from living donors. Later in the 18th century, false teeth were sewn into plates with silk thread. These dentures were mostly cosmetic, as people wearing full sets of uppers and lowers had to remove them to eat. Mother of pearl, gold, silver, ivory, and other materials were tried (contrary to popular belief, George Washington's famous dentures were more likely made of ivory than wood). Steel springs were experimented with to hold the plates together. The springs often slipped, and the dentures would pop out.

The first porcelain dentures were made ca 1770 by Frenchman Alexis Duchâteau. They were more comfortable and appealing than earlier versions but heavy. In 1791, Duchâteau's former assistant Nicholas Dubois de Chemant patented "de Chemant's Specification," which would make "artificial teeth either single, double or in rows or in complete sets and, also, springs for fastening or affixing the same." He began peddling his porcelain paste the next year, paving the way for today's natural-looking crowns and light acrylic dentures. ■

THE SPINE

THE TWISTS AND TURNS OF THE FLEXIBLE BACKBONE

THE SPINE IS A CHAIN of small, but tough, drum-shaped vertebrae, or bones, that extends from the base of the skull down to the hips. Also known as the backbone or vertebral column, it has multiple functions: supporting the body while also protecting the spinal cord, the nervous system's main conduit to the brain.

Along with other bones and muscles, the spine provides strong support for the head and trunk, and, at the same time, allows the back to bend so that the body can perform a wide range of twists and turns. It also provides areas of attachment for the muscles that hold the back upright.

COMPOSITION

Each vertebra is linked to the next by symphyses, which are cartilaginous joints. The individual amount

SPINAL BONES

THE SPINE CONTAINS 26 bones divided into three vertebrae regions (from the neck down): cervical, thoracic, lumbar, followed by the sacrum and coccyx (or tailbone). The upper vertebrae provide flexibility for the body to move, especially the seven cervical (neck) and five lumbar (lower back) vertebrae. The five vertebrae of the sacrum and the four of the coccyx are fused together.

The Vertebrae

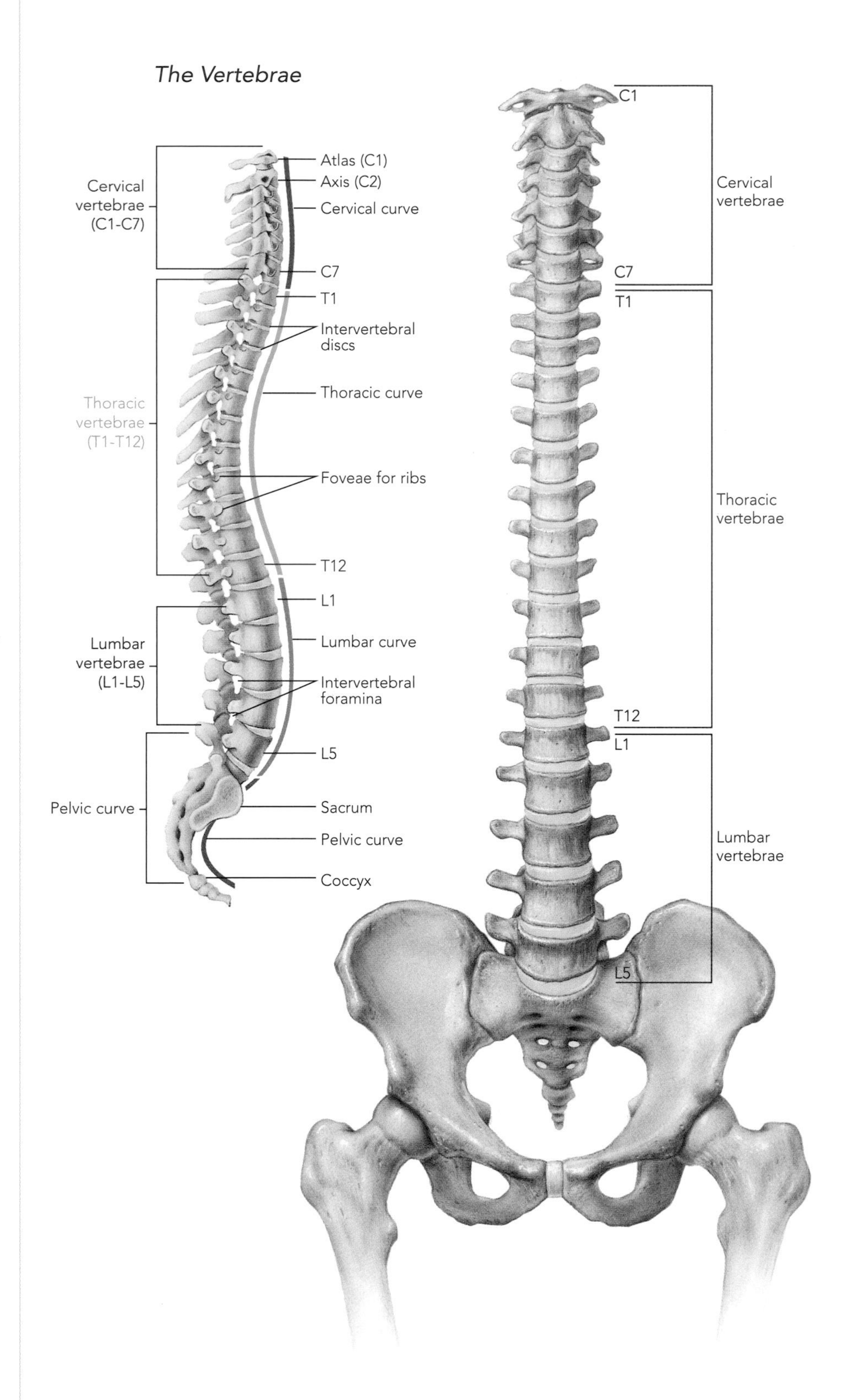

SEE ALSO: Chapter Nine, "The Spinal Cord," PAGE 246

of movement between a vertebra and its neighbors is actually very small, but altogether the vertebrae allow a large range of movement. The spine can bend in many directions. In general, the movements that can occur between vertebrae are flexion (bending forward) and extension (bending backward), lateral flexion (bending the upper body to the right or left), and rotation (the vertebrae rotate on one another in the longitudinal axis of the spine). Not all movements are equal: Humans can bend farther forward than backward, because the way the back curves limits backward movement.

The vertebrae are separated by disks of protective rubbery cartilage. These pads, sandwiched between the bones, act as shock absorbers for the spine, cushioning the bones to keep them from colliding with and grinding against each other and wearing away. During the active day, these disks are compressed, and they expand again at night when at rest. This means that we're actually slightly taller in the morning than in the evening.

The spine, pelvic girdle, ribs, and shoulder blades form the framework of the skeletal posterior. Viewed from the side, the spine is S-shaped, a design that gives it the resilience to absorb jolts when we move. The spine curves four times from anterior to posterior as it winds down the back, much like a skier slaloming down a slope. The swerves, aided by the cartilage disks between vertebrae, give the backbone spring that it would not have were it a straight rod. A curved spine also makes it easier to balance by positioning the upper body directly over the legs and better distributing the weight of the skull over the pelvic bones, which is key to walking upright. Without this stabilizing feature, humans would be top heavy.

CORD PROTECTION

A bony ring attached to the back of each spool-like vertebral body contains ridges called spinal processes. These projections link to form a tube, or canal, that shelters the delicate spinal cord, the route for all nerve signals traveling between the body and the brain. The spinal cord, on average, is 17 inches (43 cm) long and 0.4 inches (1 cm) thick. It stops growing when a child is about five years old.

Damage to the spinal column can lead to damage to the spinal cord. If the spinal cord is injured, the nerves's ability to carry messages from the brain to the muscles can be impaired or cut off completely, depending on the severity of the injury. Spinal injuries below the neck can cause paralysis below the waist, or paraplegia; neck injuries may cause quadriplegia, paralysis from the neck down.

SPINAL SECTIONS

The spine is divided into five sections: cervical, thoracic, lumbar, sacrum, and coccyx. The cervical vertebrae are the first set of seven bones that form the neck. They are also known as C1 through C7. The thoracic vertebrae make up the second set of 12 vertebrae. They form the outward curve of the spine and are known as T1 through T12 (from top to bottom). They

THE BACKBONES

SECTION	NUMBER	LOCATION	DESCRIPTION
Cervical	7 bones	Neck	Smallest, lightest vertebrae
Thoracic	12 bones	Chest (or thorax)	Medium-sized vertebrae. Feature facets to connect to the ribs
Lumbar	5 bones	Lower back	Largest, strongest vertebrae. Supports bulk of body's weight
Sacrum	1 (formed by 5 fused bones)	Upper pelvis	Fusing begins in the late teens and is complete by age 30
Coccyx	1 (formed by 4 fused bones)	Lower pelvis	Name means "cuckoo" in Latin. Its triangle shape resembles the bird's beak

are bigger than cervical vertebrae but smalled than the lumbar.

Lumbar vertebrae compose the third set of vertebrae. There are five, which are known as L1 through L5. The lumbar bones—forming the inward curve of the spine—are the largest and strongest of the vertebrae, as they have to support the most weight. The sacrum is a slightly curved triangular bone near the base of the spine, and is followed by the coccyx.

The sacrum, below the lumbar region, joins the spine to the pelvic girdle. At birth, the sacrum is composed of five separate sacral bones, but these will fuse together in early childhood to form a single bone.

The coccyx (or tailbone) is made up of four small, fused vertebrae and forms the very bottom tip of the spine. The coccyx is popularly called the tailbone, because it is believed to be the vestigial remnant of a tail that disappeared as *Homo sapiens* evolved.

Overall the spine appears flexible, allowing a wide range of movements. However, each separate region of vertebrae is slightly different and specialized according to its location. The cervical vertebrae are the smallest and lightest of the vertebrae and allow the greatest flexibility of the spinal bones, a feature that can be seen in movements of the human neck. The top two cervical bones, the atlas and the axis, allow the head both to nod up and down and to shake side to side. The atlas (C1) allows the head to move forward and back (in Greek mythology, the Titan Atlas held the world on his shoulders); the axis (C2) projects into the hollow atlas, allowing the head to rotate from side to side.

The 12 heart-shaped thoracic vertebrae—spanning the back between the lumbar and cervical regions—increase in size from top to bottom, and each bone is attached to a pair of ribs. These permit a fair range of motion with some limitations caused by the presence of the ribs.

The lumbar region of the vertebral column, commonly referred to as the small of the back, receives the most stress of all the vertabrae—and the bones there are built to take it: They are very strong, thick, kidney-shaped structures. The lumbar region also permits limited movement. Bending forward and backward is possible, but rotation is not.

The spine is capable of many types of movement, including flexion, or arching the back.

SEE ALSO: Chapter Six, "Breathing," PAGE 168

WHAT CAN GO WRONG

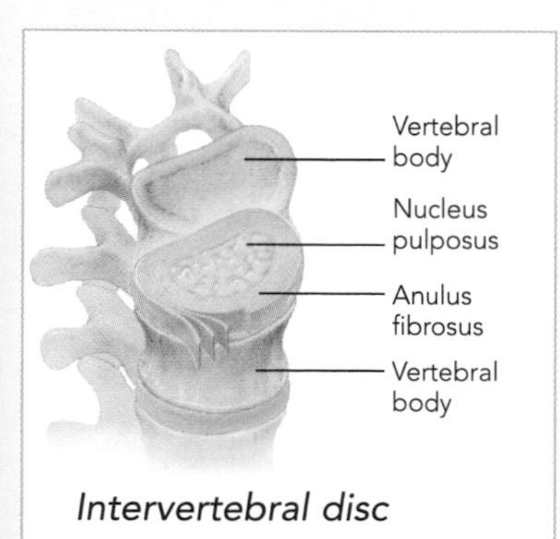

Intervertebral disc

A COMMON CAUSE of back pain is a herniated, or "slipped," disk. It can stem from injury, awkward bending or twisting, lifting heavy loads, or age-related changes. Disks, left, are small shock absorbers between the vertebrae. They have a gel-like interior and a tougher outer margin. If the exterior tears, the gel can bulge out and press on spinal nerve roots, causing pain, numbness, or weakness, opposite. A herniated disk in the neck can cause pain from the neck to the hand. When the disk is in the lower back, it may cause sciatica (inflammation of the sciatic nerve, resulting in pain from the lower spine, down the leg, into the

SPINAL PROBLEMS

Somes spines have abnormal curvatures that can cause problems. Some are congenital (present at birth); others result from disease, poor posture habits, or unequal muscle pull on the spine.

Scoliosis, which literally means "twisted disease," is an abnormal lateral curvature that occurs most often in the thoracic (midback) region. It is most common during adolescence; for some unknown reason, girls are particularly prone to developing the condition. More severe cases result from abnormal vertebral structure such as uneven lower limbs or muscle paralysis. If muscles on one side of the body are not functioning properly, those on the opposite side tug on the spine and force it out of alignment. Scoliosis is treated—with body braces or surgically—before the growth period ends to prevent permanent deformity and breathing difficulties due to limited rib motion and compressed lungs.

Kyphosis, or hunchback, is an exaggerated backward curvature of the spine in the thoracic region. It is most common in older individuals because of osteoporosis (see "What Can Go Wrong," page 78) but may also be caused by tuberculosis, rickets (typically, a childhood disorder resulting from calcium and vitamin D deficiency), or osteomalacia (abnormal softening of bones caused by disease).

Lordosis, or swayback, is an accentuated lumbar (lower back) curvature. It, too, can result from spinal tuberculosis or osteomalacia. Temporary lordosis is common in those carrying a "large load up front" such as pregnant women and men with sizable potbellies, who throw back their shoulders in an attempt to preserve their off-kilter center of gravity.

RIB CAGE

Attached to the backbone near the chest is the rib cage. This structure (also called the thoracic cage because of its location in the thoracic, or chest, cavity) consists of the thoracic vertebrae, the ribs, and the sternum (or breastbone). The ribs are the thin, flattish bones that curve around the chest to protect the heart, lungs, and other vital organs. The spongey lungs fill up most of the thorax, or chest. Part of air exchange depends upon the rib cage's ability to enlarge and reduce the size of the thoracic cavity as the lungs expand and contract.

There are a dozen pairs of ribs. Seven pairs are called true ribs: each rib is attached to the sternum by cartilage in front and curves around to join one of the vertebrae that make up the backbone. There are three pairs of false ribs. These are attached to vertebrae but not to the breastbone. In the case of ribs six through ten, each rib is attached to the rib above it by cords of costal (rib) cartilage, forming a lower edge called the costal margin.

There are two pairs of floating ribs. These are connected only to the vertebrae of the backbone. The gaps beween the ribs are called intercostal spaces, and they contain thick sheets of muscle that expand and relax the chest during breathing. The floating ribs protect organs including the liver, kidneys, stomach, and spleen. In addition to their other functions, the bones of the rib cage contain red marrow and are one of the body's major blood-cell suppliers.

foot). An MRI scan can give a definitive diagnosis and location. Conservative treatment may include moderate exercise, massage, heat therapy, painkillers, manipulation, or acupuncture. If symptoms persist, a physician may recommend an epidural steroid injection (ESI), in which a steroid is injected into the disk to shrink it and relieve pressure on the nerve. Alternatively, some of the disk's contents can be removed surgically or vaporized by a laser, treatments often done on an outpatient basis. Failing that, the disk may have to be removed; a bone graft may be done to fuse adjoining vertebrae.

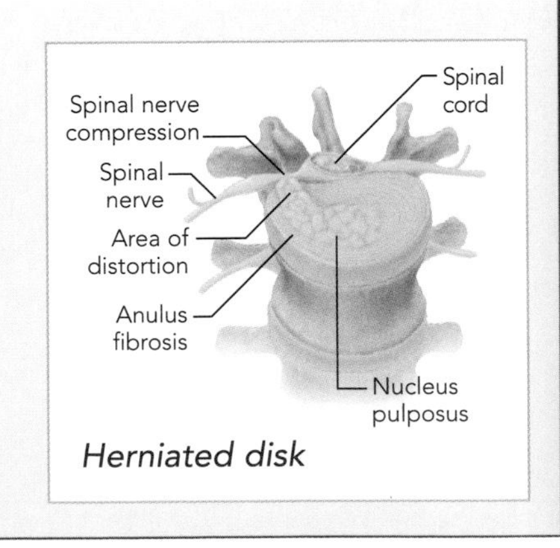

Herniated disk

HANDS & FEET

INTRICATE AND PRACTICAL

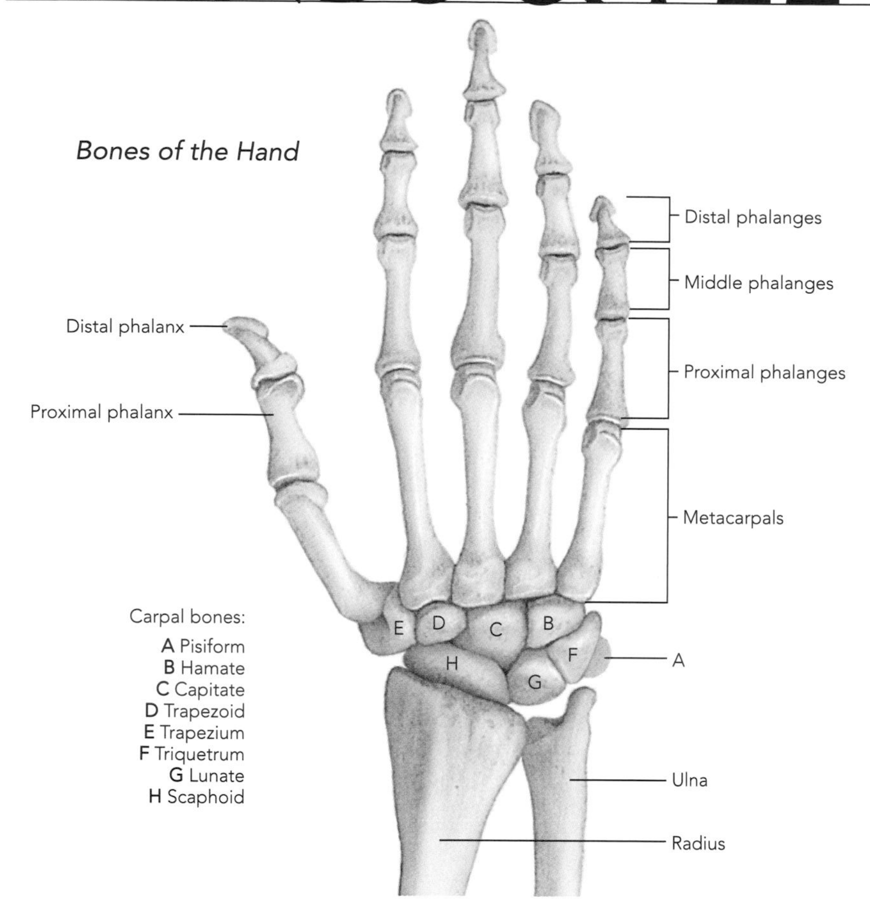

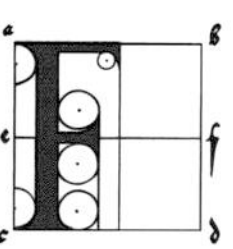

FROM THE CRADLE on, the hands and feet shape how a person interacts with the world. With the hands, we can manipulate objects—touching, feeling, grabbing the things around us. The feet bear the body's weight, taking it where it wants to go. Intricate connections between the bones and the connective tissues make these appendages highly effective tools, specialized in helping the body navigate its environment. Packed with sensitive nerve endings, they are also the place where the body takes in sensory information and relays it to the brain.

THE HAND

The hands are among the busiest, most versatile anatomic avenues in the body, together containing more than a quarter of the body's bones and a vast network of muscles. The mix of 27 bones, 37 muscles, and a highly mobile thumb in each hand allows a wide range of motion. The bones of the hand are divided into three groups: carpals, metacarpals, and phalanges. The carpals are the group closest to the arm; next, the metacarpals lie under the palm, and then the phalanges make up the fingers and thumb.

Carpals are the eight pebble-like short bones located at the end of each arm formed by the tips of the radius and ulna. The carpals make up the wrist, or carpus, and are arranged in two rows of four. Closest to the arm, proceeding from ulna to radius, are the pisiform, triquetrum, lunate, and scaphoid. The front-row carpals, which connect with the five metacarpals that form the body of the hand, are the hamate, capitate, trapezoid, and trapezium. Only the scaphoid and lunate actually articulate with the radius to form the wrist joint.

The carpals are tightly bound by ligaments. So-called gliding joints between them, however, enable the wrist bones to slide smoothly, making the wrist quite flexible, although movement at each carpal is limited.

+ HAND FACTS +

+ IN ALL, EACH hand has 27 bones.
+ There are three kinds of bone in the hands: carpals, metacarpals, and phalanges.
+ Carpals: These eight marble-size short bones are located in the wrist.
+ Metacarpals: There are five of these bones supporting the palm.
+ The fingers and thumb together contain 14 bones, called phalanges.
+ Each finger has three phalanges, except for the thumb, which has only two.
+ On average, the longest digit on the hand is the third one.

PALMS & FINGERS

The five metacarpals extend from the wrist like spokes on a wheel to form the broad metacarpus, or palm, of the hand. These miniature long bones are identified not by names but by numbers, 1 to 5, counting from the thumb to the little finger. The metacarpals attach to the carpals and the phalanges, the bones of the fingers and thumb. (The bulbous heads of the metacarpals become knuckles when the hands are clenched.) Metacarpal 1, associated with the thumb, is the shortest and most mobile of the palm bones. There's a special link between it and the trapezium carpal called a double saddle joint, which allows the thumb to flex enough to touch the tips of the other digits, giving the hand its ability to grasp and manipulate even the tiniest of objects (see "The Opposable Thumb," right).

Fingers, or digits, of the hand, are also numbered 1 to 5, beginning with the thumb, or pollux. The 14 phalanges are miniature long bones similiar to the metacarpals. Four of the fingers feature three bones: (from the top down) distal, middle, and proximal; the thumb lacks the middle. Each hand also has an assortment of sesamoid bones, small ossified nodes embedded in the tendons to provide extra leverage and reduce pressure on underlying tissue. Many exist around the palm at the bases of the digits, but the exact number varies among people.

THE OPPOSABLE THUMB

Jane Goodall extends her hand to Gregoire, a chimpanzee at the Brazzaville Zoo.

THE THUMB might be the shortest digit on the human hand, but its versatility and flexibility are a huge advantage. Its ability to swing across the palm and move toward the other fingers and touch all of them, an action called opposition, enables the hand to form a grip. Besides humans, most other primates, including great apes—such as the baboon, chimpanzee, and orangutan—also have opposable thumbs. What sets humans apart is the broad area of contact achievable between the thumb pad and pads of the other fingers, a feature that allows humans to exercise greater control. Thanks to the opposable, or prehensile (meaning adapted for grasping or seizing), thumb, humans more successfully use tools and utensils and perform multitudes of other tasks.

Several features of the human hand combine to create opposability. At the base of the thumb, a double saddle joint and associated ligaments enable the thumb to rotate an expansive 90 degrees. Joints at the base of the second, fourth, and fifth fingers allow the fingers to rotate 45 degrees in toward the thumb. These adaptations enable the precise grip between the thumb and forefinger that is key to manipulating many kinds of objects.

According to scientists, the opposable thumb and interdigital grip evolved to enable primates to take to the trees to flee their enemies or to pick fruit. Evolutionists have hailed the opposable thumb as a milestone in human evolution, the development that led to more intricate motor skills. Many researchers believe that opposable thumbs—along with walking upright—are strides made possible by more developed brains, which allow humans to intellectually distinguish themselves from other species. ■

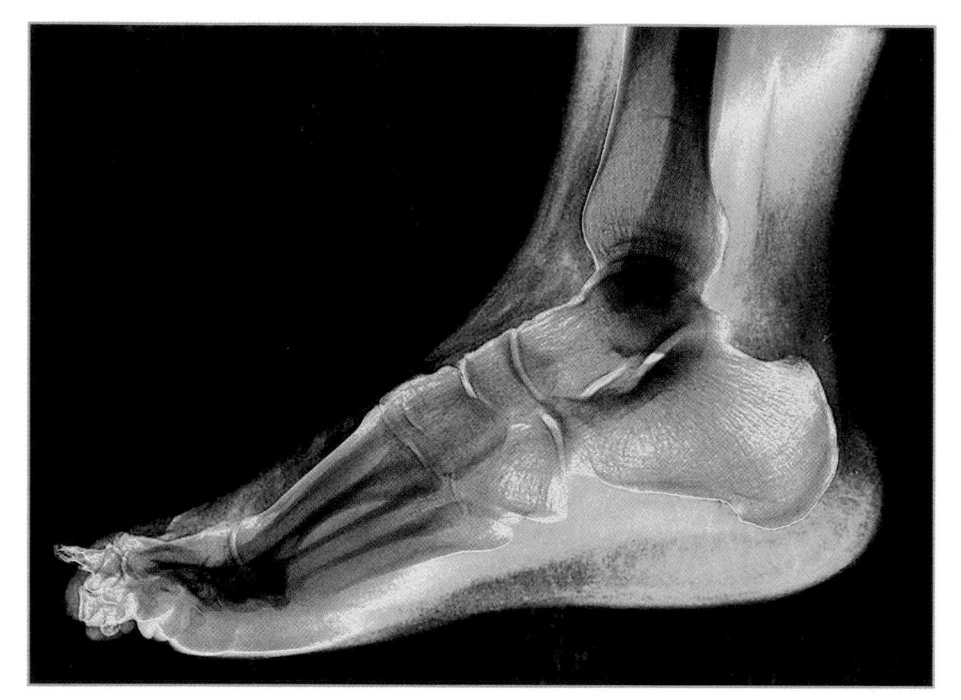

The three regions of the foot—forefoot, midfoot, and hindfoot—are all visible in this colored x-ray.

MOVEMENT

The hands' wide range of fine movements is made possible by numerous small joints, long and short muscles, and extensor and flexor tendons that allow the fingers and thumb to straighten, bend, and flex. Muscles pull with precision on multiple movable joints—14 of them, the knuckles between the finger bones and metacarpals. The knuckles at the base of each digit (between the phalanges and metacarpals) are condyloid joints, like the wrist. These joints provide greater mobility than the other finger knuckles, which are simple hinge joints that allow the fingers to bend toward the palm but not toward each other.

In a firm grip, most of the power comes from muscles in the forearm that are linked to bones by very long tendons, which can be seen in action on the back of the hand as it opens and closes. The fingers have two long flexors, fastened to the underside of the forearm. The thumb has one long flexor and a short flexor; it also has other muscles moving the thumb in opposition, making grasping possible. A fibrous band, called the retinaculum, encircles the wrist and keeps the tendons in place.

HURT HANDS

Many bones in the hands, including the wrist bones and the radius and ulna (where they connect to the wrist), are vulnerable to fractures. The two most common breaks here are Colles and scaphoid fractures. A Colles fracture is a complete crack in the end of the radius, which often occurs when the hand is flexed to brace the body during a fall. Falls can also cause a break in the bones of the palm of the hand, where the scaphoid carpal connects with the radius. This injury may initially be confused with a bad sprain and tends to heal slowly due to limited blood supply.

Broken fingers are also common, resulting from accidents (such as slamming one in a car door) to sports injuries. Luckily, taping the broken finger to a healthy neighboring finger or splinting it is usually all that's necessary for a fracture of the finger bones to heal. Surgery may be required in cases of serious damage.

THE FEET

The foot is an engineering wonder that supports body weight and acts as a lever to propel the body forward. It is both strong enough to push off with each stride, supple enough to absorb the impact of every step, and facile enough to negotiate uneven surfaces. The foot is designed in much the same way as the hand. There are three regions: the tarsals (closest to the leg), metatarsals (underneath the sole), and phalanges (the toes).

The ankle is formed where the talus articulates with the tibia and fibula (the lower leg bones). This hinge joint allows the foot to flex and to rotate when tugged on by muscles. The ankle, or tarsus—akin to the wrist (carpus)—consists of seven tarsal bones. Only one, the talus, actually helps form

SEE ALSO: Chapter Four, "Movement," PAGE 116

FOOT FACTS

+ THERE ARE 26 bones, 100 ligaments, and 35 muscles in each foot.
+ Among the bones are 14 phalanges (toe bones), 5 metatarsals (sole bones), and 7 tarsals (ankle bones).
+ The heel bone (or calcaneus) is the largest bone in the foot.
+ The knobby protrusions on each side of the ankle are actually the ends of leg bones. Each is called a malleolus: The inner one is part of the tibia; the outer one belongs to the fibula.

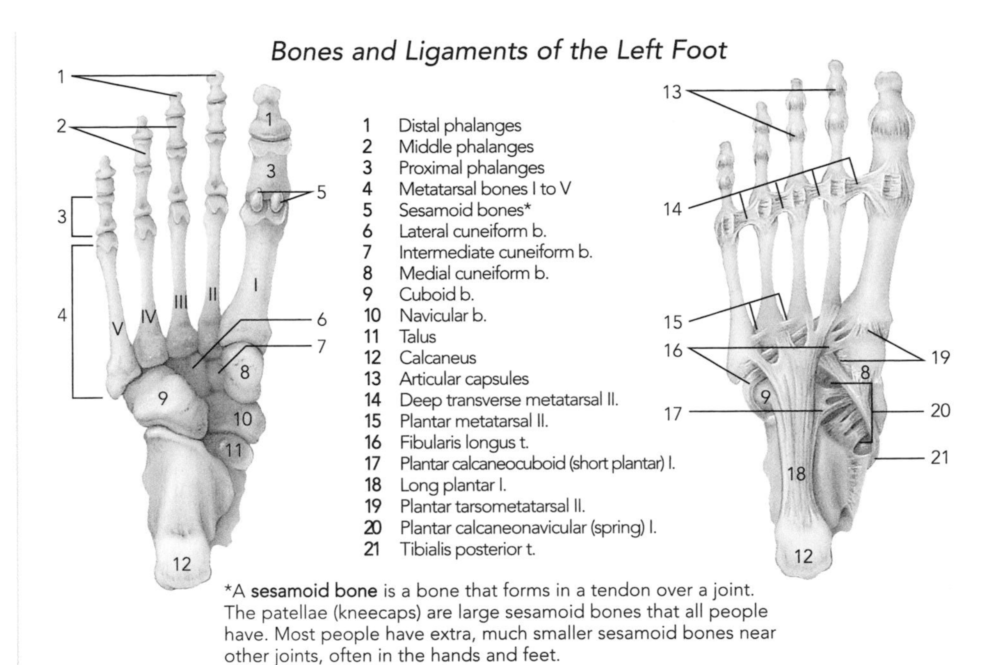

Bones and Ligaments of the Left Foot

*A **sesamoid bone** is a bone that forms in a tendon over a joint. The patellae (kneecaps) are large sesamoid bones that all people have. Most people have extra, much smaller sesamoid bones near other joints, often in the hands and feet.

the hinge-type joint. The large, topmost tarsal, it transfers weight and pressure from the leg to the foot and plays a key role in all movements involving the feet and ankles. It sits above another tarsal, the calcaneus (or heel bone)—the largest and strongest bone in the foot—which also supports body weight, and it is the site of attachment for the Achilles tendon. The Achilles connects calf muscles to the heel bone. (Calf muscles pull up on the calcaneus to bend the foot downward when moving or standing on tiptoes.)

Together, the talus and calcaneus compose the hindfoot. The midfoot contains five interlocking smaller tarsal bones. The midfoot is connected to the forefoot and hindfoot by muscles and the plantar fascia ligament, which stabilizes the foot and helps maintain its arch. The forefoot contains the metatarsal bones and phalanges, which are connected at the ball of the foot by five metatarsophalangeal joints.

Moving down the foot, there are strong bands of ligaments and tendons that pull the metatarsal and tarsal bones into three arches. These form a dome that distributes weight to different parts of the feet. About half of standing and walking weight goes to the heel bones and half to the heads of the metatarsals in the ball of the foot. The instep, or metatarsus, consists of five metatarsal bones. The metatarsals are linked to the 14 phalanges, or toe bones.

The phalanges of the toes are similiar to those of the fingers but are much smaller and, thus, less nimble. The big toe, or hallux—like the thumb—consists of only two phlanges, while the other four digits each boast three bones: the proximal, middle, and distal. (The big toe lacks the middle one.)

FLAT FEET

In flat feet, or fallen arches, there is little or no arch between the heel and toes, resulting in a sole that rests almost flat on the ground. Nearly everyone has flat feet at birth; the arches do not develop fully until the ligaments and muscles in the soles are completely formed. Some people never develop arches; others get flatter feet over time as the ligaments loosen or stretch, often due to weight gain. A person with flat feet is more prone to damage the bones of the feet, because of increased pressure on them.

Flat feet also can contribute to knee, hip, and lower-back pain, as well as a painful condition called plantar fasciitis. This inflammation of the plantar fascia is most often experienced as heel pain and stiffness when standing or walking. Plantar fasciitis can also occur when the ligament is strained from overuse or from ill-fitting shoes.

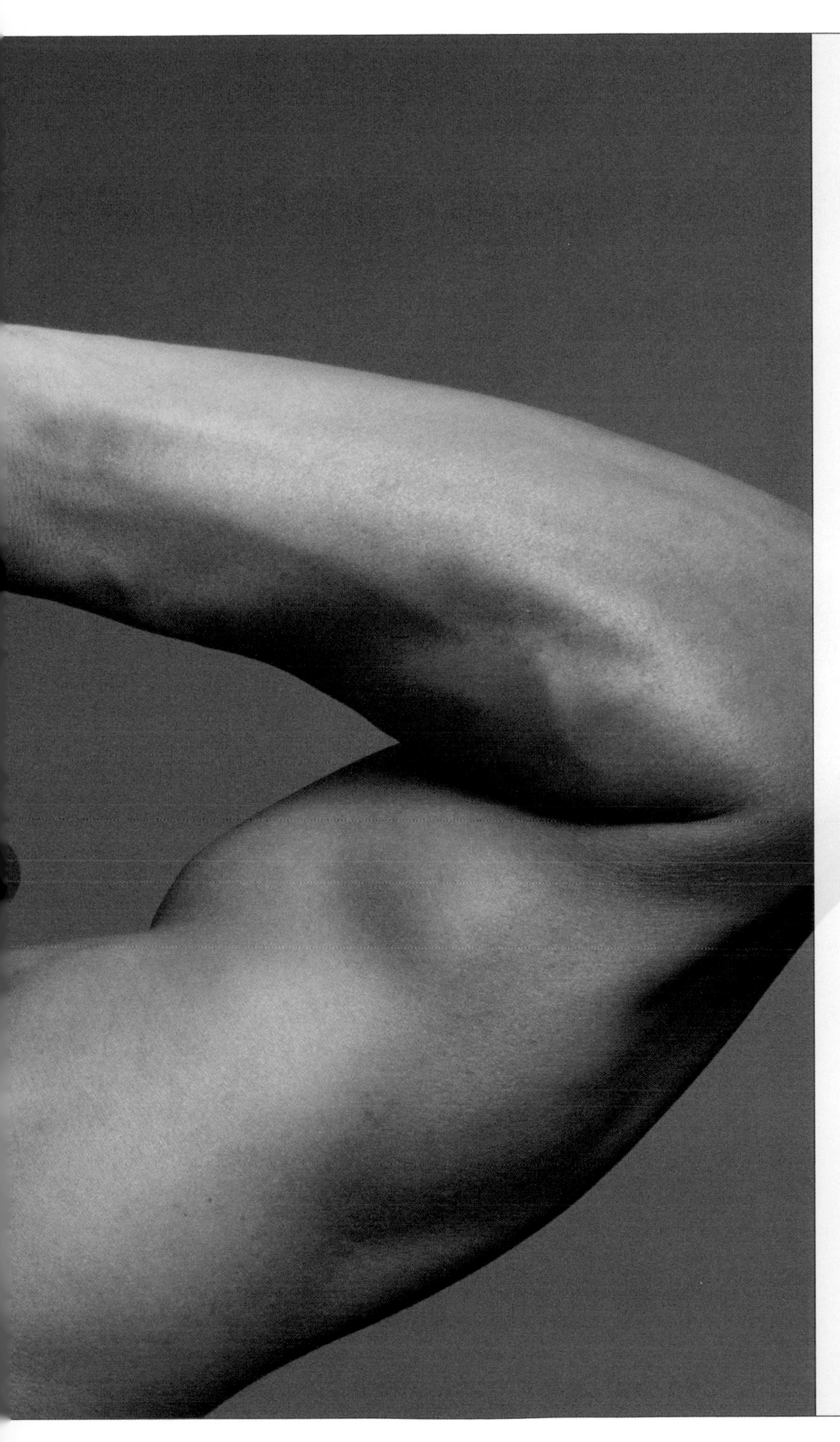

CHAPTER FOUR

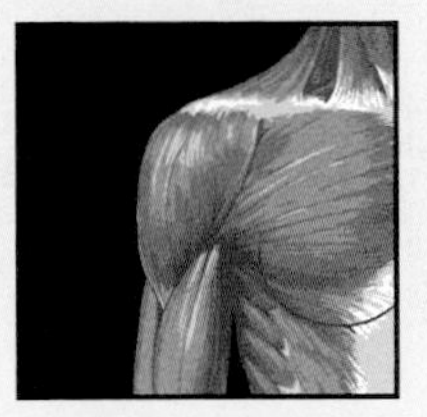

BODY MOVERS

THE WORD "MUSCLE" COMES from the Latin *musculus*. It seems that the ancient Romans thought the rippling motion of a flexing muscle resembled a "little mouse" moving under the skin. More than just the makings of a well-toned body, however, muscle tissue also constitutes the fabric of the hollow internal organs. The heart itself is composed of hardworking muscle tissue. Making up more than half the body's mass, muscles are the body's movers, shakers, and stabilizers. From pumping the blood to scratching the back, from digesting a meal to cracking a smile, muscles make it all possible.

Training with weights can increase muscle size and strength.

THE MUSCLES

GIVING THE BODY STRENGTH

THE BODY EMPLOYS muscles in almost every possible activity: sitting and standing, walking and running, smiling and frowning—and on and on. Muscles create the shape of every human body. They turn intentions—and nerve impulses—into actions. They form a strong, protective wrapping for the underlying skeletal and organ systems. Muscles hold the body's joints firmly in place so they can operate.

Muscles power the body's essential internal operating systems as well. Without muscles, the body could not breathe air or digest food. Without muscles, blood could not circulate from the heart in the core of the body, out through the limbs to the extremities, and back again. Every movement, every forceful exertion, whether voluntary or not, occurs in a body through the work of its muscles. As a result of this energetic motion of muscles, the body also stays warm.

Every muscle, no matter where it is in the body, operates through the basic principles of contraction and relaxation. Sometimes we choose to make muscles move, and this involves a deliberate combination of contracting and relaxing them. The simple action of reaching out, grasping a glass of water, and lifting it to the lips, for example, requires the voluntary and coordinated contraction and relaxation of dozens of muscles.

Night and day, whether we are asleep or awake, muscles in our bodies are contracting and relaxing on their own as well, without any awareness or deliberate thought process on our part. These involuntary muscles make the respiratory, digestive, and reproductive systems function.

MUSCLE FACTS

+ THE HUMAN BODY has about 650 distinct skeletal muscles.
+ The gluteus maximus, or buttocks muscle, is the body's largest muscle. Its primary job is to hold the trunk erect.
+ The stapedius, inside the ear, is the body's smallest muscle. Its job is to stabilize the stapes bone, which vibrates in the presence of sound waves as they enter the inner ear.

MUSCLE TYPES

The human body contains three kinds of muscles: skeletal, smooth, and cardiac. Each muscle type performs a different job. Skeletal muscles (also known as striated muscles, because of their striped appearance) make up the bulk of the body. They move voluntarily—a person can control their operation—and come in various sizes and shapes to perform many kinds of actions. Smooth muscles occur in organs and blood vessels. They contract and relax involuntarily, and they

BREAKTHROUGH

THE SPARK OF LIFE: In the late 18th century, experimentalists used primitive batteries to generate electricity to explore its effects on muscle. After Italian scientist Luigi Galvani (1737–1798) touched a charged metal wire to dissected frogs and made the frogs' legs jerk, he concluded that his experiment proved the presence of an electrical charge in the frogs' muscle tissue. Galvani's countryman Alessandro Volta (1745–1827) argued that the tissue itself simply conducted the charge. Galvani's nephew, Luigi Aldini, continued his uncle's work, experimenting on human corpses. Their results inspired the early practice of galvanism—the word coined for the use of electricity in medicine—a pseudoscience in its day but an important precursor stage leading to today's electrophysiology. We still speak, however, of a person's being "galvanized" into action.

SEE ALSO: Chapter Nine, "The Nerves," PAGE 250

motivate the essential life systems. Cardiac muscle—the muscle in the heart—is also involuntary muscle. It works differently from either skeletal or smooth muscle. From before birth until the moment of death, cardiac muscle cells contract and relax with a rhythm embedded in their cells, creating the pulsing force that circulates blood through the heart.

Of all the functions performed by muscles in the body, physical movement is the easiest to observe and understand. Every movement, from a tiny blink of the eyelid to a sprinter's initial burst of speed, happens because of the coordinated contraction and relaxation of skeletal muscle cells operating together. Lack of movement involves many skeletal muscles as well. Standing still, sitting up straight, holding a pose—all these activities require the voluntary exertion of many body muscles. These muscular exertions also generate heat, which the body needs for growth, health, and comfort.

Harder to perceive but essential to keeping a body alive and well, smooth muscles are performing tasks all the time inside the body. Smooth muscles at work in the esophagus, stomach, and intestines move food through the digestive system. Smooth muscles in the bronchi, the air tubes in the lungs, convey oxygen and other gases through the respiratory system. Smooth muscles in the lining of the blood vessels (like the veins and arteries) help maintain blood pressure in the circulatory system.

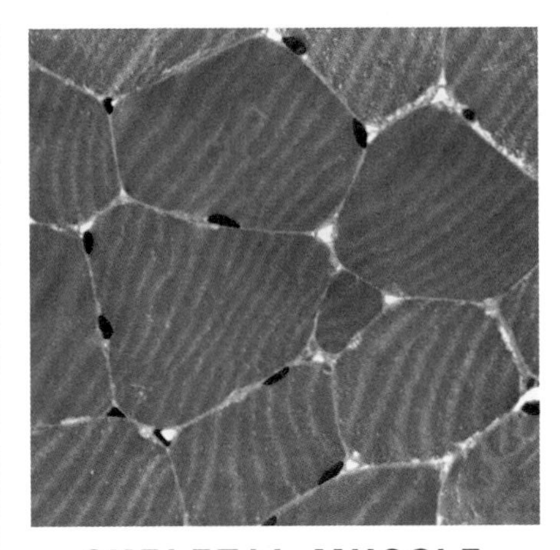

SKELETAL MUSCLE
Contractile proteins form an alternating pattern of light and dark bands called striations.

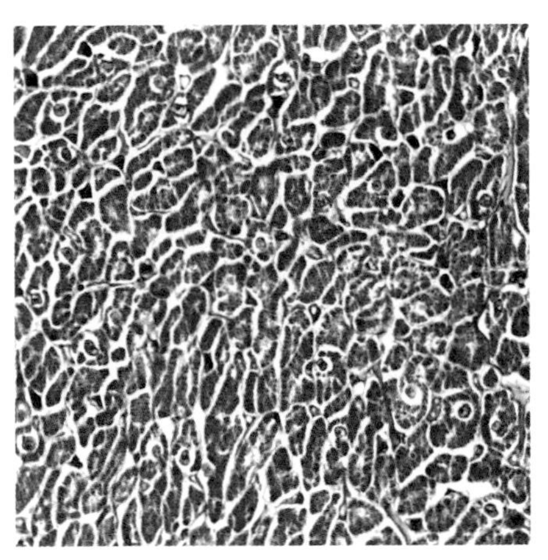

CARDIAC MUSCLE
Rich with capillaries, connective tissue (white) supplies blood to the elongated muscle cells (purple).

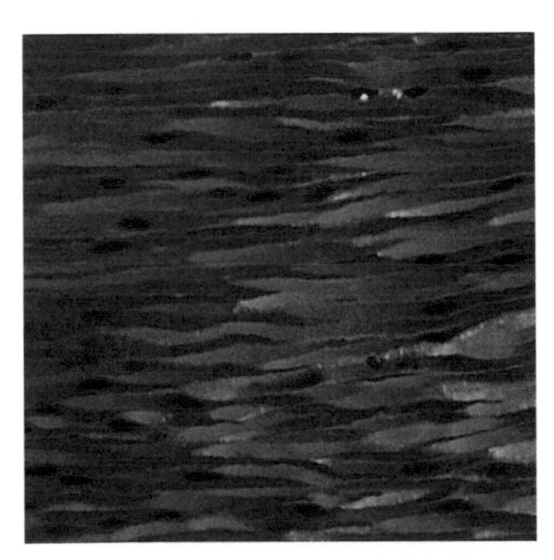

SMOOTH MUSCLE
These spindle-shaped cells, each containing a dark nucleus, are grouped in irregular bundles.

STRUCTURE

Muscle tissue, whether skeletal, smooth, or cardiac, is made up of many thousands of long, thin cells, also called fibers, working together. When a muscle contracts, some components within each fiber slide in on top of others, shortening and tightening the fiber. When a muscle relaxes, those components release and separate, lengthening and loosening the fiber. The elasticity of a muscle fiber—the ease with which it is able to contract and relax—changes according to muscle groups. It also varies among individuals and through the aging process. Regular exercise can increase muscle elasticity.

The nervous system makes skeletal muscles contract and extend. Electrical signals travel from the brain, through the spinal cord, and out to the nerves branching through the body. The signals ultimately arrive at individual muscle cells. When a neurotransmitter pings a muscle cell, that cell responds accordingly. Signals sent to a group of muscle cells elicit a coordinated set of interactions, and the muscles contract or relax, resulting in movement.

Simultaneously, neurotransmitters ping back, telling the brain what has occurred. Deliberate movements of the body involve thousands of such cross-communications between the nervous

SEE ALSO: Chapter Five, "The Heart" PAGE 140, Chapter Six, "The Stomach" PAGE 188

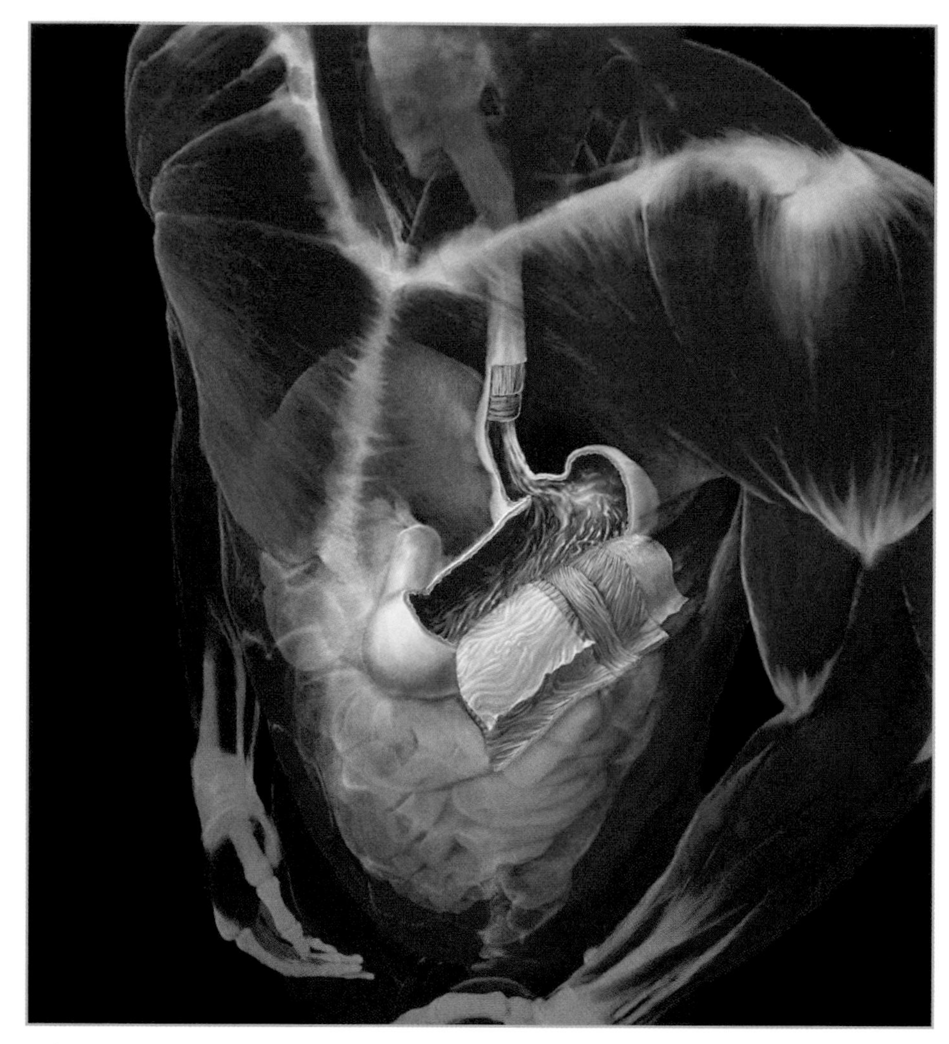

The stomach's three muscular layers are composed of smooth muscle cells.

system and skeletal muscles. It is a complex cycle of feedback and adjustment that depends on nearly instantaneous electrochemical impulses between the nervous system and the muscle tissues.

All the while, the autonomic, or involuntary, nervous system communicates in the same way, electrochemically, with the body's smooth, or involuntary, muscle cells. Without a person's knowing it or deciding to make it happen, the brain sends and receives controlling signals tending to the operation of the blood vessels, lungs, stomach, intestines, bladder, uterus, and other internal organs.

SMOOTH MUSCLE

Smooth muscle gets its name from the sleek surface of its muscle fibers, compared with the striped appearance of skeletal muscle fibers. Smooth muscle fibers are also usually organized in thin sheets, unlike skeletal and cardiac muscle cells, which are arranged in bundles. Most of the body's smooth muscle operates unseen. Smooth muscle is at work, for example, in the walls of blood vessels, in the walls of hollow organs such as the stomach, intestines, and bladder, in the airways leading into the lungs, and inside the iris of the eye.

All these body parts operate by means of the forces created as smooth muscle cells involuntarily contract and relax. Blood circulates, food is digested, urine is transported to the bladder, and the eyes respond to light or focus on objects of interest.

Groups of muscle cells often work together. In many of the body's vessels and organs, for instance, their orientation and coordinated operation make material pass through efficiently. The esophagus—the tube that sends food down into the stomach after a person swallows—is a good example because it is composed of layers of smooth muscle, one on top of another. One layer has muscle cells oriented lengthwise, and the next has muscle cells perpendicular to that, oriented like rings around the inside of the tube. The interaction of these longitudinal and

+ SMOOTH OPERATORS +

SMOOTH MUSCLE works without a person's awareness. Structures made of it or containing it include

+ veins, arteries
+ esophagus, stomach, intestines
+ ureters, urinary bladder
+ fallopian tubes, uterus
+ muscles attached to hair follicles
+ eye muscles for focusing, eye muscles for adjusting pupils to light

circular muscle groups is essential to the way that food travels down the esophagus and toward the stomach to be digested.

After a person chews and swallows, an involuntary digestive action begins in the esophagus, performed by groups of smooth muscle cells. Operating in slow waves, they contract and relax in rhythmic coordination. Cells in front of the wad of food relax, opening a pathway, while cells behind the food contract, pushing it farther down the tract. The combined effects transport the food through the length of the esophagus toward the stomach. The mechanical force of these muscle groups is so strong that food would still continue toward the stomach even if a person were standing on his head.

This process of coordinated smooth-muscle movements for the purpose of moving material inside the body is called peristalsis. It occurs not only during the digestive process in the esophagus and intestines, but in other body processes involving the ureters, the bile ducts, and the fallopian tubes.

CARDIAC MUSCLE

Cardiac muscle—the muscle that makes up the heart—is composed of a type of muscle cell different from those found in smooth and skeletal muscle. Like smooth muscle, cardiac muscle operates involuntarily. Like skeletal muscle, however, it is striated, with dark bands of color visible in the muscle cells when viewed under the microscope. Unlike skeletal and smooth muscles, the heart, although it consists primarily of cardiac muscle, is not part of the musculoskeletal system. Along with the blood vessels, it forms the circulatory, or cardiovascular, system.

Cardiac muscle cells are short and squat when compared with other muscle cells. They connect to one another with a special bonding design that makes the heart a dense, strong organ and allows tiny electrochemical impulses to pass rapidly and easily from cell to cell.

The rhythmic throb of the heartbeat happens when, in response to those impulses, all the cardiac muscle cells work together, contracting and relaxing in harmony. Muscle relaxation of the heart allows the blood into the heart organ. Muscle contraction of the heart forces the blood to leave the heart and move into the rest of the body through the circulatory system.

Some special cardiac muscle cells serve as the heart's internal pacemakers. These cells never rest. They keep up a constant rhythm of alternating contraction and relaxation, and they generate the electrochemical impulses that, almost instantaneously, transfer that rhythm to the other cardiac cells, coordinating and regulating their never-ceasing contractions.

The influence of that rhythm is felt through the entire body, moving blood out to the extremities and then returning it to the heart. You are feeling that rhythm when you take your pulse at your neck, wrist, or ankle. Although the brain and nervous system keep tabs on the heart, signaling the pacemaker cells to speed up or slow down the heart rate as the body demands, the basic motor movement of the heartbeat originates in the cardiac muscle itself.

THREE TYPES OF MUSCLE

	SKELETAL	SMOOTH	CARDIAC
LOCATION	Throughout body, primarily attached to bones by tendons	Walls of hollow organs and tubes such as blood vessels	Heart only
CONTROL	Voluntary	Involuntary	Involuntary
CELL SHAPE	Long cylindrical fiber, striated	Long, thin fiber, tapered at ends, smooth	Branched cylindrical fiber, striated

SKELETAL MUSCLE
STRONG AND FLEXIBLE

WHEN THINKING OF muscles, skeletal muscles are what usually comes to mind. Striated tissue forms distinct masses, each one a muscle. They can be big or little, thin or thick, round or flat—all shaped to perform the many diverse tasks a person undertakes in the course of a day. As the name suggests, skeletal muscles usually attach to the bones of the skeleton, where they facilitate movement. This type of muscles works voluntarily, and its actions—coordinated sequences of contracting and relaxing—move the bones at their joints.

The body's muscular system interacts intimately with the nervous and circulatory systems, as can be seen at the cellular level in skeletal muscle fiber. Skeletal muscle cells are long and cylindrical. Generally, one nerve, one artery, and at least one vein connect to each muscle cell. A network of capillaries weaves along the skeletal muscle cells. Blood flow to muscle cells furnishes oxygen and nutrients; blood flow from muscle cells removes waste products created during muscle activity.

Nerve fiber runs up and down the length of skeletal muscles. The two systems meet at neuromuscular junctions. A neuromuscular junction is a receptive spot where the axon, or nerve ending, connects into the muscle cell. There, whenever the nervous system sends a signal for a muscle to move, the nerve delivers minute amounts of biochemical messengers into the muscle to incite it to action.

TIES THAT BIND

A muscle generally is attached at one end to a fixed bone, then crosses over a joint, and is attached at the other end to the bone it moves, with the joint connecting the two bones. A muscle attaches to bone by means of a tendon, a tough, ropelike connective tissue rich in the protein collagen. (Sometimes, if the muscle is wide and thin, the connector is an aponeurosis, or flat tendon, which is sheetlike. Muscles can also attach directly to bone with their own collagen fibers.) While muscles are well supplied with blood, tendons receive comparatively little, taking their gray or white color from collagen, and contain little or no nerves.

When the muscle contracts, the tendon transfers the force to the bone, causing it to move around the joint. Whereas muscles are elastic, tendons are fairly stiff; if it were otherwise the muscle's force would be dissipated in the tendon's

Skeletal Muscle

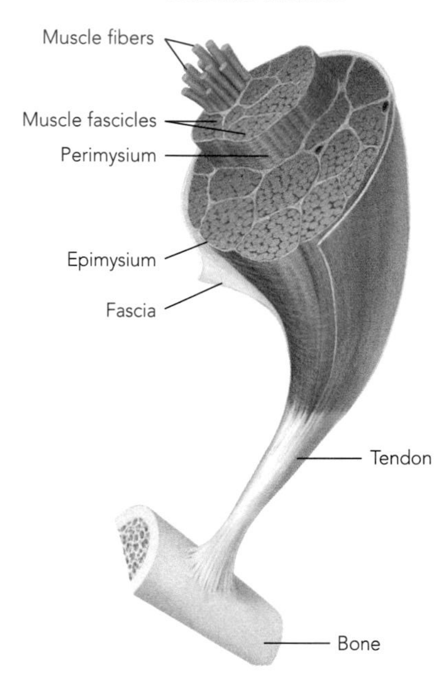

WHAT CAN GO WRONG

MUSCULAR DYSTROPHY: About one boy in 3,500 suffers from Duchenne muscular dystrophy, or DMD, a fatal genetic disease of the muscles that is expressed in males. Symptoms appear early, and afflicted boys may require leg braces (left). Researchers believe the muscle loss in DMD is caused by the lack of the protein dystrophin. Without it, muscle cell membranes tear easily, and the muscles disintegrate. Gene therapies being explored to treat DMD include injecting muscles with stem cells able to make dystrophin and prompting muscle cells to make more of a protein similar to dystrophin.

SEE ALSO: Chapter Three, "Connections," PAGE 82, Chapter Nine, "The Nerves," PAGE 250

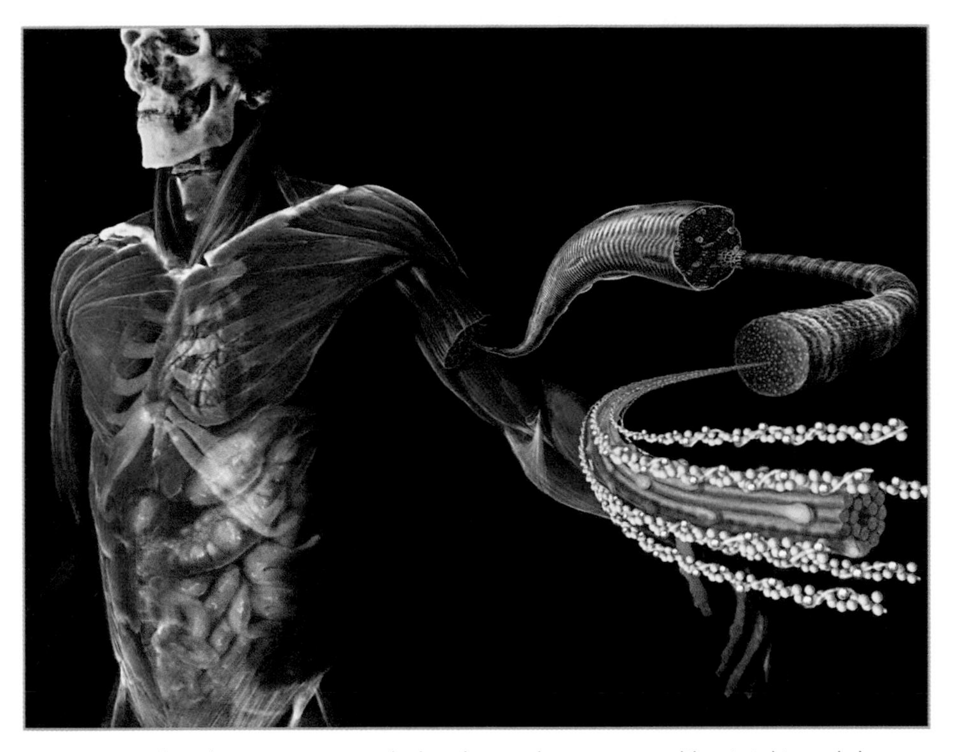

Two molecular proteins in skeletal muscle, one ropelike (pink) and the other resembling a ladder (yellow), must interact to produce contraction.

own movement. Collagen fibers originating in the muscle continue into the tendon. At the bone, the tendon grows into the bone tissue, completing the connection.

DAMAGE CONTROL

Human body tissue has the inborn capacity to regenerate when it has been wounded, but there is a limit to how well muscle mass can heal. Smooth muscle and tendons replenish themselves better than skeletal muscle. Cardiac muscle cells rarely regenerate, so physical damage to the heart usually results in scar tissue, not new heart muscle. Skeletal muscle tissue also has severe limits to how well it can build new cells.

More often, the healing of damaged muscles involves fibrosis, or the development of fibrous scar tissue. An area of tender new tissue develops at the site of the wound. New capillaries bring blood to the site. A matrix of collagen fibers and other materials knits together, filling in the gap made at the wound.

This matrix, the developing scar tissue, does not operate like the lost muscle it is replacing. Not fully infused with blood, it tends to be lighter in color than the surrounding area. Scar tissue, however, is tough and resilient, being composed almost entirely of collagen. It serves to knit together the edges of the original wound and fill them in with sturdy tissue.

MAKING A MOVE

It is the business of normal, healthy skeletal muscles to contract and relax in order to move all the parts of the body. These movements come as responses to signals sent to the cells by the nervous system.

Here, at the microscopic level, is how contraction works in a group of skeletal muscle cells. The decision to move sends a nerve impulse through the motor neuron and into the muscle cell at the neuromuscular junction. The impulse travels through the membranes wrapping each muscle fiber and into the organelles at the core of that fiber, called myofibrils. Inside every myofibril are two types of filaments, thick and thin. On the signal of the nerve impulse, the thin filaments slide over the thick filaments. The net effect is that the

WALKING

HOW MANY LEG MUSCLES does it take to walk? The answer is 12. Motor neurons, above, stimulate the quadriceps femoris, four thigh muscles; the hamstrings, three muscles at the back of the leg; the buttocks, three muscles; and two calf muscles. (Walking also involves the muscles in your feet, the muscles to keep you upright, and those to swing your arms.)

Several schools of Yoga, a system of Indian philosophy, emphasize mastery of the body as a way to attain spiritual perfection.

muscle cell shortens and bulks up. When this event occurs simultaneously in millions of muscle cells, a muscle group contracts.

At the same time, nerve signals tell a related muscle group to pull the thin filaments back away from the thick filaments—and thus relax. For every muscle group that contracts, an opposite muscle group relaxes. When a weight lifter flexes his biceps, for example, he is at the same time relaxing his triceps, the muscle group on the underside of his lifted arm. Muscles are only working when they contract; relaxation is merely the effect of the opposite muscle contracting. Thus, muscles only pull, never push.

FUELING UP

Like other living cells, skeletal muscle cells require adenosine triphosphate (ATP) to transform potential chemical energy into actual mechanical energy, the essential process of muscle metabolism. Muscle cells store only small amounts of ATP, so cells are constantly replenishing that supply, either by aerobic ("with oxygen") respiration or anaerobic ("without oxygen") respiration.

Through aerobic respiration, cells use oxygen to break down glucose molecules, resulting in energy and new ATP molecules. In anaerobic respiration, cells perform those same tasks but much less efficiently. Blood brings the necessary glucose, a sugar, to the muscle cells. When demand is high, cells may also use glycogen, a sugar stored in muscle itself.

These two metabolic processes, aerobic and anaerobic respiration, occur in living cells throughout all body systems. Aerobic respiration is the usual method of ATP replenishment in muscle cells. It takes place when a person's body is at rest or undergoing light exertion.

When a person begins to exercise more strenuously, anaerobic respiration kicks in, too. Even during moderate exercise, though, aerobic respiration provides about 95 percent of the ATP needed by muscle cells. At the highest levels of exertion, muscle cells use creatine phosphate, a molecule found only in muscle, to generate even more ATP and keep working.

With all these methods of producing energy, muscle cells actually create more energy than they need. About 40 percent of the energy is used for the work they are called on to do, and the rest is expended as heat. That is the reason why a person's body temperature rises when she exercises vigorously. Excess energy generated in the muscle cells is dumped into the bloodstream, which brings that heat to the skin's surface as

RIGOR MORTIS

THREE TO FOUR hours after death, the body lapses into rigor mortis (Latin for "stiffness of death"). The muscle cells no longer receive chemicals they need to keep up the cycle of contraction and relaxation. Without new ATP, they get stuck in a contracted position. Rigor mortis usually lasts 36 to 60 hours, then reverses as the effect subsides.

the body attempts to return to its normal temperature.

Sometimes physical effort over taxes the muscle cells. The cells need more sugar than either the blood or their own glycogen stores can supply, and so they cannot keep up ATP production to meet the continued demand. When this happens, the body undergoes muscle fatigue. Nerve stimuli still reach the cells, but the muscle fibers are unable to contract in response. The muscle cannot, and does not, do what the brain is asking it to do.

When pushed to an even greater extreme, so that they run out of ATP altogether, muscle cells undergo contracture—a state of continuous contraction. That's what happens when a person's hand seizes up in the phenomenon known as writer's cramp. The signs of rigor mortis arise from the same kind of action (see "Rigor Mortis," opposite).

WHICH TWITCH?

Just about every skeletal muscle in the body is made up of two sorts of muscle fibers: fast-twitch and slow-twitch. The type of fiber determines what sort of work the muscle can do.

Fast-twitch fibers contract rapidly but tire quickly. For energy, they use ATP and glucose but need little oxygen, protein, or fat. The fast-twitch muscles come into play for sudden, rapid movements, like jumping, sprinting, or blinking an eye. It's also the

HERMANN VON HELMHOLTZ

Hermann von Helmholtz gives a lecture shortly before his death in 1894.

HERMANN VON HELMHOLTZ (1821–1894) was one of a handful of scientists who dared, in the first half of the 19th century, to challenge the vitalist belief that a force beyond observation and measurement invigorated the living body. The physics of the human body followed principles similar to the physics of the natural world, Helmholtz argued in his landmark 1847 paper, "On the Conservation of Force." In essence, Helmholtz was applying to the human body the Newtonian law of the conservation of energy: Energy is neither created nor destroyed, but transformed from one expression into another. In Helmholtz's words, "the quantity of force which can be brought into action in the whole of Nature is unchangeable." The muscles must follow the same principle, turning chemical force into mechanical work and heat. Helmholz's application of that principle to physiology represents an essential step in our understanding of muscles. ■

fast-twitch muscles that are exerted in anaerobic exercise such as weight lifting or isometrics.

Slow-twitch fibers contract slowly, and they can keep at it for a long time. They need plenty of oxygen, and they use up protein and fat, turning their chemical energy into mechanical energy and heat. In an endurance exercise, such as running, rowing, or swimming, the body calls on the slow-twitch muscle fibers to do most of the work.

A vivid example of the two sorts of muscle fibers can be seen in the musculature of a chicken. What we call white meat is primarily fast-twitch muscle fiber, found in the breasts and wings, which exert quick bursts of movement. What we call dark meat is primarily slow-twitch muscle fiber. Found in the legs, which perform slow, steady exertion, these muscles look darker because they are infused with more blood.

Everyone has a combination of slow-twitch and fast-twitch muscle fibers, but some bodies are genetically inclined toward one or the other type. When coaches advise athletes to cross-train, they are in essence encouraging them to exercise the type of muscle fiber that is less predominant in their body. A runner might be advised to lift weights, or a weight lifter might be advised to use a rowing machine, to build up an optimal combination of strong fast-twitch and slow-twitch muscle fibers.

TONING UP

Exercise does not increase the number of muscles a person has, but it does increase the size and capability of the muscles they do have. The bigger circumference a muscle has, the more tension it can develop, the more work it

When running, changing speeds and varying the pace can work both slow- and fast-twitch muscle fibers.

SEE ALSO: Chapter Thirteen, "Building Blocks," PAGE 352

KEEPING HEALTHY

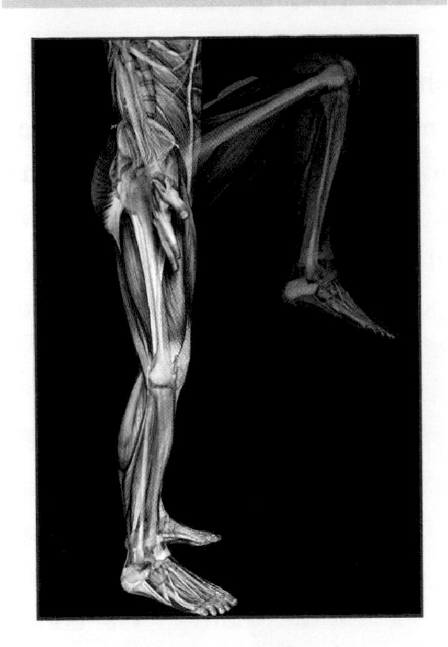

Leg lifts boost strength.

UTILIZING A VARIETY OF exercises can improve the muscles' performance. Aerobic, or endurance, exercise—slow, steady exertion such as walking, running, swimming, or dancing—boosts the capillaries bringing oxygen to muscle cells and the organelles that turn sugars into ATP and energy. The result is that muscles can work hard longer without fatigue. Muscle tone improves, too. Anaerobic, or resistance, exercise—such as weightlifting—bulks up muscle fiber and increases the connective tissue between muscle cells. It increases metabolic organelles, creates more myofibrils, and boosts the capacity of muscle fibers to store sugar. The result is bigger, more sharply defined muscles and increased strength. Most trainers suggest combining both types of exercise in a fitness plan. ■

can perform, and the longer it can sustain that work at a certain performance level. Bodybuilders are walking examples of the physical phenomenon called hypertrophy: muscle mass increased in size through concerted exercise. The opposite, atrophy, can occur as well. Muscles that are not used will shrink in size over time.

Exercise also increases the tone of muscles. We are all aware of the concept of muscle tone, which plays a large part in the modern emphasis on fitness and appearance. A body with good muscle tone holds firm, even when at ease. But how does muscle tone look at the cellular level?

Even when a muscle is relaxed, swift, tiny, involuntary contractions take place all the time. Called tonic contraction, this is a state of being we often take for granted. The neck holds the head up straight, for example, thanks to the tone of its muscles. Good posture, standing straight and tall, is accomplished thanks to muscle tone. Smooth muscles as well as skeletal muscles experience variations in tone. When digestive organs and blood vessel walls have good muscle tone, they are able to work more efficiently.

Regular exercise also increases the ability of cells to receive the oxygen that they need to accomplish work. During aerobic exercise, the constant call for oxygen by muscle cells can exceed the amount available. When a runner gets winded and gasps for air, those gasps are the body's response to its oxygen debt and its effort to bring in more oxygen to send to depleted muscle cells.

Most people who are beginning a new exercise routine have at one time or another experienced sore muscles the next day, a sure sign that the exercise-engaged muscles had not been used so much before. For a long time, muscle soreness after exercise was blamed on a build-up of lactic acid, a chemical that is released as a by-product of fast-twitch muscle activity. Excess lactic acid dissipates in less than 24 hours, however, so it cannot explain muscle soreness that continues for a day or two. Researchers now believe that delayed-onset muscle soreness, or DOMS, is caused by small-scale tearing or swelling that happens to muscles that have been overstressed. Gentle warm-up stretches prepare muscle fibers for the work ahead and can help avoid later soreness.

CRAMPS

WHEN A SINGLE MUSCLE within a group contracts suddenly and involuntarily, it's called a spasm. Tics of the eye, for example, are spasms. When a muscle spasm causes pain, it's called a cramp. Cramps can occur after exercise or while a person is sleeping. Their cause is not clearly known. The best response is to rehydrate, relax, and massage the tightened muscles.

ART OF THE MUSCLE

IT WAS THE ARTISTS OF THE Italian Renaissance, even more than the physicians of their day, who created an entirely new vision of the human body. They did so through the art of anatomical illustration.

THE FIRST MASTERS

Fifteenth-century Florentine artist Antonio Pollaiuolo (ca 1431–1498) is credited with having been the first artist of his generation to study the anatomy of the human body in the flesh, by actually dissecting cadavers and drawing the configurations of muscles and bones as he saw them revealed beneath his hands. Pollaiuolo was, wrote 16th-century artist and art historian Giorgio Vasari, the "first master to skin many human bodies in order to investigate the muscles and understand the nude in a more modern way."

By the next generation, Italian artists had come to consider the study of real human bodies essential to their work. Leonardo da Vinci (1452–1519) and Michelangelo (1475–1564) both performed dissections and created anatomical sketches that are works of art in themselves. Michelangelo's anatomical studies also clearly inform his sculptures, perhaps most magnificently in his famous statue of the biblical King David.

Leonardo, arguably as much a scientist, engineer, and inventor as he was an artist, spent years studying the human body, dissecting as many as 30 cadavers through the course of his career and producing hundreds of anatomical drawings. He began with detailed drawings of the human skull, drafting them with an architect's eye for elevation and perspective. From those, he went on to create detailed renderings of many different bones, muscles, and organs. His drawing of a fetus inside the womb is famous not only for its grace and delicacy of line but also for its accuracy. Although none of Leonardo's anatomical drawings were published in his lifetime, they were probably seen by some contemporaries.

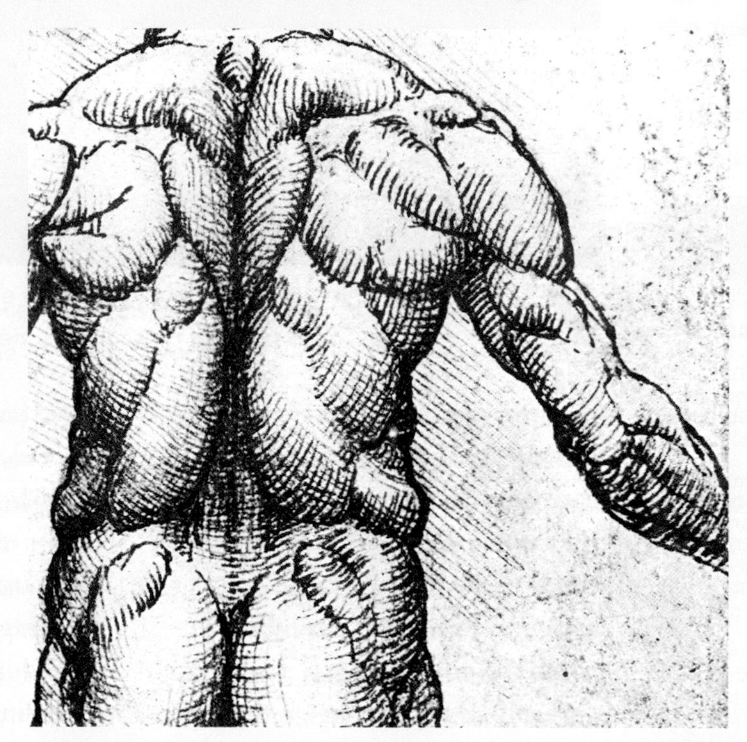

Leonardo da Vinci produced this drawing of the well-muscled upper back and arms of the human male.

BOOKS OF THE BODY

The first comprehensive, accurately illustrated volume on anatomy, *De Humani Corporis Fabrica (On the Structure of the Human Body),* was published in Basel, Switzerland, in 1543 by Andreas Vesalius (1514–1564). While the art of his volume is magnificent—Calcar, a student of the great artist Titian, may have created the images—more revolutionary for the time was Vesalius's reliance on observations of the human body instead of reference to respected ancient texts.

A lecturer at the renowned University of Padua, Vesalius performed dissections for his classes. Often, he found, the details of a dissected body contradicted statements made by the great Greek physician Galen (ca 129–216), on whose teachings—based on animal dissection—Renaissance medicine depended. Consequently, Vesalius's book outraged the Galenists and fascinated others.

Another illustrated anatomy was published in 1543: Giovanni Battista Canano's *Musculorum Humani Corporis Picturata Dissectio (Illustrated Dissection of the Muscles of the Human Body).* Like Vesalius, Canano (1515–1579) was a lecturer at the medical school in Padua and based the illustrations in his book on observations of actual bodies. His book was not as broadly distributed or as influential as Vesalius's, but it featured some elements of anatomy that his contemporary did not study. Canano represented the complicated bones and muscles of the hand, for example, and explored how valves within the circulatory system regulate the direction of blood flow through the body. His illustrations were also the first to show the action of specific muscles.

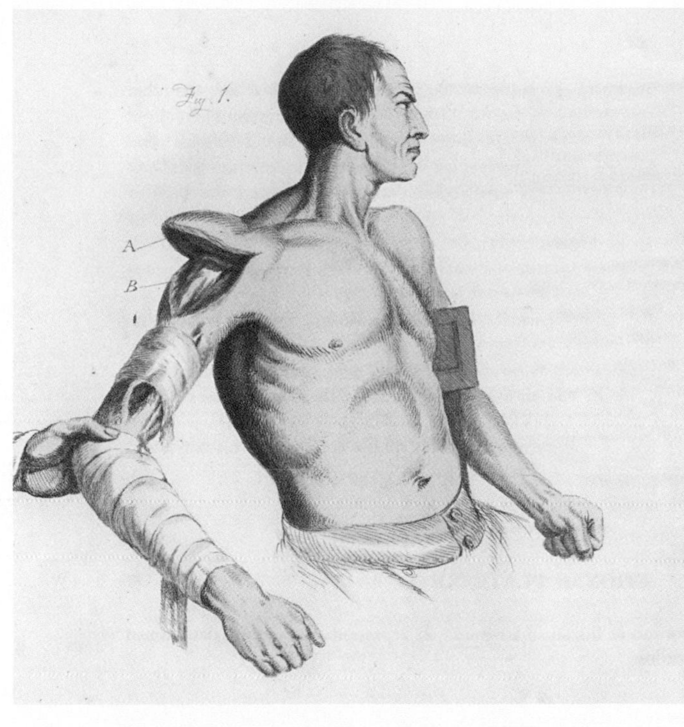

Charles Bell's instructive yet elegant illustrations show the great partnership between medicine and art.

ART OR SCIENCE?

Into the early 18th century, it was hard to distinguish between anatomy as art and anatomy as science. German anatomist Bernhard Siegfried Albinus (1697–1770), for example, mingled art and science in his volume of plates depicting the bones and muscles of the human body, published in 1747. Albinus drew full anatomical figures in exquisitely landscaped scenes. He considered that such scenery helped fill in the empty spaces of the pages and made the figures appear more three-dimensional. For him, the art was as important as the science, but with the coming of the Enlightenment, later 18th- and early 19th-century anatomists vehemently disagreed. Scottish anatomist John Bell (1763–1820), for example, bemoaned the "vitious practice of drawing from imagination" and "the subjection of true anatomical drawing to the capricious interference of the artist." His unflinchingly clinical drawings contrasted sharply with the ornamental illustrations of Albinus.

BELLS' INFLUENCE

John Bell often collaborated with his younger brother, Charles, also a physician and anatomist. Between 1797 and 1804, the two brothers published the extensively illustrated *Anatomy of the Human Body,* the illustrated *Anatomy of the Brain,* and *A Series of Engravings Explaining the Course of the Nerves.*

Charles Bell (1774–1842) outlived his older brother and continued to practice medicine into the 1830s. In 1815 he served as a surgeon on the battlefield at Waterloo, the scene of the great British victory over Napoleon Bonaparte, and then codified the lessons learned there in paintings and lectures on the effects of gunshot wounds. His tireless studies of the human face are commemorated in medical terminology today. He identified the long thoracic nerve (called Bell's nerve), which assists in breathing and stabilizes the scapula against the chest wall. Bell's palsy, a facial nerve disorder, is also named after him.

Fascinated by the way that bones, nerves, and muscles combine to create facial expressions, Charles Bell took his study full circle. In 1806 he published *Essays on the Anatomy of Expression in Painting,* a classic in the study of art history. Just as artists created the earliest anatomies, Bell shared his knowledge as a surgeon and anatomist on great works of art history.

CHAPTER GLOSSARY

ADENOSINETRIPHOSPHATE (ATP). Molecule that provides energy to living cells, essential to transforming chemical energy from food into mechanical energy and heat.

AGONIST. A muscle that drives a particular movement.

ANTAGONIST. A muscle that reverses or opposes the action of another muscle.

ATROPHY. Decrease in muscle bulk because of lack of exercise or loss of its motor nerve.

CARDIAC MUSCLE. Cells that form the wall of the heart and that contract and release rhythmically, controlled by the autonomic nervous system and often indwelling pacemaker.

CARTILAGE. White fibrous connective tissue.

CONTRACTION. Movement where filaments inside muscle cells slide over one another, making muscles denser and shorter.

CREATINE PHOSPHATE. A molecule found only in muscle tissue that helps in ATP production.

FASCICLE. Bundle of muscle fibers.

FAST-TWITCH MUSCLE FIBERS. The muscle cells responsible for quick bursts of energy.

HYPERTROPHY. Increase in muscle bulk by means of exercise.

INVOLUNTARY. Action regulated by the nervous system, but without deliberate control.

MYOFIBRIL. The organelle inside a muscle cell where filaments slide over each other to create a muscle contraction.

MUSCLE TONE. Minute, involuntary contractions in a muscle group, experienced as muscle firmness and essential to optimal health.

NEUROMUSCULAR JUNCTION. In muscle cells, the connection point between a neuron and muscle fiber.

PERISTALSIS. Rhythmic and successive muscular contractions in hollow muscular structures such as blood vessels or intestines.

RELAXATION. Movement where filaments inside muscle cells slide away from each other, making muscles longer and looser.

RIGOR MORTIS. Muscle stiffness that occurs in tissue 48 to 60 hours after death, caused when muscle cells deprived of ATP remain in a contracted position.

SARCOLEMMA. The outer membrane of a muscle fiber.

SKELETAL MUSCLE. Cells composed of striated fibers, designed to contract and release, moves voluntarily, connected primarily to bones by tendons and served by motor neurons.

SLOW-TWITCH MUSCLE FIBERS. The muscle cells responsible for steady endurance energy.

SMOOTH MUSCLE. Found primarily in vessels and hollow organs and served by the autonomic nervous system. Made of nonstriated fibers.

STRIATION. Dark bands visible in some muscle cells.

TENDON. White fibrous cord attaching muscle to bone.

middle, and the load is between the two gripping ends.

In the human body, lifting a weight with the forearm pressed against the side of the body represents a third-class lever system. The fulcrum (the elbow) sits opposite the load (the weight). In between, the biceps muscle in the upper arm exerts force against the fixed point. The load cantilevers out, and the muscle lifts it against the fixed point of the elbow.

FACIAL MUSCLES

Something as simple as a facial expression is also the product of muscular movements. A broad smile, a furrowed brow, a curled lip, a wrinkled nose—we tend to think of such facial expressions as messages of emotion. They can also be described, strictly speaking, as the combination of movements of the up to 50 bilaterally symmetrical muscles that operate just under the skin of the face.

Facial muscles are skeletal muscles, like those of the trunk and

SMILE OR FROWN

THERE'S AN old contention that it takes more muscles to frown than to smile. It's often used to encourage optimism—but is it true? A recent calculation determined that it takes 12 muscles to smile and 11 to frown. Most people smile more than they frown, however, hence the muscles for smiling are probably more fit—and operate with less effort.

limbs. Unlike the other skeletal muscles of the body, however, they do not work in opposing pairs. They stretch in a pattern of layered sheets around the eyes, nose, and mouth, and across, between, and around the bones, fat, and connective tissue of the head. Facial muscles insert into the subcutaneous tissue beneath the skin or into other muscles. Contraction and relaxation of facial muscles, which are innervated by the facial nerve, produce the myriad expressions that enliven the faces of human beings around the world.

The expressive facial muscles can be divided into groups, depending on what part of the face they serve. Over and around the scalp stretches the big occipitofrontalis muscle, divided into the frontal and occipital bellies, the scalp muscles in the front and rear of the skull.

All around and at the edges and corners of the mouth are muscles that push and pull on the lips. They not only create expressions, but also play a critical role in speech by forming the different shapes the mouth must make when forming different sounds of speech.

One large circular muscle, the orbicularis oris (from "circling the mouth"), surrounds the mouth opening. A pair of muscles, the zygomaticus major and minor, draws the corners and the upper lip up when a person smiles or bares the front teeth. The risorius muscle—its name comes from the Latin for "laugh"—attaches to the outer edges of the mouth, pulling at those corners when someone is literally feeling down at the mouth. This muscle also is ued to produce an insincere smile whose expression does not involve the eyes.

FRUSTRATION
Wrinkling the forehead in disgust is controlled by the corrugator supercilii muscles.

LAUGHTER
Recent calculations show that it takes about 12 facial muscles to crack a smile.

FROWN
Frowns are shaped by the risorius muscle, which pulls down the corners of the mouth.

An equally complex grouping of muscles works together around the eyes. Here it's the orbicularis oculi ("circling the eye") muscles that surround the orbit of the eye. Contractions of these muscles close the eyes. Above the eye, connecting the frontal bone of the forehead to the eyebrow skin, are the levator palpebrae ("eyebrow lifter") muscles. When they contract, the eyes open. Eyebrow and forehead gestures originate from the frontalis and the xorrugator supercilii ("wrinkle of the eyebrow") muscles, which angle from above the eye toward the bridge of the nose.

While the jaw muscles participate to some extent in facial expressions, their primary job is to move the lower jaw (or mandible) while chewing. They are innervated by the trigeminal nerve. Three of the four pairs of muscles—the masseter, temporalis, and medial pterygoid—operate on a vertical plane, opening and closing, hinged on the temporo-mandibular joint, also called TMJ. A fourth pair, the lateral pterygoid muscles, contracts with slight horizontal movements and works with other muscles to move the jawbone from side to side during mastications. Grouped together, these four pairs are called the masticatory muscles.

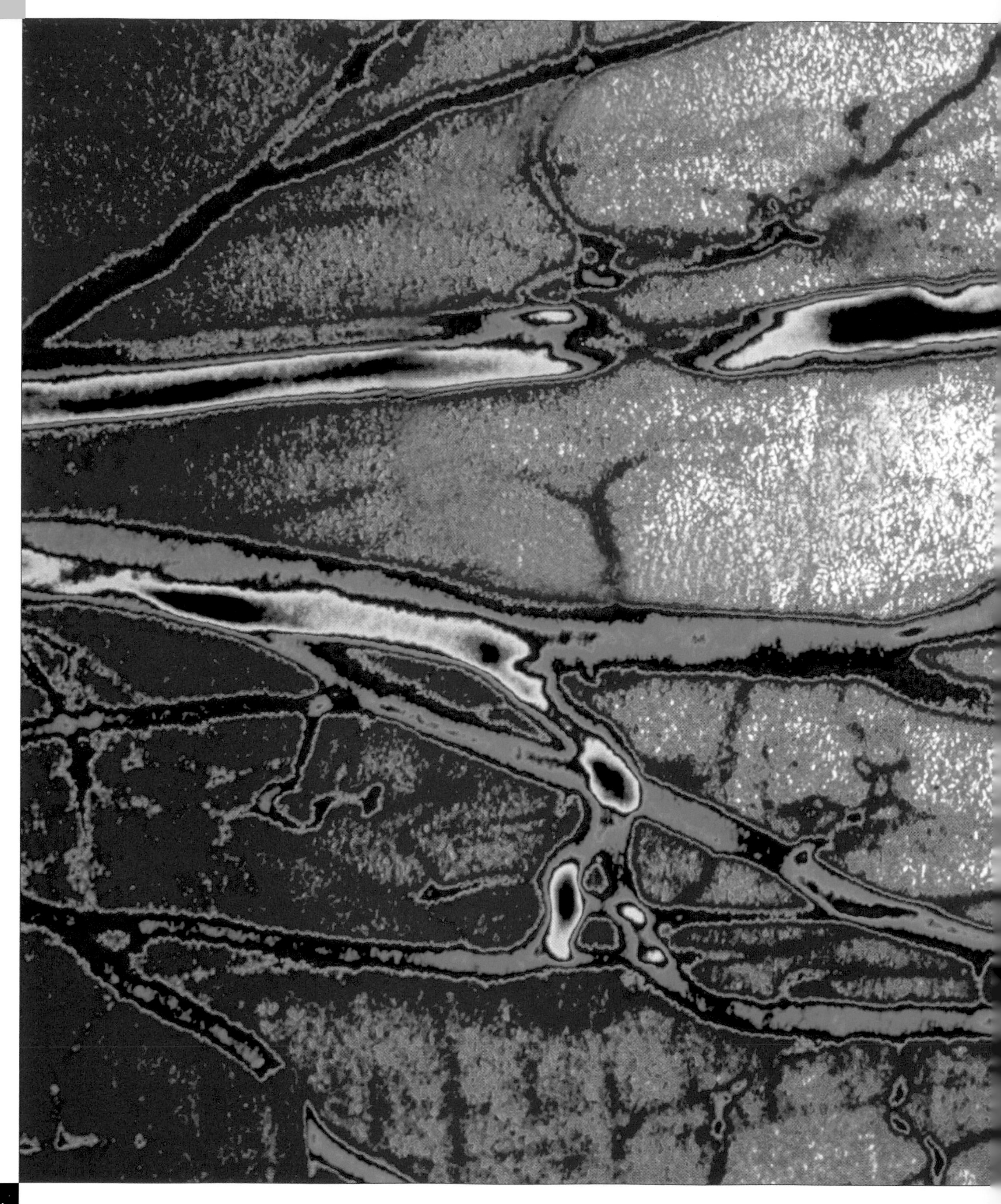

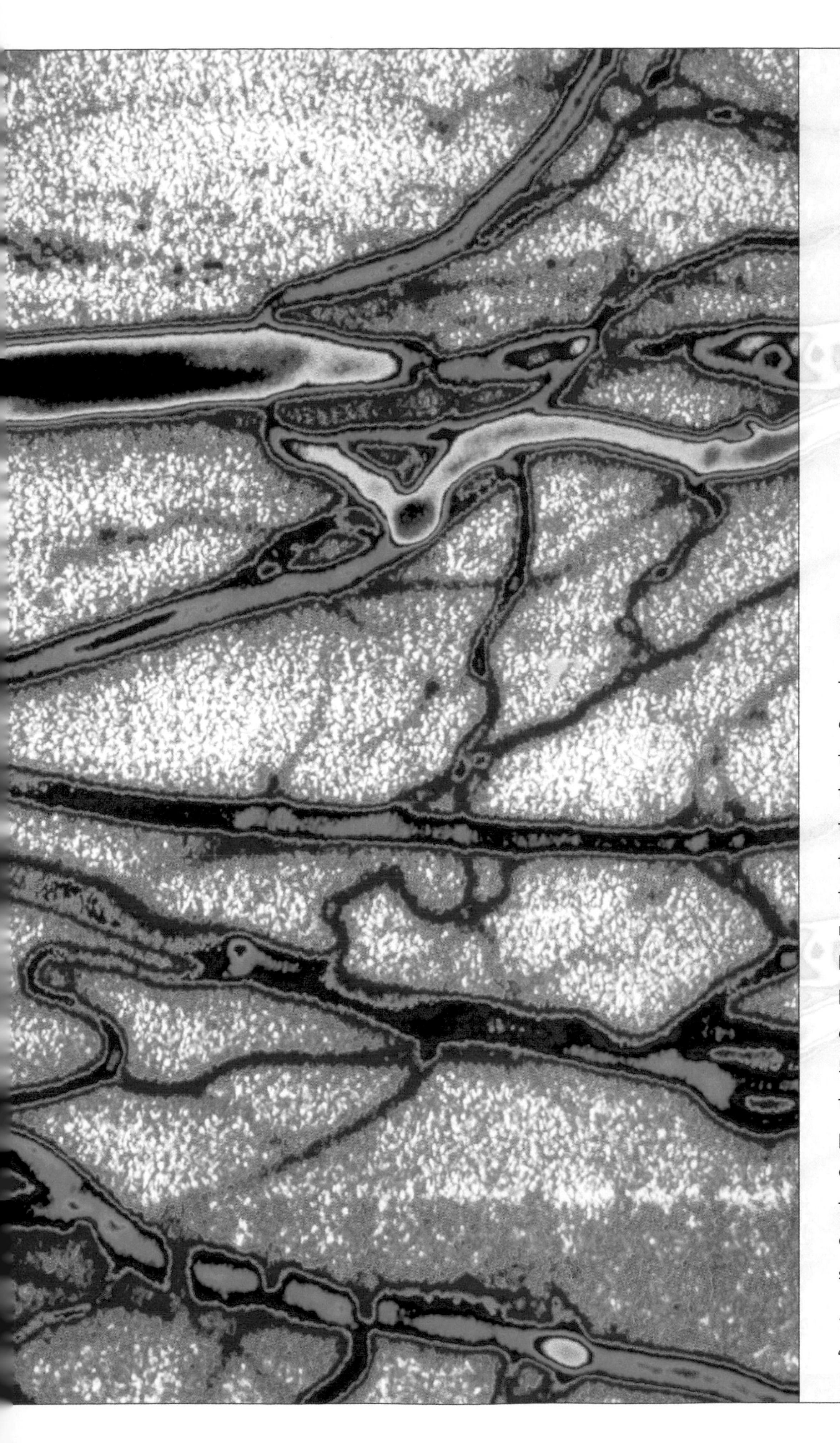

CHAPTER FIVE

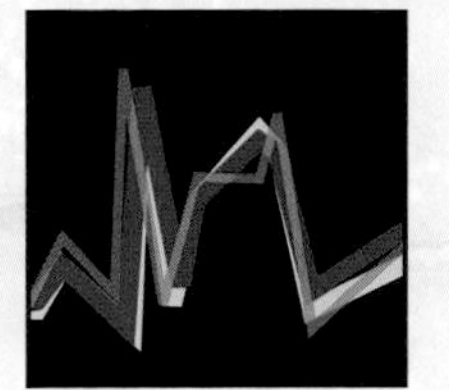

CIRCULATION

THE CIRCULATORY SYSTEM, or cardiovascular system, is the first system to develop. It nurtures and protects every cell in the body by pumping blood through a network of vessels that, laid end to end, would measure 60,000 miles (96,500 km). About once a minute, the beating heart propels the body's entire blood supply, about 5 quarts (5 L) in an adult male, through every tissue, where it performs countless vital chemical transfers before returning to the heart. The steady pulse of blood through the body is synonymous with life.

The branching structure of veins is apparent in this false-color image.

THE SYSTEM

MOVING BLOOD THROUGHOUT THE BODY

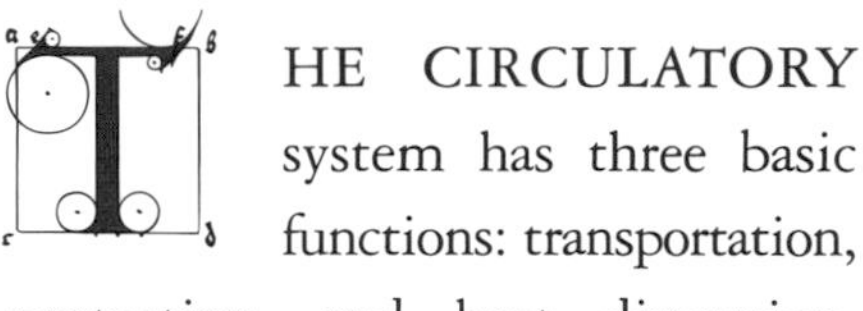

THE CIRCULATORY system has three basic functions: transportation, protection, and heat dispersion. Blood carries oxygen, nutrients, and hormones to all the body's tissues and totes away carbon dioxide and other waste products. It also contains a wide variety of immune-system cells that fight off viruses, bacteria, and other invaders and foreign substances. Platelets and clotting agents in the bloodstream speed to the site of injuries and seal them off in minutes. And blood disperses heat around the body, a lesser-known but still important job, helping to cool the brain and hardworking organs in the chest and abdomen and keeping the body from overheating during exercise.

CIRCULATING PARTS

The three main components of the cardiovascular system are the heart, blood, and blood vessels. The heart is the central engine of circulation. A fist-size, roughly cone-shaped lump of muscle, it nestles between the lungs in the middle to left part of the chest. Guided by electrical signals generated within its own muscle cells, it begins beating at four weeks after conception and does not stop until death. Its four chambers consist of two small atria on the top and two larger ventricles on the bottom. The left atrium collects oxygenated blood from the lungs and pumps it into the left ventricle, which propels it through the body. The right atrium collects oxygen-poor blood returning from the body and sends it into the right ventricle, which pumps it over to the lungs to exchange carbon dioxide gas for oxygen. The blood then returns to the left side of the heart to start the process all over again.

Connected to the heart is a closed network of blood vessels—actually, two networks. The left side of the heart pumps blood to all parts of the body except the lungs, while the right side supplies the lungs only. The largest blood vessels are the arteries—big, elastic tubes that carry blood away from the heart. Arteries branch out into smaller channels, called arterioles, which lead into the fine webbing of tiny blood vessels called capillaries that are the site of oxygen and nutrient exchange with the tissues. Capillaries, their job accomplished, merge into tiny veins, called venules, which connect to larger veins to carry blood back to the heart.

The blood that makes its regular, pulsing way through the body consists of a fluid holding a variety of cells. The fluid is called plasma; it's a nonliving, straw-colored liquid made mostly of water but with some nutrients, hormones, gases, and other substances dissolved in it. Plasma makes up almost 55 percent of the bloodstream. Suspended in it are the "formed elements" of the blood: red blood cells, making up almost 45 percent of the blood, and a potpourri of white blood cells and platelets, constituting less than 1 percent (but essential for life nonetheless).

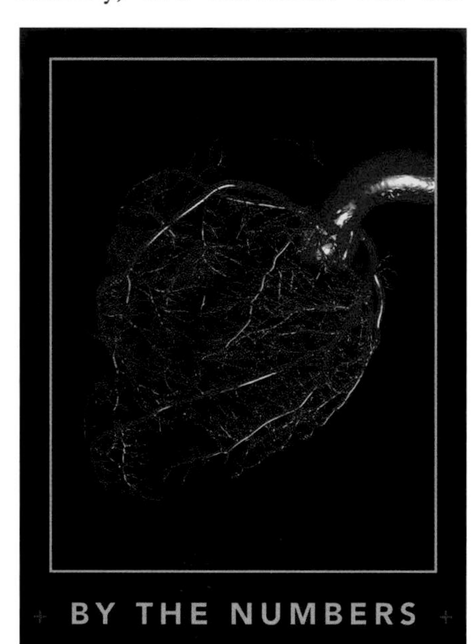

+ BY THE NUMBERS +

+ A NEWBORN BABY has about 1 cup (0.25 l) of blood in circulation.
+ All the blood vessels in an adult body, strung together, could circle the Earth two and a half times.
+ In one day, blood travels 12,000 miles (19,000 km).
+ After it leaves the bone marrow, a red blood cell makes approximately 250,000 trips around the circulatory system before dying.

MOVING THE BLOOD

Every living cell in every organ in the body must be fed by the bloodstream or die. Not all organs need

WEBLINK: Take an interactive look at Circulation at http://nationalgeographic.com/humanbody

the same amount of blood, however, and their requirements also vary depending on what the body is doing: exercising, sleeping, digesting, and so forth. The structure and feedback mechanisms of the circulatory system neatly manage these differing needs.

The total volume of blood in the body makes up about 7 percent of the body's mass, which comes to about 4 to 5 quarts (4 to 5 L) in most adults. It is in constant motion; the heart pumps all the blood through all the tissues and to and from the lungs in the course of one minute, beating, on average, 75 times.

The amount of blood expelled by each ventricle of the heart in a minute is called the cardiac output, and it is the product of the heart rate (how often the heart beats in one minute) and stroke volume (how much blood a ventricle pushes out with each beat). Typically, the ventricles pump out about half the blood they contain, or roughly 2 ounces (70 mL), with each beat.

Athletes, with bigger, more muscular hearts, may have lower heart rates but a larger stroke volume, so that their cardiac output is the same or greater than the average person's. Cardiac output changes on demand. During exercise, the heart speeds up and delivers much more blood to the needy muscles: as much as seven times the resting rate, or 32 quarts (35 L) a minute.

Arteries and Veins

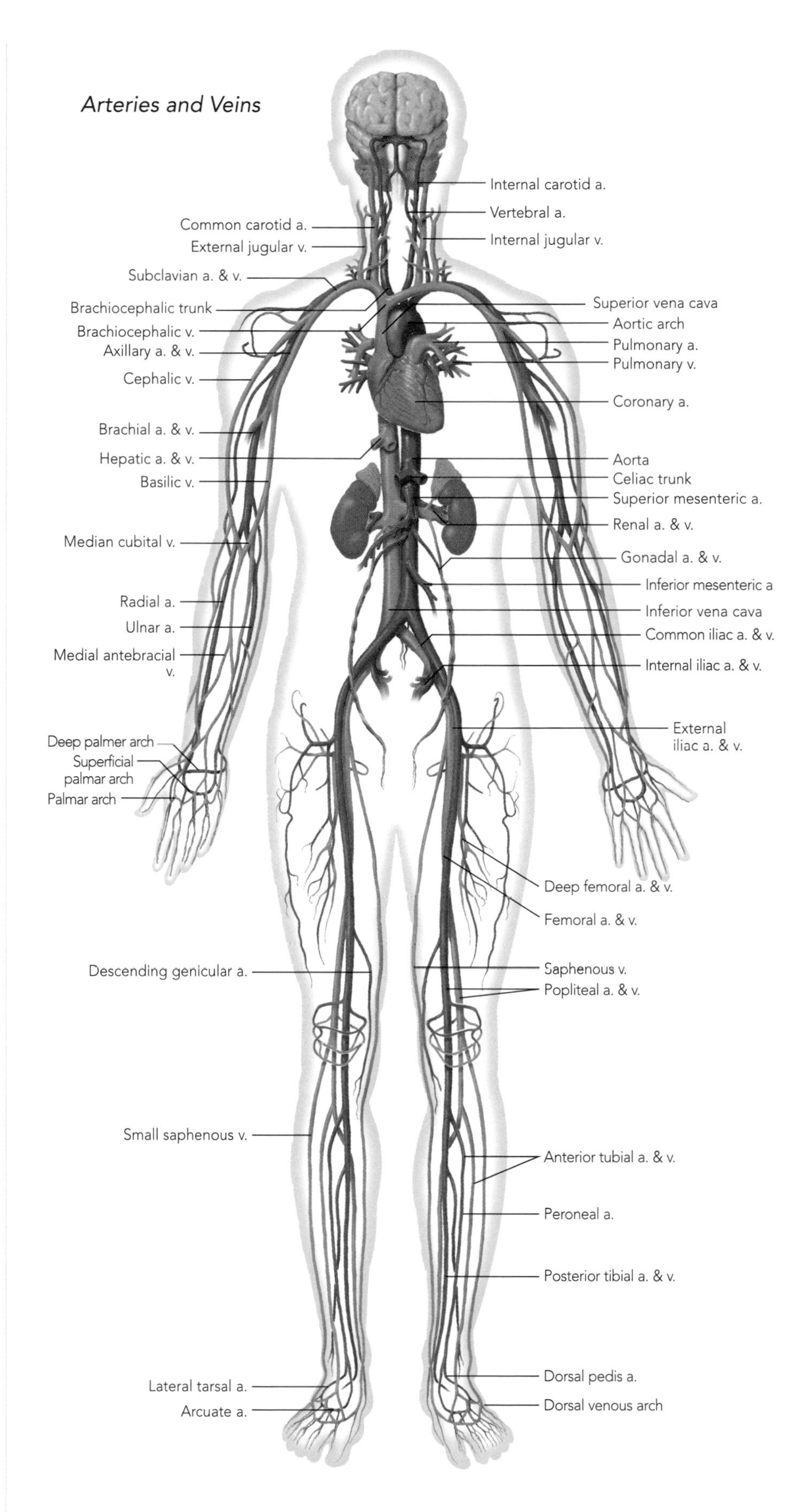

Artery is abbreviated *a*. *Vein* is abbreviated *v*.

Blood flowing through arteries and veins meets resistance, just like any fluid flowing through a tube. The amount of force needed to overcome the resistance is called the blood pressure. Blood pressure is highest in the aorta, the big artery leading out of the heart—blood from the aorta would shoot six feet into the air if the vessel were opened up when the ventricle contracts. Pressure is lowest where the blood returns to the right atrium of the heart after traveling through the body.

The highest pressure in an artery, just after the heart beats, is called the systolic pressure, which measures about 120 mmHg in healthy adults. (The unit of measure stands for millimeters, mm, of mercury, Hg being the symbol for that element, and is derived from the earliest instruments used to quantify blood pressure.) By the time blood reaches the smallest capillaries, its pressure has dropped to less than 20 mmHg.

On the return trip through the veins to the heart, the blood maintains a fairly low and unvarying pressure. Because veins don't have to withstand high pressures—in fact, they are usually only partially filled with blood—they have thinner, less muscular walls than arteries.

Blood flowing back to the heart meets little resistance, but it has to fight gravity as it returns from the arms and legs in an upright body. That's why veins in the limbs have valves: to prevent the blood from flowing back down the limb after each pulse forward. When muscles in the legs contract, they act as additional outside pumps, pressing on the veins and propelling the blood upward.

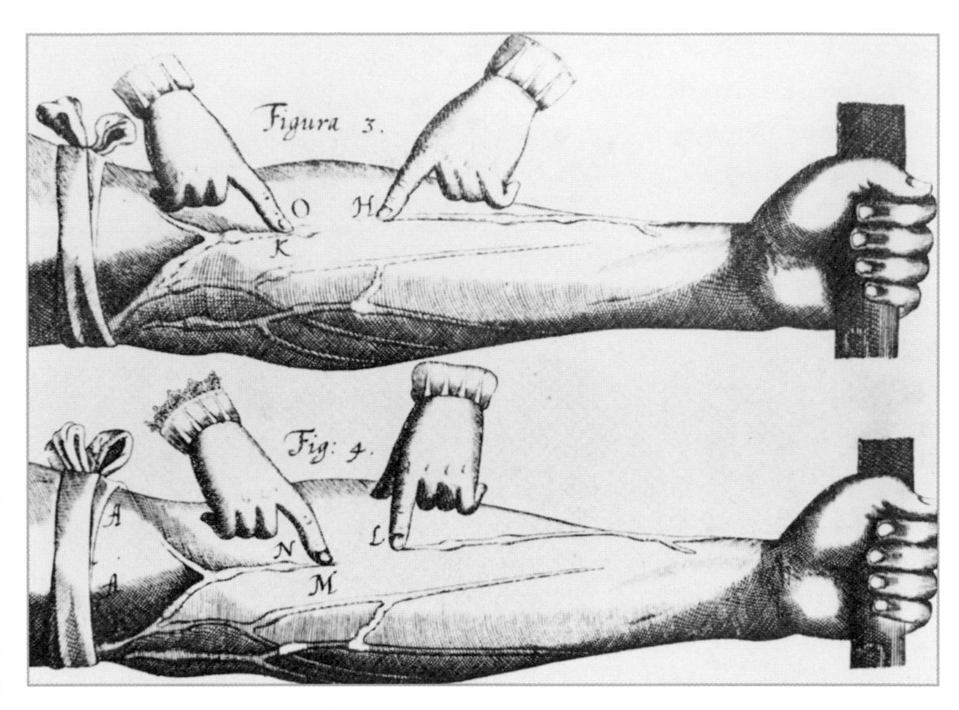

An illustration from William Harvey's book on circulation shows the way valves control the flow of blood in veins.

BLOOD SPEED

+ THE HEART PUMPS oxygenated blood through the aorta (the largest artery) at a speed of about 1 mile (1.6 km) per hour.
+ By the time it gets to the tiny capillaries, blood creeps along at about 43 inches (109 cm) per hour.

BLOOD DEMANDS

Different parts of the body have different demands for blood. At rest, the abdominal organs receive about 24 percent of the blood supply; the kidneys get 20 percent, the skeletal muscles about 20 percent, and the brain 13 percent.

As well as pumping blood to the rest of the body, the heart tissue itself needs blood—quite a lot of it—to support its ceaseless activity. Two coronary arteries arise from the base of the aorta and encircle the heart. Blood from the aorta shoots immediately into them, flowing into smaller arteries and quickly reaching capillary beds that supply the heart muscle. The depleted blood is collected by coronary veins to empty into the heart's right atrium. This blood supply pulses through the coronary arteries only when the heart is relaxed; when the heart contracts, the arteries are too compressed to carry blood. The heart makes up only 0.5 percent of the body's weight but needs 5 percent of the body's blood.

SEE ALSO: Chapter Seven, "Absorption," PAGE 194

Two organs, the lungs and the liver, have their own special circulation design. The lungs receive deoxygenated blood from the right side of the heart via the pulmonary artery. After the blood flows through the lungs' capillary beds, giving up carbon dioxide and picking up fresh oxygen, it returns to the left side of the heart via the pulmonary veins.

The liver receives about 25 percent of its blood from the hepatic artery, part of the regular systemic circulation flowing from the aorta. The other 75 percent arrives via the portal vein, which carries blood from the intestinal tract and spleen. This nutrient-rich blood is modified and detoxified by the liver before being passed on to the rest of the body.

Short-term stresses can redirect the blood supply to the tissues that need it most; during exercise, for instance, more blood flows to the skeletal muscles and skin and less to the digestive organs. The body manages this through a complex feedback system among the blood vessels and nervous system. Little receptors in the blood vessels monitor how much the vessels are stretching and the chemical composition of the blood; if a tissue needs more blood (and the oxygen it contains), these receptors send impulses through the nervous system that trigger increased respiration and dilate the blood vessels, bringing more blood and oxygen into that area.

DISCOVERING CIRCULATION

Pioneering British doctor William Harvey looks over a gentlemanly patient.

UNTIL THE RENAISSANCE, FEW people disputed the teachings of the Greek physician Galen (ca A.D. 129–216) about the heart and blood. There were two types of blood, made by the liver and taking separate courses, he taught. Dark, venous blood constantly made from digested food was sent to the tissues to be absorbed. Bright-red arterial blood joined "pneuma," a spiritual essence the lungs inhaled to spread vitality. Blood moved by innate heat and a pulsing of the vessels.

But William Harvey (1578–1657) changed all that. Harvey attended Cambridge and the university in Padua, where Galen's ideas were being challenged. Back in England, Harvey conducted experiments that convinced him Galen was wrong. Observing the beating heart in animals, he saw it as a pump, expelling blood with each contraction. Measurements of the blood leaving each ventricle made it clear the liver could not make so much so fast. In 1628, Harvey published his findings: "Since all things . . . show that the blood passes through the lungs and heart by the force of the ventricles, and is sent . . . to all parts of the body . . . it is absolutely necessary to conclude that the blood . . . is impelled in a circle, and is in a state of ceaseless motion." A contemporary reported, "[A]fter his Books of the Circulation of the Blood came out . . . he fell mightily in his Practize." But Harvey lived to see his theory accepted. Today it is considered one of the greatest contributions to medicine. ■

The normal amount of blood in the body is more than enough to keep all the organs perking along. When the heart can't deliver enough blood, however, a person can quickly go into shock, a condition characterized by falling blood pressure, pale and clammy skin, and a rapid and thready pulse.

A number of crises can reduce blood flow: the blood loss of a traumatic accident, a heart attack, an excessive loss of fluid from diabetes or vomiting, or even an allergic reaction that causes the blood vessels to dilate, causing the blood pressure to fall.

Without enough blood to deliver oxygen to organs throughout the body, cells and tissues can become damaged and begin to die. In these situations, the body reacts, when it can, by increasing heart rate, dilating or constricting various blood vessels, and retaining water—but these responses can't compensate completely for a blood loss of more than 10 to 20 percent. A person in shock needs emergency medical care.

Curling like the tendrils of a tropical plant, abdominal arteries branch off the central abdominal aorta in this colored arteriograph.

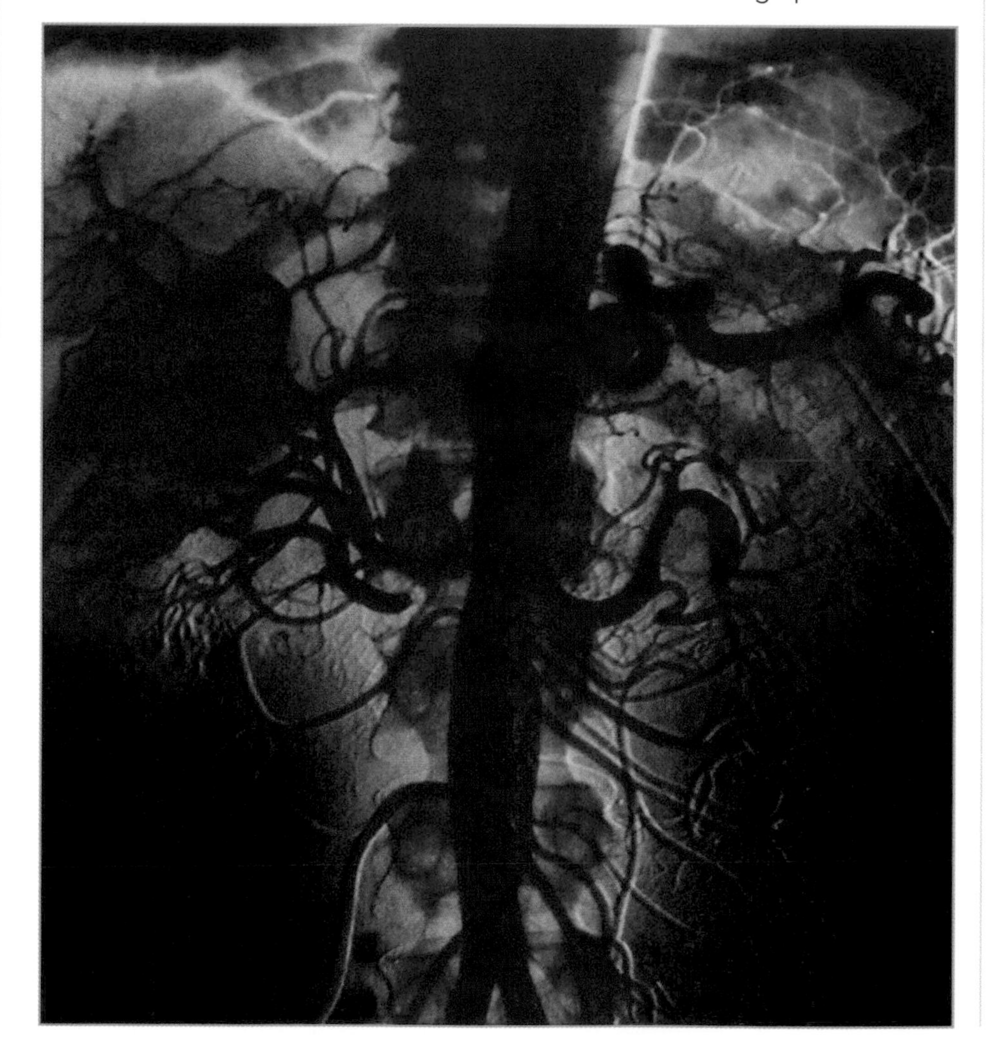

KEEPING HEALTHY

Compression socks can prevent thrombosis.

BEDRIDDEN PATIENTS AND airplane passengers on long trips may share an unfortunate side-effect of immobility: deep-vein thrombosis. The condition is caused by blood flowing slowly or pooling in the deep veins in the lower legs or thighs of an unmoving person. The blood may not be able to wash away clotting factors, and a blood clot (thrombus) can result, blocking circulation. The limb becomes tender, swollen, hot, and red. The clot may break away and travel to the brain—causing a stroke—or to the lungs, heart, or other areas. Doctors may prescribe anticoagulants to the bedridden; people prone to the condition can wear compression stockings to prevent pooling. Plane or car passengers on a long trip should take periodic breaks to stand up and walk about, lest they suffer "economy-class syndrome." Another option is to do leg stretches to keep blood moving. ■

CHAPTER GLOSSARY

ANASTOMOSIS. An end-to-end joining of blood vessels (or lymphatic vessels or nerves).

ANGIOPLASTY. A technique for opening space within a clogged artery usually by temporarily inflating a balloon inside the artery.

ARTERIOLE. A tiny, muscular artery that carries blood to capillaries.

ARTERIOSCLEROSIS. A group of diseases characterized by stiff, thickened arterial walls.

ARTERY. Blood vessel that carries blood away from the heart.

ATHEROSCLEROSIS. A disease characterized by the buildup of plaque on arterial walls.

ATRIOVENTRICULAR (AV) NODE. A specialized mass of electrically conducting cells located between the heart's atria and ventricles.

ATRIOVENTRICULAR VALVES. Valves between the atria and the ventricles.

ATRIUM (pl. atria). Either one of the heart's two upper chambers.

AUTORHYTHMIC FIBERS. Muscle cells in the heart that produce electrical signals without outside stimulus.

CAPILLARY. A microscopic blood vessel that connects arterioles to venules; the site for interchange between blood and tissues.

CARDIAC CYCLE. The events of one complete heartbeat.

CARDIAC OUTPUT. Blood pumped by a ventricle in one minute.

DIASTOLE. The relaxation phase of the cardiac cycle.

ECHOCARDIOGRAM. An image of the heart's structure and function produced by ultrasound.

ELECTROCARDIOGRAM. A recording of the heart's electrical activity.

EMBOLISM. A blood clot that forms in a blood vessel and travels to another part of the body.

ENDOCARDIUM. The smooth, innermost layer of the heart wall.

ENDOTHELIUM. The inner lining of many body structures, including the heart and blood vessels.

EPICARDIUM. The thin membrane covering the heart.

ERYTHROCYTE. Red blood cell.

FIBRILLATION. Rapid, uncoordinated contractions of heart fibers.

FIBRIN. An insoluble protein formed during blood clotting.

FIBRINOGEN. A clotting factor in plasma; converted to fibrin.

HEMOGLOBIN. A substance in red blood cells that transports oxygen.

HYPERTENSION. High blood pressure.

LEUKOCYTE. White blood cell.

MITRAL VALVE. The valve between the left atrium and left ventricle.

MYOCARDIAL INFARCTION. A heart attack consisting of damage to or death of an area of heart muscle due to inadequate blood supply.

MYOCARDIUM. Muscular wall of the heart.

PERICARDIUM. Double-layered membrane that encloses the heart.

PLAQUE. Deposits of accumulated substances, such as cholesterol, fats, and other materials, on the inner lining of the artery wall.

PLASMA. The fluid component of the blood.

PLASMIN. An enzyme in blood plasma that dissolves the fibrin in blood clots.

PLATELETS. Fragments of cells, found in blood, that aid in clotting.

PROTHROMBIN. An inactive blood-clotting factor that is produced by the liver. It can be changed into thrombin.

SEMILUNAR VALVES. The valves between the ventricles and the aorta or pulmonary trunk.

SEPTUM. The wall dividing the left and right chambers of the heart.

SINOATRIAL (SA) NODE. A specialized mass of cardiac muscle cells in the right atrium; generates the electrical current that causes the heart to contract.

STENOSIS. The narrowing of a blood vessel or heart valve.

SYSTOLE. The contraction phase of the cardiac cycle.

THROMBIN. Enzyme that induces clotting by converting fibrinogen to fibrin.

THROMBUS. Blood clot.

TRICUSPID VALVE. Valve between the right atrium and right ventricle.

VEIN. Blood vessel that carries blood from the body to the heart.

VENTRICLE. One of the two lower chambers of the heart.

VENULE Small vein that leads from the capillaries to the larger veins.

THE BLOOD

MANY COMPONENTS MANY TASKS

ALTHOUGH IT IS A fluid, blood is in fact a tissue—a connective tissue whose individual cells float suspended in liquid plasma. The cells perform a variety of critically important jobs in the body, and blood's mix of cells and chemicals is a key indicator of the body's overall health.

BLOOD COMPONENTS

Blood plasma, a pale yellow liquid, makes up the largest part of blood by volume, about 55 percent. Plasma itself is more than 91 percent water. The rest of it consists mainly of proteins produced by the liver that help to maintain the proper levels of water in the bloodstream and help blood to clot when necessary. Plasma also carries small amounts of nutrients picked up from the digestive tract, electrolytes such as salt and potassium, respiratory gases such as oxygen and carbon dioxide, and waste products including urea and ammonia. It transports heat as well, keeping the body at a steady overall temperature.

Floating in the plasma are the formed elements of the blood, the blood cells, making up 45 percent of the bloodstream. These come in three types: red blood cells, or erythrocytes; white blood cells, or leukocytes; and platelets, which are cell fragments.

Red blood cells dominate the bloodstream, making up more than 99 percent of its formed elements. These cells have one job, and one job only: to carry respiratory gases to and from the body's tissues. Hemoglobin molecules within the cells contain a red pigment that gives blood its color. Platelets, the next most common formed element, are merely cell fragments, but they have the important task of repairing rips in blood vessels and promoting blood clotting. The five varieties of white blood cells account for only a tiny percentage of the blood's cells, but they play a protective role in fighting off viruses, bacteria, toxins, and other foreign invaders.

BLOOD COUNT

COMPLETE BLOOD count, or CBC, analyzes three blood components:
+ White blood cells: A high count can indicate an infection or allergy.
+ Red blood cells: Too few red blood cells may indicate anemia; too many may impede blood flow. The test measures number, volume (hematocrit), and total hemoglobin.
+ Platelets: Measures clotting ability.

Blood cells don't live long, so they must constantly be replaced. Most are produced in the bone marrow, a spongy connective tissue within the bones. In a fetus, all the bones are involved in making blood, but in an adult, only certain larger bones—among them the ends of the arm and thigh bones, the skull, breastbone, pelvis, and ribs—produce the cells.

Bone marrow contains stem cells, undifferentiated cells that are able to develop into many different types of blood and immune system cells. These stem cells divide, proliferate, and specialize into red blood cells and different forms of white blood cells within the marrow; the cells then enter the bloodstream via blood vessels in the bone. Lymphocytes travel to lymph glands to finish developing before entering the blood, but the other cells enter the bloodstream fully formed.

BLOOD WORK

Blood has three main jobs in the body, the first being transportation. Blood is the river that carries oxygen from the lungs and nutrients

SEE ALSO: Chapter Three, "The Bones" PAGE 76, Chapter Twelve, "Innate Defenses" PAGE 322 from the digestive system to all the body's tissues so that cells can perform their metabolic tasks. Blood also takes away carbon dioxide and waste products; the lungs exhale the carbon dioxide and the kidneys dispose of other wastes in urine. The bloodstream also carries hormones from the glands and organs that produce them to their target tissues.

Both the plasma and the cells of the bloodstream perform regulatory jobs as well. Blood absorbs heat from warmer areas of the body and releases it in cooler tissues, keeping body temperatures even. Blood itself is a little hotter—about 100.4°F (38°C)—than the average body temperature. It is also slightly alkaline and contains proteins that help to regulate the pH level in cells.

The blood is also the body's guardian, one of its main avenues of defense against a host of invaders. White blood cells and antibodies float throughout the bloodstream, destroying viruses, bacteria, parasites, and other pathogens that attack by the millions. Platelets and proteins in the plasma repair damaged tissues and in a remarkably short amount of time build clots that seal up cuts and tears.

You replenish your blood supply every time you drink a glass of water. Plasma, which makes up more than half the bloodstream, is more than 90 percent water taken from the digestive system. Dissolved within it, and making up the other 9 percent of its volume, are more than 100 other substances that keep the body working. Most are proteins that exert an osmotic pressure to keep the proper balance of water within the bloodstream versus water within the tissues. Other proteins in plasma include fibrinogen, made by the liver, which helps blood to clot, and antibodies that fight pathogens. Plasma also transports nutrients—such as glucose from carbohydrates, vitamins, minerals, and amino acids—from the digestive system to feed all the body's cells. It carries away wastes for disposal and distributes enzymes involved in cellular reactions and hormones that help to control metabolism and growth.

Even though plasma by itself is a pale, yellowish liquid, blood is red because it is thronged with red blood cells (erythrocytes), which make up between 40 and 50 percent of blood by volume. Between 4 and 5 million red blood cells crowd into every cubic millimeter of blood. Just 7.5 micrometers in diameter, these flattened disks are the only cells in the body to lack a nucleus; that organelle is ejected from the cell as the cell matures. As a result, red blood cells are depressed in the center, looking rather doughnut-like. They are also extremely flexible, which allows them to squeeze through the narrowest of capillaries.

Each red blood cell holds roughly 250 million molecules of the protein hemoglobin. Hemoglobin

RED BLOOD CELLS
Lacking a nucleus, red blood cells have a hollowed-out shape that maximizes surface area for exchange of gases.

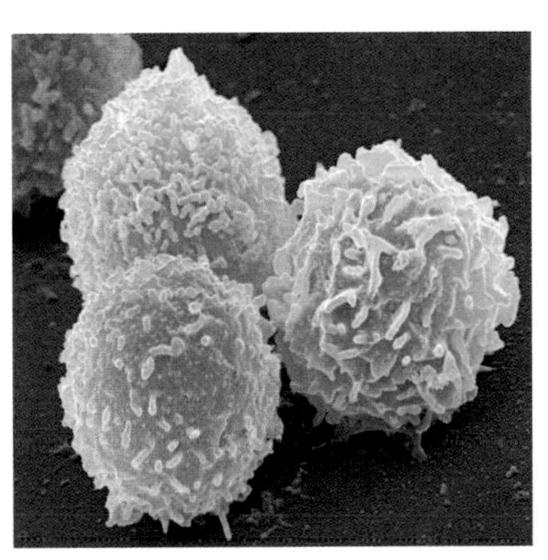

WHITE BLOOD CELLS
Mainstays of the immune system, two lymphocytes (left) cozy up to a neutrophil (right).

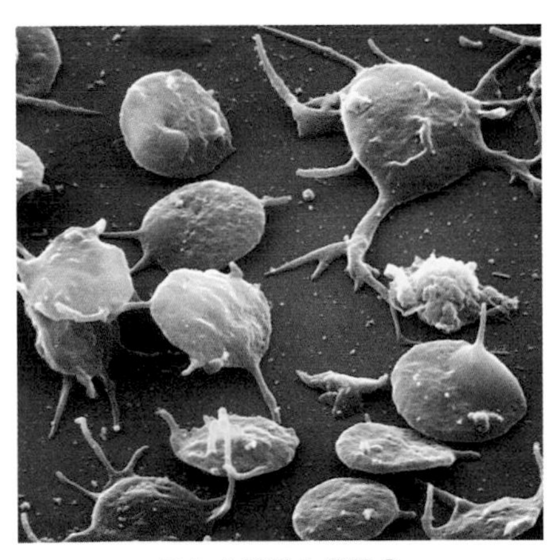

PLATELETS
Activated platelets develop fingerlike pseudopods as they prepare to form a platelet plug.

BLOODLETTING

A Renaissance-era phlebotomist adjusts a patient's humors.

BLOODLETTING has long been a method used by physicians. A common method of phlebotomy, or bloodletting, was venesection: A vein was opened and a pint or two of blood drained off. It still is used to treat some rare disorders such as hemachromatosis, an excess of iron. Removing some iron-rich blood restores balance.

Another method, the leech, has quietly returned to legitimate medical practice. From ancient times through the 19th century, healers used the blood-sucking bite of the medicinal leech *(Hirudo medicinalis)* to drain blood, basing the therapy on the ancient Greek notion that illnesses derived from an excess of certain "humors" in the bloodstream. In recent years this three-jawed, segmented worm has returned because of the useful combination of chemicals in its saliva. Containing both an anticoagulant and an anesthetic, the leech's saliva promotes blood flow in reattached limbs; the animal's bite can also remove blood pooled in tissues after a skin graft. When the leech drops off, the anticoagulant it has injected continues to work for hours, preventing clotting and helping to improve circulation. In the United States, leeches now are approved by the Food and Drug Administration for use as medical devices, typically in skin grafts or in reattaching small body parts such as ears or noses. ■

molecules contain iron atoms that bind loosely to molecules of oxygen as blood passes through the oxygen-rich tissues of the lungs. When the blood reaches the capillaries, the concentration of oxygen in the blood is higher than it is in the surrounding tissues; therefore, the hemoglobin releases the oxygen, which diffuses through the thin capillary walls (see "Blood Vessels," page 150), moving naturally from an area of higher concentration to an area of lower concentration. The hemoglobin also picks up between 10 and 20 percent of the carbon dioxide that is a waste product of cellular metabolism and trucks it back to the lungs for disposal.

Lacking a nucleus or most other organelles, red blood cells last only about 120 days before dying at a rate of about 2 million per second. The spleen and liver remove the dead, ruptured cells from circulation; the body then thriftily recycles many of their components, such as the iron and the amino acids in hemoglobin.

Compared with red blood cells, white blood cells, or leukocytes, are few and far between in the blood—about 5,000 to 10,000 per cubic millimeter—but their importance outweighs their scarcity. Soldiers of the immune system army, white blood cells float through the bloodstream until they detect chemical signals from tissues that need protection. Then they leave the blood through the

SEE ALSO: Chapter Twelve, "Innate Defenses," PAGE 322, "Adaptation," PAGE 328

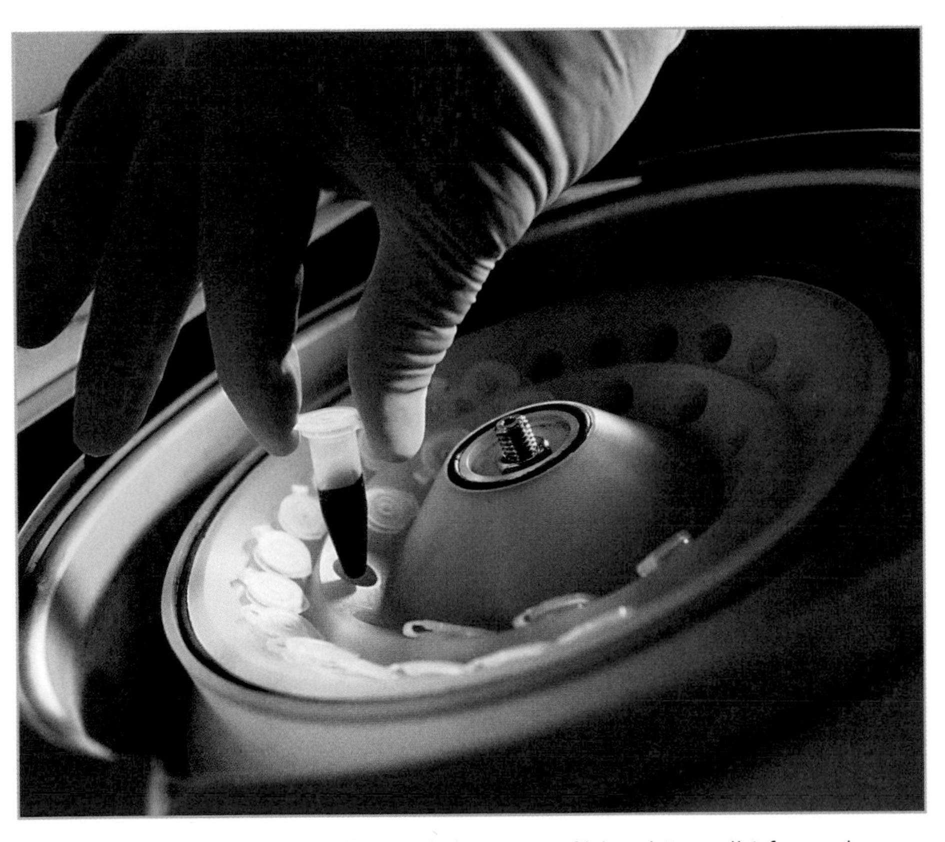

A centrifuge separates the formed elements of blood (its cells) from plasma.

capillary walls and travel through the cells to mount an immune or inflammatory defense. Most live for only a few days.

There are five types of white cells that travel through the blood; they are classified either as granular or agranular according to the presence of little grainlike vesicles in their cytoplasm. Granulocytes include neutrophils, basophils, and eosinophils. Neutrophils, the most commonly found kind of white blood cells, attack and kill bacteria and some fungi. Eosinophils kill parasites (such as flatworms), while basophils, the rarest sort of white blood cell, release histamines during allergic reactions. They also attract other white cells to areas of infection in the body.

Lymphocytes and monocytes are the two types of agranulocytes. Lymphocytes, the second most common form of white blood cell, come in several varieties that are powerful members of the immune system, attacking everything from viruses to cancer cells. Upon entering the body's tissues, monocytes change into macrophages, which can then eat bacteria and dead or damaged cells.

More numerous than white blood cells, about 150,000 to 400,000 per cubic millimeter, platelets are the third type of "formed element" in the blood. They can't be called cells, because they are merely fragments of huge cells called megakaryocytes that form in the blood marrow. The mature megakaryocytes splinter into thousands of pieces that then enter the bloodstream as platelets. Roughly disk-shaped, they have no nucleus but contain an array of chemicals that allow them to find damaged blood vessels, form clots, and repair tissues.

The blood vessels are among the most vulnerable parts of the body. Were it not for blood's amazing rapid-healing properties, even a tiny cut in a vein or artery could quickly drain the body's entire blood supply.

HEALING POWER

When a blood vessel is damaged or severed, a multistage, fast-acting defense mechanism springs into action. Chemicals are released by the injured tissue and by the platelets. These substances, as well as signals from pain receptors, stimulate a vascular spasm: The muscles of the blood vessel wall constrict, reducing blood flow. Meanwhile, the exposed collagen fibers in the damaged tissue prompt platelets, which normally slide smoothly

AS RED AS BLOOD

HUMAN BLOOD CHANGES color as it circulates, but it is always red. After hemoglobin picks up oxygen in the lungs, the blood is bright red; when the oxygen has been unloaded in the tissues, the blood turns a dark maroon. Veins look blue because the overlying tissues filter out red light at that depth.

along, to adhere to the wound. As they do this, the platelets change shape, becoming spiky, sticking to one another, and releasing a host of chemicals. Some of the chemicals summon more platelets to the injury; others bolster the vascular spasm reaction, while still other chemicals cause the newly arriving platelets to become sticky themselves. Soon, the accumulating platelets have formed a plug that temporarily seals the break in the blood vessel.

As this is happening, the injured tissues and the platelets are also stimulating a complex cascade of reactions that in a matter of minutes will lead to a blood clot. Substances on the surface of the accumulating platelets and others released by the injured tissue trigger a chemical reaction in the blood that forms prothrombin activator. This substance transforms a plasma protein called prothrombin into the enzyme thrombin. Thrombin, in turn, changes another plasma protein, fibrinogen, into loose, insoluble threads of fibrin. These sticky strands gather at the site of the platelet plug and trap platelets and other blood cells in a mesh of fibers—a clot, or thrombus, that stops blood from leaking through the damaged vessel.

Over the course of an hour or so, fibrin threads attached to the damaged area tighten, squeezing serum (plasma without clotting proteins) out of the clot and pulling the torn edges of the blood vessel together. The platelets release yet another chemical, a growth factor that helps the tissues of the damaged blood vessel repair themselves.

Once the tissue is on the way to being healed, the clot is no longer needed. Yet another plasma protein, plasminogen, takes care of the problem. Within two days after the clot has formed and healing begins, the plasminogen starts to release an enzyme called plasmin

Heir to the Russian throne Alexis Romanov (front center) suffered from hemophilia, a family ailment.

SEE ALSO: Chapter Two, "Healing," PAGE 54, Chapter Twelve, "Innate Defenses," PAGE 322

WHAT CAN GO WRONG

HEMOPHILIA is a rare, inherited disorder in which people are born without a necessary clotting factor in their blood. Hemophiliacs may bleed spontaneously, or extensively after minor traumas; they may also suffer painful, damaging bleeding in the joints, such as in the knees, elbows, or ankles. The two main types of the disorder are hemophilia A, or classical hemophilia, in which clotting factor VIII is missing (accounting for more than 80 percent of cases) and hemophilia B, linked to a missing factor IX. Both types are sex-linked conditions: They occur almost exclusively in males because the genetic mutation that accounts for the clotting factor is found on the X chromosome. (Women are protected because they almost always inherit a matching X chromosome without the mutation.) Women and men may be asymptomatic carriers of these kinds of hemophilia.

that eventually digests the fibrin strands and dissolves the clot.

Blood clots are lifesavers, but they can also be killers if they accumulate or if they break loose to travel through the blood system. Ordinarily, several natural anticoagulants in the blood—among them heparin (produced by immune system cells), antithrombin, and activated protein C (APC)—prevent this by dissolving clots. Sometimes these factors are not enough to prevent an unwanted clot. If blood is flowing too slowly, or if the walls of the blood vessels develop rough spots due to arteriosclerosis or inflammation, then large, fixed clots that block blood flow partially or completely can develop. If this happens in the blood vessels that feed the heart tissue, it leads to a heart attack (see "When Hearts Attack," page 146).

If a clot breaks loose and floats through the bloodstream—at which point it's called an embolus—it can get stuck in a smaller blood vessel and cause an embolism, or obstruction. An embolus that becomes trapped in the lungs is called a pulmonary embolism, a life-threatening condition; one that blocks a blood vessel in the brain can cause a stroke.

Certain drugs, ranging from aspirin to synthetic heparin, are anticoagulants that help to prevent embolisms in susceptible patients. Blood banks prevent clotting in donated blood by adding substances that remove calcium, which is needed for the sequence of reactions that causes clotting.

Bleeding disorders represent the opposite problem: too little clotting. Bone marrow diseases or damage, such as that brought on by radiation, can result in too few platelets being created. Without enough platelets, the blood vessels soon begin to leak blood from innumerable tiny breaks throughout the body. Bruises and tiny purplish blotches caused by the bleeding show up on the skin. Liver damage—such as that caused by cirrhosis or vitamin K deficiency—can also bring on bleeding because the liver makes clotting factors that are released into plasma. Perhaps the most familiar example of a blood clotting disorder is hemophilia, a genetic condition in which even a minor trauma can become a life-threatening event (see "What Can Go Wrong," opposite).

BLUEBLOODS?

LOBSTERS HAVE blue blood (when it is exposed to oxygen), but people never do. The expression "blue-blooded," meaning aristocratic, comes from Spain, where in past centuries the oldest families of Castile claimed to be distinguished from their mixed-blood compatriots by their pure *sangre azul*, or blue blood. In actuality, their veins just showed up more conspicuously blue against their fair skin.

Strands of fibrin trap red blood cells in a developing clot.

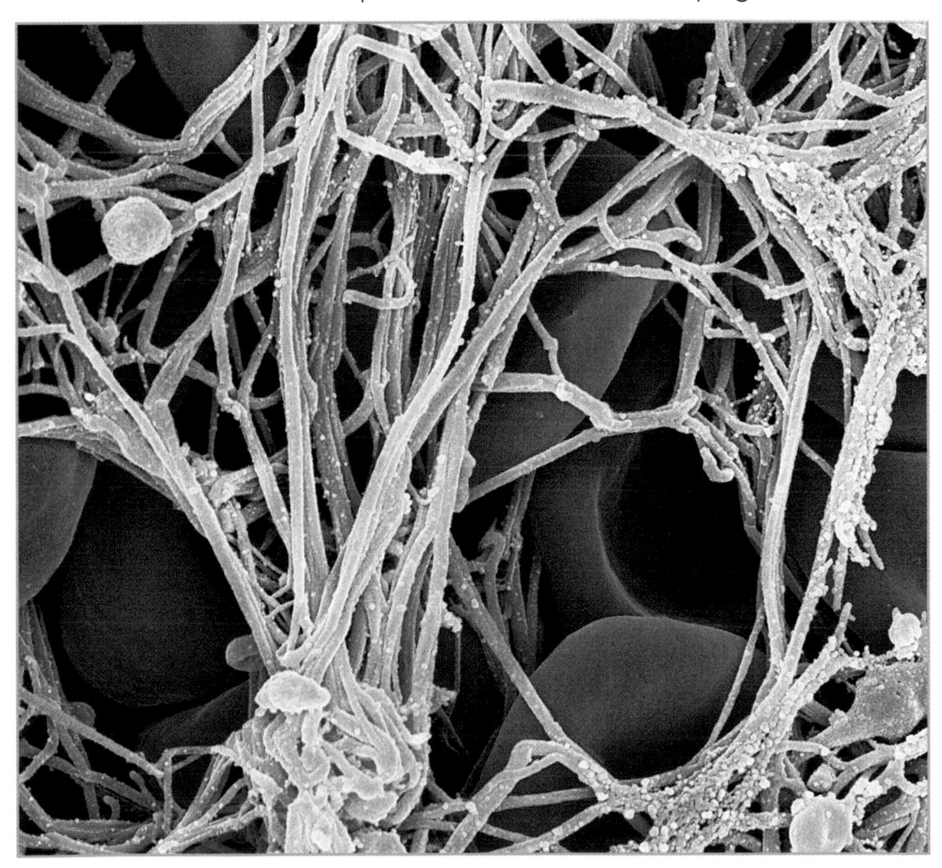

Many bleeding disorders have a seemingly simple solution: a blood transfusion. And in fact, transfusions of whole blood and packed blood cells are common, lifesaving procedures, but they are not as straightforward as they seem because of the existence of incompatible blood types. Before the discovery of blood types (see "Breakthrough," opposite), a few physicians had experimented with animal-to-human and human-to-human blood transfusions, but with little success. They attributed their patients' deaths to the rapid clotting of the transfused blood.

Like most other cells in the body, red blood cells have specialized proteins—called antigens—on their surfaces. These antigens are like ID tags, marking the cells as belonging to the individual that produced them. When someone receives a blood transfusion from an incompatible donor, the body's immune system detects the foreign antigens and attacks the cells, making them clump together and eventually burst. The ruptured cells release hemoglobin into the bloodstream, which can carry the protein to the kidneys and shut them down.

GROUP THERAPY

There are at least 30 common varieties of antigens on red blood cells, resulting in at least 30 different blood groups within the human population; about 100 other blood groups exist in small groups of related people. The two most important blood groups from a transfusion standpoint—the ones most likely to cause a major reaction—are the ABO group and the Rh group. The body typically doesn't react strongly to the other blood groups, although some may cause a reaction if a person is repeatedly transfused with them.

Every unit of donated blood is tested for blood type to prevent transfusion reactions.

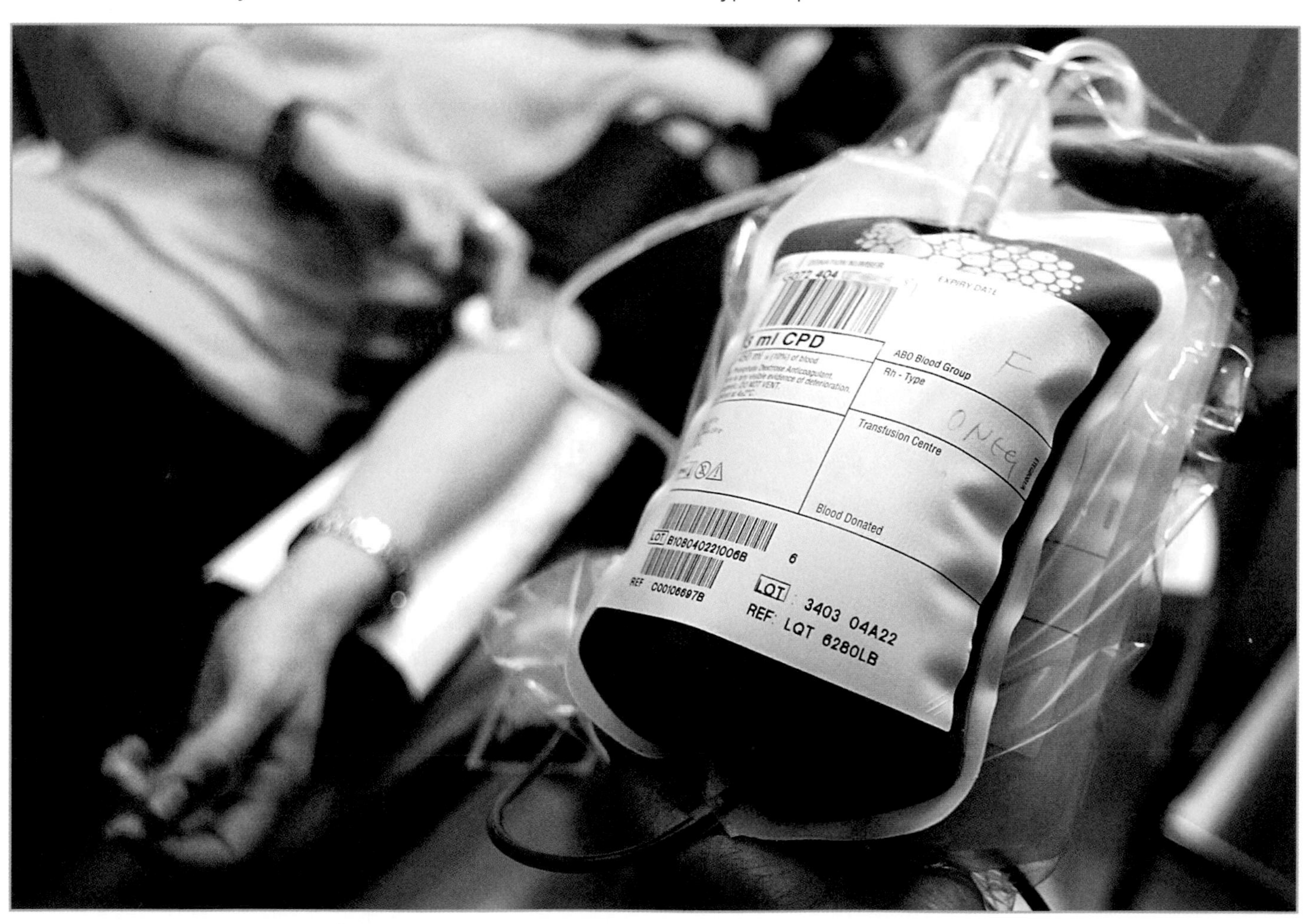

The ABO groups consists of four types—A, B, AB, and O—identified by their antigens. Type A blood has antigen A, type B has antigen B, type AB has both antigens A and B, and type O has neither.

The Rh blood group (so named because the first of its antigens, or factors, was discovered in rhesus monkeys) has at least eight types of antigens, but only one, antigen D, is important in transfusions. Blood that contains the Rh antigen D is called Rh positive (Rh+); about 85 percent of the U.S. population is this type. Blood lacking the antigen is Rh negative (Rh-). ABO type and Rh factor are not linked—they are inherited separately—but they are typically reported together, such as A+ or O-.

Everyone can receive a transfusion of his own blood type as well as type O, which is called the universal donor. Those with AB blood, called the universal recipient, can receive all types of donated blood. The Rh factor must also be matched between donor and recipient so that an Rh- patient doesn't receive an Rh+ transfusion. An adverse reaction doesn't happen the first time Rh+ is transfused into an Rh- person, but after it has occurred, the recipient's body builds up antibodies to the foreign Rh type. The next time he is transfused with that type, the antibodies will attack the transfused blood cells and rupture them.

BLOOD TYPES: GIVING AND RECEIVING

BLOOD GROUP	ANTIGENS	ANTIBODIES	CAN DONATE TO	CAN RECEIVE FROM
AB	A & B	None	AB	AB, A, B, O
A	A	B	A & AB	A & O
B	B	A	B & AB	B & O
O	None	A & B	AB, A, B, O	O

Rh blood types are not only important for blood transfusions; they are also important for expectant mothers and can cause serious problems in fetuses or newborns. If an Rh- woman is carryng a fetus with the Rh+ blood type, leakage of only a small amount of fetal blood into the mother's circulation can cause the mother to form Rh antibodies. While these Rh antibodies do not develop quickly enough to affect this fetus, they cause miscarriage, stillbirth, or serious health problems for subsequent Rh+ children. In the United States, blood typing is a routine part of prenatal care. Rh- pregnant women typically receive injections of Rh immune globulin before and after delivery to prevent formation of Rh antibodies.

BREAKTHROUGH

BLOOD TYPES: In 1901, Austrian biologist Karl Landsteiner (1868–1943) discovered that the blood serum of some people would attack the blood of other people, and that this transfusion reaction was a natural, immunological property. He and colleagues soon identified four basic blood types—A, B, AB, and O—based on this reaction. In 1907, the first transfusion cross-matching donor and patient blood was performed at Mount Sinai Hospital in New York. In 1937, Landsteiner, now an American, and colleague Alexander Weiner discovered the Rh factor, the other important blood type in transfusion reactions.

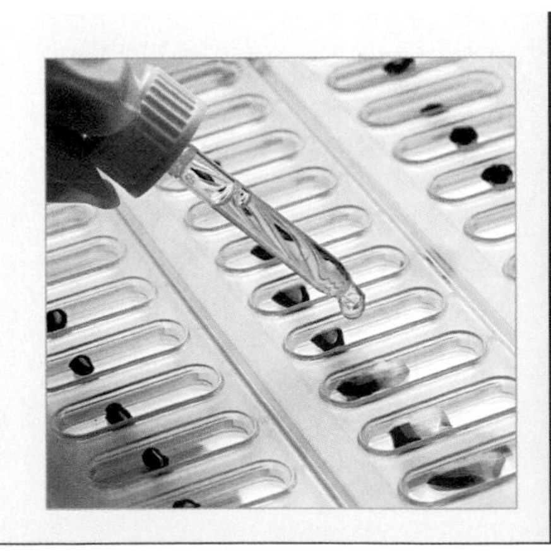

THE HEART

A POWERFUL BEAT

PRESS TWO FINGERS firmly between the fifth and sixth ribs (from the top) on your left side. The steady thump-thump-thump you feel is your beating heart. Contracting and relaxing 100,000 times a day, 35 million times a year, almost 3 billion times in the average lifetime, the heart tirelessly pumps blood throughout the body, keeping every cell oxygenated, fed, protected, and functioning.

AROUND THE HEART

The heart is lodged between the lungs and on top of the diaphragm in a central cavity in the chest, called the mediastinum. It sits behind and just left of the breastbone, its flattened top below the second rib and its pointed apex, at the bottom, behind the fifth and sixth ribs on the left. The size of a large fist, it is hollow and light, weighing about 8 ounces in adult women, 10 ounces in adult men.

Surrounding and protecting this vital organ is the pericardium,

HEART FACTS

+ IN THE COURSE of an average lifetime, the heart will pump about 48 million gallons (182 million L) of blood through the body.
+ Anger and fear increase the heart rate by 30 to 40 beats per minute.
+ Grab a tennis ball and squeeze it tightly: That's how hard the beating heart works to expel blood.

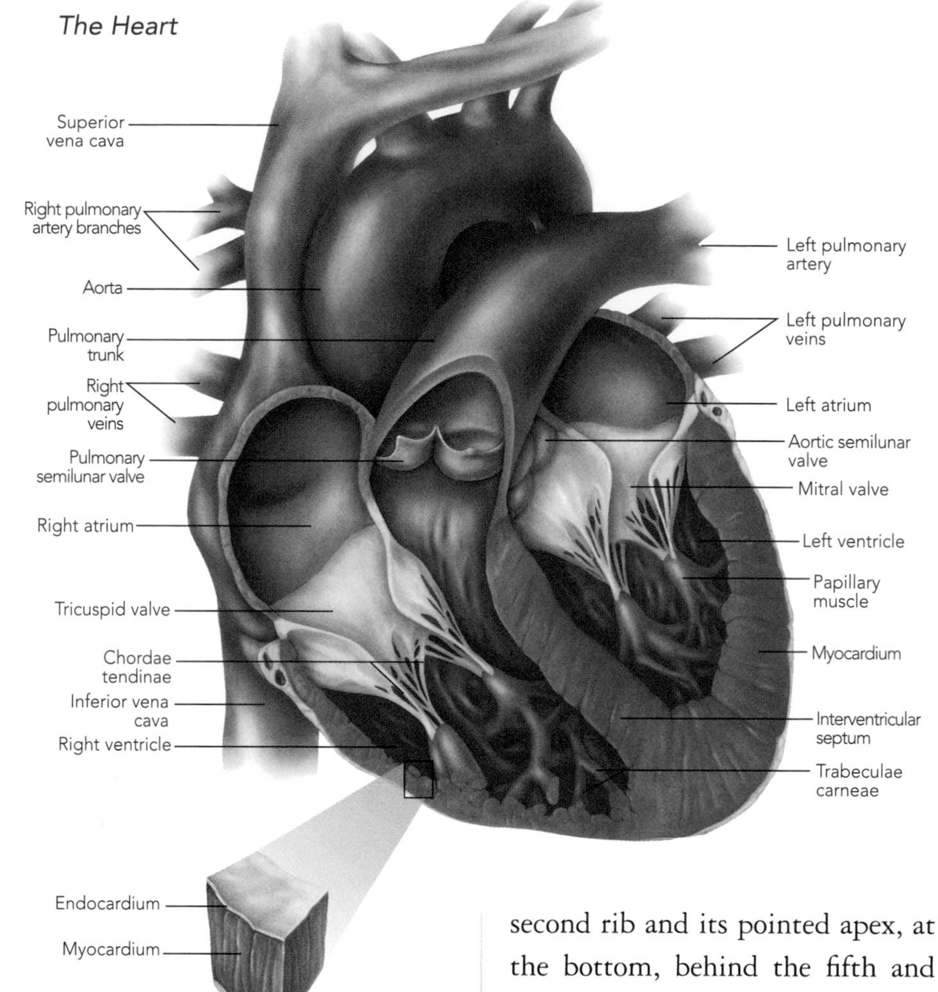

BREAKTHROUGH

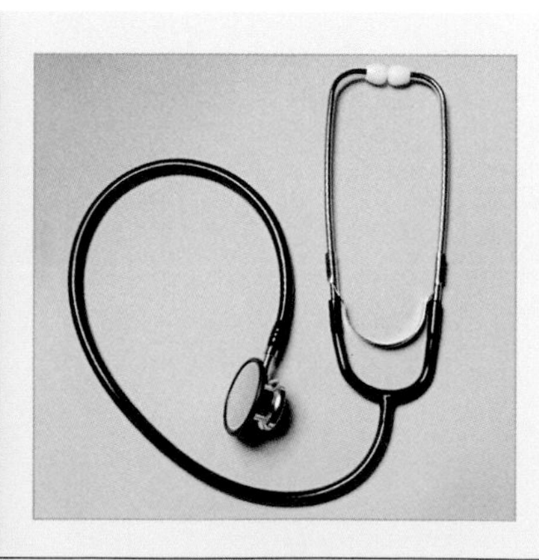

OBSERVING THE HEART: Three inventions improved the ability to detect the heart's actions without opening the chest. In 1816, physician René Laennec (1781–1826) invented the first: the stethoscope, left. Reluctant to put his ear to the chest of a female patient, he rolled papers into a tube and "perceived the action of the heart . . . more clear and distinct." He built a monaural stethoscope and published the design. In 1903 came the electrocardiograph, opposite. Physiologist Willem Einthoven's (1860–1927) device measured electrical current in the heart. A silver-coated quartz string was attached to the skin with an

WEBLINK: See the Heart at work at http://nationalgeographic.com/humanbody

HEART AND SOUL

Below: Aristotle, seated on the left, instructs his pupil, Alexander the Great.

THE Greek philosopher Aristotle (ca 384–322 B.C.) influenced medicine for centuries by teaching that the heart was the body's most important organ and the seat of its "vital spirit." Dissections of chicken embryos showed him that the heart was the first organ to be formed. After three days, he wrote, "the heart appears, like a speck of blood, in the white of the egg. This point beats and moves as though endowed with life, and from it two vein-ducts with blood in them trend in a convoluted course." The brain, he believed, served mainly to cool the body. The heart, however, was the source of heat, and the soul was part of that heat, flowing through the body. The heart was also the central sensory organ, collecting input from the peripheral organs through the blood vessels. Thought and emotion then arose from these perceptions. Today, we still speak of things as being "heartfelt." ■

a double-walled membrane. The outer wall is a tough, inelastic sac that attaches to the diaphragm at the bottom and wraps around the heart, fusing to the arteries where they exit the heart. The inner wall is a thin, double-layered membrane called the serous pericardium that clings tightly to the heart. A slippery fluid between its two delicate layers eases friction as the heart expands and contracts. If the membrane becomes inflamed, a condition known as pericarditis, the layers may rub together painfully, and fluid buildup may occur.

HEART PARTS

The heart itself is a muscular little four-chambered machine. Its walls have three layers. The outermost, the epicardium, is the same membrane that makes up the inner layer of the serous pericardium; it consists of thin connective tissue. In most adults, it contains quite a bit of yellowish fat.

The middle layer, called the myocardium, is where the hard work happens. It is made primarily of cardiac muscle, a specialized kind of involuntary muscle tissue. The branching cells of cardiac muscle are connected at spots called intercalated disks. There electrical signals pass from one cell to another. Held together by connective tissues, the cells form muscle bundles that spiral around the heart.

Among the cardiac muscle cells are autorhythmic fibers, specialized cells that are self-excitable, which means they can contract rhythmically without outside stimulation. In fact, individual fibers grown in a cell culture will beat spontaneously. The heart's inner layer is called the endocardium. This white connective tissue lines the chambers of the heart and covers its valves.

BREAKTHROUGH

electrode; as current passed through, it quivered. A photographic plate registered its movements. By 1911, readings from the "string galvanometer" were used to diagnose heart disease. The third advance, the echocardiogram, was the first medical use of ultrasound. It was invented in the 1950s by cardiologist Inge Edler (1911–2001) and physicist Carl Hellmuth Hertz (1920–1990). Using a sonar machine from a shipyard, they recorded the echoes of Hertz's heart. In 1954, their "reflectoscope" recorded movement of the heart walls. By 1956 their reflectoscope could be used to diagnose valvular diseases.

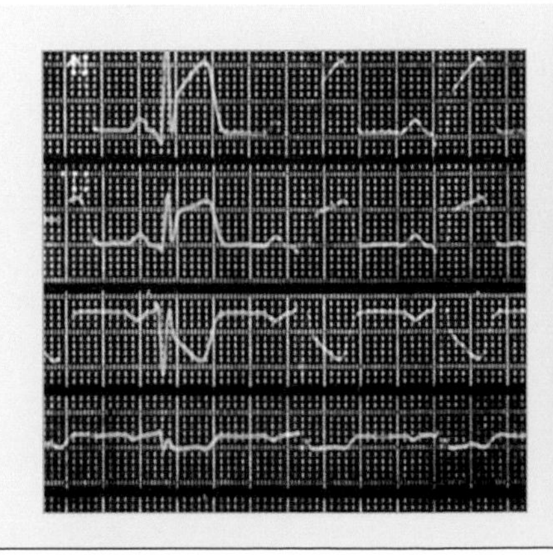

THE HEART AT WORK

The four-chambered heart is an efficient and durable piece of muscular machinery. It can be thought of as two side-by-side pumps: The left side collects oxygen-rich blood from the lungs and pumps it to the rest of the body, while the right side collects oxygen-poor blood from the body and pumps it to the lungs for oxygenation.

The chambers consist of two atria on top (atrium meaning "entry hall") and two ventricles below (ventricle meaning "little belly"). The atria are separated by a thin wall, the interatrial septum. The atria are the collecting rooms for returning blood; with each beat, they pump this blood through valves into the larger and more muscular ventricles, which pump it out of the heart. The ventricles are separated by a partition as well, the interventricular septum.

Three veins—the superior vena cava, the inferior vena cava, and the coronary sinus—feed oxygen-poor blood into the right atrium. As the heart contracts, it squeezes this blood into the right ventricle through the tricuspid valve (so called for its three leaflike parts). When the heart relaxes, the valve closes, preventing the blood from flowing back into the right atrium.

The right ventricle, which makes up most of the visible surface of the heart when seen from the front, pumps the blood it receives into the pulmonary trunk, which divides into two pulmonary arteries that lead to the lungs. A pulmonary valve between the right ventricle and the pulmonary trunk guards against backflow of the blood into the heart.

Bright-red blood returning from its refreshing trip to the lungs enters the left atrium through four

Coronary arteries (blue) spread across the surface of the left ventricle (red).

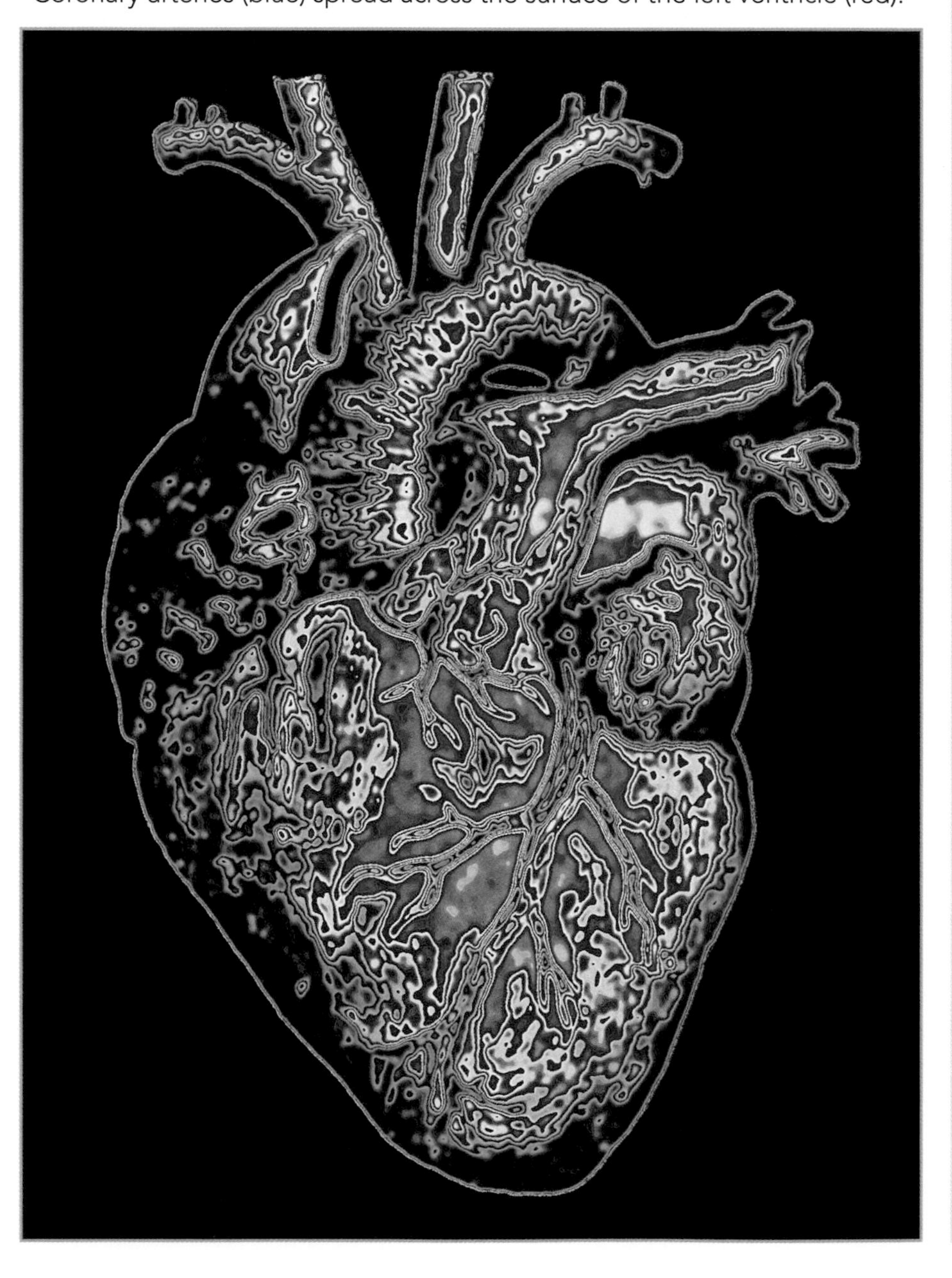

ANCIENT CULTURES

ANCIENT EGYPTIANS believed the heart was the source of the soul, memory, emotions, and wisdom. The ancient Chinese, on the other hand, understood the heart to be the center of circulation. By the second century B.C., The Yellow Emperor's Manual of *Corporeal Medicine* noted that the heart was a pump that moved nutritious blood through the body in a circle of blood vessels.

pulmonary veins. As the heart contracts, this blood passes into the left ventricle through the two-leaved mitral valve—so called because its two parts look like a bishop's miter.

The left ventricle forms the apex, the heart's pointy bottom, located toward the back of the body. Its half-inch-thick, muscular walls are three times as thick as those of the right ventricle. The hardest-working part of the organ, the left ventricle is responsible for pumping blood through all of the body's blood vessels except those leading to the lungs. With a powerful contraction, it expels oxygen-rich blood through the aortic valve into the aorta, the body's largest artery, which branches off to carry it into the rest of the body. (A little blood from this ventricle also feeds into the coronary arteries to supply the heart muscle itself.)

VALVES

The heart's four valves are the guardians of its circulation. The two between the atria and the ventricles are aptly called the atrioventricular valves. When the contracting muscle of the heart presses the blood against them, it forces them open; as the heart relaxes, the valves return to a tightly closed position. Their leaves typically don't get everted, or pulled inside-out into the atria, because they are connected to the ventricle by slender cords called chordae tendineae (the "heart strings" of popular song).

SAVING BLUE BABIES

Vivien Thomas developed many surgical techniques and instruments.

ON NOVEMBER 29, 1944, Drs. Helen Taussig (1898–1986) and Alfred Blalock (1899–1964) and surgical associate Vivien Thomas (1910–1985) stood over a baby in an operating room at Johns Hopkins Hospital in Baltimore, Maryland. Eileen Saxton, 15 months old and 9 and a half pounds, suffered from Tetralogy of Fallot, a congenital heart defect that features an underdeveloped artery leading to the lungs. These infants were called "blue babies" because not enough blood reached their lungs, depriving them of oxygen and making the skin appear bluish. Often they would die.

Taussig, one of the few female pediatric cardiologists at the time, had studied blue babies for years using a fluoroscope, an early x-ray technique. She suggested to Blalock, the chief of surgery at Hopkins, that connecting the subclavian artery to the pulmonary artery could greatly increase blood flow to the lungs. Blalock asked Thomas, his longtime head of surgical research, to work out the technique using dogs as subjects.

The team worked to perform the operation. The baby survived and gradually picked up a healthy pink hue. In the years to come, thousands of children underwent the operation. Blue-baby mortality dropped from 20 percent to under 5 percent. In 1971, surgeons from around the country honored Thomas, who had no college degree, for his contributions, and in 1976 he was appointed to the medical school faculty. ■

Between the ventricles and the big arteries—the pulmonary trunk and aorta—are the semilunar valves, named for their three crescent moon–shaped leaves. Pressed open by each ejection of blood from the ventricles, they close when the heart relaxes, preventing blood from reentering the ventricles. No valves stand between the incoming veins and the atria; as the heart contracts, the entrance to those veins is compressed so that very little blood flows back into them.

Typically, bloodflow through the heart is quiet and smooth, producing the characteristic heartbeat sounds (see "Lub-Dup," page 149). Sometimes, however, doctors will hear additional sounds, like clicking or whooshing noises. These murmurs may be "innocent," but in many cases, particularly in adults, they are the sound of blood flowing through malfunctioning valves. Diseases, congenital defects, or scarring can damage the valves so that they don't open completely (stenosis) or close completely (incompetence). Either condition makes the heart work harder as it pumps blood through narrowed openings (in stenosis) or pumps some backflowing blood again and again (in incompetence). Mitral valve prolapse, in which the mitral valve leading into the left ventricle is weakened, is a common example, and its effects can range from benign and unnoticeable to severe. In serious cases of valve damage, doctors can put in a new valve; some replacement valves are taken from pig hearts (and treated to prevent rejection) or human cadavers, and some are synthetic.

SHOCK TO THE HEART

IN 1952, Dr. Paul Zoll (1911–1999) attached electrodes to the chest of a man whose heart had stopped because the natural pacemaker signal was blocked at the AV node. Two-millisecond electric shocks from the electrodes passed into the heart, restarting it. Zoll's innovation led to the development of the portable defibrillators widely used today.

ELECTRICITY

The heart is powered by electricity. The signal begins in the sinoatrial (SA) node, also known as the heart's natural pacemaker, located near the top of the right atrium. This tiny piece of tissue is made of autorhythmic fibers, specialized heart cells that take in and expel calcium and other elecrolytes to regularly change their electric charge. The electrical impulse produced by the cells of the SA node spreads through the walls of the atria, moving from one cardiac cell to the next via gap junctions, passageways that connect two cells. The signal makes the muscle cells of the atria contract in a sequenced fashion, pressing blood into the ventricles.

As the blood enters the ventricles, the electrical impulses from the atria converge on a second junction, the atrioventricular (AV) node, located in the interatrial septum just above the tricuspid valve. The electrical signal pauses for a tenth of a second at the AV node while the ventricles fill with blood. Then it shoots out through the atrioventricular bundle (AV bundle, or bundle of His), autorhythmic cells that run down through the interventricular septum. The AV bundle splits into right and left branches, which in turn lead

BREAKTHROUGH

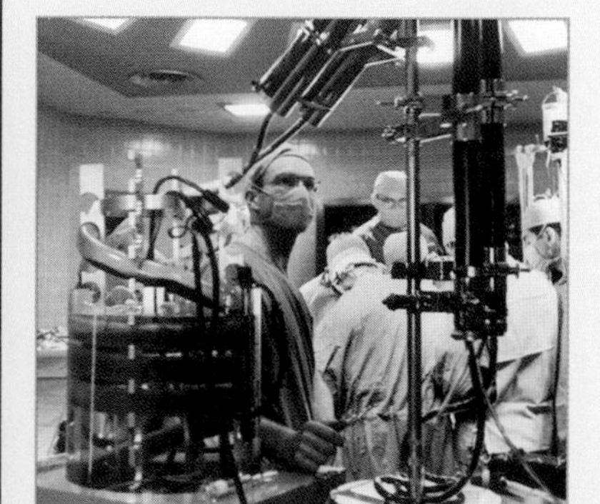

BEFORE THE 1950s, open-heart surgery was an extremely tricky business as it involved cutting into a still-beating heart. Prolonged operations to repair heart valves or to bypass blocked coronary arteries were virtually impossible to perform. After the death of a young patient with blocked lung circulation in 1930, American surgeon John Heysham Gibbon (1903–1973) became determined to build a machine that could completely take over the work of the heart and lungs during surgery. Working with technician Mary Hopkinson, who later became his wife, Gibbon pursued the development of

SEE ALSO: Chapter Four, "The Muscles," PAGE 106

into strands of specialized cells called Purkinje fibers that fan out through the ventricle walls. Electrical currents from these bundles and fibers spread through the muscular ventricles, causing them to contract and pump blood into the arteries. The whole signaling process lasts less than a second.

CURRENT DETECTION

The electrical currents generated by the heart can be detected all over the body and convey a surprising amount of information. Electrodes placed on the arms, legs, and chest can pick up the signals and transmit them to an electrocardiograph machine, which translates them into the rounded and spiky waves of an electrocardiogram, or ECG (sometimes called EKG from the German *Electrokardiogram*).

Three waves appear, one after the other, with each beat of the heart. The first, the P wave, represents the atria contracting. The second wave, the tall peaks and valleys of the QRS complex, represents the ventricles contracting. The third wave, the T wave, shows the signal repolarizing as the ventricles relax.

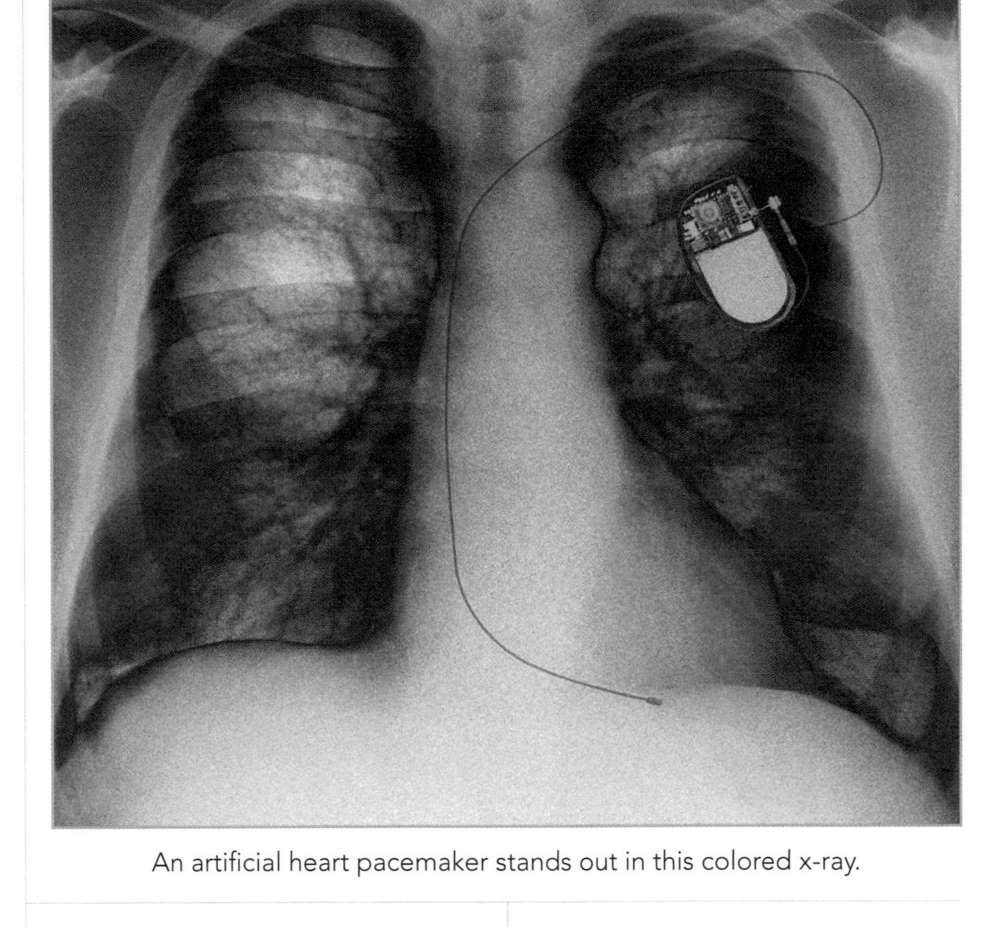

An artificial heart pacemaker stands out in this colored x-ray.

In a normal ECG, these waves repeat in a predictable pattern. Abnormal ECGs can yield clues to the location of a heart problem. Large P waves, for instance, might mean an enlarged atrium; an unusually tall R peak in the QRS complex could point to enlarged ventricles, while a flattened T wave could mean insufficient oxygen to the heart (as in coronary artery disease). Loose, random ups and downs indicate ventricular fibrillation, a serious condition in which the SA node loses control of the heart rhythm and the organ is taken over by rapid, irregular contractions. The heart can sometimes be restored to normal rhythm by shocking it with a defibrillator.

BREAKTHROUGH

the machine for two decades, funded in the 1940s by Thomas J. Watson, chairman of IBM. The Gibbons found that blood cells passing through a machine were easily damaged, and the blood was liable to clot. The researchers therefore passed the blood in a thin sheet over a layer of film to oxygenate it and added screens and filters to block air bubbles and clots. By 1952, the machine was working successfully in dogs; in 1953, Cecelia Bavolek became the first patient to successfully undergo heart surgery while her blood was being shunted through the heart-lung machine. It was still a risky endeavor in its early days—three of the first four patients to use the machine did not survive the surgery—but with further improvement, the heart-lung machine, opposite, became a standard addition to heart surgery and a piece of equipment in every hospital's cardiac surgery unit.

WHEN HEARTS ATTACK

HEART ATTACKS DON'T begin in the heart: They start in the slender coronary arteries that feed the heart muscle. When those arteries become stiff, thickened, and clogged—a condition known as arteriosclerosis—they can no longer deliver enough blood to the heart. Starved of oxygen, the heart tissue begins to die; the weakened heart may falter and stop beating normally. All too often, the first sign of this disease is its last one—a fatal heart attack. Coronary artery disease is the leading killer of adults in the United States, accounting for more than 650,000 deaths a year.

Although other diseases or traumas can stop the heart from beating properly, the artery-clogging condition known as atherosclerosis (one form of arteriosclerosis) is the most common cause of a classic heart attack. Doctors aren't sure how atherosclerosis starts, but it seems to begin when the inner lining of an artery develops a rough or injured spot, maybe due to inflammation, high blood pressure, carbon monoxide from cigarette smoke, or high LDL (low-density lipoprotein) cholesterol, among other factors. Cholesterol, fats, calcium, cellular waste products, and platelets gather at the injured spot, forming a plaque that narrows the blood vessel.

When this happens in coronary arteries, the reduced blood flow that results can starve and weaken the heart muscle, a condition called myocardial ischemia. It can happen "silently," without symptoms, or it can bring on the squeezing sensation of angina, especially during exercise, when the heart needs more blood than usual.

Even more dangerous is a complete blockage of a blood vessel. Sometimes a blood clot forms at the site of the plaque; if the clot breaks loose, it can be moved by the bloodstream into a narrower vessel, walling it off completely. In a blood vessel feeding the brain, it causes a stroke; in the arms or legs, it results in tissue damage; in coronary arteries, it brings on a myocardial infarction, or heart attack.

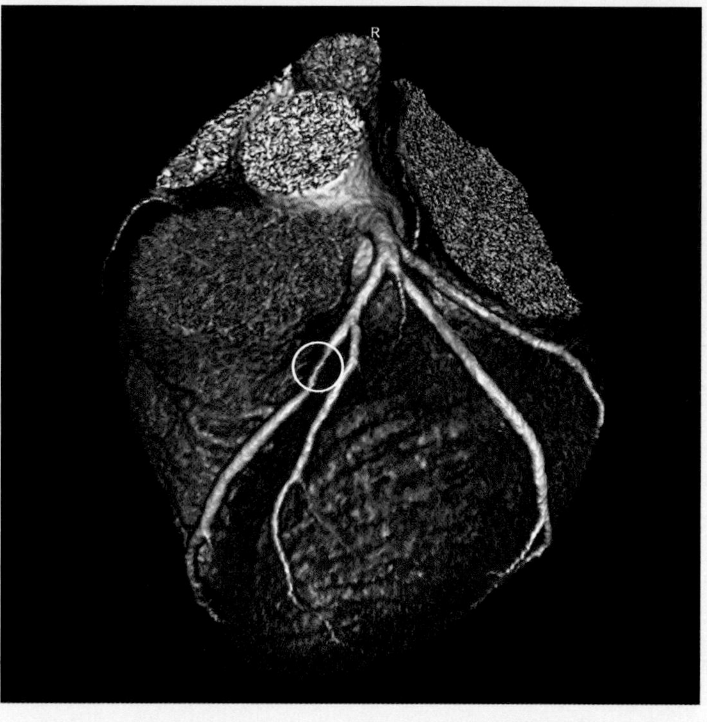

Computed tomography images allow a non-invasive, 3-D view of the heart; this one shows a narrowed artery.

SIGNS & SYMPTOMS

Some heart attacks strike silently, leaving victims (if alive) unaware that they even occurred. More often, there is some kind of chest discomfort: The symptoms can simulate bad indigestion, a tight band around the chest, or pressure so intense that it feels like an elephant is sitting on the chest. Pain can radiate into the arms, shoulder, jaw (usually the left side more than the right) or torso. Other symptoms—somewhat more common in women than in men—include nausea, shortness of breath, profuse sweating, and dizziness. The stereotypical scene in which someone clutches his chest, drops to the floor, and dies is unusual, but because they are expecting similarly dramatic symptoms, many patients wait too long to call for emergency assistance. Those who don't wait—who reach a hospital within an hour of the first symptoms—have the best chances for survival.

An infarction kills heart muscle and can trigger ventricular fibrillation and death. Hospital personnel who suspect a heart attack will give the patient clot-dissolving drugs and may perform emergency surgery, such as angioplasty, to open up clogged arteries, or coronary bypass surgery, which uses blood vessels from other parts of the body to bypass blocked coronary arteries. After the infarction, the damaged heart muscle will be replaced with scar tissue within six or eight weeks, leaving the heart permanently weakened. This can lead to heart failure.

TREATMENT

Treatment for the disease can mean drugs or surgery, or both. Cholesterol-lowering drugs will, as their name indicates, drop blood cholesterol levels substantially, helping to keep plaque from building up. Beta blockers are medicines that ease strain on the heart by preventing epinephrine from attaching to beta receptors and pumping up the heart rate. Nitroglycerine dilates coronary blood vessels, while digitalis increases cardiac contractibility and, therefore, cardiac efficiency.

If exercise, diet, and medication are not enough to keep blood vessels open, surgery might be needed. Heart surgeons might bypass blocked coronary arteries by attaching a blood vessel, taken from elsewhere in the patient's body, to the aorta at one end and an unblocked portion of the coronary artery at the other. They may also keep blood flowing by inserting a balloon into the blocked artery, inflating it temporarily to squash plaque against the artery walls, and then deflating it. Or they may insert a stent, a latticelike tube of steel, into the artery to keep it open. Some stents are coated with drugs that reduce the odds that the artery will clog up again; questions have arisen about whether these stents lead to an increased risk of clotting, so studies are underway to check on their efficacy and safety. At least one recent study suggests that, in many cases, blood-thinning drugs can be as effective as stents in keeping arteries open. Decisions regarding the specific choices listed above are best made by an informed patient and physician. Ongoing studies may well change current thinking.

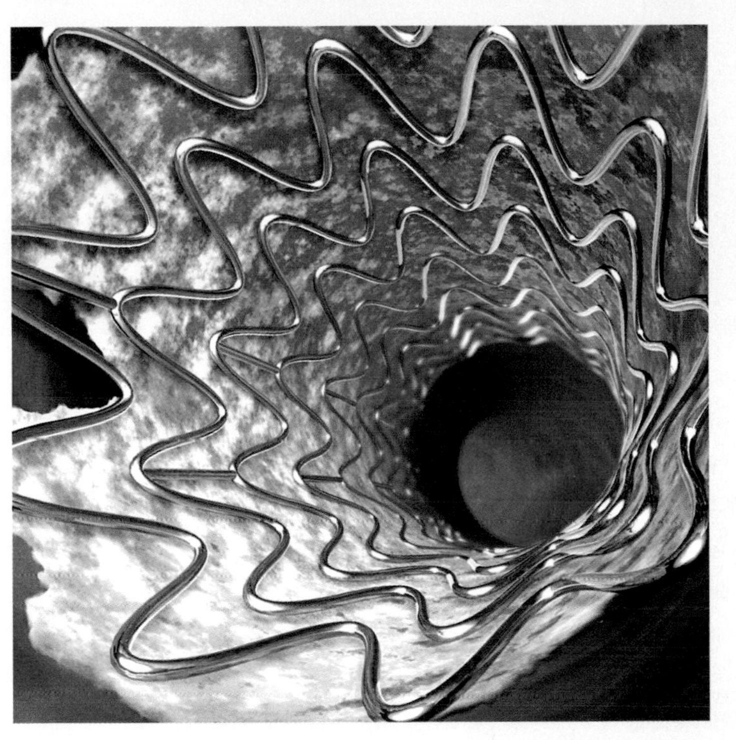

Seen from inside the blood vessel in this illustration, a metal stent holds open an artery.

RISKS & PREVENTION

The risk factors for coronary artery disease include some that can be controlled and others that cannot. Controllable factors include weight, blood pressure, smoking, diabetes, blood cholesterol (particularly, high levels of LDL), and exercise. With regard to weight, the distribution is significant, with excess pounds around the belly being more dangerous. Factors that cannot be changed include age (risk increases for men over age 45 and for women over age 55) and a family history of early heart disease.

Preventing coronary artery disease, as much as it is possible, involves removing the controllable risks—keeping weight within normal limits, exercising, refraining from smoking, and controlling blood pressure and cholesterol levels. Although the human heart itself is durable, it is as dependent as the rest of the body on the health of the blood vessels that nurture it.

BLOOD VESSELS

PATHWAYS AND PASSAGES

WITHIN ITS CLOSED loop, the entire circulatory system encompasses thousands of miles of blood vessels, reaching virtually every cell in every tissue in the body. Three types of blood vessels—arteries, veins, and capillaries—make up this intricate road map. Arteries carry blood away from the heart; capillaries, the tiniest pathways, carry it into the tissues; and veins take it back to the heart.

All blood vessels except the smallest are made of three layers of tissue, called tunics, surrounding the blood-containing space known as the lumen. The innermost layer, the tunica interna, is lined with the endothelium that coats all the inner surfaces of the cardiovascular system, including the heart's. The middle layer, the tunica media, consists of elastic and muscular tissue; the outer layer, the tunica externa, is made mainly of elastic and collagen fibers. The larger blood vessels are fed by their own vessels within their walls.

ARTERIES

Arteries come in two varieties. Elastic arteries are the biggest; ranging from about one-third of an inch to an inch in diameter, these seven

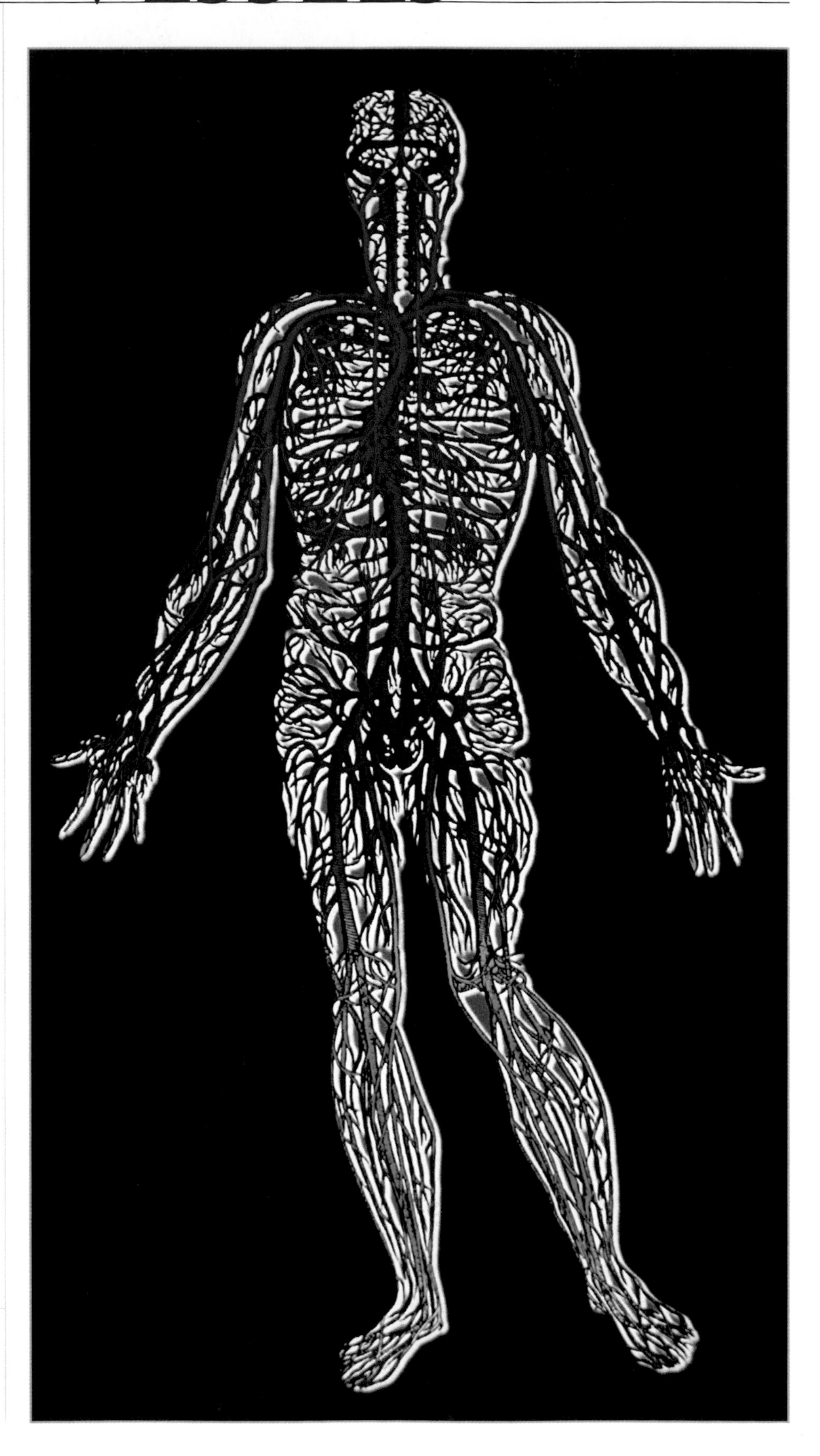

A computer-enhanced version of a 16th-century illustration depicts the body's venous system.

WEBLINK: Learn more about Blood Vessels at http://nationalgeographic.com/humanbody

arteries include the three parts of the aorta, the two common carotids, and the two pulmonaries. Their elastic walls expand and recoil as blood is propelled out of the ventricles. Because elastic arteries carry blood to the medium-size arteries, they are sometimes called conducting arteries.

The blood then enters the medium-size arteries with their thick, muscular walls. These vessels carry blood to the various organs and limbs; the smooth muscles in their walls are capable of constricting or dilating to restrict or increase blood flow as needed. Blood flows into smaller and smaller arteries, reaching the arterioles, tiny vessels with thin walls, before entering the capillaries. Arterioles can also constrict or dilate to temporarily feed or starve the capillary beds.

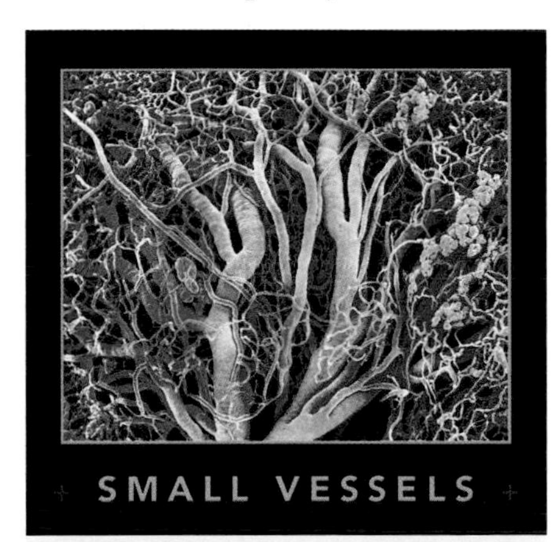

SMALL VESSELS

CAPILLARIES, the smallest blood vessels in the body, above, have an average diameter of 9 µm (about 4 millionths of an inch), much finer than a human hair. A capillary's lumen, the open space within it, is just large enough for individual red blood cells to slip through in single file.

ALEXIS CARREL

Greatly magnified, a cross-section of an arteriole reveals the red blood cells inside.

IN 1894, AN ANARCHIST stabbed French president Sadi Carnot in the abdomen. Unable to repair his severed portal vein, doctors were forced to watch Carnot die. This helplessness angered Alexis Carrel (1873–1944), a young French surgical resident in Lyon. Working first on human cadavers and then on dogs, Carrel experimented with methods for suturing the ends of blood vessels, a technique called vascular anastomosis. He was famously skilled with his hands, having learned embroidery in his youth. Soon, Carrel could cleanly stitch together a carotid artery in two and half minutes.

In 1904 Carrel accepted a position at the University of Chicago and began working with a young colleague, Charles Guthrie, on anastomoses. Together they published 21 papers in 22 months, laying out techniques for joining blood vessels, reattaching limbs, and transplanting organs, including the heart. In the next few years, Carrel described mitral valve repair and coronary artery bypass techniques. He was far ahead of his time: Most of his advanced methods had to wait until the mid-20th century for reliable blood transfusions, heart-lung machines, and other mainstays of modern medicine before they could be used on living humans. Nevertheless, his work was recognized as groundbreaking and paved the way for much of modern surgery. In 1912 Carrel was awarded the Nobel Prize in physiology. ■

CAPILLARIES

As tiny as they are, the microscopic capillaries are arguably the most important parts of the circulatory system, for they are the vessels that do the work for which the system was built. Known as exchange vessels, the capillaries transport nutrients, wastes, gases, chemicals, and even heat between the blood and the interstitial fluid that fills the spaces between the cells of tissues.

The narrow, open space within the capillary is only wide enough for individual red blood cells to pass. Capillaries also have very thin walls, consisting only of one layer of endothelial cells and a basement membrane. Some, called fenestrated capillaries, have pores or windows in their endothelium that allow for rapid entrance and exit of nutrients and hormones.

Even more porous are the sinusoidal capillaries found only in specialized tissues such as the bone marrow; their Swiss cheese–like walls allow large molecules and even blood cells to pass through. The capillaries branch through the tissues in capillary beds, networks of ten to one hundred capillaries apiece. Little sphincters control the flow of blood into the capillary beds, contracting and relaxing five to ten times a minute in response to the needs of the area.

Diffusion from higher to lower pressure pulls oxygen and carbon dioxide, glucose, amino acids, hormones, and many other substances through the gaps between the cells in the capillary walls. A few substances (such as insulin) squeeze through the walls using a process called transcytosis, in which they are first enclosed within little vesicles. At times, large volumes of fluid and solutes are driven by bulk flow pressure from capillaries into tissue or vice versa. If too much fluid floods into the interstitial space, the tissues will swell, a condition known as edema.

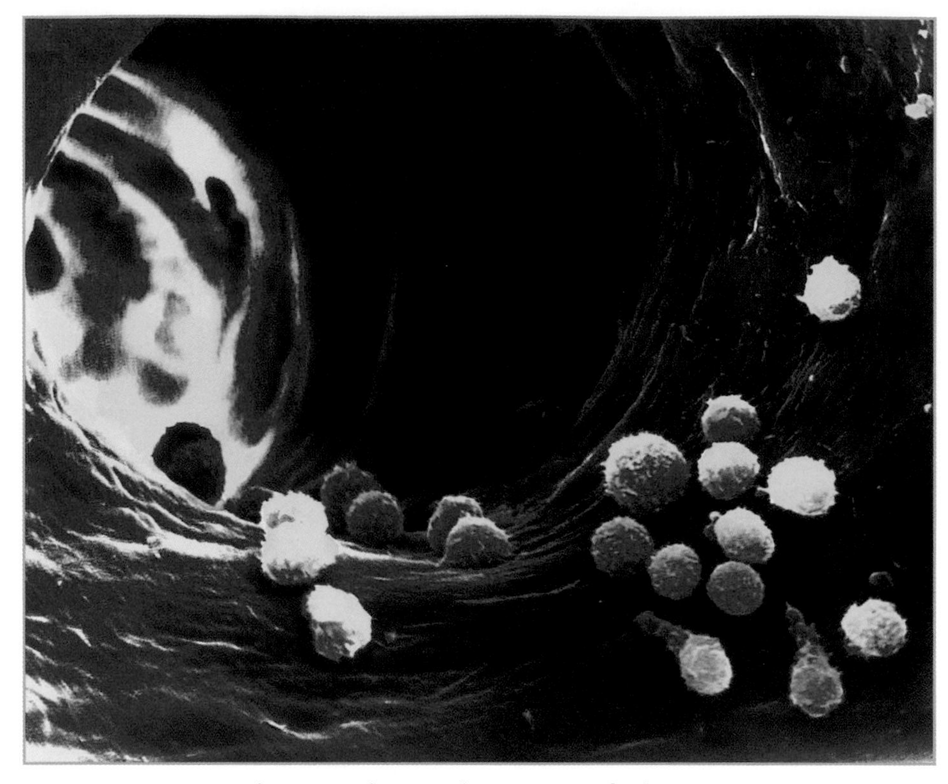

Lymphocytes cling to the interior of a liver artery in this false-color, scanning electron micrograph image.

SEE ALSO: Chapter Six, "Breathing," PAGE 168

WHAT CAN GO WRONG

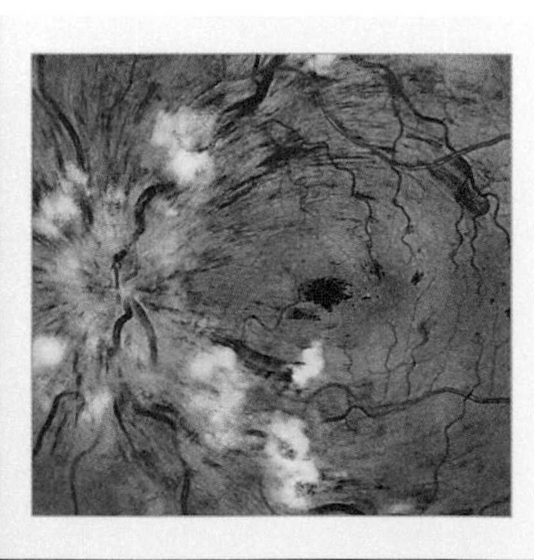

HYPERTENSION, or high blood pressure, is known as the silent killer. This (usually) symptomless condition affects about 30 percent of all Americans and more than 40 percent of African Americans. It is a leading risk factor for heart disease and stroke. A small increase in blood pressure might not seem harmful, but the added stress it inflicts on the heart and blood vessels can be deadly. Blood pressure readings are typically classified as follows: Normal falls at or below 120 mmHg systolic and 80 mmHg diastolic; prehypertension at 120–130 systolic and 80–89 diastolic; stage 1 hypertension at 140–159 systolic

VEINS

After releasing their oxygen, nutrients, and other chemicals and picking up wastes and other cargo from the tissues, the capillaries merge to form small veins called venules, which drain into larger veins that carry the deoxygenated blood back to the heart.

Blood flowing through veins is at a relatively low pressure and veins are accordingly thinner-walled and less muscular than arteries. Some veins, particularly in the arms and legs, contain valves that prevent the blood from flowing backward as it is pumped upward toward the heart. Skeletal muscles in the legs help to pump the blood toward the heart as well by constricting around the veins and milking the blood upward through the valves. Leaky valves, sometimes damaged by stress, can lead to swollen varicose veins that hold pools of backflowing blood.

PRESSURE & PULSE

Blood pressure varies greatly from place to place within the circulatory system. During a physical, blood pressure is measured in the brachial artery in the upper left arm, with the arm held level with the heart. The first reading an examiner takes after relaxing a blood pressure cuff (which corresponds to the first sound heard through the stethoscope) is the systolic pressure, the pressure that the blood exerts on the big arteries as it leaves the contracting ventricles; the reading at which the sound suddenly becomes faint is the diastolic pressure, the pressure in the arteries as the ventricles relax (see "What Can Go Wrong," below.)

Blood pumping through the elastic arteries creates a visible wave with each heartbeat—the pulse. The pulse can be felt in any artery close to the body's surface (see chart above) by pressing lightly on the skin. The pulse beats at the same rate as the heart, about 70 to 80 times a minute at rest.

WHERE TO FEEL THE PULSE

ARTERY	LOCATION
Superficial temporal artery	Temple, next to eye socket
Facial artery	Lower jaw on a line with the corners of the mouth
Common carotid artery	Neck, next to larynx
Brachial artery.	Upper arm, underside of biceps
Femoral artery	Upper, inner thigh
Popliteal artery	Behind the knee
Radial artery	Bottom half of wrist
Dorsal artery of foot	On top of the instep

WHAT CAN GO WRONG

and 90–99 diastolic; and stage 2 hypertension at 160 or higher systolic and 100 or higher diastolic. High blood pressure damages blood vessels, left. It causes the arterioles to constrict and the heart to work harder. The heart, particularly the left ventricle, enlarges and weakens. Blood vessel walls thicken, further constricting blood flow and promoting coronary artery disease and kidney damage. Vessels in the brain can weaken and burst, causing a stroke. Hypertension can be caused by kidney disorders or tumors of the adrenal gland, but in most cases the primary cause is unknown. Risk factors, however, are clear: a family history of the disorder, excess weight, smoking, a high-fat or high-salt diet, lack of exercise, high stress, and high alcohol intake. Drugs that help shed excess water and that dilate the blood vessels can help treat this common, but very dangerous, condition.

CHAPTER SIX

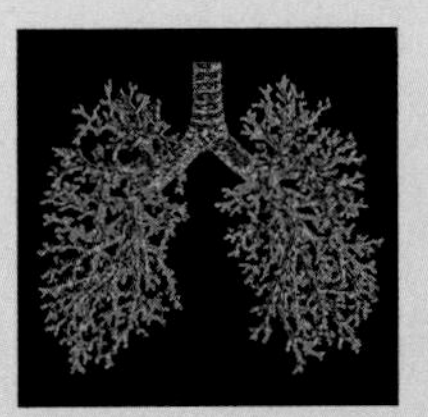

VITAL EXCHANGE

"LIFE AND RESPIRATION ARE complementary," noted the eminent British physician William Harvey in the 17th century. "There is nothing living which does not breathe nor anything breathing which does not live." This is certainly true of the human animal; the respiratory system and its close partner, the circulatory system, work together to ensure that life-giving oxygen is brought into the body while the waste product, carbon dioxide, is removed. Every cell in the body depends on this exchange for survival. It is a process that is as vital as it is automatic.

After this swimmer holds his breath for about a minute and a half, he will have to surface—or die.

RESPIRATION

THE BREATH OF LIFE

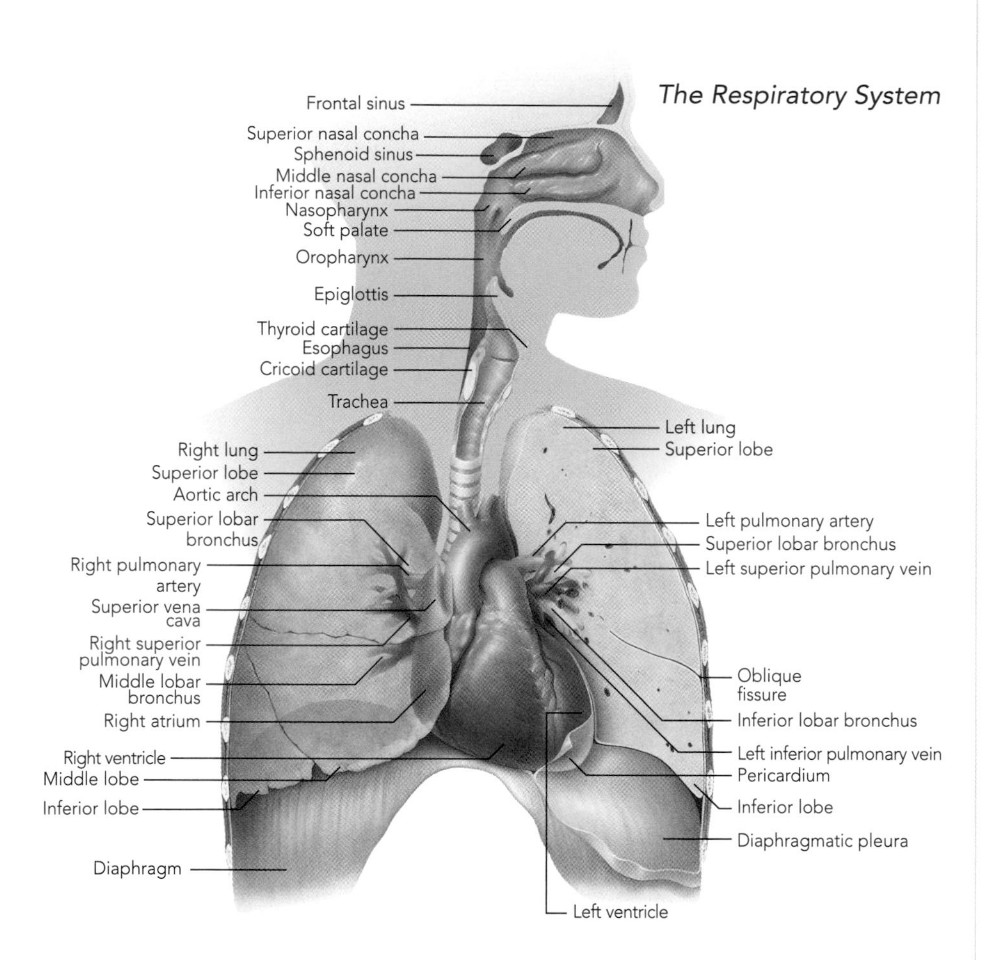

THE HUMAN BODY'S essential link to the outside world is the respiratory system. Immersed in a bath of air, the body takes in oxygen with every breath. Each of its trillions of cells depends on this infusion for its survival: Without oxygen, brain cells begin to die within five minutes, and other tissues expire within hours.

In biological terms, however, respiration is more complex. It encompasses the whole process by which oxygen is delivered to the cells and used to break down glucose from food, releasing energy. Respiration also removes the toxic waste product of that process, carbon dioxide, with each exhalation. Other respiratory tasks include moving air across the vocal cords for speech and pulling smells into the nose.

The respiratory system can be divided into two zones, conducting and respiratory, according to their function. The conducting portion includes all the tubes and channels that carry gases in and out of the body. It begins with the nose and nasal cavity, the space within and behind the nose. Below the nasal cavity is the oral cavity, the inside of the mouth; the nasal and oral cavities are connected by the pharynx (throat). At the base of the pharynx lies the larynx (voice box), which leads into the trachea (windpipe). This passageway branches like an upside-down tree, with one branch, or bronchus, extending into each lung. These primary bronchi subdivide into secondary bronchi, one for each lobe of the lungs. The secondary bronchi branch into tertiary bronchi, which split repeatedly into ever smaller tubes, or bronchioles, some no larger than 0.02 inch (about half a millimeter) across.

RESPIRATORY ZONE

The respiratory zone is made up of the tissues in the lungs where gas exchange occurs. It consists of the

TAKE A BREATH

+ AT REST, MOST people breathe about 8 to 16 times a minute.
+ An inhalation lasts about two seconds; an exhalation, about three.
+ Stress can cause hyperventilation: abnormally fast or deep breathing. This action reduces carbon dioxide and increases oxygen in the blood, lowering blood acidity and making the person feel light-headed.

WEBLINK: Take an interactive view of Respiration at http://nationalgeographic.com/humanbody

smallest, respiratory bronchioles and the alveoli, delicate air cavities that are attached—like clusters of grapes—to them via alveolar ducts. Capillaries cover the alveolar sacs, picking up oxygen and delivering carbon dioxide through the porous alveolar walls.

The lungs themselves are made up of both conducting and respiratory structures. These elastic, spongy organs consist mainly of air spaces, connective tissue, and blood vessels. They occupy most of the chest cavity, with the heart nestled next to the left lung and connected to both lungs via the pulmonary arteries and veins.

Both lungs are covered by a fluid-filled double membrane, the pleura. Beneath them, a dome-shaped muscle called the diaphragm contracts and relaxes to expand and relax the lungs. Several areas in the brain, known collectively as the respiratory center, collaborate to control the motions of breathing (pulmonary ventilation), relieving us of the task of consciously breathing in (inhalaton or inspiration) and out (exhalation or expiration) every moment.

CONDUCTING ZONE

The most visible parts of the respiratory system, the nose and mouth, are the entryways for the conducting zone. The nose plays a key role in drawing in air, moistening it, and filtering out foreign (and possibly harmful) substances. It also houses receptors that enable the

PERCUSSION

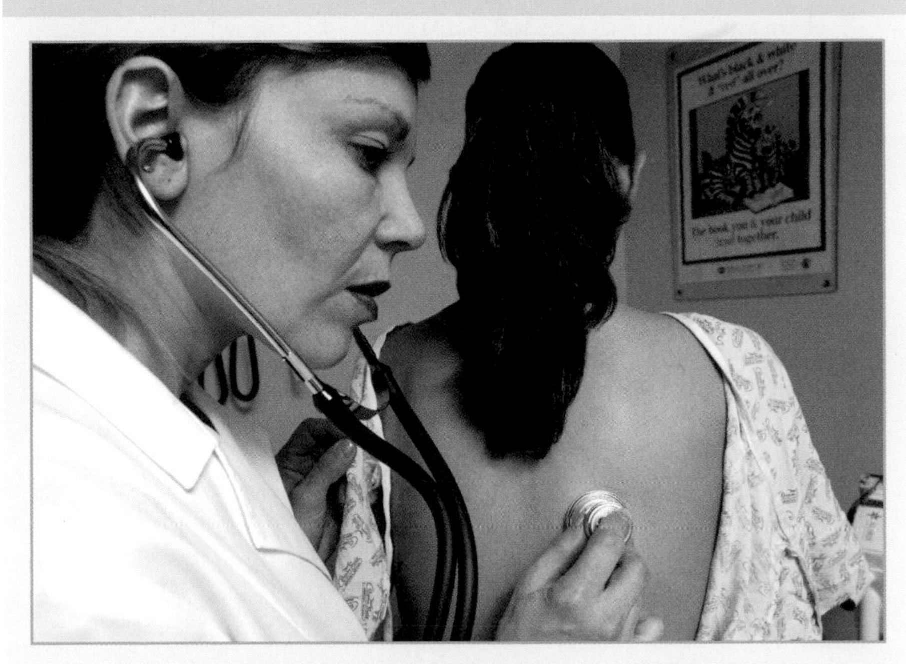

The stethoscope has replaced percussion in some doctors' physical exams.

IT'S A FAMILIAR hallmark of the physical exam: The physician bends over your chest and, using the fingertips, taps various spots, listening intently. This simple technique, called percussion, has been used successfully since the late 18th century to aid in the diagnosis of pneumonia, tuberculosis, emphysema, asthma, and other lung disorders. Yet as easy and as accurate as it is, percussion was not widely accepted until almost 50 years after it was first described.

The method's inventor, Josef Leopold Auenbrugger was born in 1722 in Graz, Austria. He was the son of an innkeeper, and as a youth, he learned how to tap on the sides of wine barrels to find the level of the wine inside. Later, as a physician in Vienna, he tried the technique on the human chest. Rapping sharply with the tips of his fingers held together, he distinguished categories of sound: the resonant sound of healthy lungs, like "the stifled sound of a drum covered with a thick woolen cloth;" the dull sound of congested lungs, as in pneumonia or fluid in the pleural spaces; and an unusually hollow resonance, as from emphysema and injuries that introduce air into the pleural cavity. Auenbrugger could accurately diagnose a wide variety of lung diseases, judge their severity, and predict the possible outcome.

In 1761 he published his description of percussion in a paper, and there it languished. Only when the technique was expanded upon by Napoleon's physician, Jean Nicolas Corvisart des Marets in 1808, did it gain general medical acceptance. Auenbrugger himself died of pneumonia in 1809; he is said to have correctly predicted the day of his own death. ■

sense of smell and adds resonance to the voice.

Within and behind the external nose (which is made of bone and flexible cartilage under the skin) is the nasal cavity. The area just inside the nostrils is called the vestibule; lined with coarse hairs (vibrassae) that trap incoming particles, it is the body's first line of defense against dust, pollen, and microbes. The nasal cavity is divided into right and left sides by the nasal septum, made of cartilage; at the top, just behind the bridge of the nose, is the olfactory epithelium, a membrane holding receptors for smell.

Mucous membrane containing cells with hairlike projections (cilia) lines the nasal cavity. Mucous glands under the membrane secrete about a quart of sticky mucus a day. The mucus traps dust and attacks invading germs with antibacterial enzymes. The cilia then sweep the mucus and contaminants into the pharynx, where they can be swallowed into the stomach and destroyed by stomach acid. Occasionally, irritating particles tickle nerve endings within the nose, triggering a sneeze that expels the invaders.

Above and around the nasal cavity are the sinuses, cavities in the skull that are lined with mucous membrane and are thought to

WHEN IT'S COLD

COLD, CRISP WEATHER can cause an unpleasant side effect: a runny nose. The cold air slows the thousands of moving cilia within the nose, allowing mucus to drain out the nostrils and not down the throat. The tendency of moist breath to condense in cold temperatures compounds the effect, adding droplets of water to the whole mess.

A hiker releases a cloud of moisture-laden breath on a cold, crisp day.

SEE ALSO: Chapter Nine, "The Senses," PAGE 256, Chapter Twelve, "Innate Defenses," PAGE 322

aid in warming and moistening inhaled air and to resonate during speech. Most people become aware of them when they are inflamed from allergy or infection, a painful conditon called sinusitis.

Beneath the nasal cavity is the oral cavity, the space between the lips and the throat. More important for digestion than for respiration, it serves as an alternate passageway for air into the lungs. It is lined by mucous membrane but is not as efficient at moistening and filtering air as the nose. The pharynx connects both cavities to the larynx and trachea below. Muscular and funnel-shaped, it conducts air and food, houses the tonsils, and creates resonance for speech.

RESHAPING THE NOSE

Before-and-after photos show the results of an early nose job performed in France.

RHINOPLASTY—the nose job—is one of the oldest plastic surgery techniques on record. As early as 400 B.C. or so, the Indian physician Sushruta described his technique for repairing mutilated noses (an injury commonly inflicted by bandits or as a punishment). Sushruta recommended cutting a plant leaf in the shape of the desired nose and using it as a pattern to cut skin from the forehead or cheek. "Attach it [the skin] to the nose tip," he wrote, "and quickly join it with perfect sutures."

The use of rhinoplasty to repair mutilated noses continued through the Renaissance and then gained strength in the 19th century, when the advent of anesthesia made the procedure easier on the patient. Around the same time, people began to have the surgery for aesthetic reasons; in 1891, for instance, the American physician John Roe described straightening a woman's bumpy nose by making an incision between the nasal bone and upper cartilage. Today, rhinoplasty usually involves making a small incision, separating the skin of the nose from its underlying bone and cartilage, and then reshaping that framework. More than 200,000 nose jobs are performed each year in the United States. ■

THROAT ZONES

The pharynx is divided into three sections: the nasopharynx, behind the nasal cavity; the oropharynx, behind the oral cavity; and the laryngopharynx, which connects to the larynx and the esophagus, the tube that leads to the stomach. Two auditory, or Eustachian, tubes lead from the nasopharynx to each middle ear, helping to equalize air pressure. Ciliated cells in the nasopharynx propel mucus (containing dust and microbes) down the throat. The back of the nasopharynx also contains the adenoids (pharyngeal tonsil). This oval lump of tissue, like tonsils in the throat, contains lymphoid tissues and, as part of the immune system, helps to trap and fight bacteria.

Below the nasopharynx, the oral cavity opens into the oropharynx through the fauces, the opening seen at the back of your opened mouth when you look at it in the mirror. The oropharynx holds two pairs of tonsils: the lingual tonsils at the base of the tongue and the palatine tonsils visible at the back of the mouth, one on each side. Occasionally, and especially in children, the palatine tonsils are overwhelmed by a bacterial or viral infection and become inflamed, swollen and sore. This condition is known as tonsillitis; the tonsils may be removed in intractable cases.

The laryngopharynx, the lowest section of the throat, leads to the larynx, where the airways and digestive passages take separate paths. The esophagus carries food to the stomach, whereas the larynx takes air to the trachea and lungs.

Just two inches long, the larynx plays a large role in social interactions. Within the larynx are the vocal cords; with the mouth and other parts of the respiratory system, they make speech possible. The larynx is more than a voice box, however. It also serves as a passageway for air and as a guardian mechanism that directs food down the esophagus. The entire larynx rises when you swallow, as you can feel if you put your hand on the front of your throat.

Short as it is, the larynx is relatively complex. Its wall consists of nine pieces of cartilage connected by muscles and ligaments. The epiglottis, a large, spoon-shaped piece of cartilage, drops down over the top of the larynx when you swallow, keeping food out of the airways. Below it is the thyroid cartilage, whose prominent midline ridge is visible as the Adam's apple—larger in men due to hormones released during puberty. The ring of cricoid cartilage below the thyroid cartilage attaches to the top of the trachea; it is the point of entry for an emergency tracheostomy. Most important for speech is the pair of arytenoid cartilages at the top and back of the cricoid cartilage. These wedge-shaped structures attach to the vocal cords and to the muscles of the pharynx that move them.

The versatile larynx can produce sounds ranging from a whisper to the train-whistle scream, like the one immortalized by Janet Leigh in *Psycho.*

CREATING SPEECH

The vocal cords (or vocal folds) are two bands of tight, elastic ligament

WHAT CAN GO WRONG

IN THE 1990s, longtime singing star Julie Andrews realized that her famous multi-octave voice was growing hoarse and losing its range. She had developed a common singer's ailment: vocal cord nodules, benign callus-like growths that prevent the vocal cords from vibrating normally. Nodules face each other, one on each cord. Other benign lesions include polyps (softer, growths that usually appear singly) and cysts (masses that often form beneath the cords). Lesions are thought to result from chronic strain of the voice by singing too loudly or shouting, or from overuse while sick. Symptoms include hoarseness, vocal fatigue, and loss of singing range and suppleness. In Andrews's case, surgery to remove the nodules was unsuccessful. Typically, doctors prescribe voice rest and voice therapy before suggesting surgery, but the microsurgery is usually beneficial.

covered with white mucous membrane and attached to either side of the larynx. When relaxed, they form a V-shaped opening, allowing air to pass freely into the windpipe. When muscles in the larynx contract, the V closes, leaving a narrow space between the cords. Exhaled air rushing past the tensed cords makes them vibrate, producing "voiced" sounds.

The pitch of those sounds depends on the tension and speed of vibration, which ranges from about 80 to 200 times a second in adults and can exceed 300 times per second in children; the faster the vibration, the higher the pitch. Larger vocal cords, such as those in men, produce deeper sounds. If the cords are relaxed, air passing through them produces little to no sound, and at a very low pitch. The throat, mouth, nasal cavity, and sinuses all contribute to the voice's resonance. Muscles of the face, tongue, and lips help to articulate the sounds of speech like vowels and consonants.

Most people don't appreciate the complex mechanics of speech until they lose it; laryngitis, an inflammation of the larynx, can make the vocal cords swell until they cannot vibrate, resulting in hoarseness or complete voicelessness. Professional vocalists spend years studying and working with their vocal cords, the resonating chambers in the head, the lift of the soft palate, and other areas of voice production to achieve their high degree of vocal control.

Below the larynx is the trachea, or windpipe. About four to five inches long, it carries air in and out of the bronchi, the two main passageways leading to the lungs. Like the nose and larynx, it is lined with mucous membrane and ciliated cells, which sweep dust and other particles up toward the pharynx and away from the lungs. Highly irritating particles can be expelled at speeds of up to 100 miles per hour in a cough.

Sometimes, though, a piece of food or other foreign object gets stuck in the trachea, blocking the airway and cutting off air supply. In the Heimlich maneuver, an energetic thrust up into the diaphragm forces air out of the lungs and through the trachea, dislodging the blockage. When an immovable obstruction occurs above the larynx, doctors can perform a tracheostomy, cutting into the trachea just below the cricoid cartilage to create a new opening that allows air to flow in and out of the lungs.

SNORING OCCURS when air passages are blocked by a stuffed nose, bulky tissue from excess weight, large tonsils or adenoids, throat and tongue muscles relaxed by alcohol or medicine, a long soft palate, or a long uvula. Snoring may alternate with sleep apnea, when breathing intermittently stops. This dangerous condition should be evaluated by a doctor.

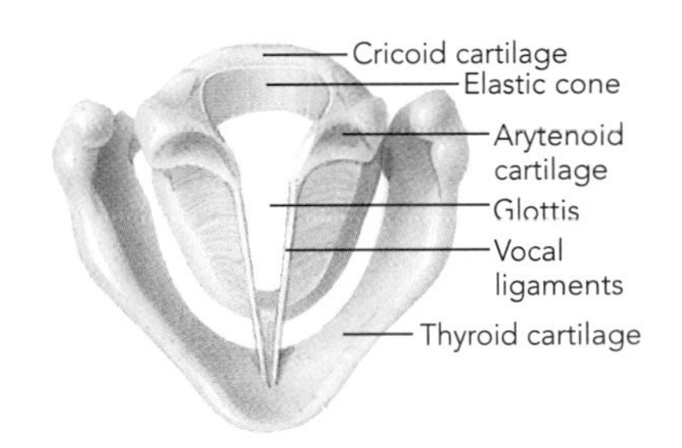

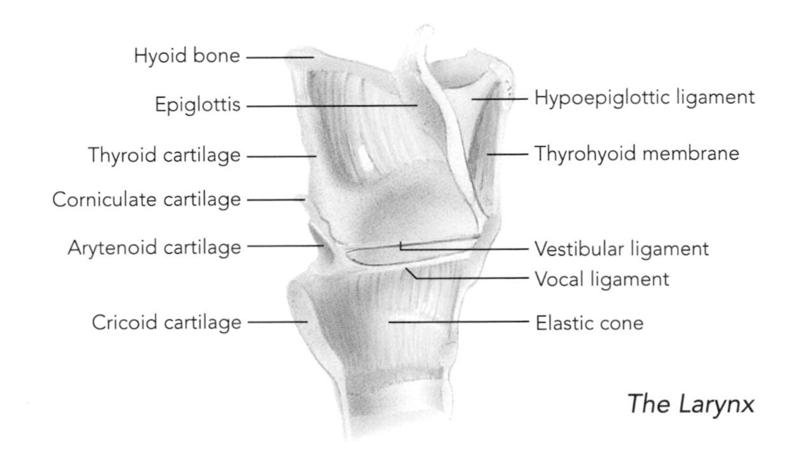

The Larynx

THE LUNGS

A HARDWORKING PAIR

THE LUNGS ARE THE central organs of respiration. They are so light that they float in water. In fact, the word *lung* comes from an early Germanic word for "light." In older British usage, *lights* is a synonym for *lungs,* as in the expression "punch your lights out." The well-known lightness of the lungs reflects what is perhaps their key characteristic: They are filled with millions of air pouches, or alveoli, that provide an immense amount of surface area for moving gases both in and out of the bloodstream.

Pale pink at birth, the lungs gradually darken to gray or even black with age as they pick up carbon particles. These two cone-shaped, spongy organs occupy most of the chest cavity within the ribs: The tip, or apex, of each can be found just above and behind the collarbone, while the base rests on the diaphragm above the lowest pair of ribs. Between the lungs, in an area called the mediastinum, are nestled the heart and its attendant veins and arteries, along with the esophagus, trachea, bronchi, and some lymph nodes and nerves.

The "root" of each lung is the bundle of bronchi, lymphatic and pulmonary vessels, and nerves that enters through the hilum, an indentation in the central portion of the lung. The left lung is about 10 percent smaller than the right because it contains the cardiac notch that curves around the heart. The right lung, however, is shorter than the left, because the liver presses the diaphragm up under its base.

The lungs are divided into sections, or lobes: three in the right lung and two in the left. Each lobe is supplied with air by its own bronchus and each has its own artery and vein. The lobes are subdivided into segments and then further divided into lobules, about 130,000 in each lung. These polyhedral structures give lungs their slightly lumpy, tiled appearance.

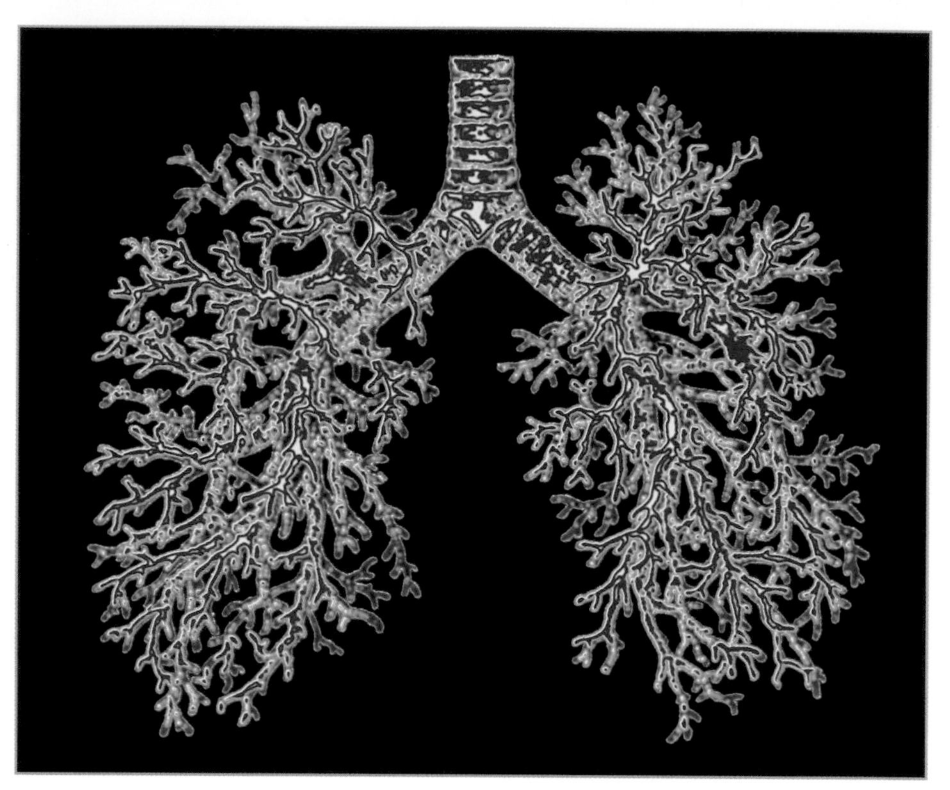

In this enhanced image of a cast, two bronchi branch off from the trachea and divide repeatedly into the upside-down-tree structure of the lungs.

LUNG FACTS

+ TOGETHER, THE LUNGS weigh just over 2.5 pounds.
+ More than 2600 gallons (10,000 liters) of air pass through the lungs each day.
+ The body loses about 11 ounces of water every day in the form of exhaled water vapor; people may lose a small amount of weight in their sleep due to this water loss.

DEFENSE SYSTEM

The lungs are versatile defenders against airborne invaders: dust, pollen, viruses, bacteria, and other unwanted substances. Specialized cells in the lung tissue form an

WEBLINK: See the Lungs in action at http://nationalgeographic.com/humanbody

important part of the immune system. Hairlike cilia on some cells sweep invaders out of the airways; others detect and pick up foreign matter in the air sacs (see "Breathing," page 168).

Encasing each lung is a thin, double-walled covering called the pleura or pleural membrane. The outer wall of the membrane, called the parietal pleura, clings to the inside of the chest (thoracic) wall and to the surface of the diaphragm. The inner layer, the visceral pleura, covers the lung's surface. The plurae produce a slippery pleural fluid that fills the narrow cavity between the two layers, allowing the membranes to slide easily over each other. The surface tension of the fluid makes the two membranes stick together. When the diaphragm contracts, it pulls the outer pleural membrane with it, which in turn pulls the inner membrane and with it, the lungs, so that they expand.

The pressure within the pleural cavity is less than atmospheric pressure, which also helps to keep the lungs expanded. If air gets into the pleural cavity (say, if it is pierced by a broken rib), the pressure within it equalizes with the atmosphere and the lung collapses. Doctors treat this condition, called a pneumothorax, by inserting chest tubes to suck the air out, reinflating the lung.

Excess pleural fluid is constantly being pumped from the membranes into the lymphatic system.

DISCOVERING WHAT LUNGS DO

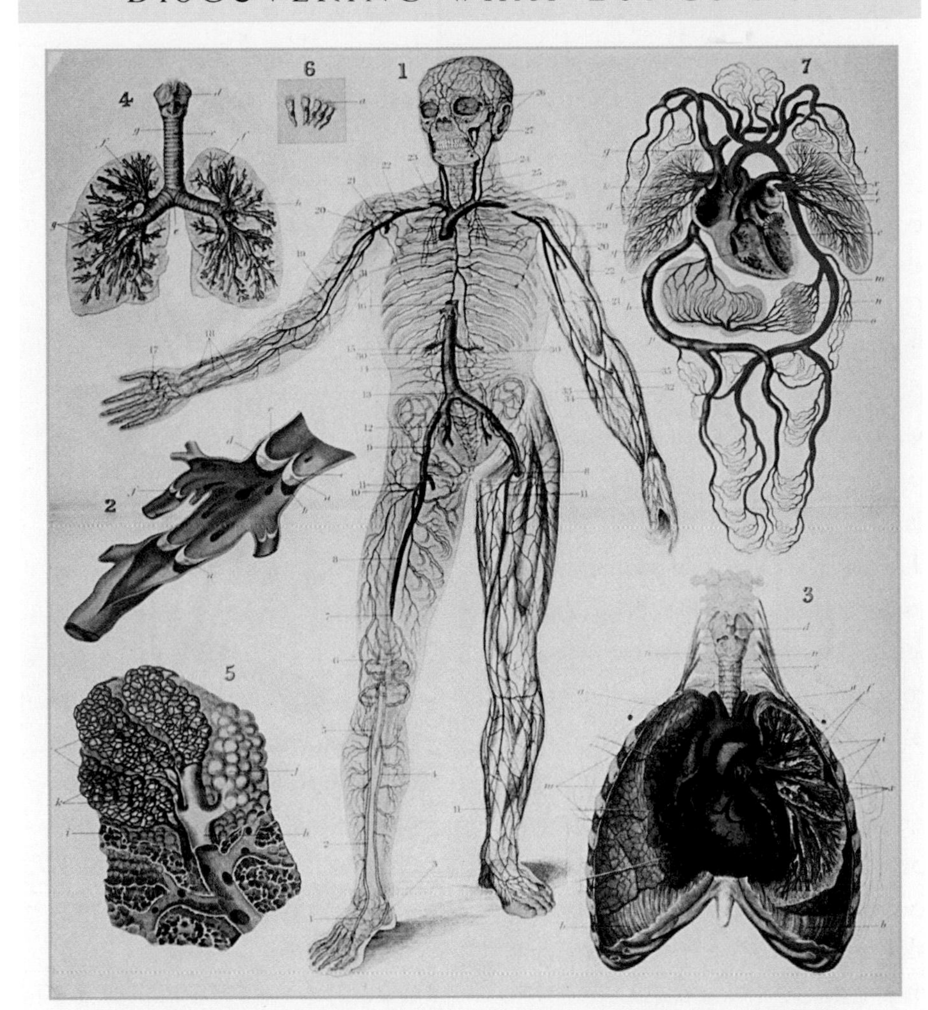

Anatomical art from the 1850s depicts the lungs and circulatory system.

UNTIL THE DISCOVERY of the element oxygen, physicians struggled to understand the purpose of the lungs. Early on, they saw that the organs expanded and contracted; the second-century Greek physician Galen believed this action helped to regulate heat within the body. The theory of the lungs as thermostat persisted with some refinements through the Renaissance. Leonardo da Vinci noted that "impurities or 'sooty vapors' are carried back to the lung by way of the pulmonary artery, to be exhaled into the outer air." In the 17th century, the British anatomist William Harvey outlined the interrelationship of the respiratory and circulatory system, but it was not until 18th-century French scientist Antoine-Laurent Lavoisier discovered the gas he named oxygen that the essence of respiration became clear. "Breathing animals are active combustible bodies that are burning and wasting away," he wrote and reported that air provided the oxygen for this combustion through the lungs. ■

Occasionally the pleural membranes become inflamed, a condition known as pleurisy. In pleurisy's early stages, the two membranes rub together painfully; in its later stages, they may produce too much fluid, which presses on the lungs.

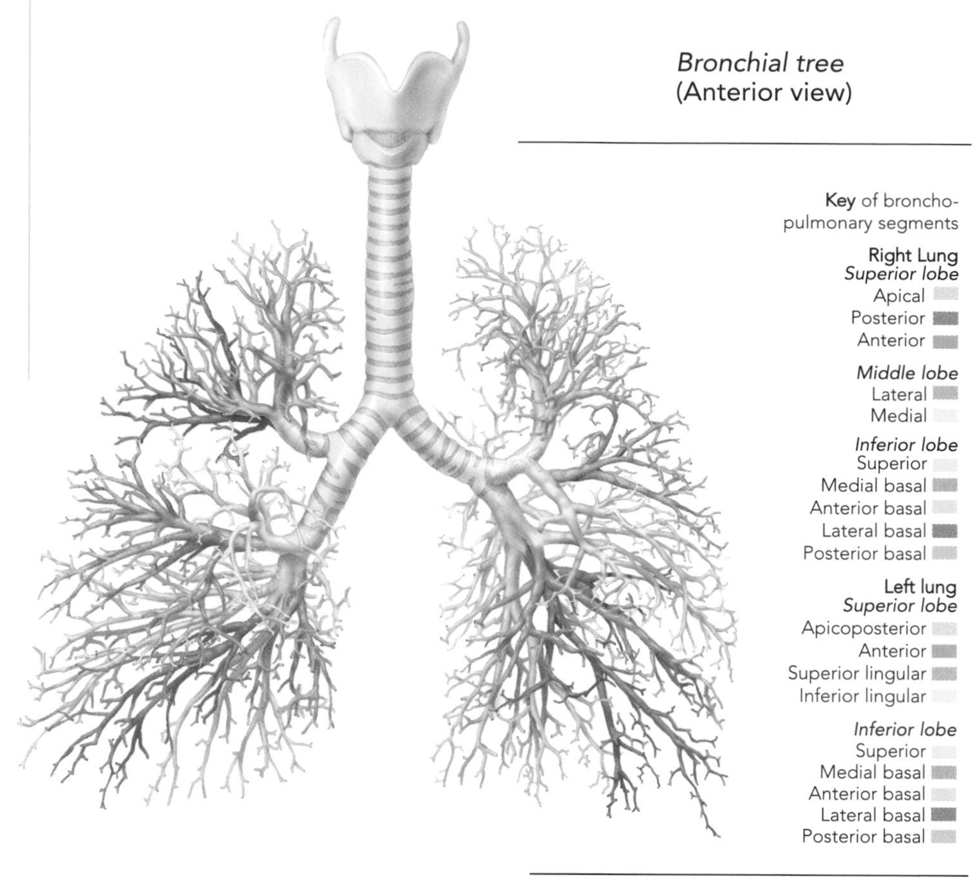

TREE OF LIFE

The lungs enclose a branching network of airways known, appropriately, as the bronchial tree. The trunk of that tree is the trachea, the central airway leading from the larynx. At the trachea's final ring of cartilage, it divides into two large branches, the primary bronchi, one leading into each lung.

Where the two bronchi divide at the trachea is an internal ridge, called the carina, lined with sensitive mucous membrane; any tickling or irritation of the carina leads to violent coughing. The right primary bronchus is shorter, wider, and more upright than the left; inhaled objects—such as food when choking—are more likely to take that path. They typically end up lodged in the right lung rather than the left.

The larger bronchi, like the trachea, are ridged with cartilage and lined with ciliated mucous membrane. The cilia move ceaselessly to propel foreign particles up the airways and eventually out the mouth or down the esophagus into the stomach. As the bronchi become smaller, stiff cartilage gives way to elastic, smooth muscle and the tissues change from ciliated, mucus-

SEE ALSO: Chapter Five, "Blood Vessels," PAGE 150

WHAT CAN GO WRONG

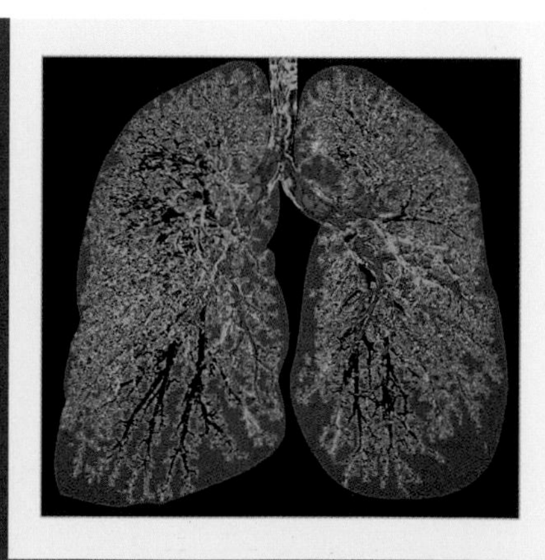

EMPHYSEMA (the term means "inflation") is a disorder that damages healthy lungs, left, by destroying alveolar walls, forming abnormally large air spaces, opposite, that cannot deflate properly to expel air during exhalations. With less respiratory surface area, less oxygen also diffuses into the bloodstream. As a consequence, people with emphysema become rapidly breathless with mild exercise. They begin to cough or wheeze and may compensate for a decreased oxygen intake with rapid inhalations. Cigarette smoking is the most common cause, but polluted air and industrial contaminants can also secreting cells to a simpler, single layer of cells without cilia.

The primary bronchi divide again in the lungs into secondary (lobar) bronchi, one for each lobe of the lungs. They branch again into tertiary (segmental) bronchi and then into smaller and smaller bronchi and bronchioles until, after about 16 divisions, they taper into the thousands of so-called terminal bronchioles, each one not much wider than a needle.

During exercise, hormones relax and dilate these small airways, or bronchioles, allowing more air to flow through them. But their elasticity has a negative side, as well: Allergic reactions and asthma attacks can close off the bronchioles, choking off the body's supply of oxygen (see "Gasping for Air," page 172).

BLOOD CONNECTION

Surrounding the entire bronchial tree and covering it like a fine webbing of vines are the veins and arteries that feed into and out of the lungs. No two systems in the body are more closely wedded than the respiratory and the circulatory. Pulmonary arteries from the right side of the heart pump oxygen-poor blood into the lungs, where it flows through branching blood vessels along the bronchi until it reaches the capillaries that cover the alveoli. Carbon dioxide then diffuses from the capillaries into the alveoli as oxygen diffuses from the alveoli into the capillaries. The oxygenated blood returns to the left side of the heart via four pulmonary veins.

+ BLACK LUNG +

COAL WORKERS' pneumoconiosis is a scourge of miners, who inhale coal dust into their lungs. Dust particles scar and thicken tissues, causing shortness of breath and, in some cases, emphysema or even heart failure. There is no cure. The disease is less common since the Federal Coal Mine Safety and Health Act of 1969 reduced the coal dust in mines.

GAS STATION

Reaching out from the terminal bronchioles are the tiniest twigs on the bronchial tree, the respiratory bronchioles. Slender tubes called alveolar ducts connect these bronchioles to bumpy alveolar sacs containing a cluster of delicate, cuplike alveoli, each about 0.008 in. (0.02 cm) wide. These are the central respiratory structures of the lungs, the site of gas exchange between the respiratory and circulatory systems. The estimated 300 million alveoli that fill the lungs—each one barely visible to the naked eye—provide a vast amount of surface area for supplying oxygen to the bloodstream and removing carbon dioxide.

Far thinner than tissue paper, the wall of each alveolus consists of a single layer of cells. Slender capillaries surround the alveoli, each

WHAT CAN GO WRONG

harm the lung tissue. Certain enzymes produced naturally by the body, called proteases, are believed to attack the connective tissue of the alveoli unless blocked by protease-inhibiting molecules, also normally made by the body. Cigarette smoke apparently impairs these inhibitors and may increase the destructive proteases. Emphysema cannot be reversed, but its progress can be halted by removing the cause—for example, by quitting smoking. Careful exercise, oxygen therapy, bronchodilators, lung volume reduction surgery, or even lung transplants can ease breathing in emphysema patients.

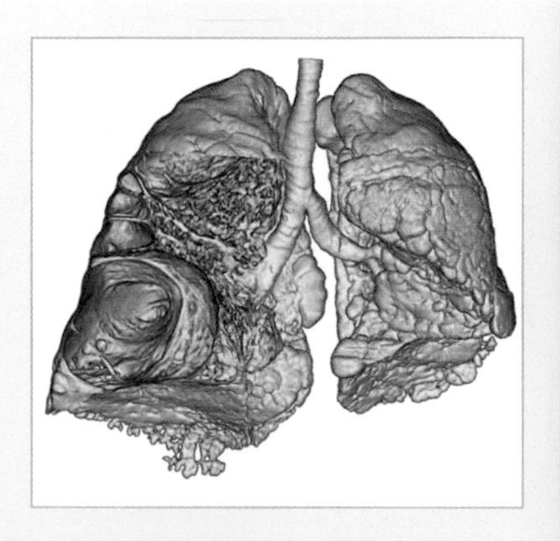

one fused to the surface of an alveolus, forming a shared wall known as the respiratory membrane. Only one-sixteenth the diameter of a red blood cell, the membrane allows gases to diffuse easily between the air sacs and the bloodstream.

SURFACTANT

On the alveolar side of the respiratory membrane, the wall consists of two types of cells. Type I cells, the most common, are the main portals for gas exchange between the alveoli and the capillaries.

The other kind, called Type II cells, releases alveolar fluid, which coats the surface between the cells and the air. In this fluid is surfactant, a slippery, detergent-like substance that lowers the surface tension of alveolar fluid. Without it, the water molecules in the fluid would cling together, pulling away from the air and ultimately collapsing the delicate alveoli.

Surfactant is not made in the lungs until the final two months of fetal development. Its absence can be deadly because babies born without surfactant, as with infants delivered prematurely, struggle to breathe. Until the discovery and analysis of surfactants in the 1950s, respiratory distress syndrome (RDS) was the leading cause of death in premature infants. It caused the deaths of some 10,000 babies a year in the United States alone. Today, pregnant women going into premature labor can be injected with steroids to encourage the fetus to begin developing surfactant on its own. Premature infants with RDS are treated with positive-pressure respirators and natural or artificial surfactants to aid their breathing. Since the development of these treatments, deaths from RDS have decreased to about 1,000 a year.

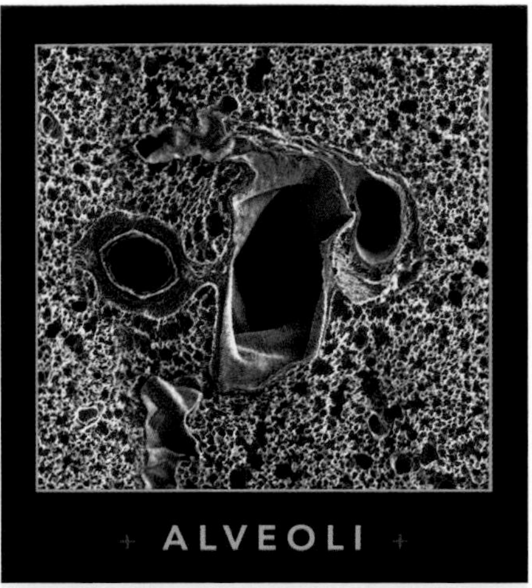

+ ALVEOLI +

+ LUNGS CONTAIN about 170 alveoli (in purple above) per cubic millimeter.
+ If stretched flat, the surface of the alveoli would cover about 861 square feet (80 square meters), roughly the area of a handball court.
+ Humans are born with about one-fifth their adult alveoli; they usually reach the full number by age 8.

Swimming freely through the fluid along the lungs' alveolar walls are millions of macrophages, cells that eat and destroy invading microbes and foreign particles. These vigilant guard dogs of the immune system help keep the alveoli remarkably clean. When their task is accomplished, the macrophages themselves are swept up into the bronchial tubes and out of the lungs.

The airways, with their easy access to the bloodstream, can provide a direct pathway for medicine that can be inhaled in a fine mist of droplets. Many asthmatics are familiar with nebulizers that deliver medicine to dilate the bronchial tubes; nebulization can also carry antibiotics directly into the lungs.

BREAKTHROUGH

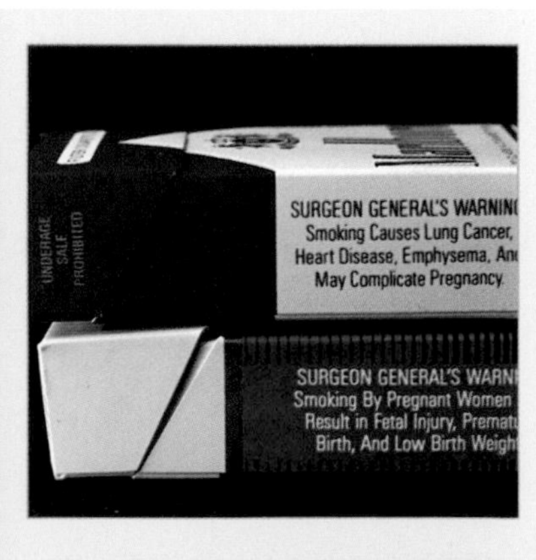

ON SATURDAY, January 11, 1964, reporters were ushered into the auditorium of the State Department in Washington, D.C., shut in without telephones, and handed a 387-page report entitled "Smoking and Health: Report of the Advisory Committee to the Surgeon General of the Public Health Service." For the first time, the U.S. government officially acknowledged what scientists had long known: Cigarette smoking damaged lungs, opposite, and caused lung and laryngeal cancer and chronic bronchitis. After 90 minutes, reporters were allowed out of the auditorium to spread the news to the public. It

CHAPTER GLOSSARY

ALVEOLUS (pl. alveoli). One of the tiny, thin-walled air cavities of the lungs; alveoli cluster in alveolar sacs around a common air passageway.

APNEA. Brief cessation of breathing. Can be caused by airway obstructions or problems in the part of the brain that controls breathing.

BRONCHIOLES. The small, final branches of the bronchi within the lungs that lead to the lobules and alveolar sacs.

BRONCHUS (pl. bronchi). Branching air passageway; primary bronchi lead from the trachea into each lung and then split repeatedly into smaller bronchi.

CILIA. Tiny, hairlike projections from a cell that can move the cell or move substances along its surface.

COMPLIANCE. The ability to stretch or yield.

CONDUCTING ZONE. All the air passageways of the respiratory system that do not play a role in gas exchange, such as the pharynx, trachea, and bronchi.

FAUCES. The narrow opening from the mouth into the pharynx.

GLOTTIS. The vocal cords and the space between them.

HEMOGLOBIN. A part of red blood cells, made of protein and pigment, that transports oxygen and some carbon dioxide.

LARYNX. The respiratory passageway between the pharynx and the trachea, containing the vocal cords; also known as the voice box.

MEDIASTINUM. The portion of the chest cavity between the lungs, containg the heart, trachea, bronchi, esophagus, and other structures.

NEBULIZATION. The reduction of medicine to a fine spray; often used for lung disorders such as asthma.

PHARYNX. The respiratory and digestive passageway that starts behind the nasal cavity and extends to the larynx and esophagus.

PLEURA. The double-layered membrane that covers the lungs and lines the chest cavity; pleural fluid fills the narrow space between the two layers.

PNEUMOTHORAX. A condition in which gas enters the pleural cavity through a rupture or puncture.

PULMONARY VENTILATION. Breathing; The rhythm of inhalation and exhalation.

RESPIRATION. The process of supplying the body with oxygen and ridding it of carbon dioxide. External respiration involves bringing oxygen into the lungs and then into the bloodstream; internal respiration is the exchange of gases between the body's cells and the blood.

RESPIRATORY CENTERS. Brain regions in the medulla oblongata and pons of the brain stem that control the rate and depth of breathing.

RESPIRATORY MEMBRANE. The tissue joining an alveolus and a capillary, through which gases pass.

RESPIRATORY ZONE. Microscopic structures within the lung—primarily the alveolı and their bronchioles—that are the site of gas exchange.

SURFACTANT. Slippery fluid produced by cells in the alveoli. Reduces the surface tension of water and prevents the alveoli from collapsing.

TRACHEA. Air passageway from the larynx to the bronchi; also called the windpipe.

BREAKTHROUGH

became one of the leading stories of the year. Despite a vigorous defense of smoking from tobacco companies, the U.S. Congress quickly passed laws to require health warnings on cigarette packs, opposite), ban cigarette advertising on television, and require further annual reports on cigarettes and health. The campaign to end smoking, jump-started by the 1964 report, dropped smoking rates among U.S. adults from 42.4 percent in 1965 to 20.9 percent in 2005, a new low. Increased awareness of the dangers of secondhand smoke has prompted bans on smoking in many workplaces and public spaces.

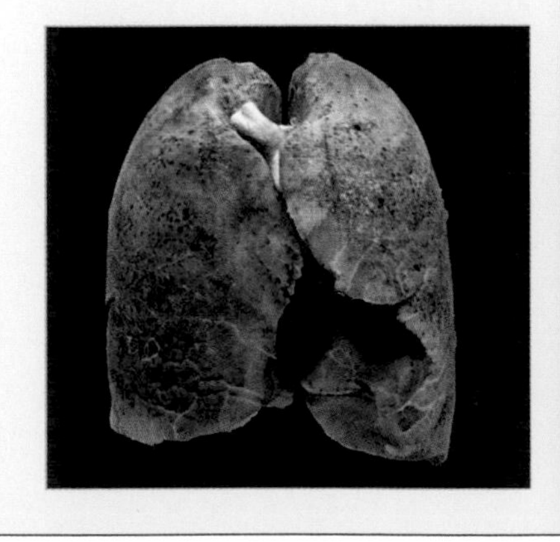

BREATHING

THE MECHANICS OF INHALING AND EXHALING

A cutaway model of the muscles used in respiration exposes the intercostals around the ribs and the dome-shaped diaphragm.

THE STEADY, EVEN in-and-out motion of breathing—called pulmonary ventilation, in medical terms—relies on an elegantly simple mechanism of pressure changes within the chest. It also depends on a complex feedback system between the body and the brain.

The chest, or thoracic cavity, is enclosed by the rib cage and the intercostal muscles, which are located between the ribs. The diaphragm, a dome-shaped muscle, forms the floor of the thoracic cavity. When relaxed, the diaphragm curves up under the lungs. When it contracts, it pulls down and flattens out.

Within the thoracic cavity, the lungs are attached to the chest wall and to the diaphragm by the double-layered pleural membrane. Pleural fluid fills the thin cavity between the layers; the fluid holds the layers together while letting them glide freely over each as the lungs expand and contract. Also helping the layers hold together is the low pressure inside the pleural cavity: normally at about 756 mmHg (millimeters of mercury), it is below the atmospheric pressure of 760 mmHg.

+ SNEEZING +

SNEEZES HAVE GENERATED their share of "urban legends."

+ Your heart stops for a moment when you sneeze. Not true: Your heart rate might change briefly, but your heart does not stop.
+ You can't keep your eyes open during a sneeze. If you do, your eyeballs will pop out. Wrong: It's difficult to do, but some people can manage it—and keep their eyeballs safely in their heads.
+ A woman won't get pregnant if she sneezes immediately after sex. Not so: One sneeze, or several, will not discourage or dislodge 300 million determined sperm.

SEE ALSO: Chapter Four, "The Muscles," PAGE 106

BALANCING ACT

The air pressure inside the lungs themselves varies, depending on whether a person is inhaling (breathing in) or exhaling (breathing out). Just before each inhalation, air pressure inside the lungs is the same as air pressure outside. The lungs are, essentially, at rest.

During an inhalation, the diaphragm contracts, pulling the floor of the chest down. At the same time, the external intercostals pull the ribs up and out. This increases the volume within the chest cavity and, with it, the volume of the lungs, which are attached to the chest wall.

A basic rule of physics, Boyle's law, states that at a constant temperature the pressure of a gas varies inversely with its volume: The greater the volume for a given amount of gas, the lower the pressure. Thus, as the lungs expand, the air pressure within them drops until it is about 1 to 3 mmHg lower than the air pressure outside. This pressure difference is enough to trigger an inflow of air to equalize the pressure; typically, about 500 milliliters (a little more than 2 cups' worth) restores the balance.

During an exhalation, the respiratory muscles relax. The diaphragm regains its rounded shape and the rib cage returns to its narrower contours. The lungs recoil, the volume within them dropping, which increases the air pressure inside until it surpasses the pressure outside the body. Air from the lungs (now containing carbon dioxide that has diffused from the capillaries) is then expelled as the lungs again equalize the pressure inside and out. Not all the air leaves the lungs during exhalation, however; a residual amount remains, keeping the alveoli open and the lungs inflated.

This is the process of quiet inhalation and exhalation, the sort of breathing one does at rest. During exercise, while singing, or sometimes when a person is suffering from an obstructive pulmonary disease (such as asthma, emphysema, or pneumonia), respiration becomes a more active process and uses additional respiratory muscles. In forced inhalation, muscles in the neck, in the front of the chest, and around the spine help to pull the rib cage out. Abdominal and internal intercostal muscles work together to force exhalation.

Ease of breathing is affected both by airway resistance and by lung compliance, a measure of flexibility. Air flowing through bronchi and bronchioles faces resistance from the walls of the passageways, just as liquid flowing through plumbing meets resistance in the walls of the pipes. The narrower the passageway, the greater the resistance. Some diseases—asthma, for instance (see "Gasping for Air," page 172)—thicken the walls of the airways and increase resistance to airflow, making breathing more

RESPIRATORY ACTS

YAWN	A deep inhalation, right, with the jaws wide open; cause unknown
CRY	Very similar to laughing, with different sounds and facial expressions
SNEEZE	Exhalation muscles contract, forcing air through the nasal passages
LAUGH	An inhalation followed by a number of short exhalations; the glottis stays open and the vocal folds vibrate
HICCUP	Spasms of the diaphragm followed by short, sharp inhalations; the glottis closes, making a sharp sound with each inhalation
COUGH	A deep breath followed by a closing of the glottis; then a sharp blast of air from the lungs forcing the glottis open and leaving through the mouth

SEE ALSO: Chapter Five, "The Heart," PAGE 140, Chapter Nine, "The Brain," PAGE 234

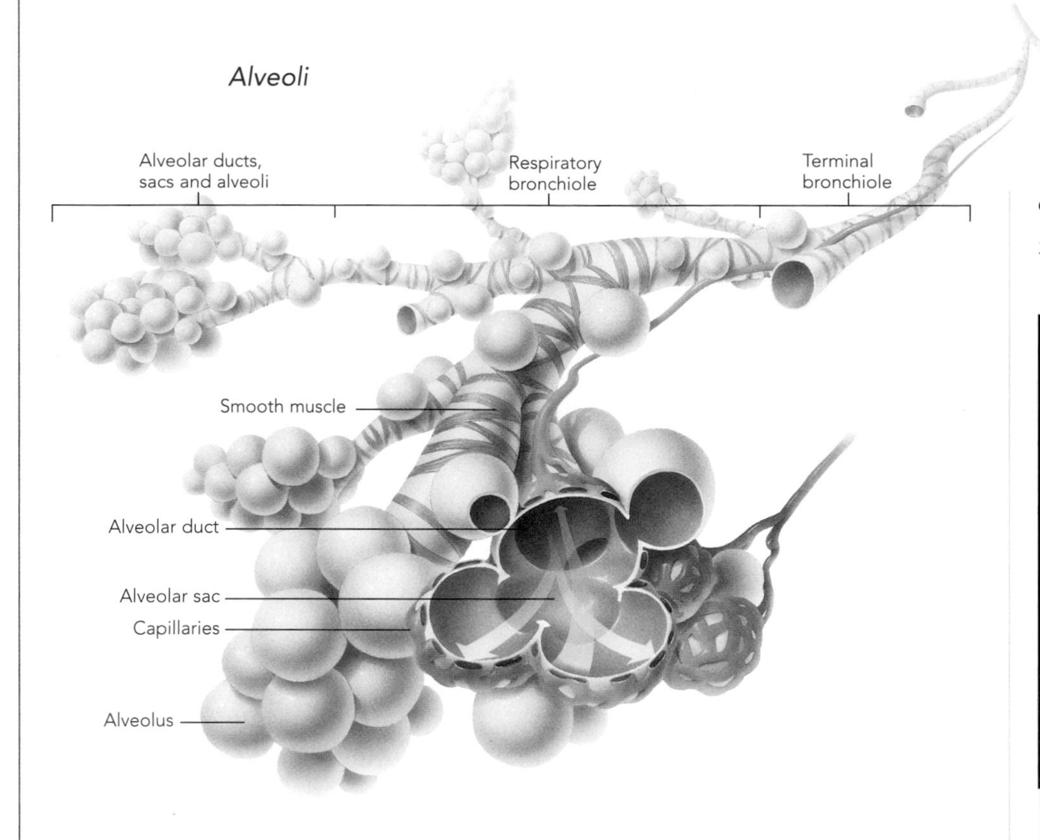

difficult. Some illnesses, such as tuberculosis or pulmonary fibrosis, scar or stiffen the lung tissue, decreasing compliance and making breathing more difficult. A few, such as emphysema (see "What Can Go Wrong," page 164), weaken the tissue, which also makes it less elastic and reduces the flow of oxygen into the blood.

Each breath draws air down through the bronchi and bronchioles into the alveoli. Here, in the air sacs, occurs the central process of external respiration, in which oxygen enters the bloodstream and carbon dioxide leaves it. Internal respiration describes the exchange of gases between the body's bloodstream and its cells and tissues.

In internal respiration, blood pumping through the body gives up its oxygen to the cells, which use it to break down glucose for energy, releasing carbon dioxide as a waste product. The oxygen-poor, carbon-dioxide-rich blood returns to the heart and is pumped through the right atrium and ventricle and then into the lungs. There, blood vessels in the presence of low oxygen concentrations constrict and drive the deoxygenated blood toward oxygen-rich areas, so that there is always a close match between blood that needs oxygen and alveoli that have it. (The opposite happens in most of the body's other tissues; arterial blood is directed toward, not away from, oxygen-poor areas.)

Once the oxygen-starved blood has reached the capillaries around the alveoli, oxygen and carbon dioxide diffuse in opposite directions through the respiratory membrane. The gases are driven

A HELD BREATH

YOU CANNOT HOLD your breath until you die. Respiratory centers in the brain stem are constantly monitoring the level of carbon dioxide in your blood, a level that rises steadily as you hold your breath. When carbon dioxide reaches a critical amount, the brain overrides your conscious control and forces your body to take a breath.

Most people can hold their breath for a minute to a minute and a half, on average. The world record for breath-holding belongs to Austrian Herbert Nitsch, who held out for 9 minutes 4 seconds while floating in water. Don't try this at home: Swimmers who hyperventilate before submerging, to lower their blood carbon dioxide and hold their breath longer, risk drowining by passing out in the water from lack of oxygen.

to do this by differences in pressure: Every gas exerts its own pressure, and each will move from an area of greater pressure to an area of lower pressure until it reaches equilibrium. Normally, oxygen in the alveoli exerts a pressure of 104 mmHg, while that in the pulmonary arteries is only 40 mmHg, so alveolar oxygen diffuses rapidly through the respiratory membrane between the alveoli and the capillaries until both sides have equal oxygenation—a process that takes about three-quarters of a second.

Meanwhile, carbon dioxide travels in the opposite direction from the blood to the alveoli. Although the difference between carbon dioxide's pressure in the blood and its pressure in the alveoli (45 mmHg to 40 mmHg, respectively) is much less than that of oxygen, it diffuses more rapidly because it's much more soluble in plasma (the noncellular fluid portion of circulating blood) and alveolar fluid.

A little (about 1.5 percent) of the oxygen entering the bloodstream from the lungs dissolves in the plasma. The rest latches on to hemoglobin, a protein found in red blood cells, turning it bright red. Carbon dioxide, meanwhile, enters the lungs from the bloodstream, where some (about 20 percent) was bound to a different portion of the hemoglobin molecules, and the rest was either dissolved in the plasma or riding along in the form of bicarbonate ions made from carbon dioxide and water.

Other gases can occasionally hijack this efficient mechanism. For example, hemoglobin has an affinity for carbon monoxide, and if exposed to this colorless, odorless gas will bind to it quickly, blocking out oxygen. People who breathe in carbon monoxide (say, in gases from a fire) become starved of oxygen and can die without treatment. In many hospitals, treatment involves hyperbaric therapy, in which the patient is held in a sealed chamber and breathes 100 percent oxygen at high pressure, forcing oxygen back into the tissues.

THE BRAIN'S ROLE

The many patterns of breathing—the quiet in-and-out of the resting body, the vigorous panting of exercise, sobbing, laughter, gasping with cold—are controlled by a network of respiratory centers in the brain stem and receptors elsewhere in the body. Although it is possible to consciously regulate breathing for short periods, control soon reverts to the respiratory centers, which are responsible for sending the right amount of oxygen to the right place at the right time.

Two closely related groups of neurons in the brain stem—the inspiratory center and the expiratory center—control the basic rhythm of inhalation and exhalation. In the typical cycle of quiet breathing, the inspiratory center sends a series of signals, lasting two seconds, to the diaphragm and external intercostals (rib muscles), prompting them to contract, expanding the chest cavity and drawing air into the lungs. Then the center becomes inactive for three seconds; the diaphragm and intercostals relax and the lungs recoil, exhaling air.

During forceful breathing—while exercising or singing—the inspiratory center also activates the expiratory center, which prompts the internal intercostal and abdominal muscles to

BREAKTHROUGH

THE IRON LUNG was once a lifeline for people with respiratory paralysis from polio, right, and other causes. The 1885 prototype, the "pneumatic cabinet," was a body-size box enclosing the patient. An attendant moved a rubber membrane, and changes in air pressure inflated and deflated the patient's lungs. A vacuum-pump-operated iron chamber that left the patient's head outside followed by 1929. "It looked more like a torture chamber than anything else," noted an observer. In the late 1950s, small positive-pressure ventilators began replacing the big machines, signaling the end of an era.

GASPING FOR AIR

ASTHMA IS A COMMON, chronic, and, for now, incurable respiratory disease. About 20 million Americans are afflicted with asthma, including about 3 to 7 percent of the adult population and 9 to 14 percent of children. Every year, 4,000 Americans die from it. Asthma's incidence is growing, with boys more likely than girls, and blacks more likely than whites to develop it. Incidences among adults are about equally divided between men and women. The condition is found disproportionately in city dwellers. Most frustrating is that the exact cause is unknown, although recent research has produced some promising leads.

The symptoms of asthma wax and wane in most people, but a typical episode is marked by wheezing, whistling, or noisy breaths; coughing; tightness in the chest; and shortness of breath. In the most severe attacks, the untreated asthma sufferer will become so short of oxygen that he or she may lose consciousness.

HISTORY

The word *asthma* comes from the Greek word meaning "to gasp." Until the 18th century, Western medicine typically treated asthma attacks as an imbalance of the four humors (yellow bile, black bile, blood, and phlegm) with asthmatics considered to be excessively phlegmatic. In the 18th and 19th centuries, physicians were able to study the disorder more closely in hospitals and apply their new understanding of anatomy and lung function to the disease. They began to understand the link between asthma and allergens and pollution: Many asthma sufferers, along with tuberculosis patients, found relief in moving to clean, dry, desert environments. By 1900, physicians also began to prescribe epinephrine to relax the airways during an asthma attack.

Aided by modern instruments, physicians today have developed a fairly clear idea of what happens in the lungs during an asthma attack. Asthma's symptoms are caused by a narrowing of the bronchi, brought on both by inflammation—swelling—of the tissue in the airways and by a tightening of the smooth muscles in the bronchial walls. Bronchial cells may also produce more mucus than usual, further clogging the passageways. Air struggling to flow through the narrow airways will make a whistling sound. The airways also become more sensitive, and are likely to tighten up quickly in reaction to small quantities of an irritant.

A boy uses a handheld inhaler for relief. Asthma is the most common chronic disorder in young people.

TRIGGERS

Many things can trigger an asthma attack, and they vary from person to person and episode to episode. Among the triggers are exercise (particularly in cold weather); allergens, including pollen (from trees and grass), mold, animal dander (from the skin, hair, or feathers of animals), dust mites in house dust and cockroach excretions; indoor and outdoor pollutants such as cigarette smoke, car exhaust and other air pollution, industrial pollutants, and paint fumes; medicines, including aspirin, NSAIDs (nonsteroidal anti-inflammatory drugs, such as ibuprofen), or beta-blockers (medications used to treat high blood pressure and certain heart conditions); respiratory infections such as a cold or the flu; and stress.

Genes play a key role in asthma. A child with one asthmatic parent has about a 20 to 30 percent chance of developing the condition based on heredity alone; with two asthmatic parents, the odds rise to 60 to 70 percent. The disorder is prevalent in children—it is the most common chronic disorder in young people, accounting for the most absences from school—and, unlike other chronic diseases, it is increasing in frequency among the young. Fortunately, more than half of children with asthma grow out of the disease by adulthood.

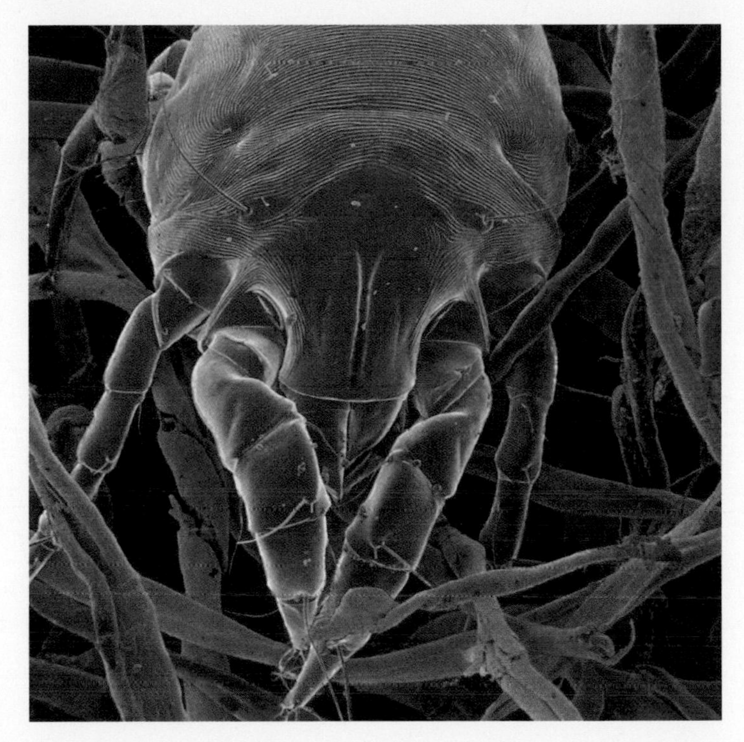

Dust mites (greatly magnified here and shown in false color) are common allergens, found by the millions in every house.

The underlying causes of asthma—the reasons why the airways become inflamed in the first place—are still being investigated. Until recently, researchers focused primarily on the muscular tightening of the airways, but now the interaction of the immune system and inflammation has taken a front seat in asthma research.

Studies suggest that many cases of asthma may begin with an immune reaction either to an infectious agent such as a virus, or to a foreign body such as a pollutant or mold. The immune system mobilizes lymphocytes, which react to these invaders in two ways. They may call up increased numbers of allergy cells called mast cells, which release irritating chemicals that cause inflammation. Alternatively, they may stimulate the production of antibodies, particularly Immunoglobulin E (IgE), an antibody that also produces chemical irritants.

TREATMENTS

Asthma treatments fall into two separate categories: quick-relief drugs taken when wheezing begins and controllers, which are taken every day to prevent attacks. The task of rapid relief falls to bronchodilators, which relax the muscles of the airways. Many of these are derived from synthetic epinephrine, a hormone that in its natural form is produced by the adrenal glands. A bronchodilator is typically delivered as a spray from a handheld inhaler, in measured doses. For more severe attacks, a patient may also use a nebulizer, which delivers the medicine in a fine mist through a mouthpiece or mask.

Long-term controllers typically mean corticosteroids, which are anti-inflammatory medicines. In recent years, the ranks of controllers have grown to include leukotriene modifiers. Leukotrienes are chemicals released by mast cells and leukocytes in the blood. They bind to receptors in the bronchial tubes and cause swelling that can persist for weeks. Leukotriene modifiers will either bind to those receptors themselves, blocking the inflammatory chemicals, or inhibit the production of leukotrienes. The good news about asthma is that it is usually controllable. Almost all asthmatics learn quickly how to manage the disease and find that it does not get in the way of an active and healthy life.

KEEPING HEALTHY

A good, long laugh can burn calories.

IS LAUGHTER the best medicine? In his 1979 memoir *Anatomy of an Illness*, Norman Cousins wrote that he could alleviate a painful spine condition for two hours ten minutes by laughing at Marx Brothers films and TV comedies. Researchers have since made other claims for laughter as therapy. Laughter burns about 50 calories every 10 to 15 minutes. It may block damaging chemicals released during stress and boost the immune system. Critics point out that any distraction can reduce pain and stress; and to burn significant calories, one would have to laugh for hours. People who can laugh while ill may also be innately healthier than others or have a stronger social support system. By itself, then, laughter may not be the best prescription, but it can still be good medicine. ■

contract during an exhalation, adding oomph to each expelled breath. Other sites in the brain stem—the pneumotaxic area and the apneustic area—help to coordinate the exact timing of inhalations and exhalations as well.

The brain's cerebral cortex controls voluntary breathing, though it can be overruled by the respiratory centers if oxygen runs low. Emotional and pain stimuli travel from the hypothalamus, at the base of the cerebrum, to the respiratory centers, prompting rapid breathing when angry, for example, or gasping when in pain.

The job of monitoring oxygen and carbon dioxide levels in the blood falls to chemical receptors both in the brain stem and on the walls of the aorta and the carotid arteries. If the amount of carbon dioxide in the blood rises, or if the amount of oxygen falls, these receptors signal the inspiratory center to increase the breathing rate, bringing in more oxygen and expelling more carbon dioxide until the blood gases return to normal levels.

Hyperventilating—breathing unusually rapidly and deeply—can build up oxygen levels and drop carbon dioxide in the blood to the point where the chemical receptors no longer send signals to the respiratory area. People who hyperventilate may be able to hold their breath an unusually long time, which sometimes lowers their oxygen level until they faint.

None of the finely tuned mechanisms that regulate breathing is needed until the moment of birth. Until it takes its first breath outside the womb, the fetus does not use its lungs at all; they are collapsed and filled with amniotic fluid (quickly drained at birth). Instead, the fetus picks up all the oxygen it needs through the umbilical cord from the mother's placenta. The moment of birth sees dramatic changes in the newborn baby. With oxygen from the mother cut off, carbon dioxide levels build up rapidly in the baby's blood. This activates the respiratory center in the brain stem, which signals the respiratory muscles. The baby takes a deep, gasping breath and cries: the first sign of a pair of healthy lungs.

EXERCISE EFFECTS

Several systems come into play at once during exercise. Just the anticipation of movement appears to activate the limbic system, a part of the brain that deals with emotions and behavior. The limbic system then signals the respiratory center to speed up breathing. As soon as a person begins to exercise, proprioceptors (specialized groups of nerve endings within tendons, joints, and muscles that monitor movements) and neurons in the brain's motor cortex also signal the respiratory area that it's time to breath more deeply.

As increased activity continues, the respiratory center responds to the decreased oxygen and increased

RESUSCITATION

A lifeguard gives mouth-to-mouth resuscitation to a downed swimmer.

CARDIOPULMONARY resuscitation (CPR) combines mouth-to-mouth resuscitation with chest compressions to revive a person who has stopped breathing and whose heart has stopped beating, as in instances of cardiac arrest or near drowning. The two techniques are intended, respectively, to supply oxygen until breathing can resume and to keep the blood transporting that oxygen to vital organs.

In 1903, American surgeon Dr. George Crile became the first person on record to successfully revive a patient by means of closed-chest compressions. In 1954, American doctors James Elam and Peter Safar showed that even the exhaled air of mouth-to-mouth resuscitation could give a patient enough oxygen for survival. By 1960, the American Heart Association began to train physicians in the techniques of CPR. Over the years, researchers refined the combination of chest compressions and breaths, recommending 30 compressions to every 2 breaths when a single rescuer was performing CPR. In recent decades many non-physicians have been trained in the technique.

Then in 2000, a study published in the *New England Journal of Medicine* suggested that mouth-to-mouth was not necessary for cardiac arrest victims. In 2007, the largest CPR study yet, involving 4,000 cases of cardiac arrest, found that patients who received chest compressions alone were twice as likely to survive without brain damage as those who also got mouth-to-mouth. The study, published in the British medical journal *The Lancet*, concluded that organs can function for several minutes on oxygen already in the body and that rescuers should focus on chest compressions to keep blood pumping. ■

carbon dioxide in the blood, as monitored by chemical receptors. All these signals together prompt deeper, faster breathing; the heart, beating rapidly, pumps more blood into the lungs to pick up the additional oxygen. These mechanisms can increase the body's ventilation by 10 to 20 times its resting rate during really vigorous exercise.

Other stimuli can speed up or stop breathing, as well. Body temperature, for instance, has a direct effect on the rate of respiration, speeding it up when the body is hot (as in a fever) and slowing it when the body is cold. A rise in blood pressure can slow respiration, while a drop can accelerate it. Sudden pain can make a person stop breathing briefly, but prolonged pain may speed up respiration. Anxiety, fear, or anger can also ratchet up breathing, perhaps as part of a flight or fight reponse that triggers the body to prepare for self-preservation.

Exposure to high altitudes, where oxygen is thin, can leave a person short of breath after even modest exercise. After a few days, the body becomes acclimatized to the altitude; oxygen uptake increases slightly but will not return to the amounts found at sea level. Skiers often suffer a mild form of altitude sickness as the body adjusts. Astronomers living at 13,780 feet (4,200 meters) on Hawaii's Mauna Kea have a blood oxygen content similar to that of patients with severe emphysema.

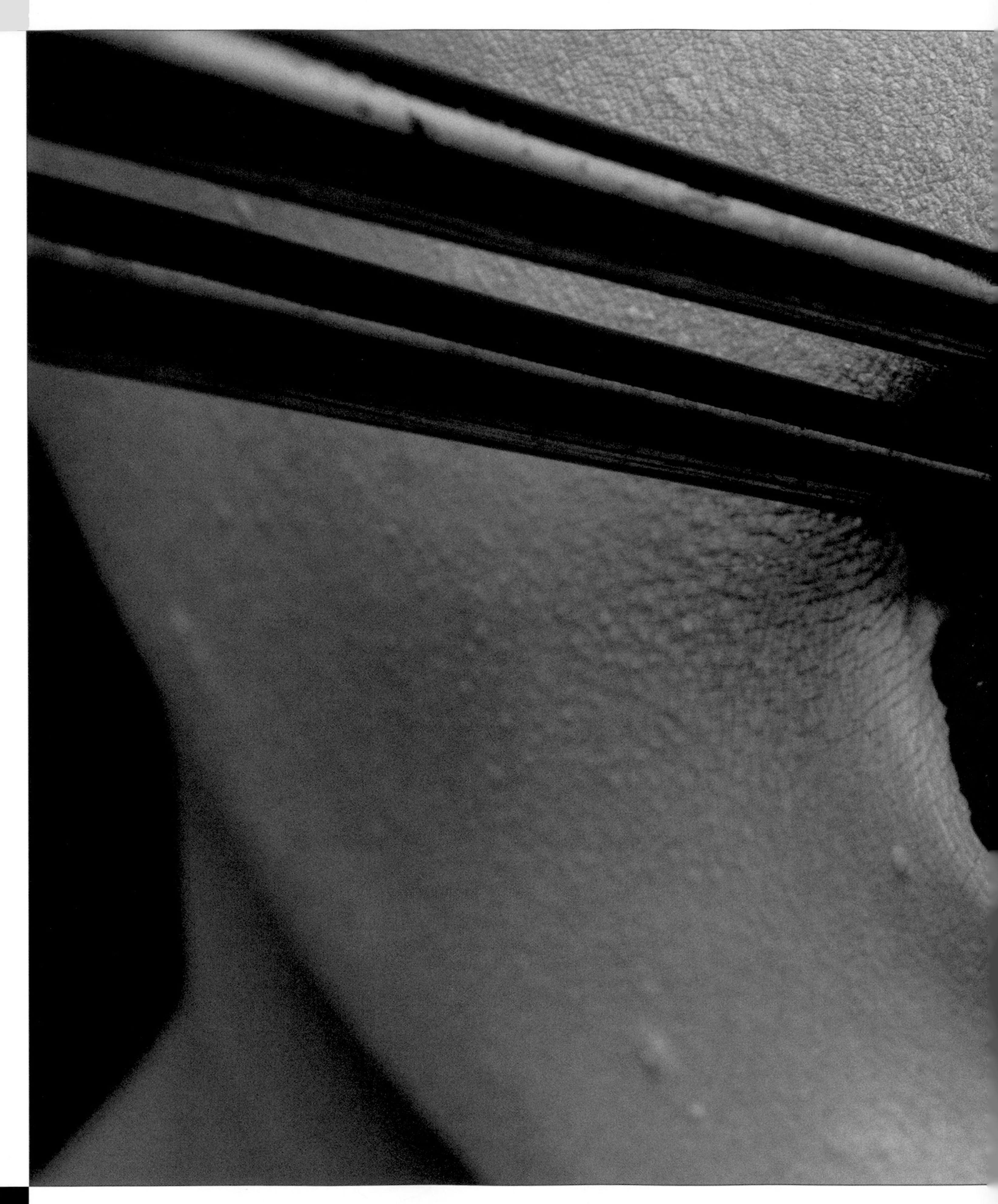

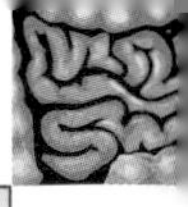

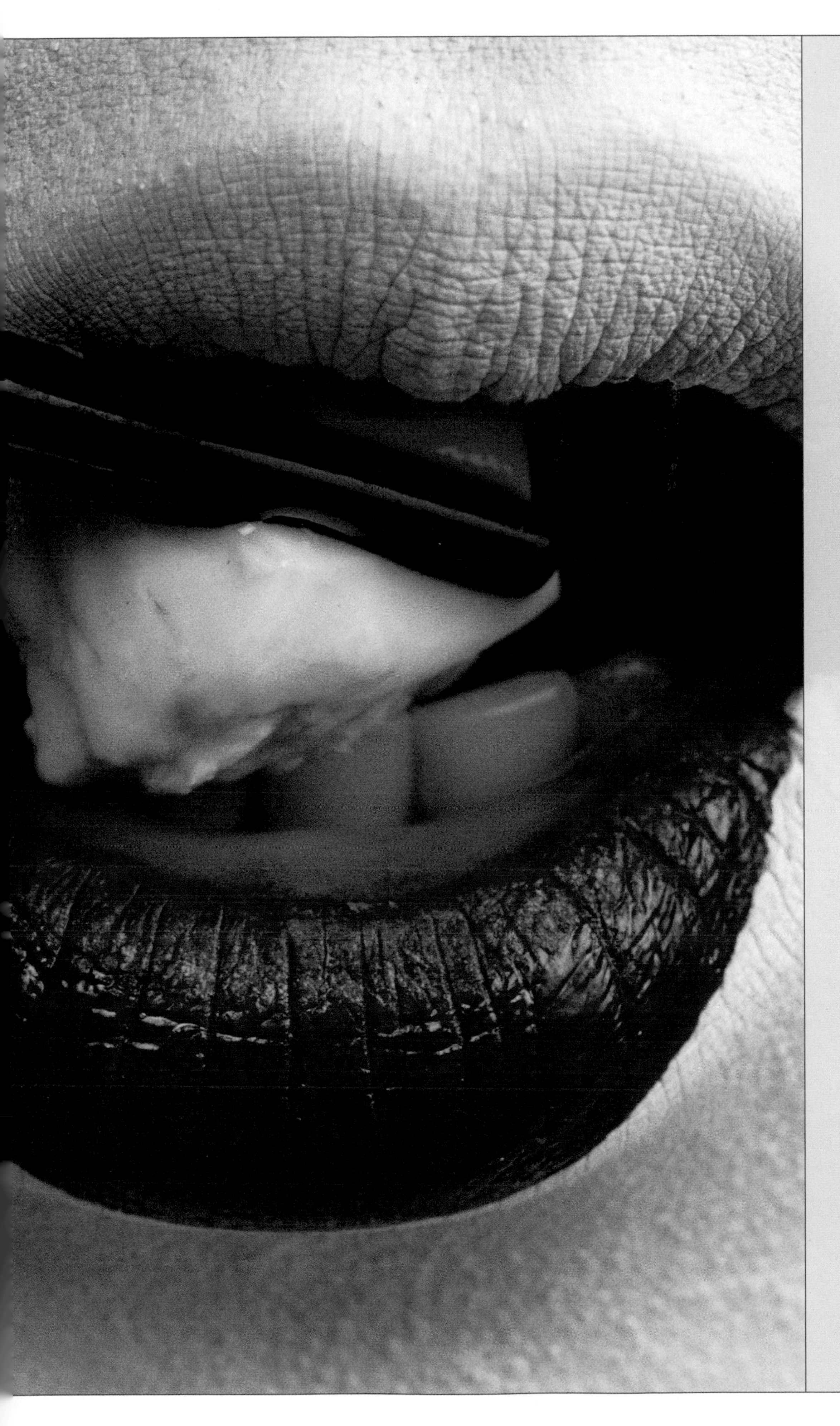

CHAPTER SEVEN

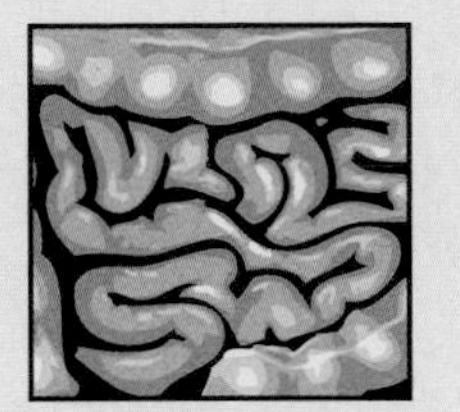

BODY FUEL

CONVERTING THE BLUEBERRY muffin you had for breakfast into fuel for your morning swim is the job of your body's digestive system. Like finely tuned plumbing, the tubes and organs of the digestive system route food through the body, in the process breaking down nutrients into smaller particles that can be passed through the walls of the intestines into the bloodstream and carried throughout the body for use in metabolism. This system then flushes out material not needed for energy or cellular growth.

Chopsticks deliver a morsel of tofu to the mouth, where digestion begins.

DIGESTIVE SYSTEM FROM FOOD TO ENERGY

THOUGH SOME PEOPLE live to eat, everyone must eat to live. Food nourishes the body by providing nutrients that are essential for the functioning of the heart, nerves, muscles, and glands. It also fuels growth, tissue repair, and the maintenance of organs and systems. In short, without food the body would soon die. The habit of eating good food is so important that food plays a major role in holidays and rituals. We celebrate life by partaking of that which gives us life, enjoying the sight, smell, and taste of food.

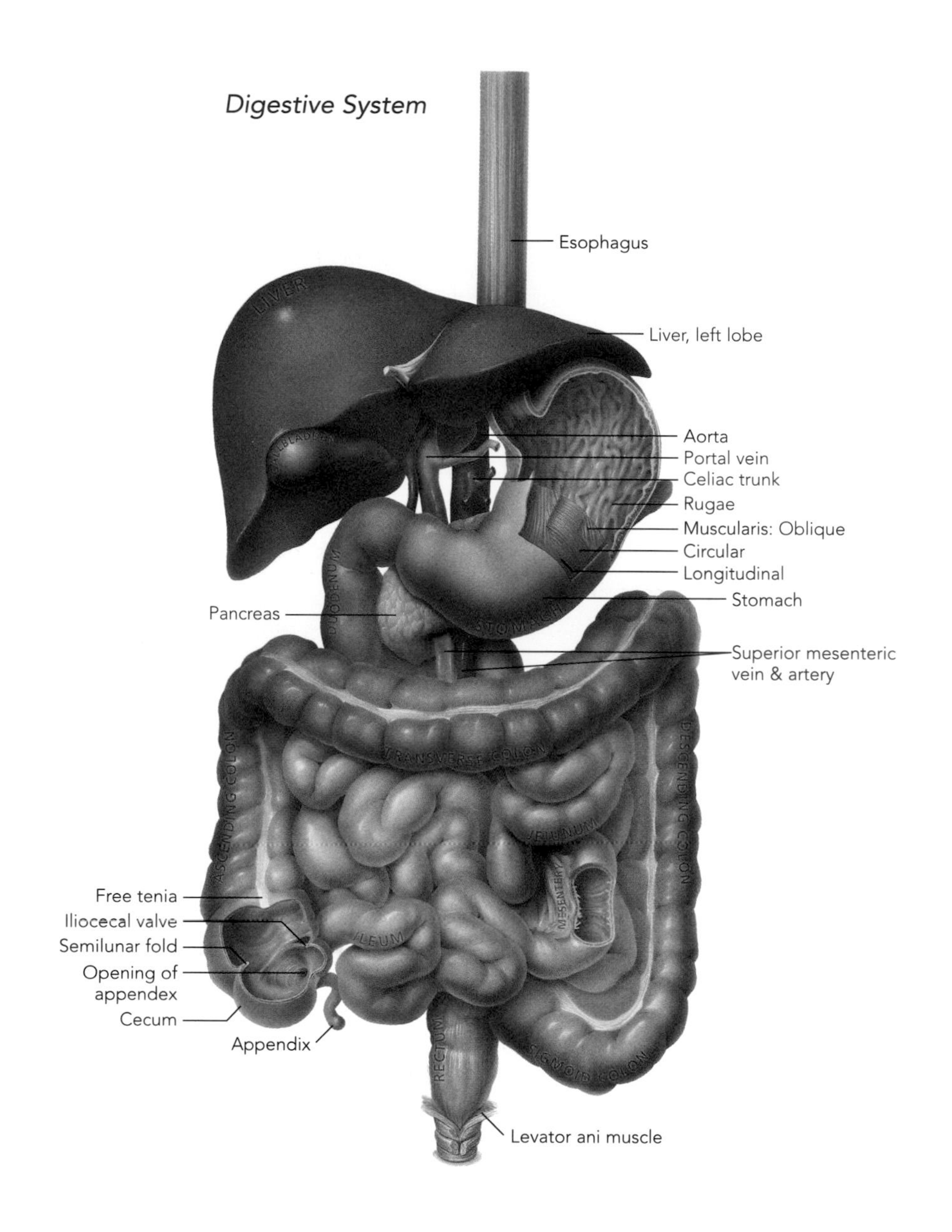

FUELING THE BODY

What does the digestive system do? We stoke the furnace by eating, but then the digestive system automatically takes over, transforming food into chemical compounds that can go through the membranes of a cell. From milk, for example, the digestive system extracts molecules of nutrients from every category it needs—proteins, carbohydrates, fats, vitamins, minerals, and water. From bread we get carbohydrates and fats, as well as some proteins and minerals.

The journey from slice of bread to molecule of carbohydrate takes place in the gastrointestinal (GI) tract, or alimentary canal, a 30-foot-long tube that begins with the lips and ends with the anus. The teeth begin the breakdown process by tearing, mashing, and grinding the food. While the tongue is helping distribute the food around, salivary glands located beside and beneath the mouth cavity secrete saliva and mucus to moisten it into a soft paste.

FOOD FACTS

+ AMERICANS EAT about 100,000 pounds of food in a lifetime; or about 3.6 pounds a day.
+ Swallowed food moves into the pharynx (throat) at 20–25 feet per second, enough to move the food deep into the esophagus.
+ The average meal takes between 15 hours and two days to pass through the entire alimentary canal.

WEBLINK: See Digestion in action at http://nationalgeographic.com/humanbody

Chewing is primarily a voluntary activity; but once the food is swallowed, the reflexes take over. The food slides down the throat and esophagus toward the stomach. The chemical transformation of food begins there. Acidic gastric juices secreted in the stomach turn the food into a soupy mixture, which is then churned until it passes into the small intestine. Hydrochloric acid secreted by the stomach is strong enough to burn through a thick rug, but a lining of mucus protects the stomach from a similar fate.

Once the food passes into the small intestine, the liver and the pancreas aid the chemical breakdown. These two glandular organs outside the GI tract are connected to it by small-caliber tubes (ducts). Situated on about stomach level, the triangular-shaped liver fits into the right side of the abdominal cavity beneath the right diaphragm. Among the liver's many functions are decomposition of old red blood cells, detoxification of alcohol and other poisons, production of clotting proteins, and storage of carbohydrates and some proteins and fats. Its main role is to produce bile salts, which help absorb fats in the small intestine. Bile is concentrated and stored in the gallbladder.

The pancreas is a large gland that secretes digestive enzymes into the small intestine. These enzymes break down proteins, carbohydrates, and fats into compounds that can be further processed within the small intestine.

The longest part of the digestive system, the small intestine measures about 21 feet in a cadaver, but about 10 feet in a living person because of a normal state of partial muscle contraction. This coiled organ folds into the lower part of the abdominal cavity. Most digestion occurs within the small intestine, aided by the liver, pancreas, and glands within the small intestine itself, which produces about one quart of intestinal juices a day. As the food is traveling through the small intestine, it is absorbed through the cells lining the intestine and delivered to the cells of the bloodstream and lymph system. To increase the surface area of the small intestine some eight times to about 2,000 square feet, it is lined with millions of tiny projections called villi. From pores in the villi, digested nutrients pass into the capillaries and thus into the bloodstream.

Like a widening river, the small intestine joins with the large intestine, or colon. This final section of the bowels frames the small

Ripe berries make a rewarding snack. Fresh berries are rich in antioxidants and vitamins, both keys to good health.

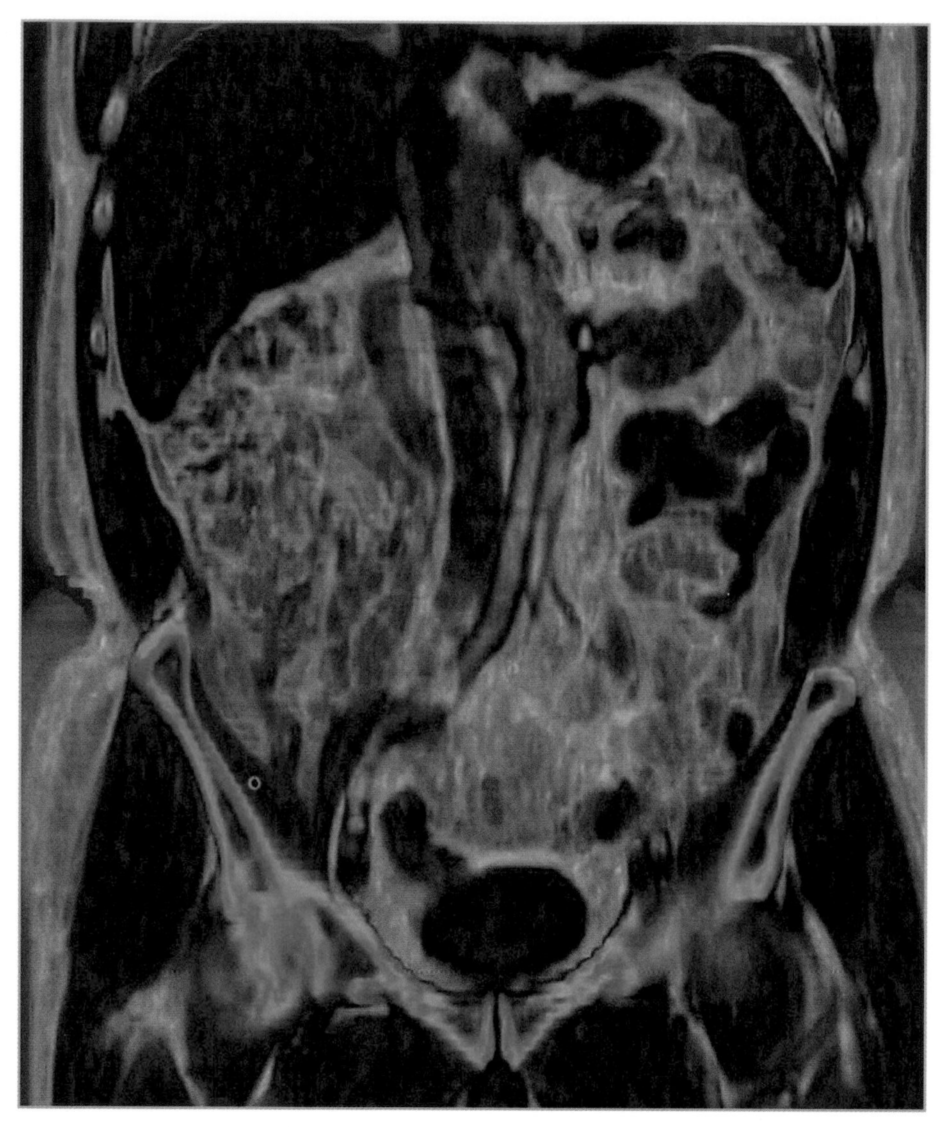

The organs of the digestive system are tightly intertwined in the abdomen, as seen here in a 3-D CT scan.

intestine, by ascending on the right side of the body, traversing the body just above the coils of the small intestine, and dropping down the left side. It then takes an S-like turn to the right as the sigmoid colon and joins with the muscular rectum, which connects to the anus, the posterior opening of the GI tract. Though it absorbs water, the main function of the large intestine is to move waste out of the body.

A miraculous processing factory, the digestive system is constantly at work. Each step in the process and each organ will be detailed in the rest of the chapter.

DIGESTIVE PROCESSES

The process of digestion has five main divisions: ingestion, propulsion and peristalsis, mechanical and chemical digestion, absorption, and defecation. Digestion begins with ingestion, the taking in of food—in other words, eating and drinking. Once food is in the GI tract, propulsion (with the help of gravity) moves it along. Propulsion begins with the act of swallowing, a voluntary or involuntary activity. From then on propulsion is accomplished by peristalsis, the involuntary contraction and relaxing of muscle walls. Through peristalsis food reaches your stomach even if you stand on your head. In addition to transporting the food along the GI tract, propulsion serves to mix the food with secretions from within the cells of the tract and the accessory organs. Some 7 quarts of water, acid, enzymes, and buffers are daily secreted into the tract.

Mechanical digestion occurs inside the mouth, when the teeth break up food into smaller units that, when mixed with watery saliva, may be swallowed. While the first part of ingestion obviously involves conscious effort—eating food—the tongue and cheeks work deliberately and automatically to move it around

HERBAL REMEDIES

HERBAL TREATMENTS have long been used for a variety of ailments. Health professionals caution that pregnant women, the very young or old, or anyone on pharmaceuticals should consult their doctor before taking herbal medicines. Some useful herbs include ginger as a treatment for motion-sickness nausea and garlic for lowering cholesterol.

SEE ALSO: Chapter Four, "The Muscles," PAGE 106

until it is ready for swallowing. Though chewing is not strictly necessary for digestion, it helps: Careful chewing of the food is beneficial to maintaining a trouble-free digestive system. The churning of food within the stomach and intestines by peristalsis also comes under the heading of mechanical digestion.

In chemical digestion large molecules of carbohydrates, fats, and proteins are broken into smaller molecules by hydrolysis (the splitting of bonds resulting in the addition of the elements of water). Chemical digestion starts in the mouth with the secretion of enzymes by the tongue and salivary glands. In the stomach and small intestine, more enzymes are introduced, some from the pancreas. A few substances within food need no chemical digestion, including vitamins, minerals, glucose, cholesterol, amino acids, and water.

Absorption—the passing of molecules, ions, and ingested and secreted fluids through the plasma membranes of epithelial cells lining the GI tract— is the ultimate goal of the digestive system. These end products of digestion move into the blood or lymph system and circulate to cells in the body.

Unabsorbed material (indigestible substances, bacteria, and cells sloughed off from the lining of the GI tract) gets shunted out of the body through defecation. All this waste exits the body through the anus and is called feces.

LOOKING ON THE INSIDE

To take interior shots, this tiny endscopic video capsule is equipped with a camera.

LIKE SPELUNKERS with headlamps, physicians need light to see inside the body. Use of light sources in diagnosis and surgery began in the early 1800s. The first instruments—hysteroscope (inserted in the uterus), laparoscope (inserted in the abdomen), and thoracoscope (inserted in the chest)—were generally rigid. By the 1930s, semiflexible endoscopes allowed physicians to see inside the stomach. In 1957 Basil Hirschowitz at the University of Michigan introduced a fiber-optic endoscope, in which thin glass fibers transmitted light throughout the length of the instrument. By the 1960s, fiberoptic endoscopes were being used in many major health facilities.

Endoscopy for the digestive system includes esophagogastroduodendoscopy or EGD (upper GI tract endoscopy), enteroscopy (examination of the small intestine), colonoscopy, and proctosigmoidoscopy (examination of the rectum and lower colon). In a colonoscopy, a flexible videoendoscope is inserted into the rectum of a sedated patient. A physician guides it along the length of the colon, looking for abnormalities such as precancerous polyps. The physician can remove polyps and retrieve them for biopsy. The risk of hemorrhage by perforation of the intestine is low compared to the benefits of early detection of colon cancer.

Colorectal cancer claims the second highest rate of cancer deaths in the Untied States, and because it spreads slowly, the American College of Gastroenterology recommends that persons 50 years or older have colorectal cancer screening tests every ten years; people with a family history of colon cancer should begin screening earlier. ■

FOOD & NUTRITION

To stay healthy, people need nutrients—chemical compounds used for energy and for cell growth, repair, and maintenance. The most essential nutrient is water: Humans require two to three quarts a day. The organic nutrients—carbohydrates, fats, and proteins—compose the major part of the diet. Carbohydrates are compounds made of carbon, hydrogen, and oxygen; they take the form of sugars and starches. Simple carbohydrates (sugars) come from fruits and vegetables; complex carbohydrates (starches) derive from breads and cereals.

Of the sugars, the simple monosaccharide glucose is the one that fuels the cells. Neurons and red blood cells depend almost exclusively on glucose for energy. A temporary shortfall of glucose can lower brain function and cause body systems to shut down. Hence carbohydrates are the body's workhorses. Although the body can rely on fats and proteins, carbohydrates provide a ready source of glucose. Current thinking calls for five daily servings of fruits and vegetables, and 6 to 11 servings of bread, cereal, or pasta.

Lipids (organic compounds not soluble in water) provide the most concentrated form of potential cell energy. Saturated fats come from meat and dairy products, unsaturated fats from nuts, seeds, and vegetable oils, and cholesterol from egg yolks, liver, and milk. Though "fat" has become the bad guy of the diet, the body needs it for many reasons: for the absorption of fat-soluble vitamins; as a cushion around organs; as insulation beneath the skin; and as an easily stored energy source.

WHAT IS A CALORIE?

A CALORIE is a unit of heat energy—the heat needed to raise the temperature of one gram of water by one degree Celsius. In dietary terms, large calories (or kilocalories) are used to measure the energy-producing potential contained in food. Depending on size and exercise routine, about 1,700 to 5,000 large calories are burned per day.

VITAMINS

	HOW VITAMINS HELP THE BODY
VITAMIN A	Maintains skin cells, helps form light-sensitive pigments, used in bone growth; found in liver and milk, also formed in GI tract from substances found in yellow and green vegetables
VITAMIN D	Necessary for absorption of calcium and phosphorus; UV light converts 7-dehydrocholesterol in the skin to vitamin D3, which is then converted to active form by enzymes in the liver and kidneys; food sources include fish liver oils, egg yolk, and fortified milk
VITAMIN E	Inhibits breakdown of unsaturated fats, involved in forming DNA, RNA, and red blood cells; dietary sources include nuts, wheat germ, seed oils, and leafy vegetables
VITAMIN K	Essential for synthesis of clotting proteins and others produced in the liver; synthesized by bacteria in large intestine, also gleaned from spinach, cauliflower, cabbage, and liver
VITAMIN C	Used in formation of connective tissues, healing of wounds and creation of antibodies, and as an antioxidant; sources include citrus fruits, tomatoes, and green vegetables
VITAMIN B1	Aids in the metabolism of carbohydrates, essential in production of neurotransmitter acetylcholine; found in eggs, whole grains, lean meat, and liver
VIATMIN B2	Functions as coenzyme in carbohydrate and protein metabolism, especially in eye cells, mucosa of intestine, and blood; sources include milk, yeast, liver, egg whites, whole grains, and legumes

SEE ALSO: Chapter Thirteen, "A Wise Diet," PAGE 360

Essential proteins come in high doses in eggs, milk, and meat. Nuts, legumes (beans and peas), and cereals also contain protein but lack important amino acids, the building blocks of protein. Leafy vegetables have low amounts of the most essential amino acids. Proteins are used in forming skin and muscle tissue and regulating enzymes, hormones, and other functions. Nutritionists suggest consuming protein equivalent to about 0.08 percent of a person's body weight. A 150-pound (68-kg) person should eat about 2 ounces (56 grams) of protein a day.

The body needs a variety of foods to get the nutrients it needs. No one food can provide all the necessary vitamins and minerals. Some vitamins, such as B, D, and K, are synthesized within the body. Since vitamins assist the other nutrients in doing their jobs, they are an essential part of the diet. Vitamins A, C, and E act as antioxidants that neutralize free radicals and thus help prevent some cancers. Though vitamin supplements may be of some help, the best source of all essential vitamins is a well balanced diet. Finally, the human body requires moderate doses of seven minerals—calcium, phosphorus, potassium, sulfur, sodium, chloride, and magnesium—as well as trace amounts of some 12 others. Vegetables, legumes, and milk are good sources. Like vitamins, minerals work in concert with other nutrients. Iron binds to red blood cells so that they can carry oxygen. Calcium, phosphorus, and magnesium provide strength and hardness to bones and teeth. Ions of sodium and chloride in the blood help to keep body fluids balanced. And potassium maintains osmotic pressure in cells and helps conduct nerve impulses.

About five daily servings of colorful, delicious fruits are part of a healthy diet.

INGESTION

THE FIRST STEP OF DIGESTION

PUTTING FOOD IN your mouth is the first step in digestion. At that point accessory organs—the teeth, tongue, and salivary glands—take over. The boundaries of the mouth, formed by the lips and cheeks, help keep food in position while you chew. The palate forms the roof of the mouth. In the anterior, or front part of the mouth, the hard palate is composed of bone covered by a mucous membrane. In the posterior part of the roof of the mouth, the soft palate is formed of muscle; at the back of the soft palate hangs a muscular process called the uvula. When we swallow, the soft palate and uvula rise to prevent food from entering the nasal cavity. To see how this works, try breathing and swallowing at the same time.

Defining the floor of the mouth, the tongue is composed of skeletal muscles covered by a mucous membrane. The top and sides of the tongue are covered with little projections called papillae, many of which contain taste buds. While we're enjoying the taste of the food, the tongue sees to it that food becomes thoroughly chewed. Lingual glands on the tongue also chip in by secreting mucus and an enzyme called lingual lipase which helps break down lipids.

When the tongue has maneuvered the food around and mixed it with saliva, the food becomes shaped into a wet lump called a bolus. Once the bolus is positioned at the back of the tongue, it can be swallowed and projected into the pharynx and then to the esophagus. Muscles within the tongue allow it to change shape—it can be

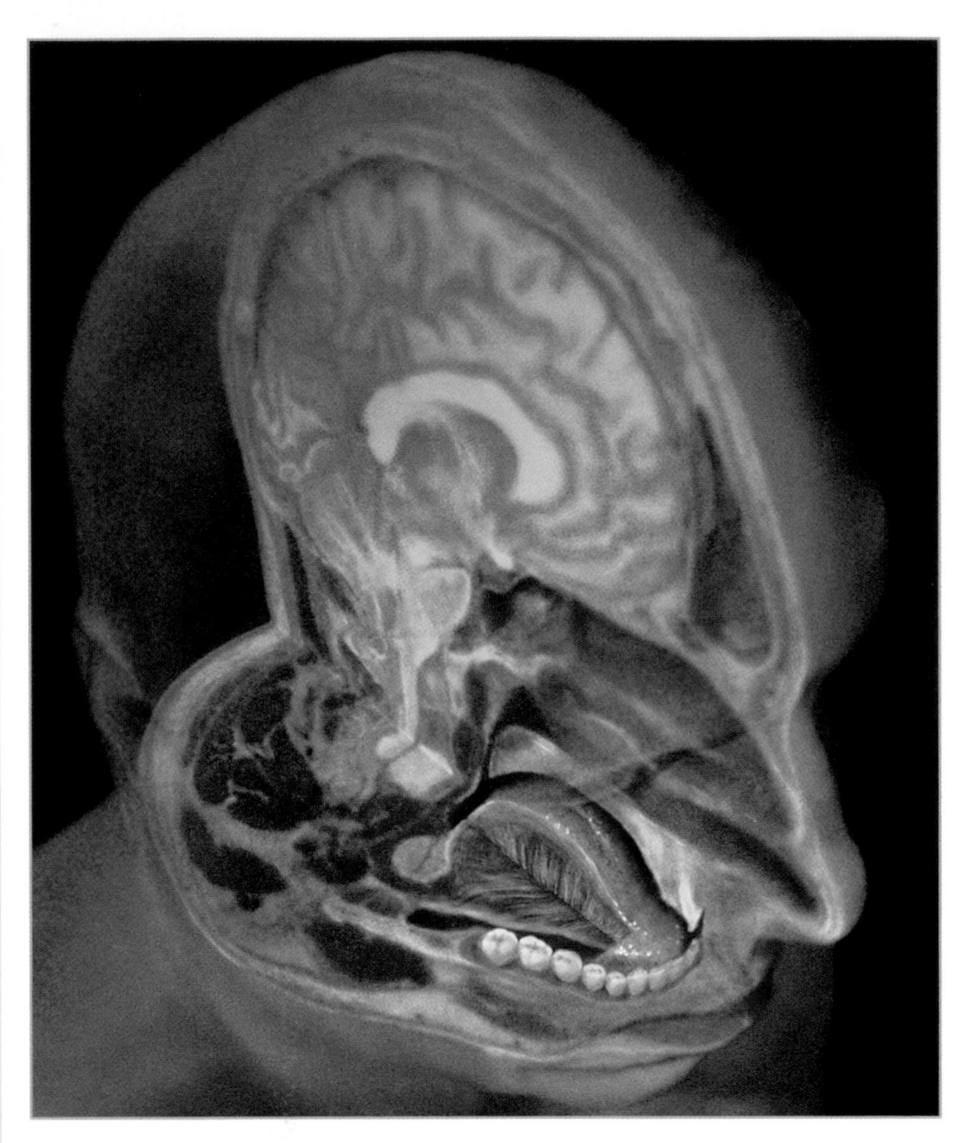

A cross section of the human head reveals the main structures involved in chewing: the tongue, the teeth, and the oral cavity.

SALIVA

ABOUT 97 TO 99.5 percent of saliva is water. The rest is an assortment of useful chemicals. Some buffer acidic foods. The antibody immunoglobulin A keeps microbes from getting into the epithelium (covering and lining tissue). Contributed mostly by the parotid glands, salivary amylase is a digestive enzyme that helps break down starch.

SEE ALSO: Chapter Three, "The Teeth," PAGE 94, Chapter Nine, "The Senses," PAGE 256 thick or thin, rounded or flat—to accommodate the needs of speech and eating. A membrane called the lingual frenulum attaches the tongue to the floor of the mouth and restricts its movement.

To moisten the food and start the chemical digestion process, salivary glands excrete saliva into the mouth. Even with no food, saliva is present in small amounts to keep the mouth wet and the teeth clean. With eating, saliva production increases. A person produces about 1 to 1.5 quarts (1,000 to 1,500 mL) of saliva a day.

The three main pairs of salivary glands lie outside the oral cavity, though some saliva is produced by glands inside the mouth. The largest glands, the parotid glands, are situated along the cheek and jawline in front of each ear, tucked between the skin and muscles. The walnut-size submandibular glands lie within the jaw. The smaller sublingual glands are farther forward, just under the tongue. Mumps is a virus that attacks the parotid glands, causing fever, malaise, sore throat, and swelling of the cheeks.

The sight or smell of food is often enough to trigger salivary flow. Just the thought of a hot waffle with butter and maple syrup might make a hungry person's mouth water. Chemical receptors in the mouth send impulses via the nervous system to salivatory nuclei in the brain stem, which then signal the salivary glands to begin production in anticipation of a meal.

PAVLOV'S DOGS

Pavlov's dogs with their keepers in St. Petersburg, Russia

THE RUSSIAN physiologist Ivan Petrovich Pavlov (1849–1936) was interested in, among other things, how secretions work during digestion. From about 1898 to 1930 he conducted a series of famous experiments on dogs from which he developed a theory of conditioned reflexes—that is, physiological responses to learned habits.

Using dogs as subjects and saliva as a measurable quantity, he found that if a ringing bell was associated with food over several occasions, the sound of the bell alone would trigger saliva flow in the dogs. Pavlov reasoned that not only simple physiological impulses, but also mental and emotional states are similarly subject to conditioned responses. Working during a time of great political upheaval in Russia did not impede Pavlov's progress in his research. When a coworker arrived late for work, saying there had been a shooting, Pavlov told him, "What difference does a revolution make when you have experiments to do in the laboratory?" On another occasion, he even took the risk of openly criticizing Communist leaders: "For the kind of social experiment that you are making," he said, "I would not sacrifice a frog's hind legs!"

As for the dogs, Pavlov was somewhat ahead of his time. In all of his experiments, he worked only with normal, unanesthetized animals. Dogs under his care typically lived full, healthy lives. In 1904 Pavlov was awarded the Nobel Prize in physiology or medicine. ■

CHEWING & SWALLOWING

Chemical digestion begins in the mouth, when salivary amylase and lingual lipase, released from the salivary glands and tongue, interact with the food. Assisting the tongue and the salivary glands, the teeth are also considered digestive organs because they participate in mechanical digestion by chopping up the food during chewing, also called mastication.

In the saliva, salivary amylase goes to work on starches, which account for most of the carbohydrates we consume. Starches are complex polysaccharides, but the bloodstream can only absorb monosaccharides, thus the starches must be simplified to be of use. Salivary amylase can break bonds between glucose units and reduce starches to di- and trisaccharides, a process that can continue—once the saliva is mixed in—for an hour after swallowing. The lingual lipase does not even begin working until it is in the stomach. There, stimulated by stomach acids, it breaks triglycerides into diglycerides and fatty acids.

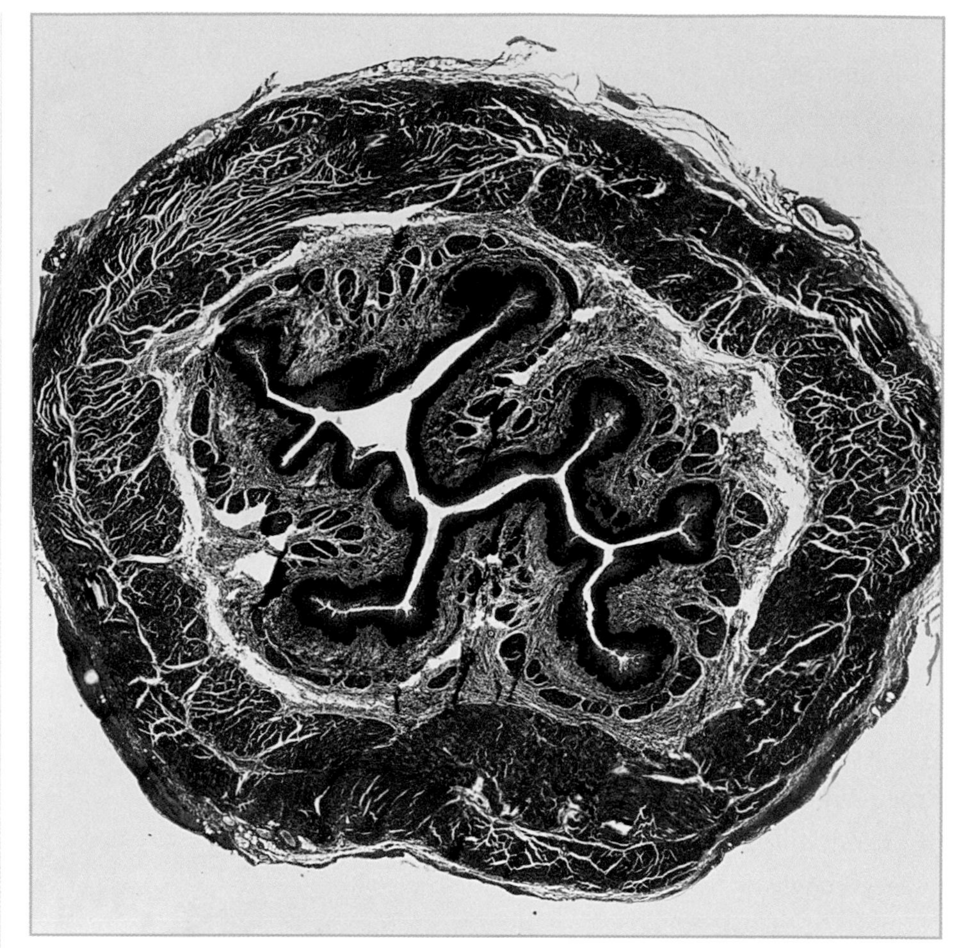

A cross-section of the esophagus shows how, when at rest, the esophagus is folded. It will expand when food enters the tube.

Down the hatch the food goes, the hatch being a funnel called the pharynx (throat). The pharynx is made of skeletal muscles lined with mucosa and is employed for breathing and digestion. During the act of swallowing, or deglutition, the bolus of food moves into the oropharynx and then involuntarily into the laryngopharynx. Repeated constriction of muscles in the pharynx moves the food down to the esophagus, which is positioned behind the trachea (windpipe).

SEE ALSO: Chapter Three, "The Teeth," PAGE 94

WHAT CAN GO WRONG

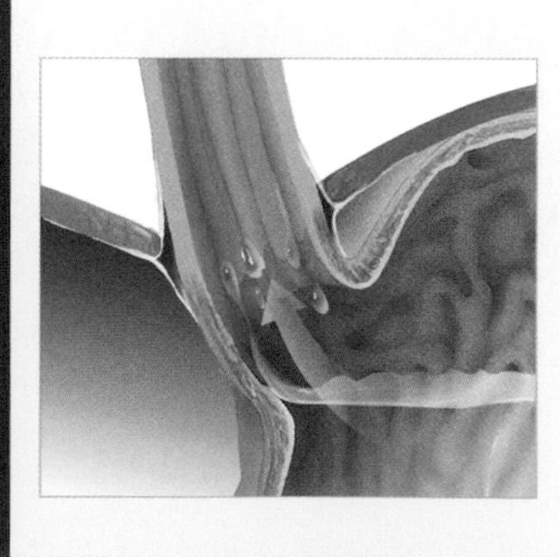

HEARTBURN: Ever feel a burning pain in your chest after a meal? You could be suffering from gastroesophageal reflux disease, which causes acidic stomach juices to back up into the esophagus, left. About one in ten adults has heartburn at least once a week; one in three has an episode every month. The problem is often related to hiatal hernia: The stomach protrudes above the diaphragm, and when the lower esophagal sphincter is in the chest (with negative intrathoracic pressure) rather than in the abdomen (with positive intraabdominal pressure), it doesn't function as well, and juices flow back up. Pregnancy, obesity,

The act of swallowing not only makes the soft palate and uvula seal off the nasopharynx; it also signals the larynx to rise and a large piece of cartilage called the epiglottis to cover the larynx (voice box) and block access to the trachea. Without this action, food would end up choking you. With the epiglottis covered, the vocal cords constrict, widening the tunnel to the esophagus. In seconds the food is through, and breathing can resume.

The esophagus is a 10-inch (25 cm) long tube that collapses when not in use so that its curving walls can almost touch each other. From the end of the laryngopharynx, the esophagus (commonly known as the gullet) passes through the diaphragm (the muscle between the thoracic and abdominal cavities) at an opening called the esophageal hiatus and then enters the top part of the stomach. A hiatal hernia occurs when a portion of the stomach pushes up through the esophageal hiatus, sometimes contributing to heartburn (see "What Can Go Wrong," below). The innermost layer of the esophagus is a mucosa made up of squamous (scaly) epithethial cells, connective tissue and smooth muscle. There are also mucous glands in this layer near the stomach. The cells protect the esophagus from the wear and tear from the daily stream of foods and liquids. Beneath this layer is a submucosa that holds connective tissue, blood vessels, and mucous glands. The outer layer is called the adventitia and is composed of connective tissue that attaches the esophagus to the surrounding muscles and dividing structures.

CHOKING

WHEN YOU swallow, your larynx rises, and the epiglottis pops down to seal off the respiratory passages. If you talk and eat at the same time, food can get stuck in the larynx. Without air getting to the lungs, the coughing reflex does no good. The Heimlich maneuver, above, has saved many chokers from death through a jab just above their navels.

The esophagus acts as an expressway, transporting food from the mouth to the stomach. Nothing is absorbed, and only mucus is excreted. Like a tollgate, a band of muscle called the upper esophageal sphincter, normally contracted, relaxes during swallowing and allows the food to move from the pharynx into the esophagus.

Peristalsis then causes a sequence of contractions and relaxations, which push the food along. The circular muscle layers above the food contract, squeezing the food down, and the longitudinal muscle fibers beneath it also contract, shortening that section and pulling the food down. The net result is a wave of contractions that both push and pull the food toward the stomach. Mucus secretions in the esophagus keep the bolus sliding along. The whole commute from the top of the esophagus to the stomach is a mere four to eight seconds.

WHAT CAN GO WRONG

and post-prandial exercise can also put abnormal pressure on the stomach and thus cause heartburn. Frequent episodes may lead to inflammation of the esophagus, right, esophageal ulcers, Barrett's esophagus and, rarely, esophageal cancer. To prevent heartburn, cut down on foods that make for an acidic stomach—coffee, chocolate, tomatoes, and fatty foods. Avoid eating just before going to bed. Some over-the-counter medications (such as antacids) may help neutralize the stomach environment, and acid-suppressing medications, which prevent the formation of acid, can save you a lot of pain.

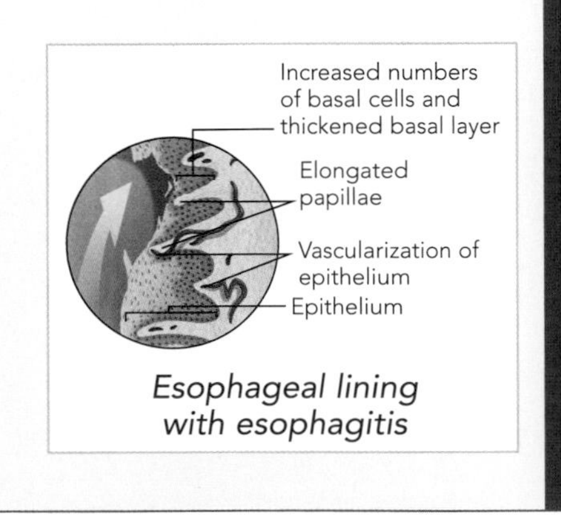

Esophageal lining with esophagitis

THE STOMACH

WHAT HAPPENS TO THE FOOD

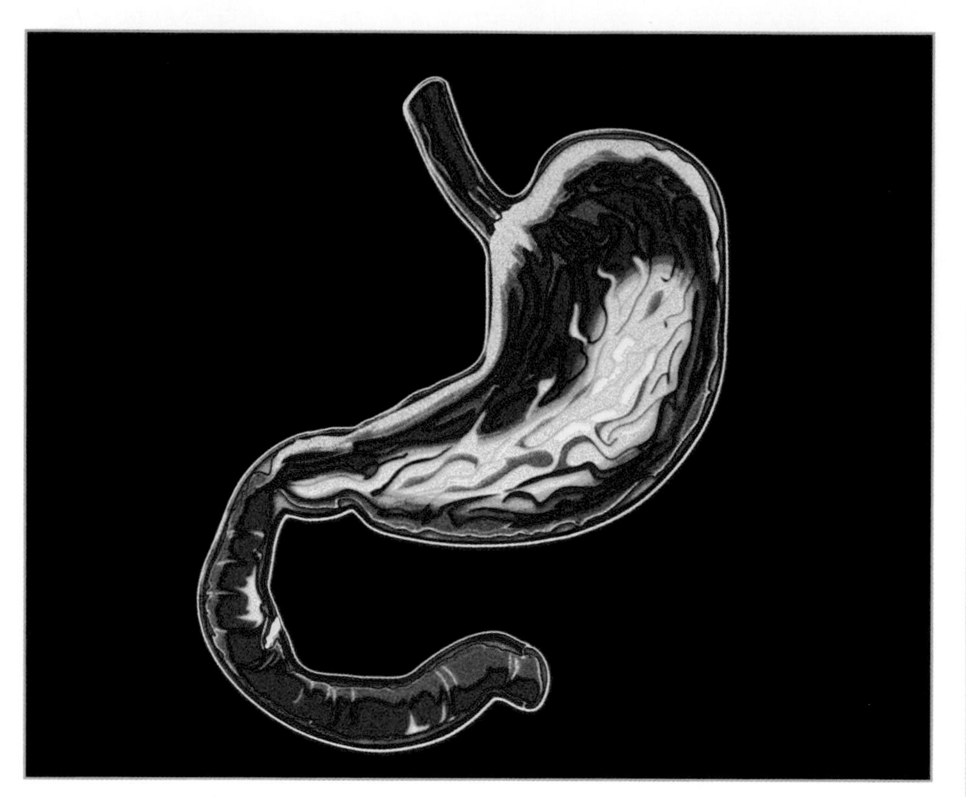

Food enters the stomach via the esophagus, upper left and exits through the duodenum, lower left, into the small intestine.

AFTER THE ESOPHAGUS, the GI tract enlarges into a flexible reservoir called the stomach. Lying below the diaphragm, the stomach is situated in the left portion of the upper abdomen. Inhalation pushes the stomach down slightly; exhalation raises it. In shorter, stouter people, the stomach tends to lie horizontally ("steer-horn shaped"); in taller, thinner people it lies more vertically ("J shaped"). In the stomach food is churned and liquefied into a creamy substance called chyme. It is then held there until ready for release into the small intestine.

ANATOMY

The adult stomach measures from about six to ten inches in length, but it is greatly expandable. An empty stomach has a volume of nearly 2 ounces (50 ml); a completely distended one can hold a gallon (about 4 liters). The empty stomach looks like a collapsed pouch, with lengthwise folds called rugae. Of the four major regions, the cardia (near the heart) lies at the entrance from the esophagus; the fundus is the large rounded portion; the body is in the middle; and the antrum is the funnel-shaped end of the stomach, which connects with the duodenum (first part of small intestine) at the pyloric sphincter. In terms of cellular anatomy, the stomach walls have four layers, or tunics. The serosa (outside layer) is composed of simple squamous epithelial cells and connective tissue. The muscularis externa contains within itself three layers of smooth muscles. Next is the submucosa, made of loose connective tissue. The mucosa (lining) is the most complex layer, composed of a number of different kinds of cells. The stomach lining is speckled with millions of gastric pits, leading deep into gastric glands that secrete gastric juice. The cells lining these pits are mostly goblet cells, which produce mucus or the stimulatory hormone gastrin. Within the fundus and body of the stomach there are more kinds of secretory cells. Mucous neck cells in the higher parts of a gland produce an acidic mucus. Parietal cells in the middle area of stomach glands

HUNGRY?

ALTHOUGH A GROWLING stomach can signal hunger, it can also be a sign of digestion. Peristaltic waves rippling through the stomach and small intestine vibrate gas and fluids in the GI tract and cause "rumbling." Hunger contractions themselves can make noise—from 10 to 20 minutes at a time. The contractions then repeat within an hour or two.

WEBLINK: Take an interactive look at the Stomach at http://nationalgeographic.com/humanbody

secrete hydrochloric acid (HCl), an intrinsic factor (needed for vitamin B12 absorption in the small intestine). Then there are the chief cells, which release an inactive form of the protein-splitting enzyme pepsin. Finally, enteroendocrine cells produce hormones that instead of going into the stomach, circulate via the blood vessels to trigger responses in other digestive organs.

INSIDE THE STOMACH

Once it has food, the stomach becomes a churning, spurting, bubbling plant of digestion. All of the digestive processes except ingestion and defecation occur here—mechanical and chemical digestion, secretion, peristalsis, and absorption. Every 15 to 25 seconds a series of peristaltic waves ripples across the stomach, sloshing its warm brew. During digestion, waves are more intense, with the strongest waves hitting the chyme as its enters the pylorus. Back and forth waves spill a little chyme into the pyloric sphincter until eventually all the chyme makes its way out of the stomach.

The stomach continues the job of starch digestion and begins work on protein and triglyceride digestion. Food stays in the fundus for about an hour, continuing to mix with salivary amylase. Then gastric juices kick in and activate lingual lipase, which dissolves triglycerides into fatty acids and diglycerides. In children the enzyme rennin acts

WILLIAM BEAUMONT

William Beaumont first saw action with the U.S. Army in the War of 1812.

WHEN AN UNUSUAL case came before U.S. Army surgeon William Beaumont (1794–1853), he realized he had been given a unique opportunity. Posted at Fort Mackinac in Michigan, he was called on June 6, 1822, to attend to a 28-year-old French Canadian voyageur who had accidentally shot himself.

Alexis St. Martin had a gaping wound in his left side: two broken ribs, a punctured abdominal wall and stomach, and a protruding lower lung and stomach. What should have been a mortal wound began to heal under Beaumont's expert care. Compresses kept food and drink from seeping out of the wound. But the wound sealed only around its edges and tightened to form an inch-wide fistula, or opening, onto St. Martin's stomach. The doctor began a series of experiments on St. Martin, who was then living in Beaumont's home as a handyman.

With a finger, Beaumont opened the edges of the fistula to access the stomach. He tied small bits of different foods to pieces of silk thread and pulled them out at intervals to observe their rate of digestion. He was able to prove that digestion was mainly a chemical process and that the gastric juice contained hydrochloric acid. He found that vegetables did not digest as well as other foods, that milk coagulated before digestion, that cold gastric juice would not digest food, and that alcohol causes gastritis. ■

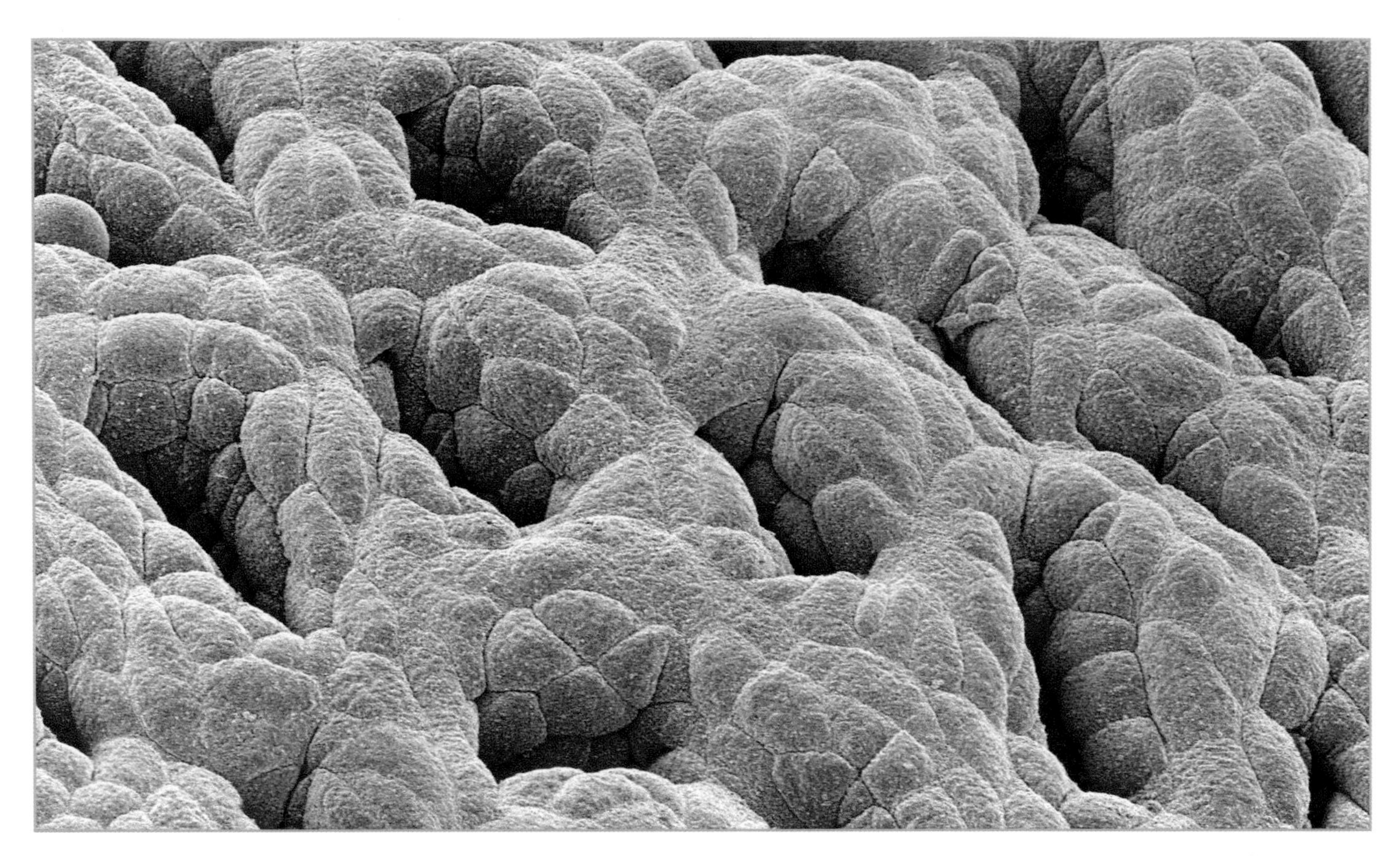

Found in the lining of the stomach's inner layer, these pits produce digestive enzymes to break down food.

on milk. Gastric lipase also breaks apart triglycerides of milk into fatty acids and monoglycerides. Parietal cells are pumping out ions of hydrogen (H+) and chloride (Cl-), which form hydrochloric acid. Chief cells secrete pepsinogen, which is not converted to the protein-digesting enzyme pepsin until it comes in contact with hydrochloric acid, but the stomach lining stays protected by a thin layer of alkaline mucus. Though most absorption takes place in the small intestine, the stomach does absorb water, ions, short-chain fatty acids, alcohol, and aspirin, and the latter two can add to problems with gastric ulcers (see "Breakthrough," below).

PHASES

Neural and hormonal mechanisms regulate gastric contractions and secretions. This stage can be divided into three overlapping phases: cephalic, gastric, and intestinal. Every day gastric glands can unleash up to 3 quarts of stomach juice—an acid strong enough to dissolve nails. The cephalic

SEE ALSO: Chapter Ten, "Hormones," PAGE 268

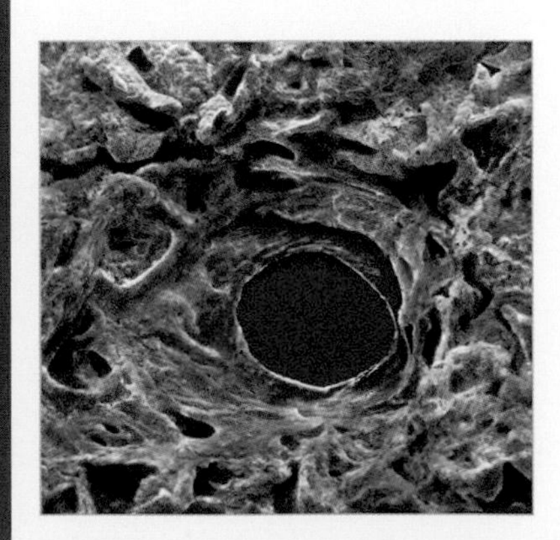

UNDERSTANDING ULCERS: Some 5 to 15 percent of Americans suffer from peptic ulcers, left, lesions in the GI tract exposed to gastric juices. The mucosa can become damaged, opposite, and cannot produce mucus to protect the stomach lining from hydrochloric acid. Symptoms include a gripping pain that starts shortly after eating and often subsides with the next meal. Worldwide, about 90 percent of duodenal ulcers and 80 percent of gastric (stomach) ulcers are caused by *Helicobacter pylori* bacteria, which are found in 50 percent of the population. The 2005 Nobel Prize for medicine went to two Australian scientists, phase occurs before food enters the stomach. Like Pavlov's dogs, we're conditioned to respond to sensory stimuli—the aroma, sight, taste, or even thought of food triggers the stomach to prepare itself. Olfactory nerves and taste buds relay signals to the brain's hypothalamus, which transmits the message to the medulla oblongata and thence to neurons in the cranial nerves. The signal stimulates gastric glands to release hydrochloric acid, pepsinogen, mucus, and gastrin. The same impulses also stimulate stomach motility. Depression, anger, and fear can inhibit this neural activity.

In the gastric phase, about two-thirds of the gastric juice is secreted. Both neural signals and hormones control this phase. The stretching of the stomach, for example, sends messages to the medulla, which in turn leads to the release of more gastric juice. The hormone gastrin plays an even larger role in secretion. Undigested proteins, caffeine, and an increasing pH level trigger the output of gastrin, which then causes parietal cells to generate more hydrochloric acid. The higher in protein content the meal, the greater will be the output of HCl and gastrin. HCl release is triggered by two other chemicals in addition to gastrin: acetylcholine (released by parasympathetic nerve fibers) and histamine (released by mucosal cells). When all three chemicals bind onto parietal cells, hydrochloric acid fountains out.

The intestinal phase is activated by the small intestine. Instead of causing secretion and movement, the small intestine slows things down so that it does not become overloaded with chyme. Chyme rich in fatty acids and glucose triggers the small intestine to release into the blood two hormones that will have an effect on the stomach: secretin and cholecystokinin (CCK). Secretin decreases the secretion of gastric juices. CCK inhibits the stomach from emptying too quickly into the small intestine. These hormones also affect the liver, pancreas, and gallbladder and their digestive operations.

After a meal, the stomach's three layers of muscles mix the chyme. Near the cardiac sphincter, peristalsis is gentle, but closer to the pylorus the waves are more vigorous. This churning is the last chance for chyme to become completely blended before entering the small intestine.

The antrum of the stomach holds only about 1 ounce (30 mL) of chyme. As the peristaltic waves wash chyme toward the partially-open pyloric valve, they end up spilling about one-tenth ounce (3 mL) of liquid and small particles out into the small intestine. Secretin and cholecystokinin inhibit motility to make sure that the small intestine does not receive more food than it can handle. It takes up to four hours for the stomach to process an entire meal.

EATING DISORDERS

WHEN A PERSON'S appetite controls are out of balance, with an obsessive fear of weight gain, he or she may have an eating disorder. They often develop during adolescence and occur most often in women. Anorexia nervosa sufferers have a mortality rate of 0.6 percent. Bulimia nervosa patients typically binge eat and then purge by vomiting.

BREAKTHROUGH

Barry J. Marshall and J. Robin Warren, who proved that *H. pylori* was the leading cause of gastric ulcers. Marshall went so far as to deliberately infect himself with the bacterium to study its effects. The bacteria can often be treated with antibiotics. Gastric ulcers are also caused by nonsteroidal anti-inflammatory drugs like aspirin, as well as by over-secretion of hydrochloric acid. Avoiding aspirin, alcohol, caffeine, and cigarette smoking can help keep the mucosa healthy and less likely to be hurt by HCl. To cut down on hydrochloric acid, medications, such as H2-blockers and proton-pump inhibitors, have proven effective.

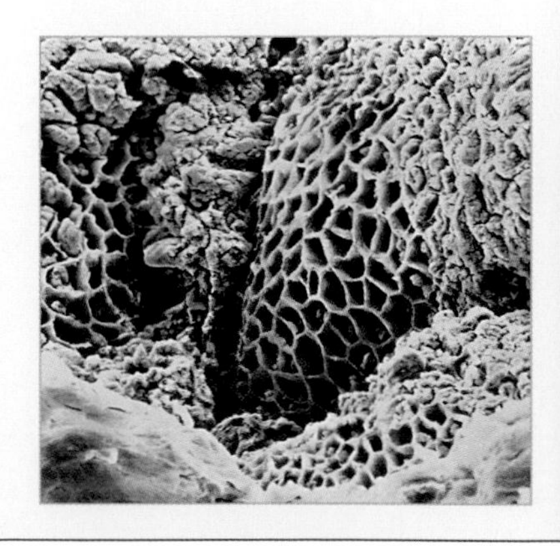

TOO MUCH WEIGHT

A HEALTH PROBLEM ON the rise in the United States, obesity afflicts one-third of the nation's adult population. Simply put, an obese person is too fat. That is, he or she has too much adipose (fatty) tissue relative to body mass. In general, a person who is 20 percent over normal weight is considered obese. To find out if you are overweight, multiply your weight in pounds by 705 and divide by your height in inches squared. The resulting number is your body mass index (BMI). A BMI of 25 to 30 indicates overweight; more than 30 is obese. Keep in mind, though, that weight is not always an accurate measurement for obesity, since a very muscular person may weigh more than a less-fit person of the same height. For an older person, who has lost muscle mass, the index may underestimate body fat. A body fat content of about 20 percent of your total weight is normal; if your fat content is more than 22 percent, however, you are carrying too much adipose tissue.

If Rubenesque figures—a term that refers to the full-bodied women portrayed by Dutch artist Peter Paul Rubens—were a modern ideal, obesity might not be the problem it is, were it not associated with a host of serious maladies: hypertension, arteriosclerosis, coronary artery disease, diabetes, pulmonary disease, arthritis, colon cancer, and varicose veins, to name but a few.

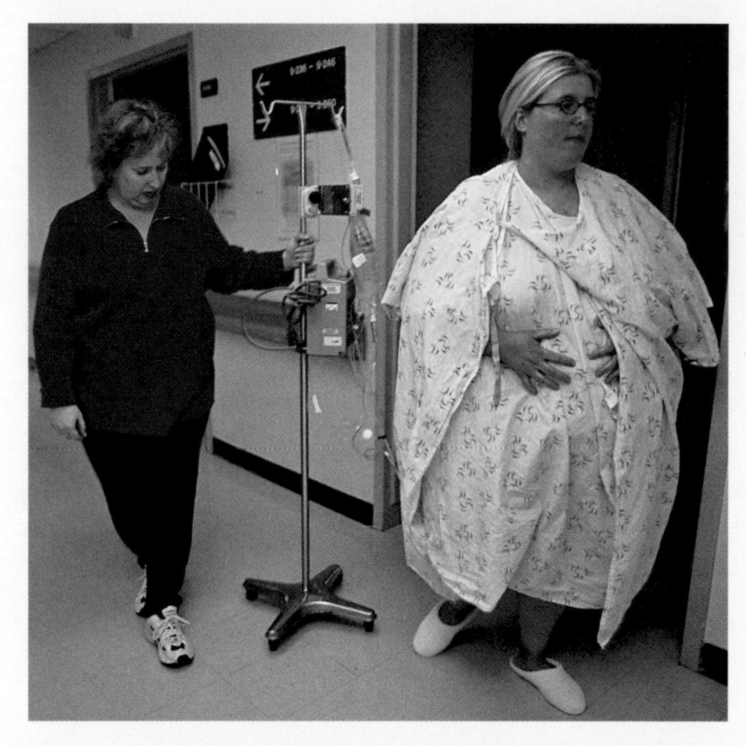

Seen before surgery, this obese woman has opted for gastric bypass surgery.

CAUSES

Why are some people too fat? Though there is no one overriding factor, there are many reasons. In general the problem lies with an out-of-balance food intake mechanism. Several influences can cause morbid obesity (obesity as a disease), including genetics, excess food intake, or bad eating habits learned young, tension-induced overeating, and trauma to the food-control center of the hypothalamus. Appetite regulation and fat-processing within the body are both contributing factors.

Among causes, genetics plays a definite role—those who inherit two obesity genes are nearly certain of having the disease. In the United States, though, only about 5 percent of obese people can blame genetics. These unfortunate people will store excess calories as fat, while others store some of them as muscle tissue.

Another main cause of obesity is childhood eating habits. Children who learn to overeat and who eat a lot of fast food, junk food, and other high-fat-content food add fat cells to their body, which are able to hold more fat as the person ages. Normal signals that tell the body that it is full are overridden in some way, possibly by stimulation from the fat cells themselves.

Another current theory holds that fat people are better at storing fat than others. Many obese people actually eat less than normal people but end up fatter because they convert food to fat and then do not easily burn it off. Another theory holds that thin people are fidgeting more than fat people and thus use more energy.

High-fat diets are particularly bad for people, because fat foods are the slowest to be exhausted as fuel. Carbohydrates are the most rapidly used. About 23 percent of carbohydrate calories eaten are tapped for metabolism, with the rest being stored, as opposed to a measly 3 percent of fat used and a walloping 97 percent of fat stored. Fat cells also beget more fat cells by growing alpha receptors, which help accumulate fat. In obese people, the lipoprotein lipase enzyme, which draws fat from the blood, is heavily produced and highly efficient.

TREATMENT

Ranging from pills to surgery, treatments vary widely in type and efficacy. Diet plans themselves come in a multitude of flavors, a new one or two popping up every year in popular books and magazines. While some of them may work temporarily, the problem is that the pounds shed too often return when the dieter begins slipping back into his or her old lifestyle. High-protein (low-carbohydrate) diets may have some benefits, but they can also be deficient in vital nutrients. Diuretics, or water pills, simply cause the dieter to lose water weight; and they can also cause electrolyte imbalance and dehydration.

Some anti-obesity drugs, created since the 1970s, have achieved moderate success. The fen-phen regimen involved the use of both fenfluramine to increase serotonin secretion (and thus increase feelings of satiety) and phentermine (an amphetamine) to stimulate metabolism. However, in 1997 fenfluramine was recalled when researchers discovered a link to problems with heart-valve closure.

Another drug, Orlistat, was approved in 1997. Orlistat blocks the action of pancreatic lipase, an enzyme that breaks down triglycerides in the intestines, so that ingested fat is never broken down and absorbed in the first place and so is moved out undigested. New drugs are being developed that bind to fat cells and increase the rate in which they are used up in the bloodstream. But all such drugs have side effects and should not be taken without the supervision of a physician.

Surgical options, such as stomach stapling and gastric bypass surgery, are sometimes effective in weight loss by giving food less chance for absorption. In one radical technique, biliopancreatic diversion, two-thirds of the stomach and half the small intestine are removed and pancreatic juice and bile are redirected so that essentially no fat is processed. The fat is simply eliminated. With liposuction, excess fat is vacuumed out. The problem is that these are major surgeries, with all the attendant risks.

After the operation, she has been able to maintain her weight at a healthier level.

The best solution to losing weight remains to be simply cutting back on fat intake and calories in general. A very low-calorie diet, coupled with a doctor-supervised exercise program (often starting with about 30 minutes of walking per day, five to seven days per week, and a gradual increase in intensity) is the best method for weight loss. After a few months, caloric intake can be increased to an optimal level.

CHAPTER GLOSSARY

ABSORPTION. The passage of nutrients from the GI tract into the blood and lymph systems.

BOLUS. Food mass that travels from the mouth to the stomach.

BRUSH BORDER. The fuzzy line created by microvilli between the lumen (interior) of the small intestine and its mucosa.

CIRCULAR FOLDS. The transverse ridges in the mucosa of the small intestine that increase the surface area for absorption.

DUODENUM. The first 10 inches of the small intestine.

ILEUM. The terminal section of the small intestine, which connects to the colon.

INTESTINAL GLAND. A gland in the intestinal mucosa that secretes digestive enzymes (also called a crypt of Lieberkühn).

JEJUNUM. The middle section of the small intestine.

LACTEAL. The lipid-absorbing lymphatic vessel in a villus.

LIPASE. A lipid-digesting enzyme, it splits fatty acids from triglycerides and phospholipids.

MICROVILLI. Miniscule projections atop epithelial cells (for example, on villi) that increase area for absorption.

MIGRATING MOBILITY COMPLEX (MMC). A peristaltic wave sequence in the small intestine, in which a wave travels a short distance before dying out, then followed by another wave.

PANCREATIC AMYLASE. A starch-cleaving enzyme secreted by the pancreas.

PEPSIN. A protein-digesting enzyme secreted in the stomach and activated by hydrochloric acid.

SALIVARY AMYLASE. A starch-cleaving enzyme secreted by the salivary glands; its action ends in the stomach.

VILLI. Hairlike projections of the intestinal mucosa through which nutrients are absorbed.

completed in the small intestine. Pancreatic enzymes convert proteins into smaller units, called peptides, each enzyme splitting the peptide bonds at a different point. Lipids are mostly digested in the small intestine, when pancreatic lipase hydrolyzes triglycerides into fatty acids and monoglycerides. The nucleic acids DNA and RNA are also digested there.

Absorption in the small intestine is the culminating process of digestion. After the long journey from mouth to stomach, the food gets drawn into the body through the epithelial cells of the small intestine. Here 90 percent of all absorption takes place. The carbohydrates have been broken down into the monosaccharides glucose, fructose, and galactose and are ready to be absorbed in the blood and lymph systems. Likewise, the proteins have been chopped up into dipeptides, tripeptides, and amino acids. The fats are now fatty acids, glycerol, and monoglycerides.

The vast chemistry lab of the small intestine has several ways of absorbing these materials: diffusion, facilitated diffusion, osmosis, and active transport. Each will be explained as we examine how the different nutrients are taken up.

The small intestine is capable of absorbing 4 ounces (118 mL) of monosaccharides every hour. Under normal conditions every carbohydrate eaten (except cellulose) will be absorbed in the body. Molecules of fructose pass into the bloodstream by facilitated diffusion, which means that, since they are too large to simply pass through membrane pores, they have to be escorted through by a transporter protein. Glucose and galactose are absorbed via active transport—they expend energy going across a concentration gradient against the normal path of diffusion; in other words, they go against the flow. Likewise, amino acids and the end products of proteins are taken in by an active transport mechanism. About 95 to 98 percent of proteins in the small intestine end up being absorbed; the rest is passed out. Half the amino acids absorbed come from food; the rest come from digestive juices and mucosal cells.

Simple diffusion takes lipids across the boundary between the small intestine and blood vessels. Adults absorb some 95 percent of all lipids; the figure is about 10 percent lower in newborns, as

+ CELIAC DISEASE +

CAUSED BY inadequate absorption in the small intestine, celiac disease stems from a sensitivity to gluten. Consumption of gluten leads to damage of the villi, which impairs the body's ability to absorb nutrients. Symptoms include weight loss, diarrhea, abdominal cramps, weakness, and stunted growth. The only cure is to eliminate gluten from the diet.

they don't produce as much bile. Short-chain fatty acids diffuse into the villi of the small intestine and move into the lacteals. Long-chain fatty acids and monoglycerides need bile salts for dissolving before they can be taken up by the villi. About 95 percent of the bile salts then get recycled through the blood and into the liver. People who do not produce enough bile salts—usually because of blocked bile ducts or liver disease—may end up passing 40 percent of their ingested fats along as waste. The fat-soluble vitamins A, D, E, and K then are not sufficiently absorbed.

The small intestine also absorbs water—about 2 gallons a day—through osmosis into the bloodstream. Only about 2 quarts come from ingested water, the rest from GI tract secretions (1 quart from saliva, 2 quarts from gastric juice, 1 quart from bile, 2 quarts from pancreatic juice, and 1 quart from intestinal juice). This leaves 1 quart, most of which is absorbed by the colon; 0.1 is excreted in feces.

THE LIVER

The versatile liver is the body's largest gland, weighing in at 3 pounds (1.3 kg). Among its many regulatory and metabolic functions are synthesis of plasma proteins, detoxification, storage of glycogen and iron, destruction of old blood cells, and synthesis of vitamin D. It produces bile, which breaks down fats.

The blood-rich, wedge-shaped liver is mostly tucked in the right rib cage, just below the nipple. It occupies the right hypochondriac and epigastric regions. The liver is divided into two main lobes: the right and left lobe. From the visceral, or back (posterior), side of the liver, the left lobe shows two additional, smaller lobes: the caudate and quadrate lobes. The right and left lobes are divided by the falciform (sickle-shaped) ligament, a part of the peritoneum, which helps suspend the liver. At the bottom, or free, end of this ligament runs the ligamentum teres (round ligament), a remnant of the fetal umbilical vein, which connects to the umbilicus, or

Lining the small intestine, the velvety villi (pink) are topped by microvilli, which are instrumental in breaking down carbohydrates and proteins.

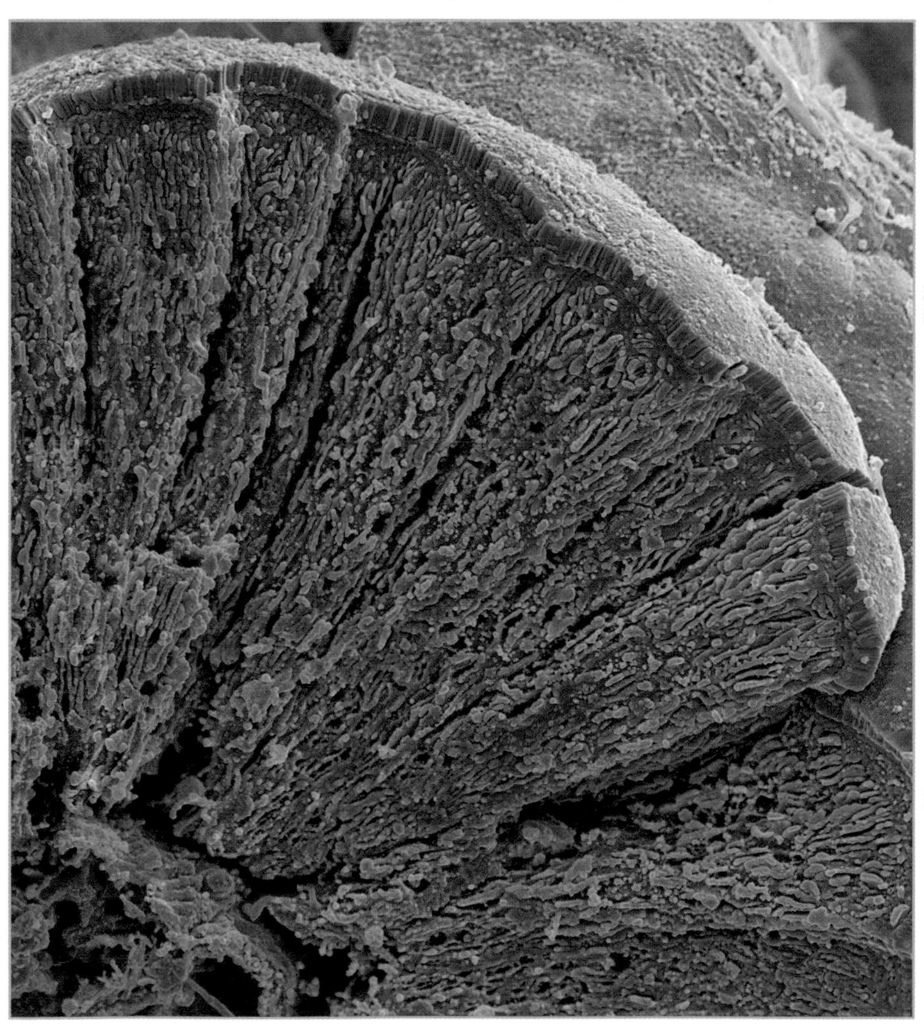

FOUR HUMORS

The four humors, from left to right: phlegmatic, choleric, sanguine, melancholy

WITH LITTLE to go on except naked-eye observation and philosophy, Western physicians from the time of the ancient Greeks to the Middle Ages practiced humoral medicine. Based on the theories of the Greek physician Hippocrates (460–370 B.C.) and later Galen (A.D. 131–200) the four humors (from the Latin for "moisture") were the description of the bodily fluids: blood, phlegm, choler (yellow bile), and melancholy (black bile).

A well-balanced person had perfect proportions of the four, but in most people one humor would predominate and dictate his or her temperament. Thus the four corresponding temperaments, or dispositions, were sanguine, phlegmatic, choleric, and melancholic. The humors were also associated with the four elements of nature—air, water, fire, and Earth. The four seasons—spring, winter, summer, and fall—were likewise aligned with the humors.

The system was a highly flexible tool for diagnosis and treatment. In the springtime, for instance, blood was said to run high, especially in the young. As a treatment, bloodletting or simply cutting back on red meat was prescribed. Colds in winter were caused by excessive phlegm and cured with hot wine. In the summers, diarrhea and vomiting were due to bile, and bile boiling in the brain led to mania. Cold baths were one attempt to help. The four principal qualities—hot, dry, cold, and wet—were matched in opposition with the humors; thus hot and moist were correlated with phlegm.

Not until the 19th century and with the full understanding of the circulation of blood did doctors declare this theory obsolete. Today we still use these terms to describe personality traits: A sanguine person is robust and confident, while a person who is slow and impassive could be called phlegmatic. If you lose your temper easily, you are said to be choleric. And the perennially sad person is often described as melancholic. Each of the four terms had additional characteristics no longer attributed to them—the choleric type, for instance, was described as yellow-faced, lean, and ambitious. ■

middle the body, and the tapered upper portion the neck. Inside, the gallbladder has a mucosa that pleats into rugae (as in the stomach) when the gallbladder is empty. The middle layer of the gallbladder wall is composed of smooth muscle fibers. When these muscles contract, the gallbladder squirts bile into the cystic duct. The outer, and final, layer of the gallbladder is formed by part of the visceral peritoneum.

A more complex organ, the pancreas (from "all" plus "flesh") produces digestive enzymes and secretes them into the small intestine. The pancreas is about 5–6 inches (12–15 cm) long and consists of a tail that abuts the spleen, a body, and a head that nestles into the curve of the duodenum.

Pancreatic juices, rich in digestive enzymes, pass into the pancreatic duct, which is joined by the bile duct just before it enters the duodenum at the ampulla of Vater and surrounds a valve, called sphincter of Oddi, that controls the flow of juice. The pancreas also has an accessory duct located about an inch above, or superior to, the sphincter that empties directly into the duodenum.

Surrounding the ducts are acini, little clusters of enzyme-secreting epithelial cells, called acinar cells. Pancreatic islets (or islets of Langerhans) among the acini have an endocrine function: They produce hormones, including glucagon, insulin, and somatostatin. The pancreas is both an exocrine gland

SEE ALSO: Chapter Ten, "Producers," PAGE 274

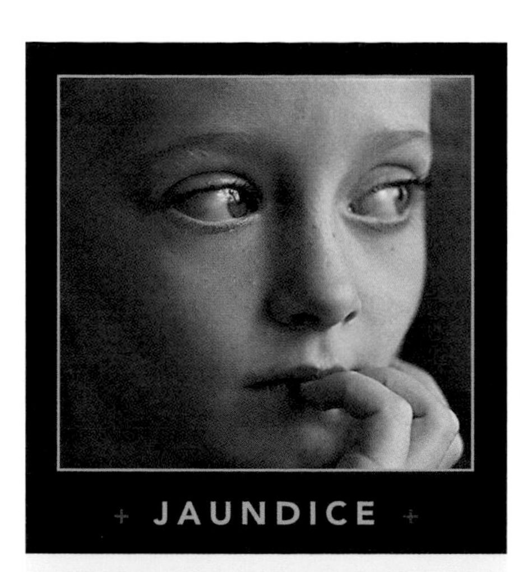

A YELLOWING of the skin, whites of the eyes, and mucous membranes are signs of jaundice, caused by too much bilirubin in the blood. Newborns often have jaundice because their livers have not matured; usually it clears without treatment. In adults it could be caused by bile duct obstruction from a gallstone, a tumor, hepatitis, or cirrhosis of the liver.

producing enzymes and an endocrine gland producing hormones.

The pancreas secretes approximately 1.2–1.5 quarts (1,200–1,500 mL) of pancreatic juices a day. A clear, colorless fluid, the juice is made of water, salts, sodium bicarbonate, and enzymes. The sodium bicarbonate helps create the proper pH for the enzymes in the intestine. Since the pancreas has no mucosa, it secretes enzymes in an inactive state so that the pancreas will not digest itself. Trypsinogen, for example, is converted to trypsin only when it is in the small intestine. Trypsin then can activate the enzymes carboxypeptidase and chymotrypsin. In addition to these protein-digesting enzymes, the pancreas secretes the carbohydrate-digesting enzyme pancreatic amylase and the chief triglyceride-digesting enzyme, pancreatic lipase. It also secretes the nucleic acid–digesting enzymes ribonuclease and deoxyribonuclease.

The secretion of pancreatic juice is regulated by both nerve impulses and hormones. Before food enters the small intestine, impulses from the vagus nerves tell the pancreas to secrete enzymes. When chyme enters the small intestine, the latter secretes the enzymes cholecystokinin (CCK) and secretin into the bloodstream; CCK makes its way back to the pancreas and stimulates the production of bicarbonate ions, while the secretin stimulates enzyme secretion. (Yes, this is the same hormone that inhibits gastric secretion.)

Just as alcohol abuse can damage the liver, it can also lead to a condition called pancreatitis. Instead of releasing the inactive trypsinogen, the pancreatic cells may release trypsin, which then starts eating away at the pancreas. Attacks can continue to occur even after treatment. Pancreatitis may also be triggered by gallstones that lodge at the common Y-connection, where the pancreatic duct and bile duct meet at the ampulla of Vater. Recurrent symptomatic gallstones call for complete surgical removal of the gallbladder, which can be done by laparoscopy.

Absorbed nutrients travel through the blood stream to the liver, where these specialized cells carry out more than 500 functions.

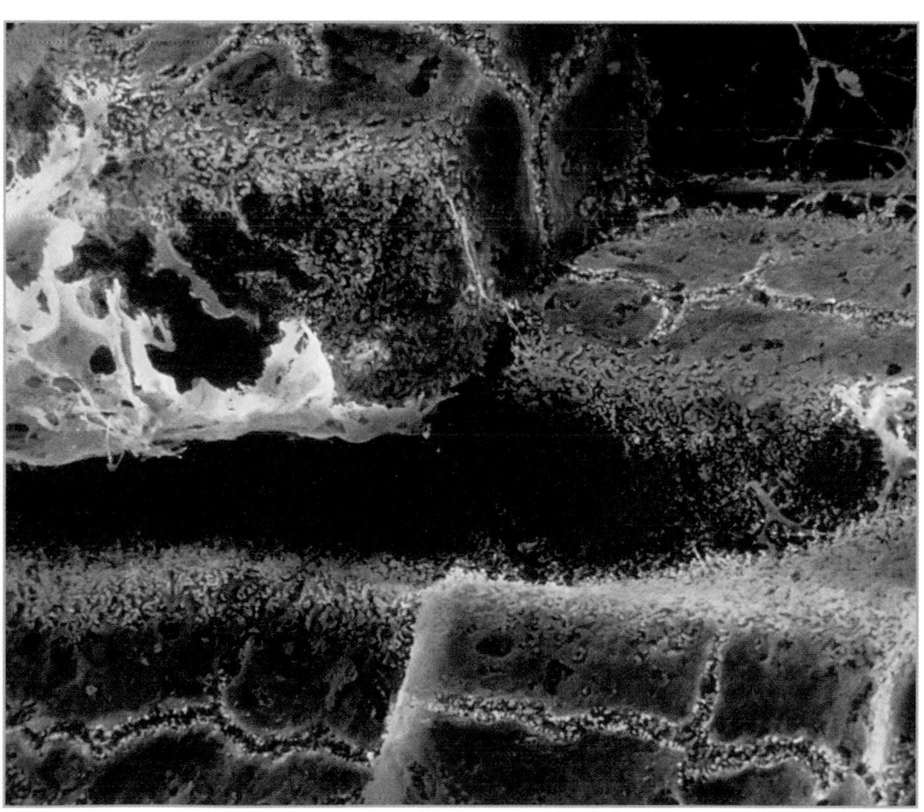

About 90 percent of water absorption has already taken place in the small intestine, but that other 10 percent absorbed in the colon—a pint to a quart—is a crucial part of the body's daily water intake. The vast majority of the colon's water absorption is through osmosis. Vitamins and ions such as sodium and chloride are also absorbed in the colon.

After some 3 to 10 hours, the large intestine has absorbed enough water from the chyme to turn it into a solid or semisolid state, and the chyme is now termed feces. What's in feces? Actually, quite a lot. Water and mucus in the feces help ease its passage; the feces also contains cellulose and other undigested food residue, inorganic salts, sloughed-off epithelial cells, and millions of bacteria.

The feces travel down the descending colon and into the sigmoid colon, where peristalsis pushes them into the funnel-like rectum. Distension of the rectal walls initiates the defecation reflex. Nerve impulses travel to the spinal cord, which sends signals back via parasympathetic nerves to the descending colon, sigmoid colon, rectum, and anus. In response, longitudinal muscles in the rectum contract, shortening the rectum and increasing rectal pressure. Voluntary contractions of the diaphragm and abdominal muscles also help in opening the internal anal sphincter. There is also an external anal sphincter, under voluntary control, that has to be relaxed in order for defecation to occur. Infants do not have control of the external anal sphincter, hence pressure on the rectum is enough to cause an infant to empty contents out through the anus. Adults also can contract their levator ani muscle to lift the anal canal. If an

A resin cast of the large intestine shows how larger blood vessels (green) supply a dense capillary bed (orange).

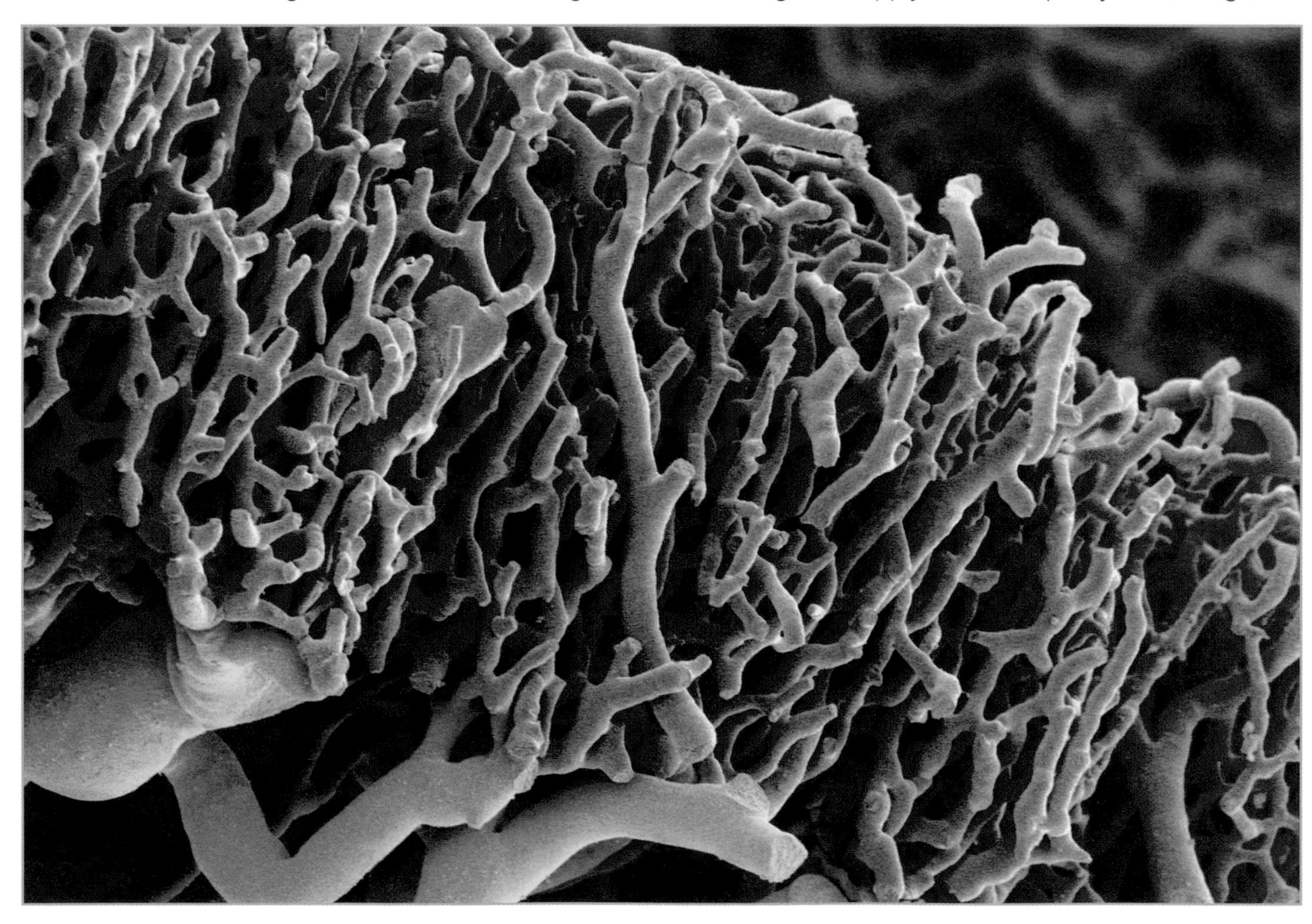

KEEPING HEALTHY

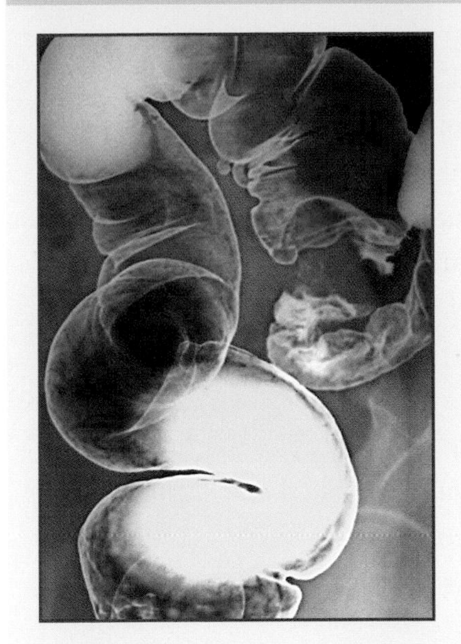

This x-ray shows colon cancer.

COLORECTAL CANCER is the third most common cancer in the United States and the second highest cause (after lung cancer) of cancer deaths. According to the American Cancer Society, there are 112,340 new cases of colon cancer and 41,420 new cases of rectal cancer per year; they will combine for a total of 52,180 deaths. Presently, some 5 to 6 percent of Americans develop colorectal cancer sometime during their lives; most people diagnosed with the cancer are over 50. Thanks to early detection, the death rate has been on the decline in the past 15 years. The most thorough type of screening is a colonoscopy, which allows a physician to examine the entire colon and remove precancerous polyps. A screening every 10 years starting at age 50 or earlier is the currently preferred strategy. ■

adult chooses not to defecate right away, the reflex impulses stop, and he or she then may wait until the next mass peristaltic movement.

COMPLICATIONS

Diarrhea (from "through" plus "flow") results when food hurries through the intestines without sufficient water absorption. The feces are then watery and increased in volume and frequency. The excessive motility of the intestine may be caused by microbes, lactose intolerance, or stress. The resulting frequent diarrhea can lead to dehydration and electrolyte imbalance.

The opposite problem, constipation, occurs when the feces remain too long in the colon and so much water is absorbed that defecation becomes increasingly difficult. Causes include failure to heed "nature's call," lack of exercise, emotional stress, and inadequate fluid and fiber in the diet. Mild laxatives may temporarily fix the problem but are not a good long-term solution.

Constipation can lead to another annoying problem called hemorrhoids ("flowing with blood"), or piles. Excessive straining during defecation puts extra pressure on the rectum, which causes rectal veins to swell. If the walls of these varicose veins dilate too much they start oozing blood. Anal itching and bleeding often indicate the presence of hemorrhoids. Over-the-counter creams are often successful in reducing the swelling; in rare cases, hemorrhoids are removed surgically. A diet high in fiber is often the best cure for both hemorrhoids and constipation. A low-fiber diet can narrow the bore of the colon and increase the strength of its contractions, which puts undue pressure on the colon walls. Weaknesses in the walls can balloon into pouches called diverticula (from "to turn aside"). Inflammation of the diverticula, diverticulitis, may cause severe abdominal pain (usually on the lower left side), fever, and vomiting. Fiber is also recommended as a guard against diverticulitis.

GAS

THE AVERAGE ADULT produces up to 17 fluid ounces of gas per day, passing it out the rectum in about 14 daily bursts. That's 48 gallons of gas—5,110 bursts—a year. It's mostly a mix of odorless vapors: carbon dioxide, oxygen, nitrogen, hydrogen, and methane. Trace amounts of sulfur released by bacteria in the large intestine cause the unpleasant smell. If you eat and drink quickly, chew gum, or smoke, you may be swallowing excessive amounts of air, which leads to gas buildup. Most of this gas is belched out; the rest burbles its way through the GI tract. Food undigested by enzymes in the small intestine is another source of gas; it moves into the colon, where bacteria break it down and release it. Carbohydrate sugars produce most of the gas.

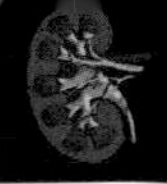

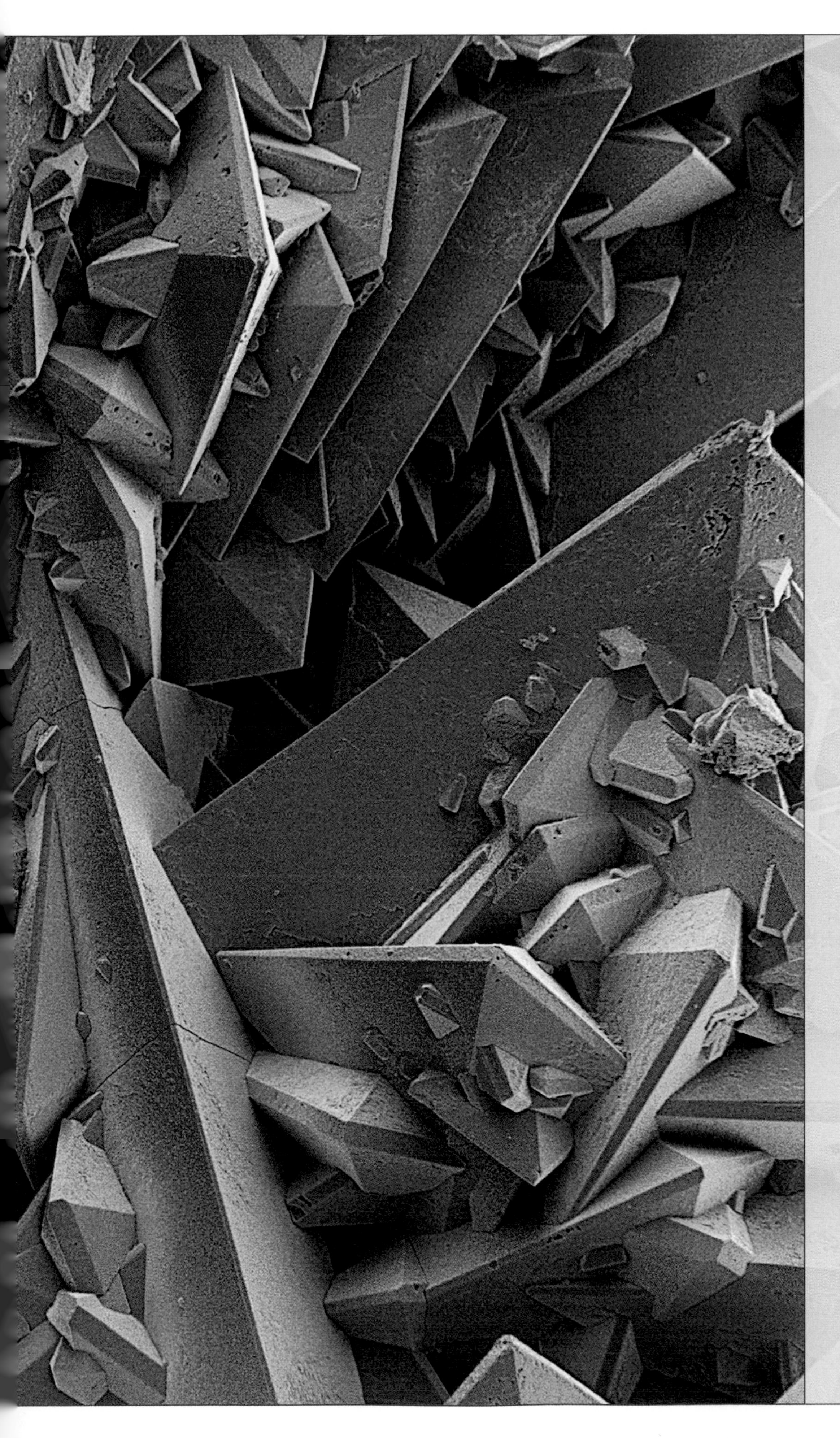

CHAPTER EIGHT

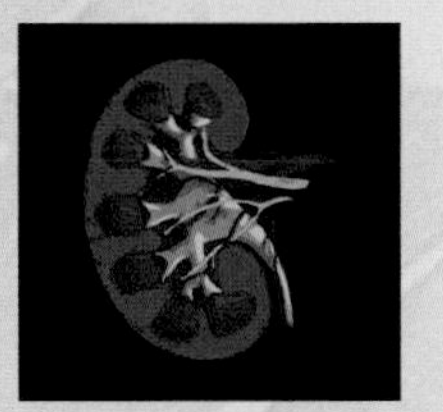

BODY CLEANSING

AN AQUARIUM WITHOUT A filter would in short order become filled with waste lethal to the life within the tank. Likewise, a person's kidneys are so vital in filtering out waste from the blood that kidney failure, if left untreated, will cause death in two to three weeks. The kidneys are constantly at work, filtering almost 200 quarts (200 L) of fluid from the blood every 24 hours. After purification, most of the fluid pulled from the blood plasma is cleaned and returned to the bloodstream; the rest is excreted as urine.

A kidney stone's jagged crystalline form under high magnification.

WASTE DISPOSAL

EFFICIENT FILTRATION

AN INTRICATE AND efficient filtration system, the urinary tract is the body's main way of keeping itself clean. The lungs and skin have a small role in ridding the body of metabolic wastes and other toxins, but the urinary system does the majority of the heavy lifting. It accomplishes this by filtering through the body's entire blood plasma more than 60 times a day. This work is carried out by the kidneys. The bladder is simply a storage reservoir, and the ureters and urethra are passageways for urine, the waste product of the kidneys.

+ CLEANING +

+ THE KIDNEYS process about 1 quart of blood every minute, for a total of about 423 gallons a day. This is equal to more than 300 times the body's blood volume.

+ By contrast, the heart pumps about four times the volume of blood daily processed by the kidneys.

+ The kidneys filter about 48 gallons of water from the blood daily. Most is reabsorbed; less than 1 percent, or about 1.5 quarts, is excreted in the form of urine.

+ Increased water intake can raise the urine volume to 3 quarts; heavy sweating can reduce it to a pint.

+ The blood plasma (blood minus the cells) is cleansed by the kidneys at the rate of 60 times per day.

The Urinary System

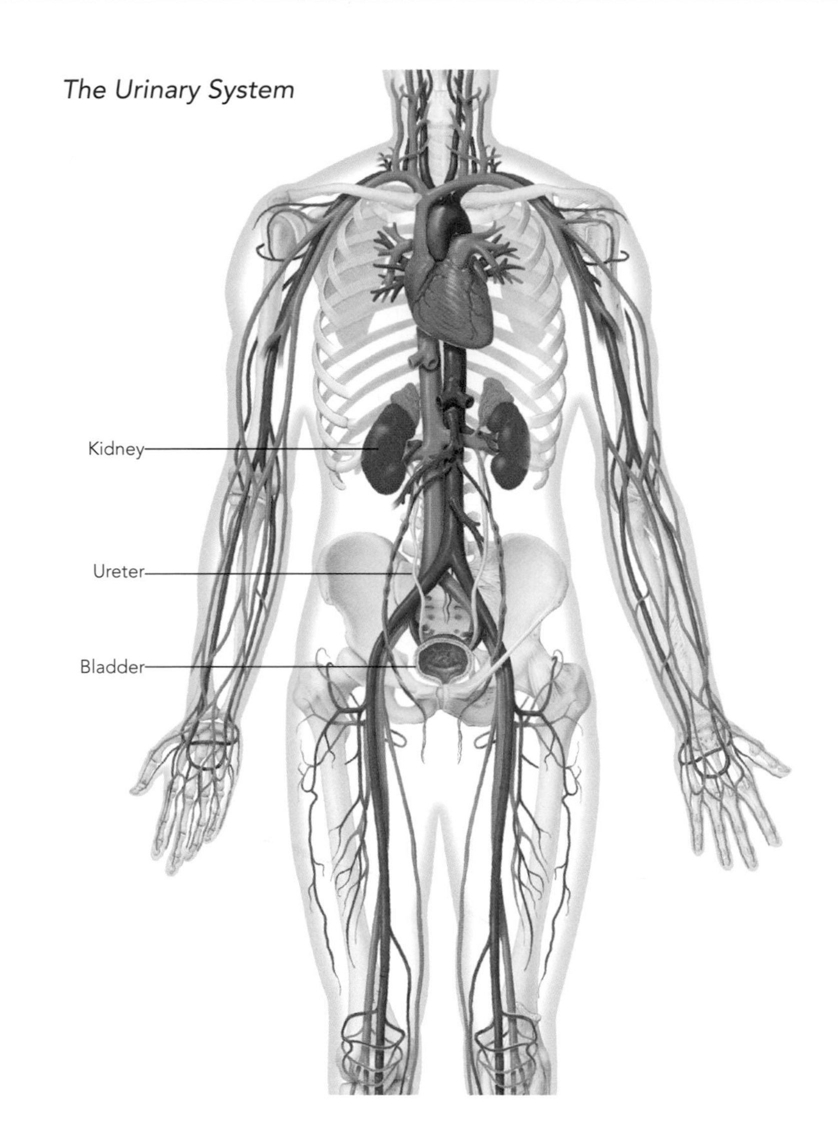

PURIFYING BLOOD

While the kidneys are making urine, they are also purifying the blood by fine-tuning its chemistry and volume. By regulating the blood's pH level and ion makeup, the kidneys keep the bloodstream and other fluids at optimal condition. The amount of salt (sodium) and water that the kidneys filter, reabsorb, and excrete from the blood regulates the blood's volume. A higher volume of blood raises blood pressure, and a lower volume decreases the pressure. By changing the volume and by secreting the enzyme renin, the kidneys can adjust the body's blood pressure as needed. The kidneys also regulate the osmotic pressure, or concentration of the blood, and all fluids in the body—that is, the amount of dissolved particles within the liquid. They do this by separately regulating the amount of water, in response to changes in the

SEE ALSO: Chapter Five, "The Blood," PAGE 132

concentration of body fluids (osmolality) and the presence or absence of antidiuretic hormone (secreted by the brain), and the amount of solutes they add to the urine. Among the solutes are calcium, potassium, sodium, phosphorus, and sulfate.

These three interrelated tasks—blood pressure, pH balance, and osmotic pressure—are absolutely essential for the functioning of the human body. By maintaining electrolyte and water homeostasis, or equilibrium, within narrow limits, the kidneys make it possible for human bodies to function on dry land, where they must conserve water and salts.

Among the other key tasks performed by the kidneys are the production of hormones, the regulation of the blood's glucose level, and the excretion of waste. The hormone calcitriol is the active form of vitamin D, and the hormone erythropoietin helps stimulate the production of red blood cells. And by creating urine, the kidneys rid the body not only of metabolic wastes such as ammonia and urea but also of foreign toxins—drugs and environmental poisons.

URINE

The waste product produced by the kidneys, urine, is a clear, pale to dark yellow fluid. The yellow tint comes from urochrome, a pigment resulting from the breakdown of hemoglobin. The more concentrated the urine, the darker its color. Occasionally the urine

URINE TELLS ALL

In 1998, Scottish sprinter Dougie Walker tested positive for the steroid nandrolene.

THE EXAMINATION of urine has long been used as a diagnostic tool in medicine and is a routine part of a physical examination. We can test urine for drug use, diseases, pregnancy, and ovulation. Testing for alcohol and marijuana, for example, are common in some workplaces, and testing for steroids and other illegal substances in athletes has kept pace with those desperate for an edge in their sport.

Diseases often are indicated by the presence of protein, glucose, bilirubin, or ketones in the urine. In addition to chemical testing, urine sediment can be examined by microscope for white and red blood cell casts (tubular structures containing cells or degrading cells), findings that could mean a tumor or kidney disease. Screening for prostate disease involves ultrasound imaging and a blood test for prostate-specific antigen. A new test for the prostate cancer gene 3 may make such screening more accurate. Diabetics can use urine tests to monitor the amount of glucose spilling into the urine when the blood sugar is very elevated, generally greater than 80–200 g/dL.

Pregnancy tests have a solution in a test tube that when mixed with urine will turn a certain color if the human chorionic gonadotropin hormone (secreted by a fertilized egg) is present. If you want to know the optimal time to try to get pregnant, use a test kit that looks for luteinizing hormone in the urine; a surge in this hormone occurs a day or so before ovulation. ■

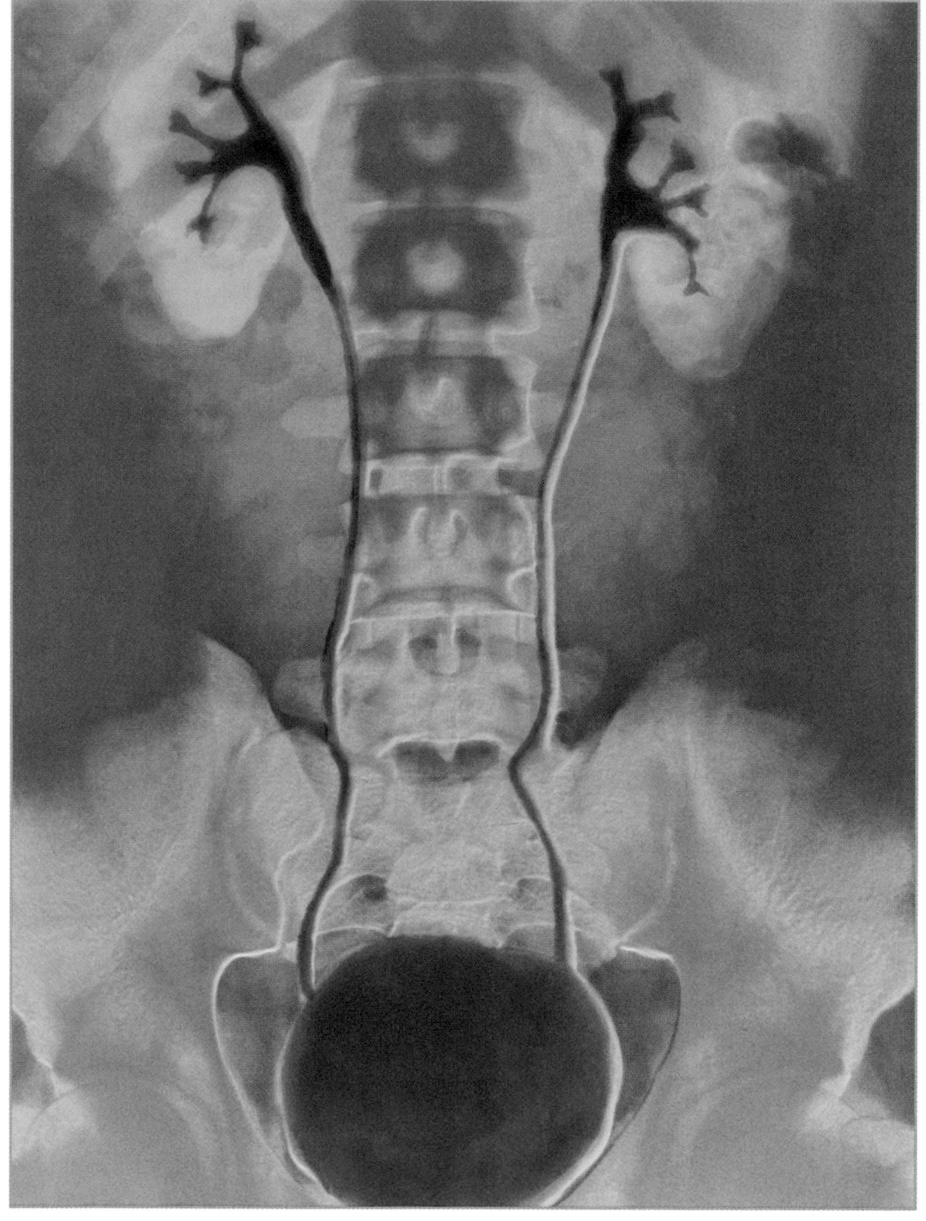

Colored red in this x-ray, the ureters connect the kidneys to the bladder.

has a pink or brownish tinge, which is usually brought about by consumption of foods such as red beets but can be due to the presence of blood or bile salts in the urine. Vitamin supplements and prescription drugs can also cause discoloration. The frothiness of urine comes from bile salts or proteinuria. Smoky or cloudy urine is often caused by an infection.

Urine has only a slight odor at first, but outside the body it begins to take on the smell of ammonia as its solutes are metabolized by bacteria. Diabetes mellitus (when out of control) lends the urine a fruity odor because of the presence of ketones. Other diseases can change the odor, as can consumption of drugs. Eating certain foods such as asparagus causes sulfur-containing amino acids to break down, lending urine a sulfurous smell.

Although urine is normally slightly acidic, with a pH of about 6 (7 being neutral), it can range into the alkaline (up to around 8) with diet or bacterial infection, or during protracted vomiting—as the stomach loses acid, leaving the blood alkaline. A diet leaning heavily on fruits or vegetables is high in potassium salts of organic acids, which when metabolized leave alkaline residues in the urine. Urine can also veer toward the acidic (down to pH 4.5)—for

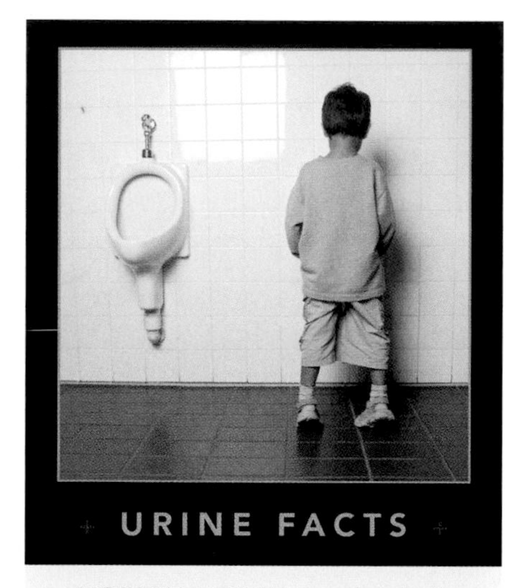

URINE FACTS

+ NORMAL URINE is nearly odorless. It develops an ammonia smell as bacteria begin to metabolize its solutes.
+ The yellow color derives from the pigment urochrome; the more concentrated the urine, the darker the color.
+ Drinking coffee or tea adds caffeine to the urine; exercise adds lactic acid.
+ The ketone bodies acetone and acetoacetic acid accumulate in the urine of untreated diabetics and those on a low-carbohydrate diet, giving the urine a fruity odor.

instance, after eating a lot of protein and whole wheat.

The specific gravity of urine varies from 1.001 to 1.035, depending on the concentration of solutes in the urine. With distilled water having a specific gravity of 1.0, urine concentration thus goes up with more solutes or down with less water. And the higher the amount of solutes per volume of water, the more likely it is that the solutes will precipitate into stones (see "Keeping Healthy," page 219).

If you drink nothing for a full day, your kidneys continue producing urine at a specific gravity of at least 1.025. Without at least 1 quart (946 mL) of water per day, the water for urine will begin coming from tissues, which can lead to dehydration. The kidneys have a limit beyond which they cannot concentrate urine. On the other hand, chronic renal (kidney) disease limits the ability to both concentrate and dilute urine.

COMPOSITION

What is urine composed of? About 95 percent of it is water; the rest is solutes. The main solute is a nitrogenous compound called urea, which is made from the breakdown of amino acids. Other waste products in urine are ions of chloride, sodium, potassium, phosphate, and sulfate. Other nitrogen-containing substances in urine include creatinine (a white, crystalline, basic compound) and uric acid (a product of nucleic acid metabolism).

URINE COMPONENTS

WATER	About 95 percent of urine is made up of water
AMINO ACIDS	Urine also has about 3 g of amino acids per day
HIPPURIC ACID	About 0.7 g of hippuric acid per day are excreted through urine
UREA	Of the remaining 5 percent, the most plentiful solute by weight. Urea accounts for more than 50 percent of all solutes and 90 percent of urine's nitrogenous components, with ammonia 2–4 percent, creatinine 3 percent, and uric acid 1–3 percent
OTHER	Of inorganic components, the daily urine contains about 6.3 g of chloride ions, 4 g of sodium ions, 2–4 g of potassium ions, up to 2.5 g of phosphate ions, up to 2.5 g of sulfate ions, and small amounts of calcium, magnesium, bicarbonate, and iron

Above: The pads on a test stick will reveal the presence and quantity of substances in urine.

There can also be found small quantities of calcium, magnesium, and bicarbonate ions.

A higher concentration of any of these urinary solutes often means there is something wrong with the kidneys or some other body system. Another indicator of disease is the presence of glucose, blood proteins, hemoglobin, red blood cells, white blood cells, or bile pigments in the urine.

Each of these abnormal components in urine points to a different probable cause. High amounts of glucose in the urine could mean diabetes. Too much protein in your urine most likely reflects glomerular inflammation (glomerulonephritis), but small amounts of protein can also be seen with high fever or prolonged upright posture. Ketone bodies in the urine point to untreated diabetes but also to fasting or starving.

Excessive bile pigments occur as the result of liver disease or a blockage of bile ducts. Hemoglobin in the urine can result from a reaction to a secondary blood transfusion, anemia, or severe burns.

Red blood cells, also called erythrocytes, in the urine might indicate bleeding in the urinary tract, which could be caused by kidney stones, a tumor, infection, trauma, or glomerulonephritis. Leukocytes, or white blood cells, often indicate a urinary tract infection or an inflammation.

THE KIDNEY

CONTROLLER OF WATER, ACIDS, AND SALTS

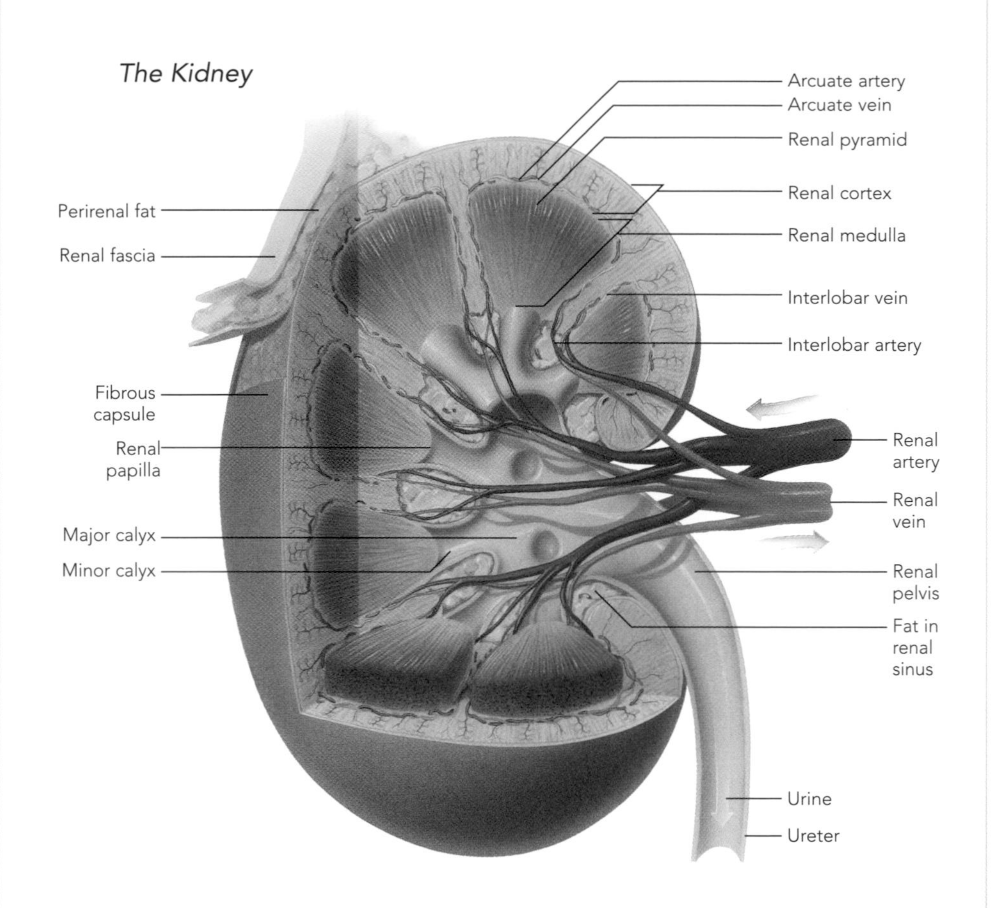

THE MAIN FUNCTION of the body's two kidneys is to filter the blood, produce a waste product called urine, and maintain the body's fluid composition. The kidneys also convert vitamin D to its most active form, and they produce an enzyme called renin, which helps maintain blood pressure. The hormone erythropoietin, secreted by the kidneys, stimulates the production of red blood cells.

Situated above the waistline, the kidneys are bean-shaped organs lying behind the abdominal cavity. The liver pushes the right kidney somewhat lower than the left. Both kidneys are partially protected by the lower part of the ribcage. An adult kidney isn't very large; it is approximately the size of a deck of playing cards.

ANATOMY

The ureter, nerves, blood, and lymphatic vessels pass through the renal hilus, a deep vertical fissure on the concave side of each kidney. Surrounding each kidney are three layers of tissue. The innermost layer is the renal capsule, a fibrous covering on the kidney itself. Next is a fatty layer called the adipose capsule, which supports and protects the kidney. And on the outside is a thin layer of connective tissue, the renal fascia, which anchors the kidney to surrounding structures.

The adipose capsule is so important that extremely thin people especially run the risk of nephroptosis, or floating kidney, in which the kidney drops and can kink the ureter, causing urine to back up into the kidney, leading to kidney damage and even failure.

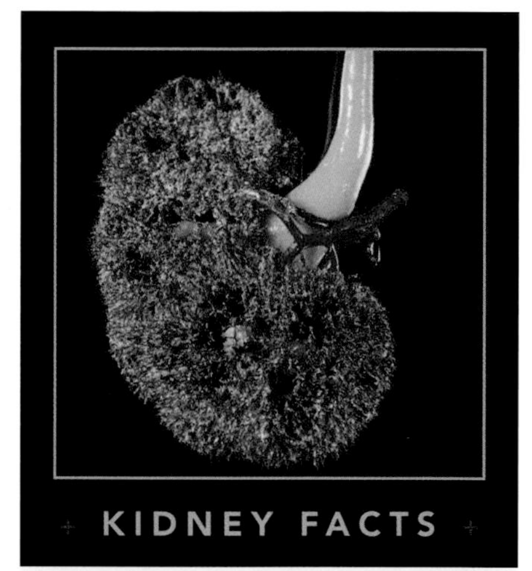

KIDNEY FACTS

+ AN ADULT KIDNEY weighs about 4.5 to 5 ounces.
+ The typical kidney is 4 to 5 inches long and about 1 inch thick.
+ The kidneys make up less than 0.5 percent of the body's weight, yet they take in 20 to 25 percent of the resting body's cardiac output and use 20 to 25 percent of the body's oxygen.

SEE ALSO: Chapter Five, "Blood Vessels," PAGE 150

Drinking enough water and staying hydrated is essential to maintaining proper kidney function.

The internal anatomy of the kidney breaks into two main regions—a smooth, reddish renal cortex and a darker inner region, the renal medulla. Within the renal medulla are 8 to 18 renal pyramids, tapering to the middle of the kidney. These pyramids are composed of bundles of tiny urine-gathering tubules. Within the cortex are nephrons, the basic processing units of the kidney. These functional units consist of a glomerulas (filtering device composed of a capillary network) and a tubule. At the end of the pyramids are urine-collecting cups called calyces, all of which drain into a collection cavity known as the renal pelvis. From the renal pelvis, the urine flows out through the ureter and to the bladder. Smooth muscles in the calyces, renal pelvis, and ureter move the urine along by peristalsis.

Since the kidneys are the blood's filters, it is no wonder that they are well supplied with blood vessels. In fact, one-quarter of the body's arterial blood is delivered to the kidneys via the renal arteries every minute. Inside the kidney, the renal artery splits into numerous segmental arteries that branch out into the kidney.

Each of the one million nephrons is hooked to an afferent arteriole (a terminal branch of the artery). These arterioles then divide into little tangled capillary balls called glomeruli. Instead of connecting to a vein, as do the body's other capillary networks, they connect to an efferent (outgoing) arteriole. Some 90 percent of the blood entering the kidney floods the cortex, where most of the glomeruli are. The veins then reverse the route of the arteries, draining each kidney via a renal vein, which connects to the inferior vena cava. The left renal artery is twice as long as the right to reach the inferior vena cava, which lies to the right of the vertebral column.

DEVELOPMENT

The kidneys begin to appear during the third week of human fetal development. Not one, but three pairs of kidneys develop in succession. The first pair begins degenerating after the fourth week and disappears by the sixth week; by the eighth week the second pair is almost completely gone as well. The pair farthest down along the fetus's intermediate mesoderm is the one that will survive. By the third month, the fetus is excreting urine into the amniotic fluid; urine then becomes the main component of the amniotic fluid.

On the other hand, renal activity slows with age. Renal blood flow and filtration rate decrease by some 50 percent between the ages 40 and 70. The kidneys diminish from a mass of about 9 ounces (260 g) in a 20 year old to 7 ounces (200 g) in an 80 year old. Kidney stones and inflammation become more common with age, as do urinary tract infections. Other ailments include dysuria (painful urination), hematuria (blood in the urine), and and incontinence (see "What Can Go Wrong," page 225).

NEPHRONS

Each kidney has about one million nephrons, the miraculous miniature filtration plants of tubes and capillaries. In each of these microscopic processing units, blood is filtered by glomeruli; the waste product, urine, is created in the tubules, which reabsorb and secrete substances through one of thousands of collecting ducts.

The interface end of the nephron is the renal corpuscle, which consists of a glomerulus (a ball of entangled capillaries) surrounded like a helmet by a glomerular capsule (or Bowman's capsule). The glomerulus is a double-layered epithelial cup, which acts as filter through which blood plasma passes. The filtrate from the blood plasma oozes from the glomerulus into the highly porous inner layer of the glomerular capsule, and the resulting filtrate then flows through the renal tubule, where it then becomes urine.

The renal tubule is about 1.25 inches (3 cm) long. The first part of the tubule—the part that exits

A detailed computer scan and colorization enhances the visibility of the kidney's intricate blood supply.

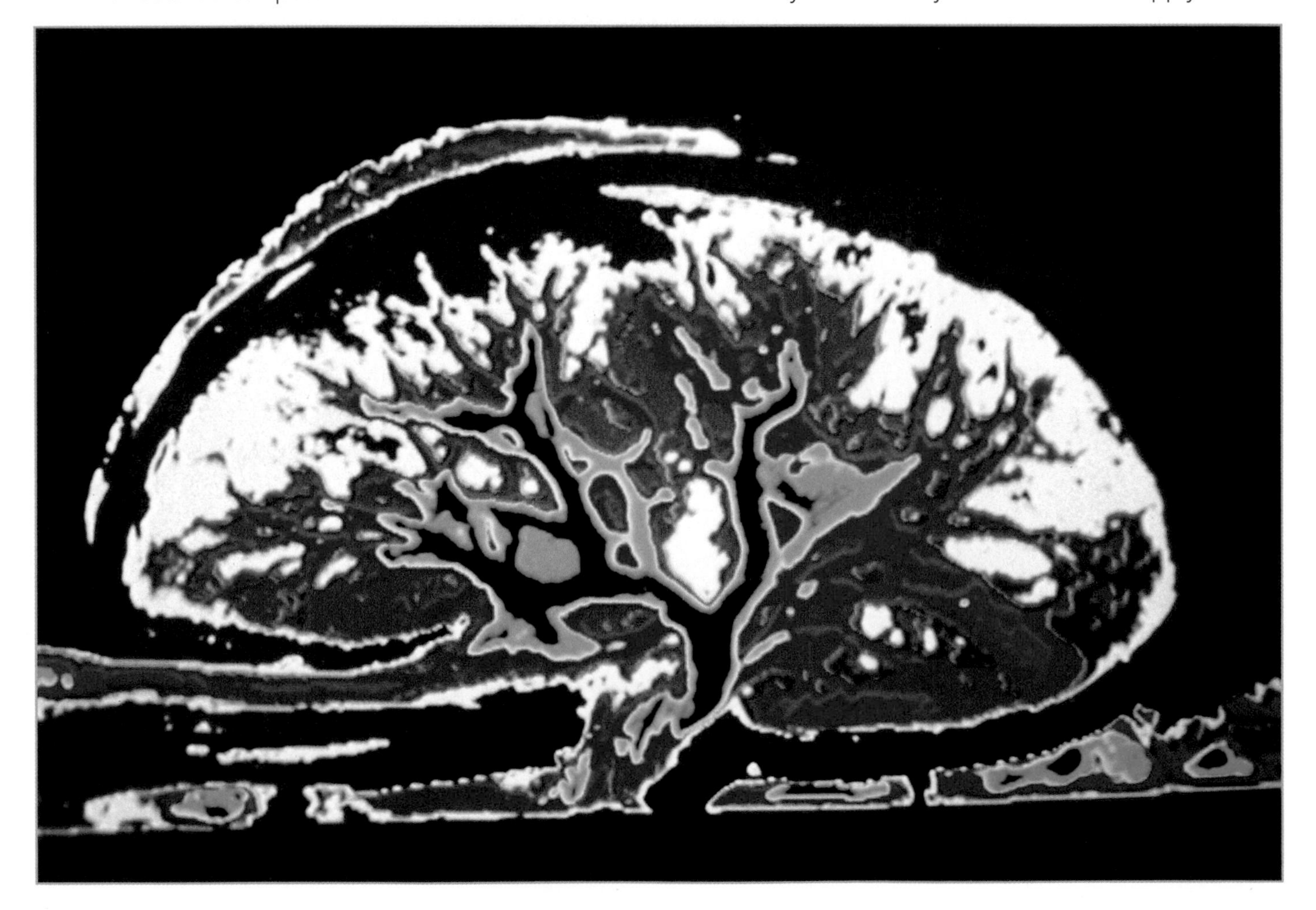

the glomerular capsule—is a twisting roller coaster called the proximal convoluted tubule. The tubule then straightens out and makes a long hairpin turn called the loop of Henle. Finally, the tubule widens and begins to twist again in the section called the distal convoluted tubule. Fluid leaves the renal tubule as urine and then enters one of the kidney's collecting ducts, which receive urine from many other nephrons.

Some 80 to 85 percent of all nephrons lie primarily in the outermost part of the kidney (the cortex), with the loop of Henle barely entering the medulla before ascending back into the cortex. The rest of the nephrons lie deep within the cortex and contain long loops of Henle that dip far down into the renal medulla.

Like little mountain creeks and streams emptying into rivers, the nephrons drain into the collecting ducts, which merge into several hundred papillary ducts. Then the papillary ducts merge into the minor and major calyces, carrying the urine down from the renal cortex to the renal medulla to the renal pelvis.

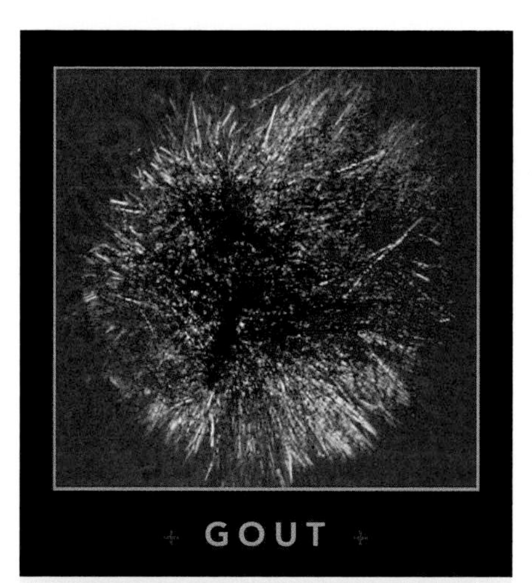

GOUT

EXCESS URIC ACID in the blood can crystallize in a joint—often at the base of the big toe. These crystals, above, cause swelling, redness, and sudden severe pain. Known as gout, this condition was once thought to be due to gluttony. Though it is more likely to affect the obese, gout can hit any person who is producing too much uric acid or not eliminating enough. Gout can be brought on by stress, injury, or diuretics taken for high blood pressure. Ninety percent of cases are in men over the age of 40. The age-old treatment colchicine (an alkaloid from the seeds of meadow saffron) is still used. Sufferers should avoid excessive alcohol and purine-rich foods: sardines, sweetbreads, and liver.

Zooming in to the cellular level of the nephron, we can see a number of interesting features. While the outside of the glomerular capsule contains a simple layer of epithelial cells, the visceral (inside) layer next to the glomerulus is composed of podocytes, cells with little feet that grip onto the endothelial cells of the capillaries. Between the feet are filtration slits, or pores, where the filtrate passes into the glomerular capsule.

The renal tubule and collecting duct have different kinds of cells. In the first part of the tubule, the proximal convoluted tubule, there are cuboidal epithelial cells. These multitasking cells both reabsorb materials from the passing filtrate and secrete substances (waste, excess ions, and drugs) into it. A brush border of microvilli (similar to that in the small intestine) increases the surface area of the tubule for the absorption of water and solutes. Remember, the tubules are carting off waste. When they secrete something into the urine, they are at the same time removing it from the blood. In short, their job is to maintain

BREAKTHROUGH

BEFORE GUSTAV SIMON (1824–1876), no one knew the body could still function with just one kidney. Although the kidney's nephrons do not heal if they become injured or diseased, they can enlarge to take on a bigger amount of blood filtering if one kidney is removed or damaged. The healthy one grows big enough to handle 80 percent of the load that two kidneys used to handle. In 1869 the German surgeon treated a woman with a urogenital fistula that could only be remedied by removal of a kidney. Simon had already distinguished himself in experimental surgery—first he proved that animals with only one kidney could still excrete urine at normal levels. His operation on the woman was the first successful removal of a human kidney. The procedure was a medical milestone, ushering in a new age of surgery that would eventually lead to kidney transplants.

A NEW HOPE

BEFORE THE MID-1950s, people with kidney failure faced a grim future. With nothing to stop the deterioration of the kidneys, and no machine on Earth capable of duplicating the kidneys' complex filtering job, patients faced an irreversible death sentence. But the development of artificial dialysis (from the Greek for "separation") and the artifical kidney gave such patients the hope for a much longer and more active life. Although commonplace today, dialysis was an amazing breakthrough that improved the lives of countless patients around the world.

FAILURE

When the kidneys begin to fail, they can no longer sufficiently filter metabolic and other wastes from the blood. As nitrogenous wastes begin to accumulate in the blood, the blood's pH turns dangerously acidic, and severe electrolyte imbalances begin wreaking havoc throughout the body. Symptoms of failure can include nausea, vomiting, edema, weakness, fatigue, uriniferous breath, and labored breathing from lung congestion. Blood tests reveal accumulation of urea (from protein metabolism) and creatinine (from muscle metabolism). The final stages of kidney failure involve convulsions, coma, and death.

Kidney failure occurs when 85 to 90 percent of the kidney's nephrons are destroyed. Nephrons may deteriorate for a number of reasons: diabetes mellitus, high blood pressure, kidney diseases such as glomerulonephritis (inflammation of and sometimes permanent damage to the glomeruli), chronic kidney infections, physical trauma, obstructed blood flow to the nephrons, and exposure to environmental toxins such as mercury or lead that poison renal tubules. Any of these can lead to chronic renal failure, an especially treacherous condition in that it often gives no clear warning until the damage is already done. In fact, in many cases there are no symptoms until renal function is down to less than 25 percent of normal.

Victims of acute renal failure might be luckier because symptoms come on more quickly—fluid retention, weakness, fatigue, and nausea indicate the need for immediate medical attention. The main sign of acute kidney failure is decreased urine output—down to 1 pint (500 mL) or less—and elevated levels of substances normally removed by healthy kidneys, such as blood creatinine and blood urine nitrogen (BUN). Causes for this kind of failure include blockage from an enlarged prostate gland or kidney stones, low blood volume, infections in the bloodstream (called sepsis), as well as toxic effects from drugs and other substances.

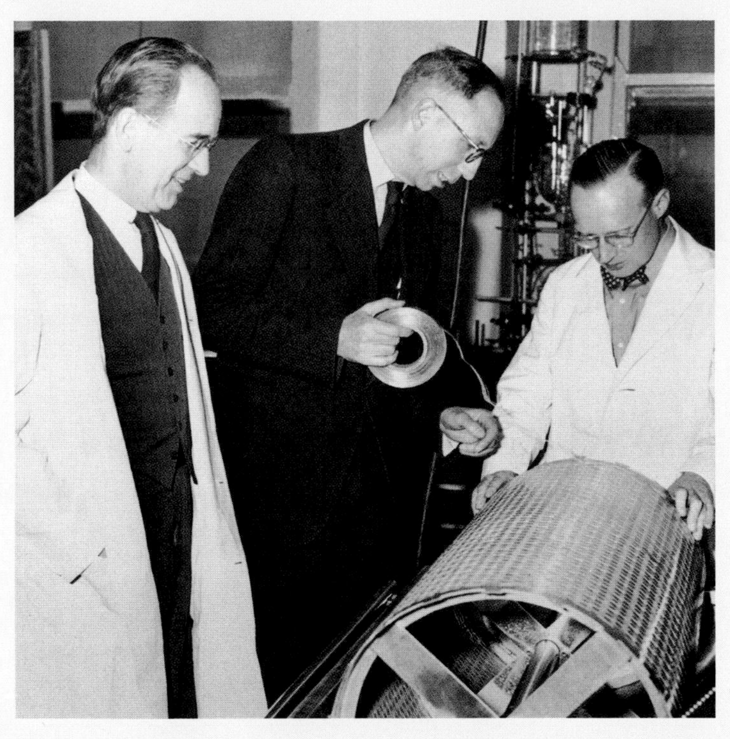
Dr. Willem Kolff (center), inventor of the artificial kidney, works with American physicians to improve it.

INVENTION

The artificial kidney changed everything for patients suffering from these conditions. Dutch physician and medical pioneer Willem Kolff

(1911–) invented the first dialysis machine in the Netherlands. While a student at the Leiden Medical School, Kolff witnessed the death of a 22-year-old patient from kidney failure. This pivotal event began his research into an artifical kidney.

During World War II, Kolff was working in a small municipal hospital in Kampen in the Netherlands. There in 1941, he developed the first dialysis machine. He made it from coarse materials gathered from a local factory. The machine consisted of a long tube of cellophane sausage wrapped around a slatted wooden drum. An electric motor turned the drum in a tank filled with dialyzing solution. Gravity drew the patient's blood through the tube as the drum spun. The waste materials in the blood filtered out through the tubing into the solution before the clean blood returned to the body.

Over the next few years, Kolff worked to improve on this initial design. By 1945, he had successfully treated his first patient: a 67-year-old woman who lived for 7 more years after treatment. Kolff worked with other physicians to improve the machine's basic design. Kolff's revolutionary machine would become a routine treatment for what was once considered a fatal condition.

DIALYSIS TODAY

Currently about 217,000 people in the United States are on some form of dialysis. People with acute renal failure may need the procedure only for a short time until the kidneys recover function. However, in cases of chronic renal failure, patients must have dialysis on a regular basis for the rest of their lives, or undergo a kidney transplant. With end-state renal disease (ESRD), the kidneys function at 10 to 15 percent or less of their normal capacity. In such cases, dialysis or kidney transplant are the only solutions.

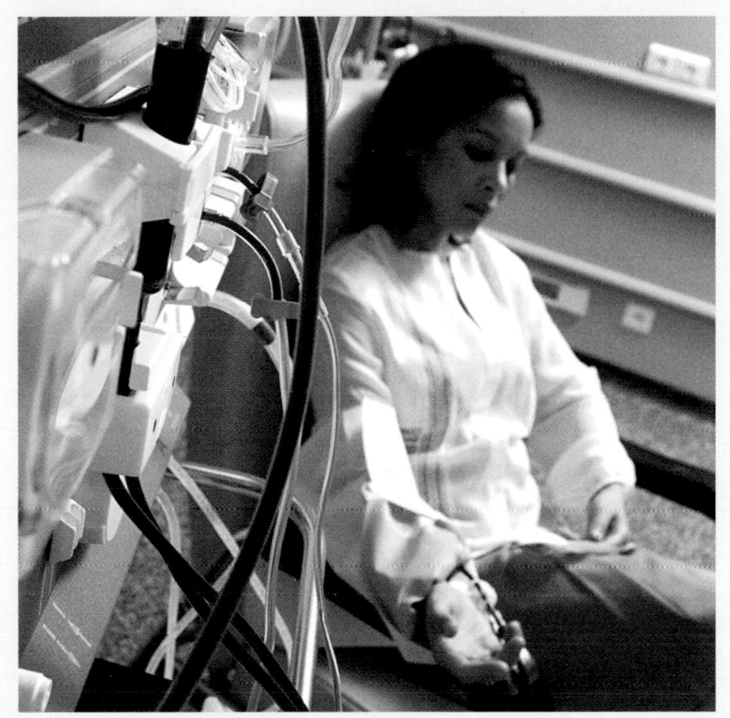

This woman receives hemodialysis several times a week to remove metabolic wastes from her blood.

The most prevalent form of dialysis, hemodialysis, works in much the same way as Kolff's original machine. It pulls waste products out of the blood by filtering it through a membrane. Tubes inserted into the patient's arm or leg form a connection between an artery and a vein and the dialysis machine. The unpurified blood runs through the machine, which filters out impurities and excess fluid. Potassium ions and nitrogenous wastes diffuse out into a solution in the machine, and bicarbonate ions and other substances are added to the returning blood. A typical hemodialysis regimen involves three weekly sessions, each lasting four to six hours.

In continuous ambulatory peritoneal dialysis (CAPD), the patient's own peritoneum (the membrane lining the abdominal cavity) acts as the filter. A catheter is inserted into the abdomen, through which the dialysis solution infuses the abdominal cavity, equilibrates with the blood, and then drains out of it. Tiny blood vessels in the peritoneal membrane do the filtering job the kidneys normally do.

Using this method, it can take an hour or longer for the toxins and fluids to be completely removed, and the procedure must be done four to five times a day. Patients can perform CPAD at home. A continuous cycling peritoneal dialysis machine (CCPD) does the same thing as CAPD—cycling dialysis solution in and out of the peritoneal cavity—while the patient sleeps.

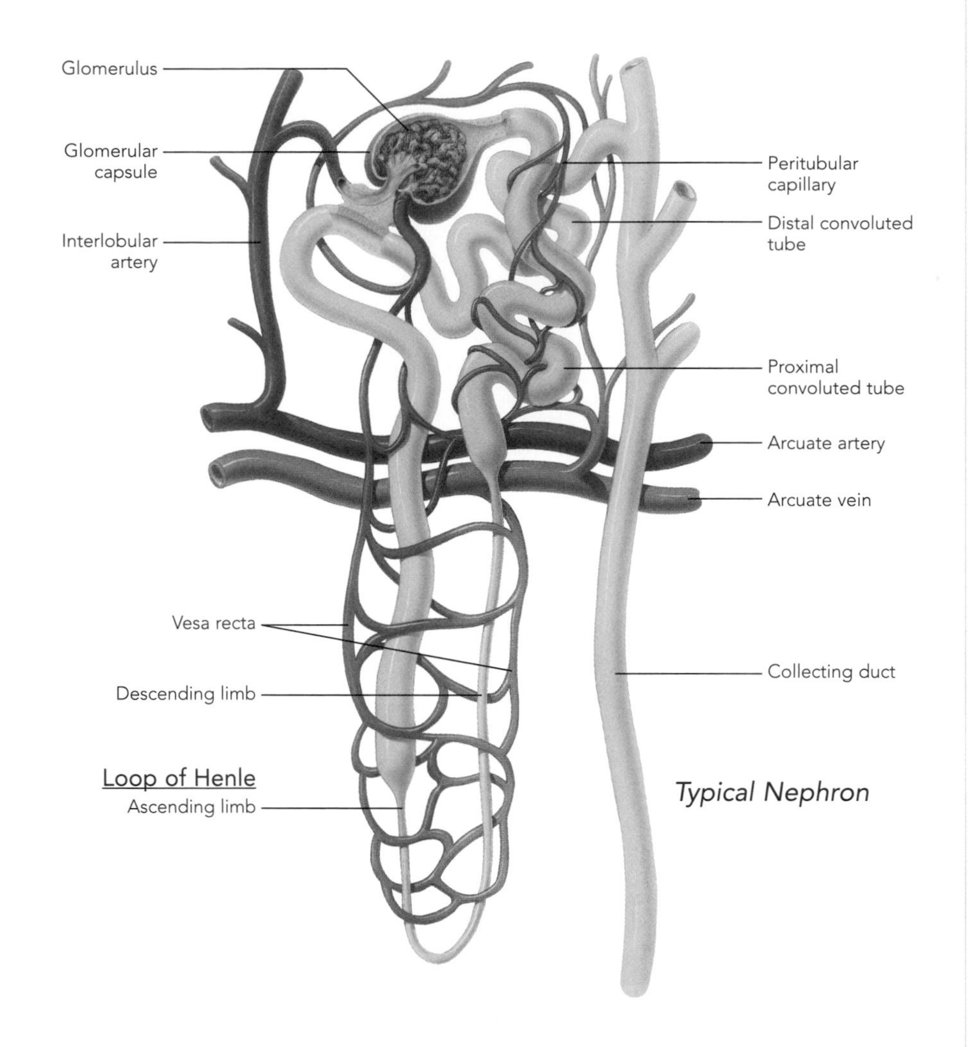

Typical Nephron

balance (homeostasis) by excreting waste and keeping salt, water, and acid and base in balance.

Simple squamous epithelial cells compose most of the loop of Henle; they are water permeable. Then the thicker, ascending part of the loop contains cuboidal and low columnar epithelial cells. And finally, the distal convoluted tubule has principal and intercalated cells. At this point, much more secretion than absorption is occurring. The principle cells are involved in balancing water and salt, and the intercalated cells help maintain the blood's pH. The urine drains into the collecting and papillary ducts, which are lined by columnar cells.

KIDNEYS AT WORK

The three basic processes carried out by the nephrons and collecting ducts are glomerular filtration, tubular reabsorption, and tubular secretion. First, water and solutes filter from the blood plasma into the glomerular capsule. Then, in the renal tubules, some 99 percent of the water, as well as some solutes, are reabsorbed into capillaries. While the filtrate flows, cells lining the tubules secrete wastes from the capillaries, further purifying the blood.

The kidney's nephrons filter wastes from the blood plasma and turn it into urine in several stages. The first is glomerular filtration: the blood plasma filters from the glomerulus into the glomerular (Bowman's) capsule. How much blood plasma actually gets through? Only about 16 to 20 percent of what comes in via the arterioles goes on and becomes filtrate in the nephron. And of that amount, more than 99 percent ends up being reabsorbed by the tubules and ducts. So of the 40 gallons (150 L) of glomerular filtrate that daily passes through a woman's kidneys (more for men), only about 1 to 2 quarts (1 to 2 L) ends up as urine.

The initial filtrate is far closer to blood plasma than urine in its chemistry. The junctions between the capillaries and the podocytes (footlike cells) of the glomerular capsule have a three-layer meshing that does not allow blood cells, platelets, and most plasma proteins to pass through. These "good things" remain in the bloodstream while the rest, the filtrate, enters.

The kidneys maintain a steady glomerular filtration rate (rate at which filtrate is made) of about 3.5 fluid ounces (105 mL) per minute in females and 4.25 fluid ounces (125 mL) per minute in males. They do this by controlling the rate that blood flows into

SEE ALSO: Chapter Five, "The Blood," PAGE 132

and out of the glomerulus and by adjusting the glomerular capillary surface area.

The filtrate passes from the glomerular capsule into the first part of the renal tubule, the proximal convoluted tubule where reabsorption begins to take place. Remember that 99 percent of the filtrate still has to be reabsorbed before it can be called urine. Most of the reabsorption happens in the proximal convoluted tubule. In addition to water, a number of solutes are reabsorbed, including (in order of quantity) chloride ions, sodium ions, bicarbonate ions, glucose, potassium ions, urea, uric acid, and proteins. Notice that some urea is reabsorbed by the blood. Of all the solutes, though, urea has the lowest percentage of reabsorption—only 0.84 ounces (24 g) out of 2 ounces (54 g) per day (44 percent). The other solutes are absorbed in percentages varying from 91 to 100 percent.

Since 20 percent of the body's entire volume of blood is filtered through the glomeruli into the tubules every 45 minutes, the reabsorption that takes place there keeps it from being drained of fluid. As in the glomerular corpuscle, there are three layers of membranes through which the water and solutes have to pass: the luminal and basolateral membranes of the tubule cells and the endothelium of the peritubular capillaries (the capillaries that return the water and solutes to the bloodstream). Using active or passive transport mechanisms, the substances cross this barrier by

This cross-section of kidney tissue reveals the tubules of the nephrons surrounding the glomerulus.

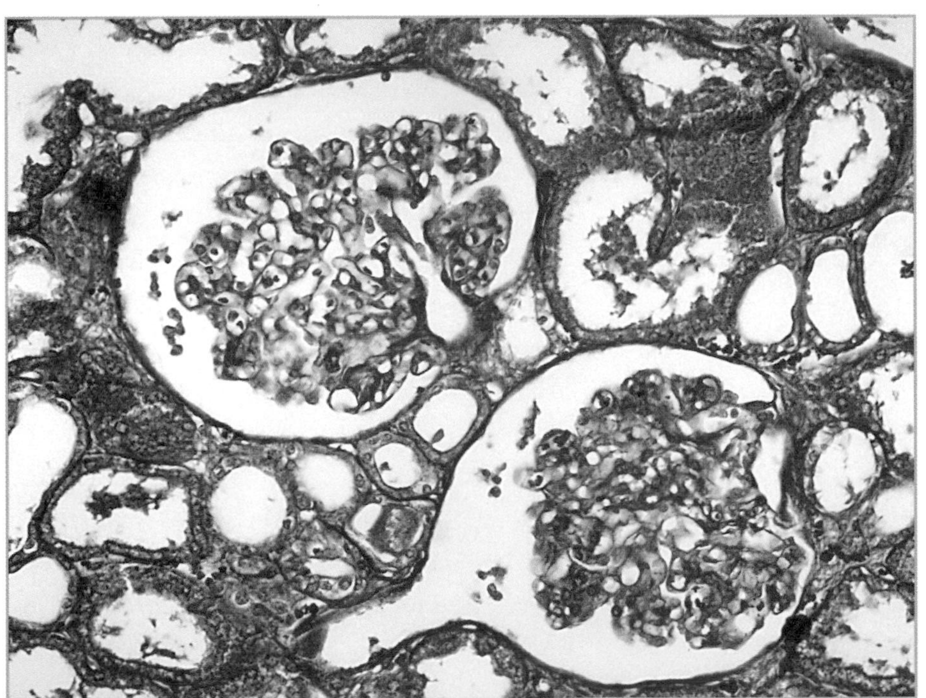

KEEPING HEALTHY

An endoscope holds a kidney stone.

A SEARING PAIN that comes in waves, moving from the back to the abdomen. The worst pain of my life." Descriptions of having a kidney stone relate what it is like to have a sharp-edged pebble travel down the ureter, into the bladder (where the pain often subsides), and then through the urethra. About one million Americans between the ages of 20 and 40 are treated each year for a kidney stone. Some 10 percent of men and 5 percent of women will have a kidney stone by the age of 70. When the urine becomes too concentrated, solutes such as calcium salts and uric acid can crystallize into little mineral deposits or stones. While 90 percent of stones pass without help, some require surgery or the newer technique of lithotripsy, in which high-energy sound waves crush the stone into fragments that wash out with the urine. The best way to avoid getting them is to keep your urine diluted by drinking three to four quarts (3 to 4 L) of water a day. ■

THE FIRST KIDNEY TRANSPLANT

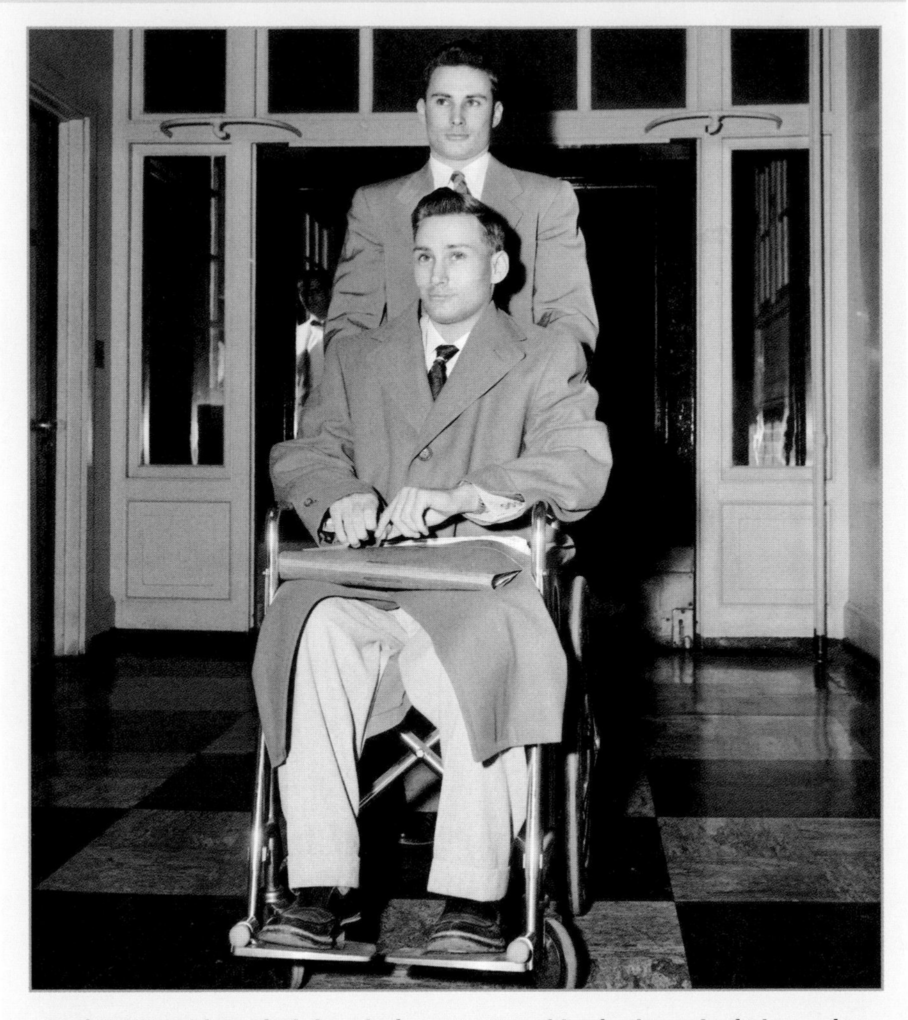

Richard Herrick is wheeled out by his twin, Ronald, who donated a kidney to him.

THE PIONEERING work of Dr. Joseph Murray, born in Massachusetts in 1919, has brought hope and years of extra life to thousands of kidney disease patients. During World War II at a Pennsylvania hospital, Murray treated soldiers who had been so badly burned that they had little unblemished skin available for grafting. He wondered why skin grafts donated by other people were often rejected by the patient's skin. He and a colleague conjectured that the closer the genetic relationship between the donor and the recipient, the longer the graft would last. Murray's research into the field of organ donation made landmark progress on December 23, 1954, when he and a team of doctors in Boston transplanted a kidney from a 23-year-old man into his identical twin brother. The operation became the first successful long-term human organ transplant. The recipient lived another eight years. For his groundbreaking work, Dr. Murray was awarded the Nobel Prize in 1990. ■

a couple of different routes. They can either squeeze between the tightly packed tubule cells, or they can pass right through the tubule cells themselves. Hormones constantly regulate the amounts of water and ions that are reabsorbed in any given time period.

The proximal convoluted tubules absorb about 65 percent of the filtrate's initial water, and the loop of Henle absorbs 15 percent. The distal convoluted tubule then takes out its share—10 to 15 percent—so that when the fluid enters the collecting duct some 90 to 95 percent of its water and solutes has been reabsorbed.

While this reabsorption is occurring, secretion is also happening. Instead of substances moving out of the tubules into the blood, the reverse occurs. The net result of this two-way process, though, is the same—the blood becomes cleaner, the urine is more loaded with waste, and fluid balance and composition is maintained.

Such substances as hydrogen and potassium ions, creatinine, ammonium ions, and organic acids pass from the peritubular capillaries through the tubule cells and into the filtrate. Except for the potassium, most of the substances are secreted in the proximal convoluted tubule, the same place where most of the absorption takes place. Some of the urea and uric acid that have already been passively reabsorbed can, via secretion, make their way through the tubules and

SEE ALSO: Chapter Twelve, "Body Attacks," PAGE 342

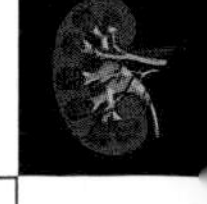

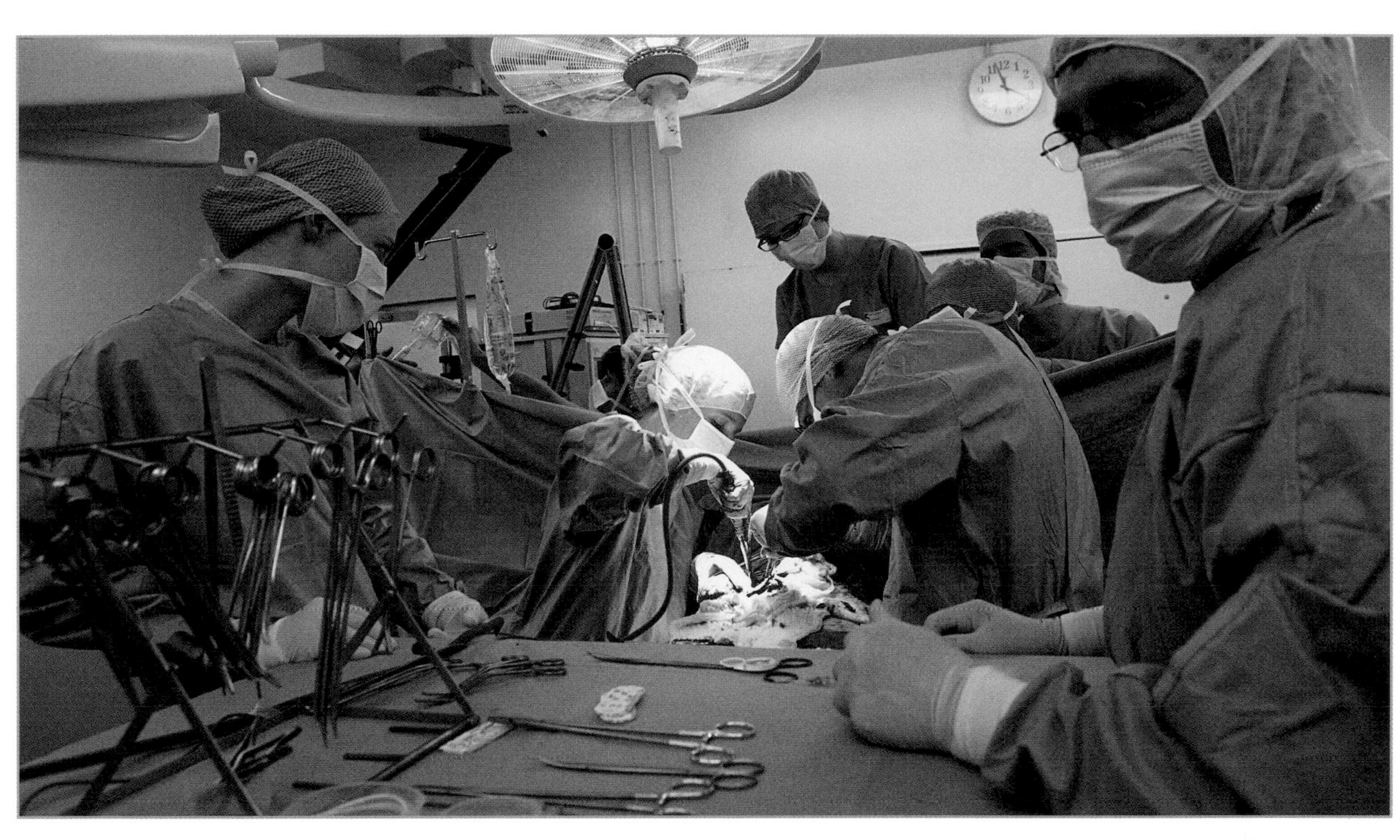

Patients in need of a kidney transplant may have a long wait. In 2007, more than 74,000 were on the waiting list.

then out of the body. And drugs such as penicillin and phenobarbital are secreted into the renal tubules for elimination.

In addition to removing undesirable substances from the blood, secretion also helps balance the blood's pH. When the blood falls into the acidic range, the tubule cells secrete hydrogen ions into the filtrate and reabsorb more bicarbonate ions. Bicarbonate, for example, which is alkaline (or basic), is retained and helps buffer the blood. When the blood's pH climbs, bicarbonate ions are secreted into the filtrate, and chloride ions (Cl-) are reabsorbed; the alkaline substances are thus passed out of the body in urine, and the blood can return to its proper pH.

CHAPTER GLOSSARY

COLLECTING DUCT. Duct carrying urine from the nephron into a duct that leads to the renal pelvis.

DISTAL CONVOLUTED TUBULE. Final section of the renal tubule.

GLOMERULAR (BOWMAN'S) CAPSULE. Double-layered helmet enclosing the glomerulus.

GLOMERULUS. Ball of capillaries in the nephron where blood enters.

NEPHRON. Functional unit of the kidney, which filters blood plasma.

PODOCYTE. Cell with footlike projections through which filtrate passes into the glomerular capsule.

PROXIMAL CONVOLUTED TUBULE. First section of the renal tubule.

RENAL CORPUSCLE. Combination of glomerulus and enclosing glomerular capsule.

RENAL CORTEX. Smooth outer husk of the kidney, where most nephrons are located.

RENAL MEDULLA. Dark red inner portion of the kidney.

RENAL PELVIS. The cavity in the middle of the kidney where urine drains into the ureter.

URETER. Duct that transports urine from the kidney to the bladder.

URETHRA. The tube through which urine moves from the bladder to outside the body.

URINARY BLADDER. A reservoir for the temporary storage of urine.

URINARY TRACT REMOVING WASTE FROM THE BODY

WHILE THE KIDNEYS are the center of action in the urinary system, they need a delivery route for transporting the urine out of the body. That's where the ureters, bladder, and urethra come in.

URETERS

After being formed in the kidney, the urine drains from the renal pelvis into the ureter. Each kidney has a ureter, a narrow tube about 10 to 12 inches (25 to 30 cm) long and about 0.04 to 0.4 inches (1 to 10 mm) in diameter that runs from the kidney to the urinary bladder. When the bladder fills, the near end of the ureter closes and prevents urine from backing up to the kidney. Occasionally the ureters do not properly seal off urine from the bladder; microbes can then enter the ureters and cause kidney infections, called urinary reflux.

The walls of the ureters are composed of three tissue layers. The innermost layer, the mucosa, has connective tissue loaded with collagen

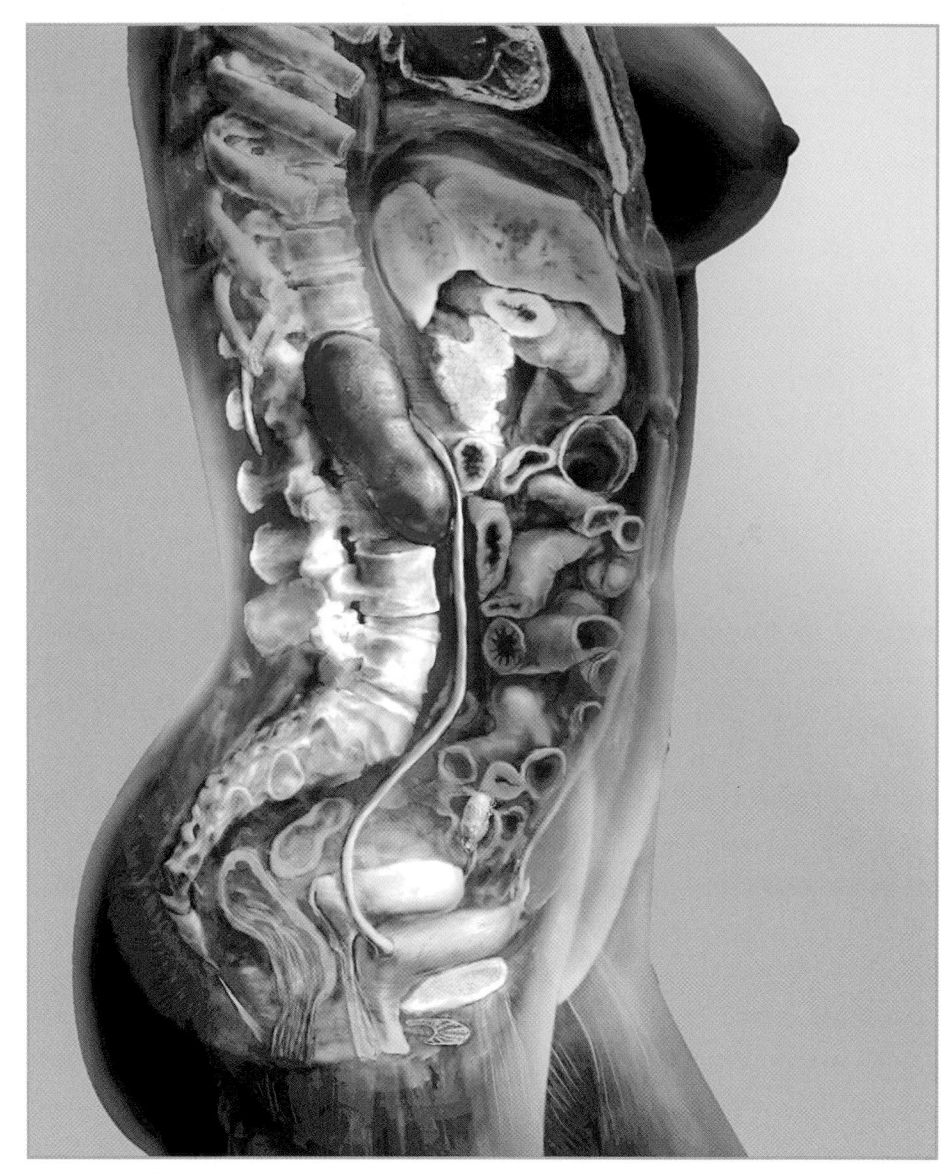

The female urinary system—from the kidneys, to the ureters, and the bladder—can be traced in this 3-D visualization.

SEE ALSO: Chapter Eleven, "Male System," PAGE 294, "Female System," PAGE 300

WHAT CAN GO WRONG

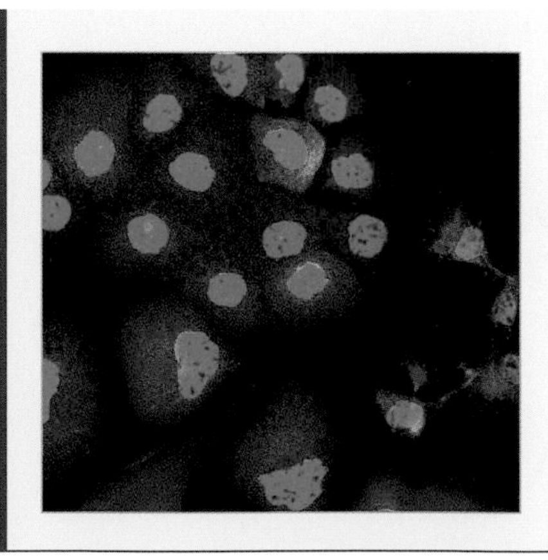

PROSTATE CANCER is the second leading cause of cancer deaths among American men (lung cancer is the first). Every year some 200,000 men are diagnosed with it; 40,000 die annually. Prostate cancer cells, left, typically spread slowly. Symptoms are similar to the common complaint of prostate enlargement and include decreased strength in urine stream (from a compressed urethra), painful urination or ejaculation, hip or back pain, or blood or pus in the urine. The American Cancer Society recommends that most begin receiving annual digital rectal exams after age 50, but those with a family history of the disease could and elastic fibers that enable the ureter to stretch to accommodate varying amounts of fluid. Mucus secreted by the mucosa keeps the urine from entering the ureter cells, preventing any alteration of their pH and solute concentration.

The intermediate layer of the ureter is the muscularis, which consists of both longitudinal and circular smooth muscle fibers. The distal, or final, third of the ureter has circular muscles sandwiched by layers of longitudinal muscles. The muscularis propels the urine by peristalsis. Depending on the rate of urine formation, some one to five waves per minute ripple along the ureters. Gravity and hydrostatic pressure also help push the urine through the ureters. The adventitia is the outer layer of the ureter. Its connective tissue not only secures the ureter into position, but also holds the blood and lymphatic vessels, as well as the nerves that service the muscularis and the mucosa.

BLADDER

Urine moves from the ureters into the urinary bladder for temporary storage. The bladder is an expandable muscular sac that is situated on the pelvic floor just in front of the rectum in men and in front of the vaginal canal and uterus in women. In men, the prostate gland surrounds the neck of the bladder at the point where it joins the urethra. The female bladder is smaller because the uterus lies just above.

+ URETHRA +

+ IN BOTH MEN and women the urethra is the passage through which urine exits the body; in men it also serves as the discharge passageway for semen.
+ The urethra measures 1.5 inches (4 cm) in women and 6 to 8 inches (15 to 20 cm) in men.
+ The flow of urine through the urethra is controlled by both voluntary and involuntary muscles.
+ Newborns void urine 5 to 40 times a day. At two months, a baby voids about 14 fluid ounces (400 mL) of urine a day; adolescents and adults void 1.5 quarts (1,500 mL) a day.
+ Awareness of urination starts at about age 15 months. Control of nighttime urination sometimes takes until age four.

When empty, the bladder assumes a pyramidal shape. When it begins filling up—which it does continuously—it becomes spherical and then takes on a pear shape as part of it distends up into the abdominal cavity. Like the stomach walls, the bladder walls have folds called rugae that allow for quite a bit of expansion.

A medium-full bladder holds about 1 pint (500 mL) of urine and measures 5 inches (12.5 cm) in length. As urine accumulates, the walls of the bladder thin out and internal pressure grows. Fully expanded, the bladder can hold 1 quart (1 L) or more, but it is also capable of bursting. In general, though, the bladder can store urine until it is released through the urethra.

Like the ureters, the bladder walls have three layers—a mucosa, a muscularis, and an adventitia. The mucosa is made up of transitional epithelium and lamina propria (or connective tissue). Under this coat is the muscularis, which in the bladder is called the detrusor ("to thrust out") muscle. The detrusor has three stacked muscle

WHAT CAN GO WRONG

begin at age 45. This exam may reveal a hard lump that could indicate the need for further tests. An enlarged, infected, or cancerous prostate may also produce higher than normal amounts of prostate-specific antigen (PSA), which a blood test can measure. Once prostate cancer has been diagnosed and its progress assessed, men have many options available for treatment, such as cryotherapy, hormonal therapy, radiation, chemotherapy, and surgery. Removal of the prostate can be successful if the cancer has not yet metastasized. Limiting testosterone's effects by medication may also decrease the growth of a tumor, right.

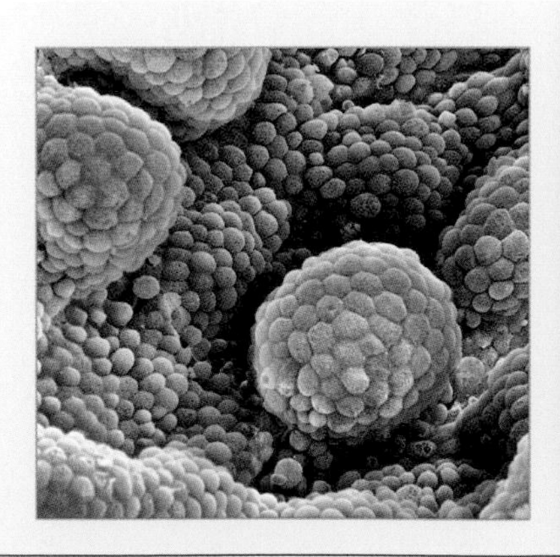

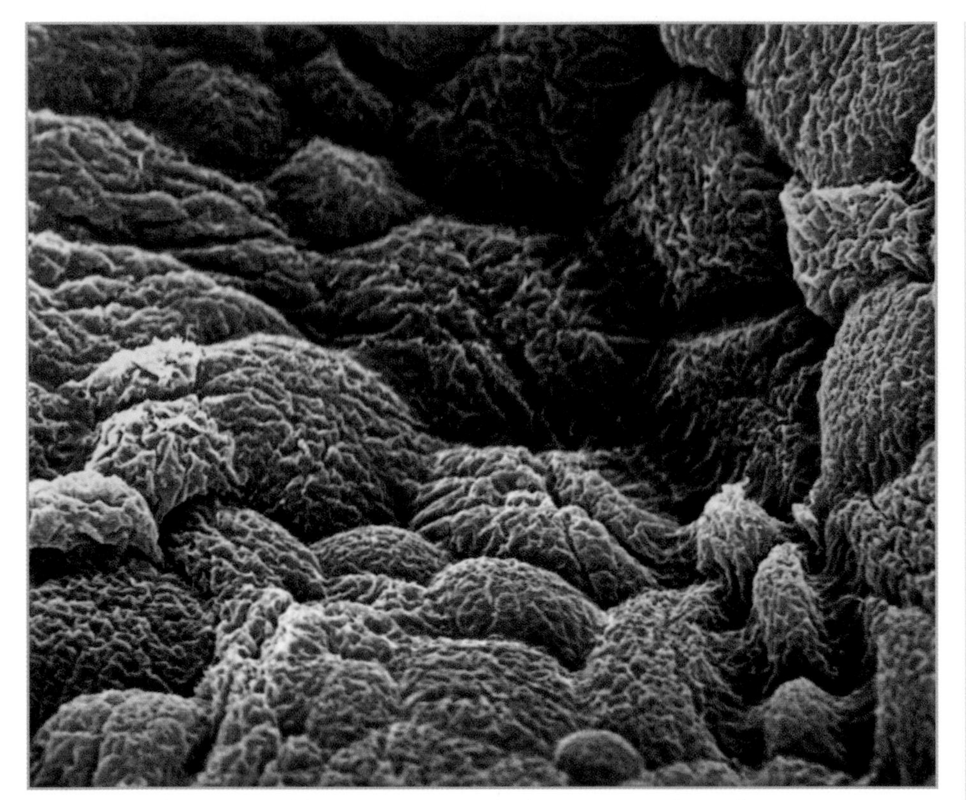

Scanning electron micography brings into sharp focus the epithelial cells that line the interior of the bladder.

layers—inner longitudinal, circular, and outer longitudinal. The connective tissue of the adventitia makes a smooth connection with the adventitia of the ureters.

The base of the urinary bladder is a triangular area called the trigone, defined by three openings—the two ureters and the urethra. Bladder infections usually occur in the trigone. The muscularis around the urethral orifice here forms an internal urethral sphincter, a valve that, since it is composed of smooth muscles, involuntarily controls the release of urine.

Just beneath this opening is the external urethral sphincter, which wraps around the first part of the urethra. This sphincter is composed of skeletal muscles and is, thus, under voluntary control. This control, in other words, allows a person to decide when he or she will urinate.

URETHRA

Urine exits the body through the urethra, a narrow-walled muscular canal that measures about 1.5 inches (4 cm) in females and 6 to 8 inches (15 to 20 cm) in males.

In women, the urethra is attached to the vaginal wall by fibrous connective tissue. Its opening, the external urethral orifice, lies between the clitoris and the vaginal opening. The shorter urethra in females and its location near the vagina and anus make it prone to infection. About 40 percent of women experience urinary tract infections at some point in their lives. Wiping back to front after a bowel movement can cause bacteria to enter the urethra or even colonize the periurethral area. Likewise, sexual intercourse can push vaginal bacteria up into the urinary tract. Infections can be localized in the urethra or move up into the bladder or kidneys. Fever and frequent and painful urination are typical symptoms; antibiotics are usually successful in treating such infections.

In men, the much longer urethra is more complicated. The male urethra discharges both urine and semen, which contains sperm. The first inch is called the prostatic urethra and is surrounded by the prostate gland, which contributes semen from the vas deferens and ejaculary duct. Next, the membranous urethra runs for almost an inch from the prostate gland, through the urogenital diaphragm, and on to the penis. Then for about the next 6 inches (15 cm), the spongy, or penile, urethra extends the length of the penis, opening at the external urethral orifice.

The composition of the urethra is similar in females and males. The urethral wall is composed of a mucosa and a superficial muscularis of circular smooth muscles.

In males, the prostatic urethra has ducts that carry secretions from the prostate gland and seminal vesicles and sperm from the vas deferens into the urethra. Some of these secretions neutralize the acid

SEE ALSO: Chapter Four, "The Muscles," PAGE 106

environment of the female reproductive tract and increase sperm motility. Then, along the spongy urethra, ducts of the bulbourethral (Cowper's) glands secrete a lubricating mucus during sexual arousal, as well as an alkaline material to buffer the urethra's acidity before ejaculation.

URINATION

The final step in the excretory process is the actual elimination of the waste product, urine. A number of specialized muscles and tissues make this action possible. What happens during urination, or micturation? Put most simply, the bladder is emptied.

What seems the simplest of daily activities is more than a mindless duty for the body. In the absence of any disease, the bladder can hold about 17 fluid ounces (500 mL) or more urine for two to five hours.

But after accumulating about 7 to 14 fluid ounces (200 to 400 mL) of urine in the bladder, stretch receptors there begin to send nerve impulses to the sacral (pelvic) region of the spinal cord to stimulate the micturation reflex. Impulses going to the brain also stimulate the urge to urinate. The more frequent the bladder's contractions occur, the more urgent the signal to start the micturation reflex will be.

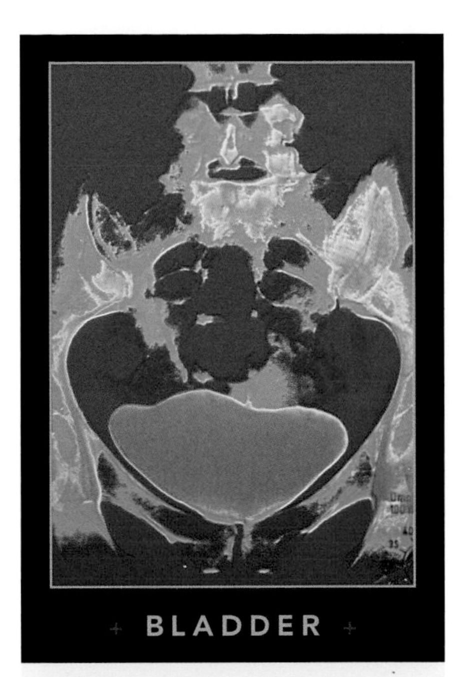

+ BLADDER +

+ A MODERATELY FULL urinary bladder, above, holds about 17 fluid ounces. The bladder is capable of holding 1 quart or more of urine.
+ The base of the bladder, the trigone, is defined by the openings for the two ureters and the urethra.
+ Men with enlarged prostate glands that limit bladder outlet often retain more than 5 fluid ounces of urine in their bladder after trying to empty it.

The impulses cause contractions of the bladder's detrusor muscle, while relaxing the internal urethral sphincter. Remember that the external urethral sphincter is under voluntary control. When a person decides to urinate, neural signals will inhibit the normally contracted external urethral sphincter and thereby allow it to relax. Then the micturation reflex takes over and urination can happen.

If a person consciously decides not to void the bladder, the contractions should subside within approximately one minute. In addition to the external urethral sphincter, the levator ani muscle on the pelvic floor (the same muscle used in defecation) helps control the urination impulse. When another 7 to 10 fluid ounces (200 to 300 mL) of urine have collected in the bladder, the micturation reflex will begin all over again. This process will keep repeating until the bladder is so full that voluntary control is overridden, and urination occurs.

WHAT CAN GO WRONG

URINARY INCONTINENCE affects about 10 million U.S. adults. Coughing, exercise, and pregnancy can cause stress-related incontinence. Smoking, nerve damage, drugs, disease, and depression can also cause loss of control. The bladder becomes less flexible with age and may lose some ability to hold urine. Treatments include exercises to strengthen the pelvic muscles. Increased nighttime urination may be helped by changing sleep or medication routines. Some patients may need prostate gland surgery or a surgically implanted sphincter. Herbal remedies, such as Saint-John's-wort, right, are being studied.

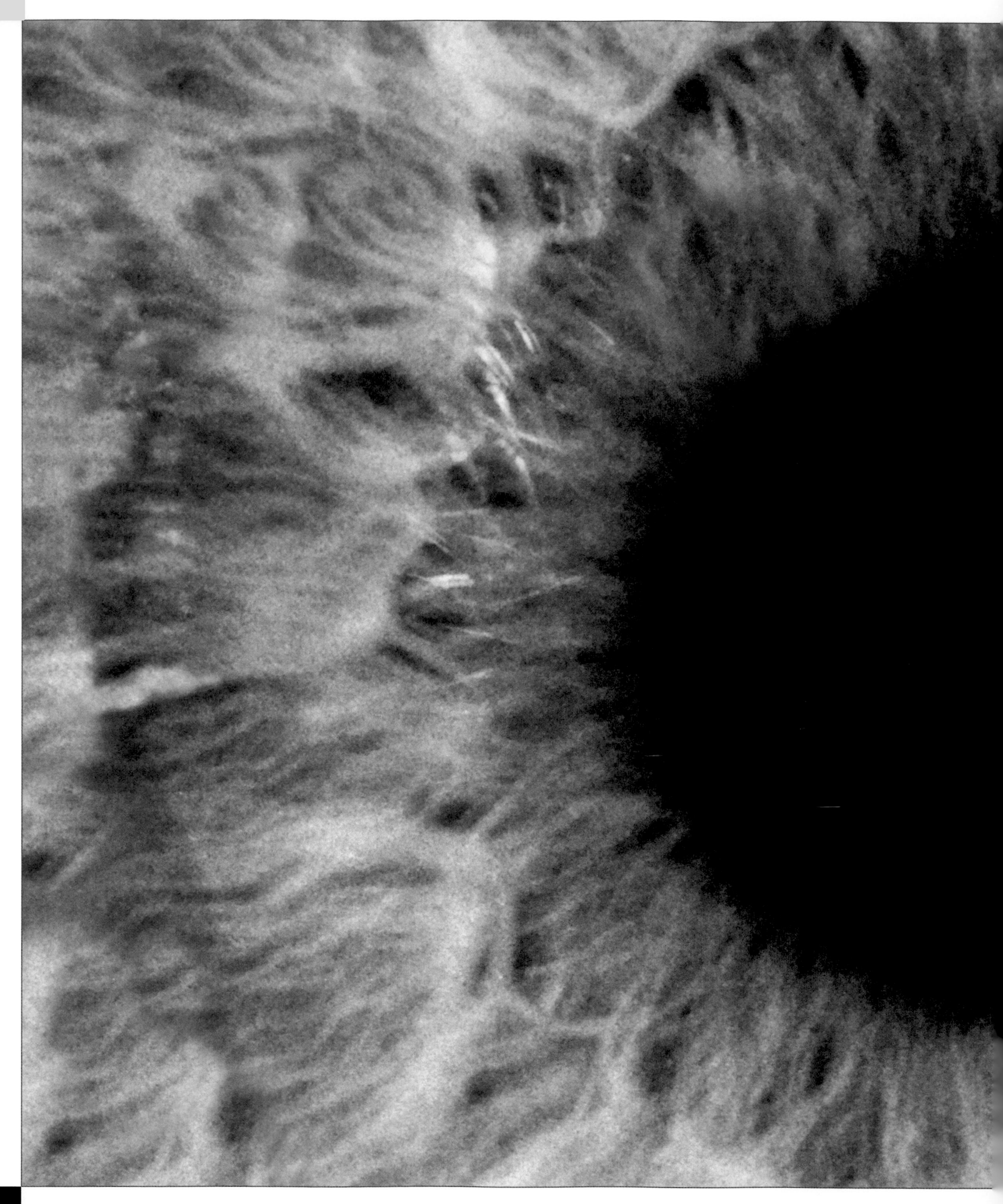

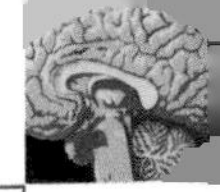

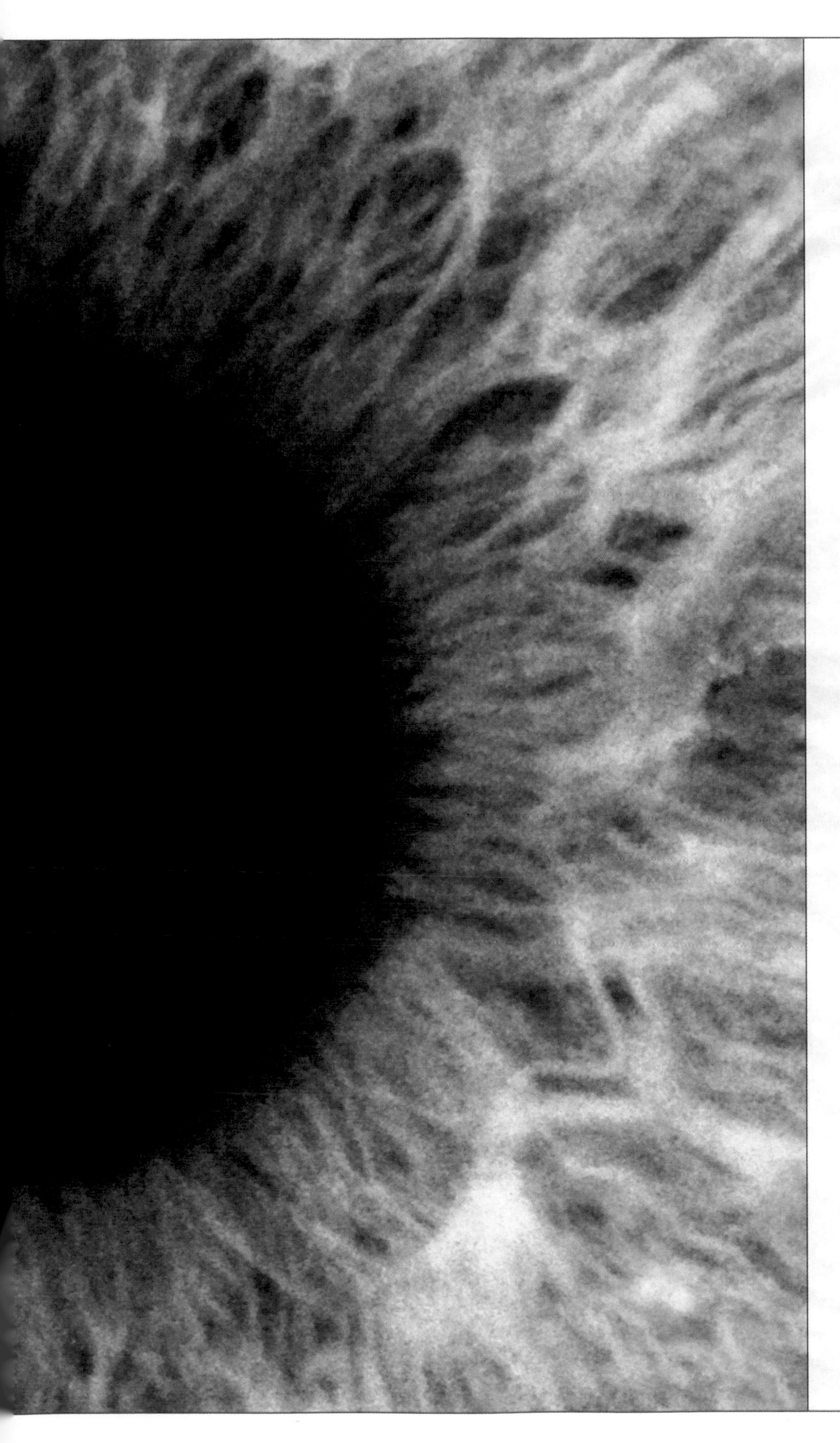

CHAPTER NINE

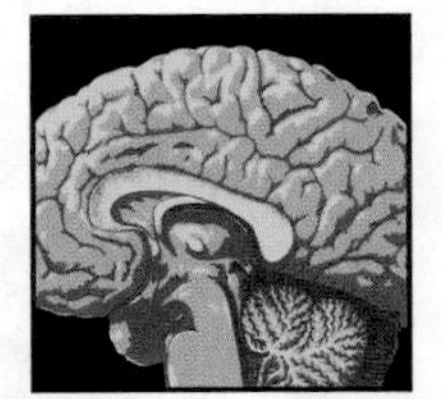

CONTROL CENTER

THE NERVOUS SYSTEM IS both an intricate communications network and a control center for the body. Every moment of every day, billions of nerve cells relay countless electronic messages along their lines to and from the brain, bringing in information, allowing the brain to interpret it, and then telling the body what actions to take. The nerves, the spinal cord, and the brain must all work perfectly together to help the body navigate safely through the world.

A gateway to the brain, the pupil channels light to millions of nerves in the retina.

NERVOUS SYSTEM

COMMAND CENTER

SPREADING THROUGH every tissue in an intricate electrical web, the nervous system is the body's central controller and communicator. Its billions of neurons, or nerve cells, connect with one another in a network of intricate complexity. Along with the endocrine system, the nervous system dictates and coordinates every action the body takes. Its sensory nerves deliver information about the outside world, and its primary organ, the brain, is the seat of the mysterious processes of human consciousness.

What does this all look like, anatomically speaking? It literally starts at the top with the brain—arguably the most important organ in the human body. Enclosed in the skull, the brain is a double handful of pinkish gray, mushy tissue containing billions of neurons and an even greater number of support cells, called glial cells.

Twelve pairs of cranial nerves branch off from the base of the brain to the head and upper body. Many more nerves run from the brain down through the spinal cord in the center of the vertebral column. Thirty-one pairs of spinal nerves extend from the spine to the muscles and glands of the body. Nerves divide as they fan out through the body and, in some places, converge in clusters

The Nervous System

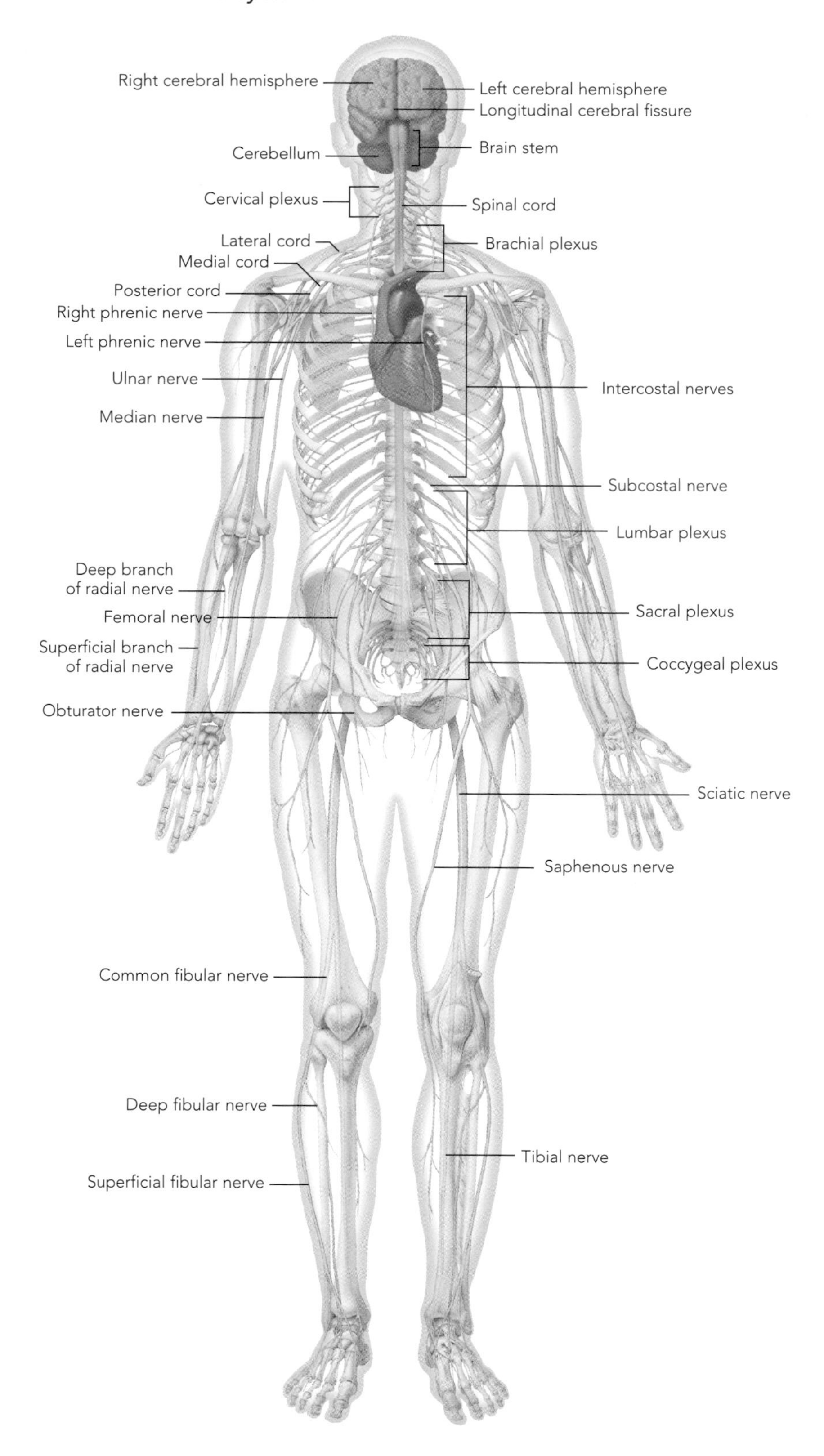

WEBLINK: Learn more about the Nervous System at http://nationalgeographic.com/humanbody

called plexuses, as in the gastrointestinal tract where they serve as mini-controllers of digestion.

A COMPLEX SYSTEM

The nervous system has three connected functions. The first is input: Sensory nerves receive information from outside and inside the body. These nerves gather data about changes in light, sound, touch, and other stimuli in the environment as well as inside information about the status of internal organs, muscles, blood chemistry, and myriad other changeable states of the body. They then convert this information into signals sent to the brain or spinal cord.

The second nervous system job is integration: The central portion of the system receives and interprets sensory input and makes decisions about what should be done.

This leads to the nervous system's third function: response. The brain or spinal cord sends a signal—a motor output—that dictates an action in some part of the body. Millions of signals every second tell the body to blink, or increase its heart rate, or utter a syllable, or release epinephrine, or perform countless other actions, most of them unnoticed by the conscious mind.

Neuroscientists typically divide the nervous system into two main parts and a number of subdivisons. The two primary parts are the central nervous system (CNS) and the peripheral nervous system (PNS). The central nervous system consists of the brain and spinal cord—the command and sensory integration centers of the body and the seat of thought, memory, and emotion. A communication network, the peripheral nervous system consists mainly of nerves that carry signals to and from the CNS.

The peripheral nervous system is made up of three parts: motor, sensory, and autonomic. The motor division transmits orders from the CNS to the muscles and glands, telling them to take action. The motor division includes the somatic nervous system, also called the voluntary nervous system because its actions fall under conscious control.

The sensory division carries impulses from the sensory organs and the visceral (or internal) organs to the central nervous system. The autonomic (or involuntary) nervous system carries signals to areas of the body rarely under conscious control, such as the smooth muscle, cardiac muscle, and glands. It regulates processes that we don't actively think about, such as respiration and digestion.

The autonomic nervous system can itself be divided into two parts: the sympathetic and parasympathetic divisions. These motor pathways generally have opposing effects on the body and its systems. The sympathetic division usually mobilizes body systems. For example, the sympathetic division increases the heart rate while the parasympathetic division inhibits and slows it down (see "The Nerves," page 254).

NERVES

Two kinds of cells make up the body's nervous tissue: neurons and neuroglia (also called glial cells). Five to fifty times more common than neurons, the small neuroglia have the unglamorous but essential role of support and protection in the nervous system. Unable to carry electrical impulses, the six types of neuroglia cells—four in the central nervous system and two in the peripheral nervous system—

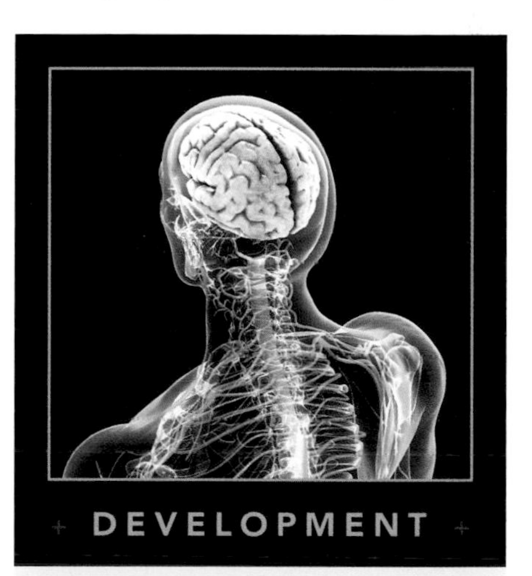

+ DEVELOPMENT +

IN THE WOMB, the brain takes longer than most organs to develop.
+ 4 weeks: Neural tube, basis of the central nervous system, has formed.
+ 5 weeks: Forebrain, midbrain, and hindbrain form at one end of neural tube; spinal cord encompasses rest.
+ 13 weeks: Cerebrum has begun to form; brain stem, cerebellum appear.
+ 26 weeks: Cerebrum folds begin.
+ Birth: All structures are formed, but neural networks continue to develop throughout childhood.

SANTIAGO RAMÓN Y CAJAL

Ramón y Cajal was an artist as well as a neurologist.

SANTIAGO RAMÓN Y CAJAL (1852–1934) was an artistic, energetic Spanish physician who became the first person to accurately describe the anatomy of the neuron and the fine structure of the nervous system.

Born to a country doctor in Petilla, Spain, Ramón y Cajal was apprenticed to a butcher and a shoemaker first before he finally began studying medicine at the University of Zaragoza. By 1875 he had been named an auxiliary professor of anatomy at the University of Zaragoza, where he began a lifelong study of the body's tissues as seen through the microscope.

In 1887, a colleague showed Ramón y Cajal samples of brain tissue that had been stained for better viewing through the microscope by the pioneering Italian neurologist Camillo Golgi. Golgi's studies had led him to believe that the nervous system consisted of a network of continuous fibers. Inspired by Golgi's samples ("sharp as a sketch with Chinese ink," Ramón y Cajal wrote), the Spanish doctor went on to study the nervous tissue of the brain, the spinal cord, and the retina, developing his own techniques for staining cells. Through careful preparation and observation, he saw clearly that what Golgi believed to be uninterrupted strings of nervous tissue were in fact strands made up of separate cells that met, but did not touch, at narrow gaps.

Ramón y Cajal's meticulous and beautiful drawings of the various cells in the tissues he studied were widely published and became the basis for modern neuroanatomy. In 1906, Golgi and Ramón y Cajal jointly received the Nobel Prize for physiology or medicine. Ramón y Cajal continued to contribute to neurology for the rest of his life, in his last years becoming director of the Cajal Institute, founded in his honor by King Alfonso XIII. ■

provide nutrients and a supporting framework of tissue for their neuronal cousins. Some protect brain cells by killing invading microbes and clearing away dead cells. Unlike neurons, mature neuroglia can divide and reproduce. They fill in areas of the brain where neurons have been destroyed.

The neuron—nerve cell—is the basic functioning unit of the nervous system, notable for its ability to conduct electrochemical messages from one part of the body to another. Like cardiac muscle cells, neurons are long-lived, surviving the length of a human lifetime. However, mature nerve cells are unable to divide and reproduce, and they cannot be replaced if they are destroyed, which is why damage to the brain or spinal cord is often irreversible. Neurons are hungry cells, demanding oxygen and glucose for survival. After cardiac arrest, people can suffer brain damage because of the relatively brief loss of oxygen to nerve cells.

The central portion of these highly specialized cells ranges in diameter from 5 to 140 microns, just barely visible to the naked eye. Most neurons consist of three parts: the cell body and two kinds of nerve "processes" or fibers. These fibers are the dendrites—short, branchlike extensions reaching out from the cell body—and an axon—a long, thin, electrically excitable projection.

The cell body contains a nucleus and typical cell organelles. Its

SEE ALSO: Chapter One, "The Cell," PAGE 14

surface, the plasma membrane, sometimes serves as a receptive surface for messages from other neurons. Dendrites extend like leafless trees from the cell body; some neurons have just a few thick dendrites, while others have hundreds. These are the main receivers for signals arriving from other nerve cells.

Axons are the most remarkable parts of these remarkable cells. Arising from the cell body, these thin fibers can range in length from just a fraction of an inch to up to several feet long (as from the spine to the toes). Each axon finishes up in a spreading axon terminal with up to 10,000 or more branches, and each branch ends in a rounded bulb or knob.

Most axons are covered with a myelin sheath, a fatty coating that insulates the fiber. It also increases the traveling speed of nerve impulses along its length. Two kinds of neuroglia—Schwann cells in the PNS and oligodendrocytes in the CNS—wrap around the axons to form this protective sheath; between the glial cell wraps are little gaps called nodes of Ranvier.

Nerve impulses travel from one neuron to the next through one-way junctions called synapses, the points where axons' terminal branches reach the dendrites of other neurons. Most neurons are covered with hundreds—or even thousands—of synapses, so that they are able to receive signals from many different neurons. Neurons can also make synaptic connections with a few other kinds of tissue, such as skeletal muscle or glandular tissue, allowing nerves to trigger action in those tissues directly.

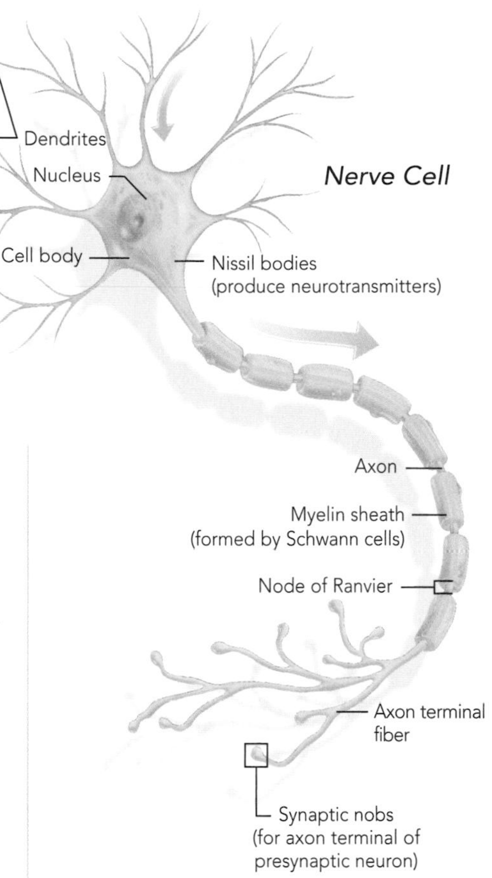

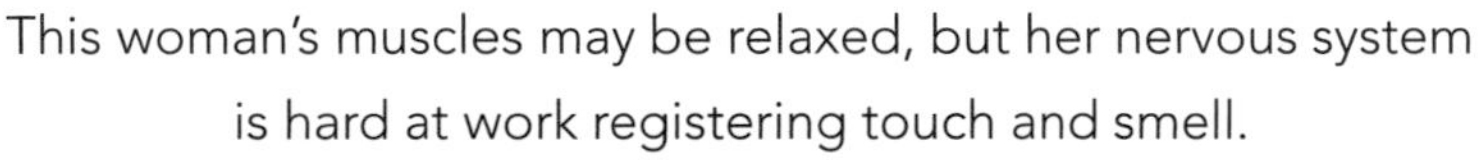
This woman's muscles may be relaxed, but her nervous system is hard at work registering touch and smell.

NEURONS

Neurons come in all sizes and many shapes, but they can be classified into three groups according to the number of dendrites and axons extending from them. Most, including almost all in the brain, are multipolar neurons, with several dendrites and one axon extending from the cell body. A few (such as in the retina) are bipolar, with one main dendrite and one axon. Some are unipolar: A single process extends from the cell body and divides into a T, one end branching out into dendrites and the other into axon terminals. Some unipolar neurons are found in the peripheral nervous system.

SELECTED NEUROTRANSMITTERS

NEUROTRANSMITTER	WHERE IT IS RELEASED	EFFECTS
Acetylcholine, shown right	Junctions in skeletal muscles, autonomic motor endings, motor cortex of brain	Muscle contraction; in the brain, involved in memory, sleep, attention
Norepinephrine	Brain, peripheral nervous system	Regulates arousal, moods, heart rate, blood pressure
Dopamine	Brain, peripheral nervous system	Involved in movement, reward, hormones, psychosis
Serotonin	Brain stem, cerebellum, pineal gland, spinal cord	Involved in sleep, mood, depression, anxiety, appetite
GABA (gamma-aminobutyric acid)	Hypothalamus, cerebellum, spinal cord, retina	Inhibits neurons
Glutamate	Brain, spinal cord	Excites neurons; involved in learning and memory
Endorphins	Brain, pituitary, spinal cord	Inhibit pain; natural opiates
Nitric oxide	Brain, spinal cord, adrenal gland, nerves to genitals	Affects digestion, messaging, male sexual response

Neurons can also be grouped according to their jobs. Sensory neurons carry impulses between the sensory receptors in the skin or internal organs and the central nervous system. Motor neurons send signals from the CNS to the muscles and glands. By far the most common are interneurons, which make up much of the brain and spinal cells; these are primarily responsible for integration of signals and for decision-making.

Regardless of shape or place in the nervous system, every neuron can send an electrical impulse down its length and transmit it to another cell. The neuron's plasma membrane contains tiny tunnels called ion channels that allow electrically charged atoms or small molecules to flow in and out of the cell. When the cell receives an impulse from another neuron, the impulse triggers a tiny voltage change (an action potential) on the surface of the receiving neuron. This voltage change then whizzes down the length of the axon at speeds of up to 65 miles per hour. A neuron can fire repeatedly and rapidly, up to a thousand times per second.

When the signal reaches the terminal bulb at the end of an axon, it usually triggers the release of chemicals called neurotransmitters that are stored there in little bubblelike vesicles. In less than one-thousandth of a second, the chemicals diffuse across the tiny space of the synapse—the gap between the transmitting axon and the receiving target cell—and bind to receptors on the target neuron or effector cell. The receptors then activate their own neuron, prompting it to fire off an impulse in turn or else inhibiting it from firing. After the neurotransmitter has done its job, it is diffused, destroyed by enzymes, or reabsorbed and destroyed by the transmitting terminal (a process called reuptake).

Scientists have discovered dozens of different neurotransmitters at work in the nervous system, with more sure to be found. Many bind to a specific receptor, like a key in

SEE ALSO: Chapter Four, "Skeletal Muscle," PAGE 110

Synaptic Connections

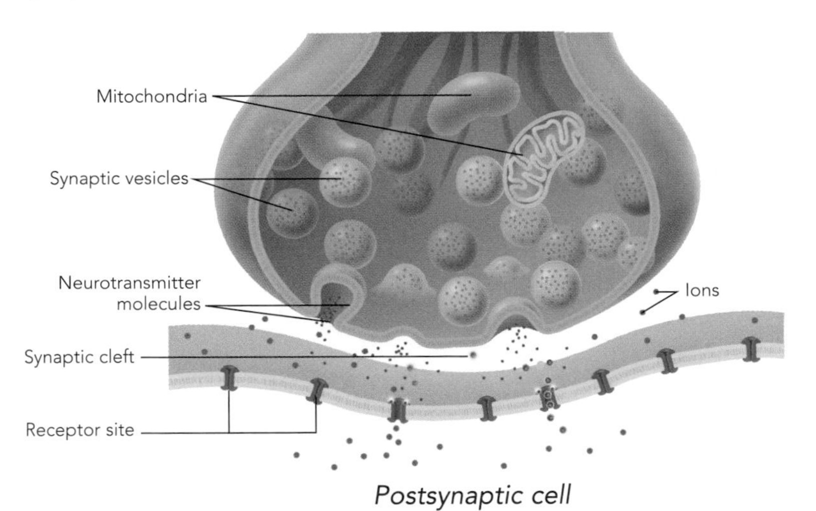

a lock, and each has a different, specialized set of effects. Research into neurotransmitters is one of the busiest areas of neuroscience, promising insights into the causes of Alzheimer's disease, Parkinson's disease, depression, schizophrenia, and many other ailments.

The first neurotransmitter was discovered in 1921 by German scientist Otto Loewi. One night Loewi had a dream that outlined an experiment to prove that nervous impulses were transmitted chemically. "I awoke," he wrote later, "turned on the light, and jotted down a few notes on a tiny slip of thin paper. Then I fell asleep again. It occurred to me at six o'clock in the morning that during the night I had written down something most important, but I was unable to decipher the scrawl. That Sunday was the most desperate day in my whole scientific life. During the next night, however, I awoke again, at three o'clock, and I remembered what it was."

In Loewi's experiment, two frog hearts were placed in connected chambers that were filled with a saline solution. Loewi stimulated the vagus nerve connected to the first frog heart, causing it to slow down; soon afterward, the second frog heart also slowed, showing that a substance was flowing from one heart to the other in the saline. This neurotransmitter, later named acetylcholine, is now known to control muscular contractions. It may also be essential to the brain's role in memory and sleep.

Other notable neurotransmitters include dopamine, which controls some movements, can cause psychosis, and actuates during some hormonal responses. The brains of people with Parkinson's disease have a shortage of dopamine; administration of levodopa, from which dopamine is made, can relieve the symptoms of Parkinson's in the short run. Serotonin, found in the brain and throughout the body, may be involved in the experiences of sleep, anxiety, and mood disorders.

Fluoxetine (Prozac) and other drugs that affect serotonin are used to treat depression and similar illnesses. Endorphins, natural opiates that act like morphine, may be released to reduce pain. And even gases can act as neurotransmitters: Nitric oxide and carbon monoxide travel from neuron to neuron by diffusing through their plasma membranes.

WHAT CAN GO WRONG

MULTIPLE SCLEROSIS (MS) is a neurological disease, often progressive, caused by the destruction of the myelin sheath that covers many nerves in the brain—damaged areas in orange at right—and spinal cord. Myelin insulates the nerves and allows impulses to be conducted normally along their lengths. MS appears to be an autoimmune disorder, in which the body attacks its own myelin. Symptoms include blurry vision, fatigue, numbness or tingling, weakness in limbs, and coordination problems. Medications exist for MS, but there is currently no cure. However, most people with MS will live active lives for decades.

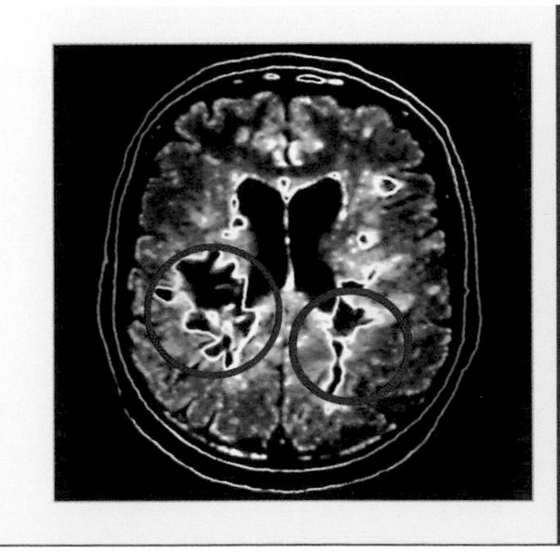

THE BRAIN

COMMUNICATION CENTRAL

"THE BRAIN—IS WIDER than the Sky," wrote Emily Dickinson. And indeed, not only can the brain encompass the idea of "sky" and more, but it is an organ of astronomical complexity. The total number of synapses in the brain—the connections within its web of neurons—is about 10^{15}, or one thousand trillion, greater than the stars in our galaxy.

The brain is the body's central controller and the coordinator of virtually all its functions, from the automatic in-and-out of breathing to the act of solving an equation. It registers sensations and orders actions based on those sensations; it is the seat of memory, intellect, emotions, behavior, sleep, and wakefulness. It directs communications and creates in humans a sense of self-awareness and identity.

All of these amazing abilities come from a wrinkled, jellylike mass weighing about 3.5 pounds. Composed of about 100 billion neurons and up to 50 trillion neuroglia, the brain sits on top of the spinal cord like a mushroom on a thin stem. Its delicate tissues are protected by layers of bone, membrane, and fluid.

The outermost, and toughest, protective covering is the cranium, the rounded bones that form the top and back of the skull. Beneath the skull are three layers of connective tissue membranes known as meninges. The external layer is the dura mater (meaning "tough mother"), a strong leathery covering. Beneath it is the loose arachnoid ("spidery") mater, connected to the next layer below. This open area is flooded with cerebrospinal fluid that serves as a suprisingly effective liquid cushion for the brain. The subarachnoid space also contains blood vessels that feed the organ. The third layer, the pia mater ("gentle mother"), is a more delicate membrane, filled with small blood vessels, that clings directly to the brain's surface.

Parts of the brain are curiously wrinkled and grooved. Its ridges are called gyri ("twisters") and its grooves are known as sulci ("furrows"). Deeper grooves, called fissures, separate major areas of the brain; the most noticeable is the longitudinal fissure that separates the cerebrum, the main part of the brain, into two hemispheres.

Curving like horns deep within the brain are fluid-filled ventricles, hollow chambers that connect to the central canal of the spinal cord. Porous capillaries hanging from the walls of each ventricle release cerebrospinal fluid into the ventricles, where it circulates into the subarachnoid space and also into the spinal cord. As well as serving as a watery buffer for the brain and spinal cord, the liquid nourishes the brain tissue and may carry chemical signals as well. If a tumor or injury blocks the flow of this fluid into the subarachnoid space, it can build up in the ventricles and put pressure on the brain tissue, a condition known as hydrocephalus.

Blood flows into the brain through the internal carotid and vertebral arteries, spreading out

WHAT CAN GO WRONG

CONCUSSION, a brain injury, occurs when the head takes a sudden blow, banging the brain against the inside of the skull. The injured brain tissue may bleed or tear. Concussions can range from mild, with no loss of consciousness, to more severe, involving unconsciousness. Patients typically suffer headaches, nausea, and dizziness in the short run, and some develop enduring headaches, dizziness, or confusion for months. Researchers are now investigating athletes, like football players and boxers, left, who after repeated concussions have developed depression and dementia in their 30s and 40s.

WEBLINK: See the Brain in action at http://nationalgeographic.com/humanbody

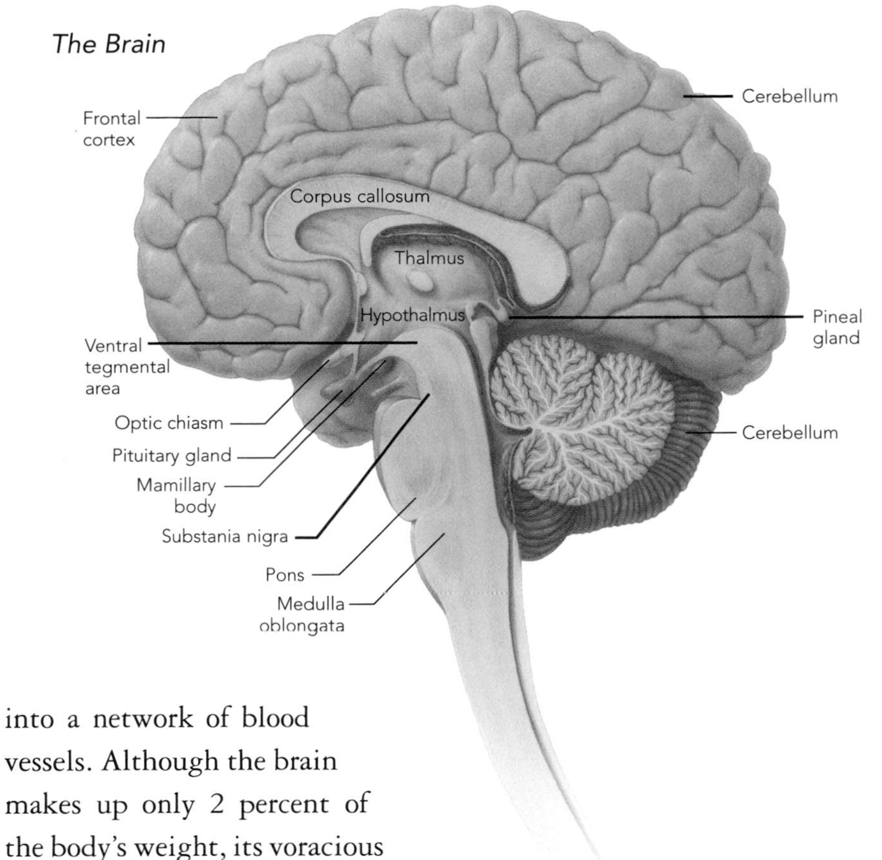

into a network of blood vessels. Although the brain makes up only 2 percent of the body's weight, its voracious neurons use fully 20 percent of the body's oxygen and glucose. When parts of the brain become active, blood flow to the regions instantly increases (and can be seen with MRI machines). If blood flow to the brain slows even momentarily, or if glucose levels drop, a person can lose consciousness.

The sensitive brain is protected from a host of harmful substances in the blood by a unique blood-brain barrier. The capillaries in the brain, unlike those in the rest of the body, are tightly sealed, their endothelial cells forming particularly tight junctions. A thick basement membrane also surrounds the little blood vessels. As a result, many of the chemicals and hormones that routinely pass through the capillaries in the rest of the body are barred from the brain. Some substances can pass through, including glucose, oxygen, and carbon dioxide. Alcohol, nicotine, and anesthetics can cross into the brain. As useful as it is, the blood-brain barrier frustrates efforts to deliver helpful drugs, such as those that attack brain tumors.

BRAIN STRUCTURE

The brain's four main areas are the brain stem, cerebellum, diencephalon, and cerebrum. The brain stem begins with the medulla oblongata, a 1.2-inch continuation of the spinal cord. The medulla contains all the sensory and motor nerves that run up the spinal cord into the brain. In the medulla, the nerves from the body's left side cross to the brain's right side, and vice versa. The medulla regulates some of the body's most basic functions, including reflex actions such as coughing or swallowing. It contains the cardiovascular center as well as the rhythmicity area of the respiratory center, controlling the pace of heartbeats and breathing.

Directly above the medulla oblongata in the brain stem is the inch-long pons ("bridge"), a relay station that connects the medulla to other regions of the brain, such as the cerebellum.

The third part of the brain stem, the midbrain, also about an inch long, connects the pons to the diencephalon. It holds visual and auditory reflex centers—areas that control pupil size, for instance, and the startle reflex.

Tucked behind the pons, under the cerebrum, is the cerebellum, two rounded hemispheres that

ON THE MOVE

AS THE BRAIN DEVELOPS in the growing embryo, different types of neurons actually scoot like tiny inchworms along temporary scaffolds of glial cells to their final destination in the brain. For instance, neurons originating near the ventricular structures pull themselves along glial fibers to the cerebral cortex. This journey of several millimeters would compare to a human walking from the East to the West Coast of the United States.

are deeply fisssured and leaflike in cross-section. The cerebellum's primary job is to coordinate movement and balance. It mediates between the muscles and the areas of the cerebrum that dictate movement. Skilled rote activities, such as hitting a baseball or playing the piano, are processed in the cerebellum.

Located in the middle of the brain and cradled by the two hemispheres of the cerebrum, the diencephalon is made of four structures: thalamus, hypothalamus, subthalamus, and epithalamus. Largest of the three is the double-lobed thalamus, which relays sensory input—perception of pain, temperature, and pressure—to the cerebral cortex and plays a role in emotions, memory, and alertness.

Below it is the hypothalamus, a small but vital part of the brain's structure. Among orchestrating other tasks, it controls the autonomic nervous system, which regulates smooth and cardiac muscles, blood pressure, and other basic body functions; it produces hormones and stimulates the pituitary gland; it regulates body temperature and feelings of hunger; and it is involved in the perception of emotions such as fear and pleasure. Behind the thalamus is the epithalamus, containing the pineal gland, which helps to regulate the sleep-wake cycle.

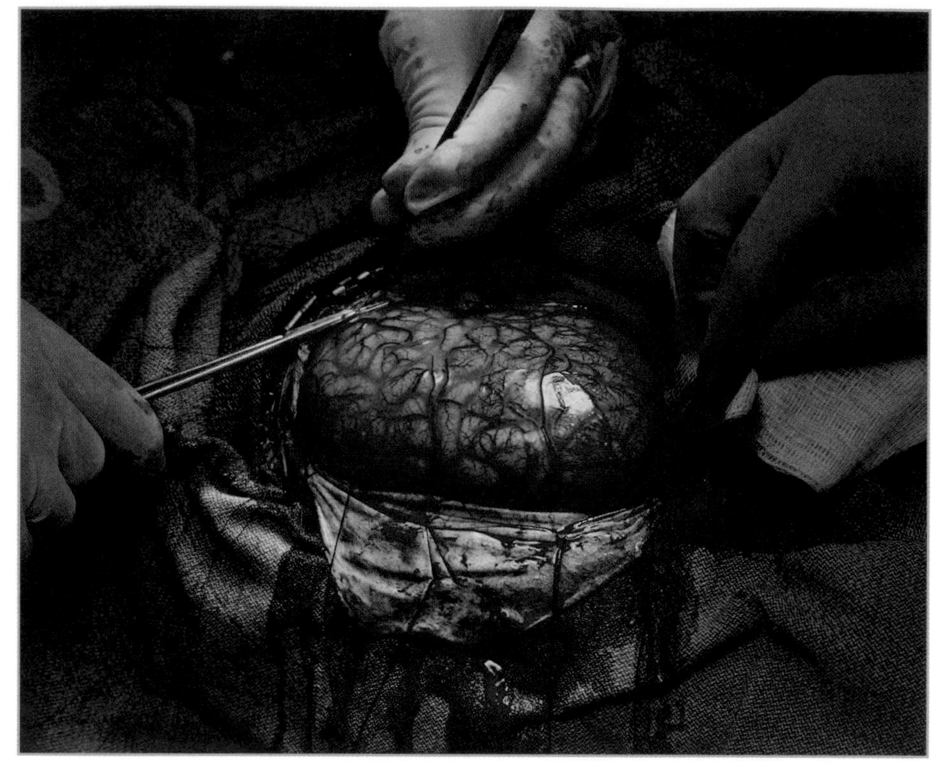

Shorn of its protective membranes, the brain is a delicate organ enriched by a network of blood vessels.

Curling above is the cerebrum, the structure most people picture as the brain. Divided into two hemipheres, the cerebrum makes up more than 80 percent of the brain's mass. It is the center of intelligence, reason, communication, memory, and imagination.

The cerebrum has three kinds of areas: sensory areas that interpret sensory impulses, motor areas that control movement, and association areas that communicate with the other regions to process information. Its thin, grayish outer layer,

SEE ALSO: Chapter Ten, "Producers," PAGE 274

WHAT CAN GO WRONG

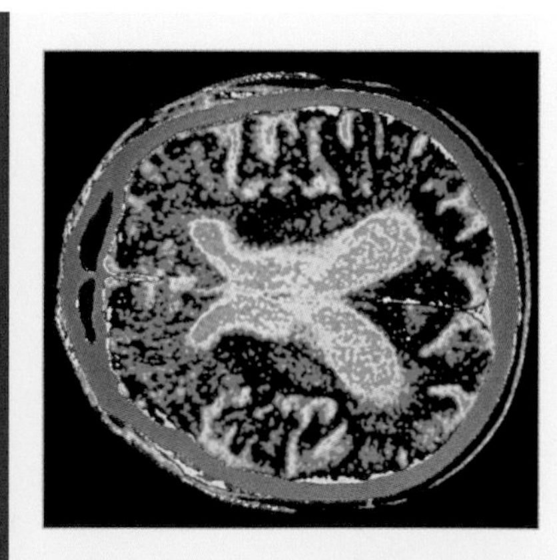

PARKINSON'S DISEASE shows the role of neurotransmitters in the brain. This disease is characterized by muscle tremors, slow movement or rigidity, and loss of balance. It begins with death or damage to cells that secrete dopamine, a neurotransmitter that mediates the signals that control voluntary movement. As the disease progresses, the ventricles enlarge as brain tissue decreases, left. Understanding dopamine's role led to the development of the medication levodopa, a dopamine precursor that helps to restore levels of the neurotransmitter in the brain—although its effects diminish over time.

the cerebral cortex, is made primarily of billions of neurons; the white matter that makes up much of its interior consists mainly of myelinated axons. Most of the brain's information processing takes place in this region. Specific functions can be mapped to specific parts of the cortex, although many regions connect with one another as well as with other parts of the brain.

Though similar in structure, the right and left hemispheres perform unique functions. Each is divided by fissures into five lobes, named for the cranial bones that overlie them. The frontal lobe is responsible for thought, speech, and skilled movements; the parietal lobe registers sensations; the occipital lobe processes visual input; and the temporal lobe, sound and music.

The two hemispheres communicate through bridges of nerve fibers. The largest of these is the corpus callosum, consisting of some 200 million nerve fibers in all.

THE MIND

All of our understanding of the brain's anatomy does not take us very far toward answering the questions: What is the human mind, and what is the nature of human consciousness?

Clinically speaking, consciousness is awareness and responsiveness to surroundings. But in a larger sense, it is synonymous with mind, more difficult to define. The mind seems to arise from an

BRAIN BELIEFS

Three giants of medicine: Galen, Persian physician Avicenna, and Hippocrates

UNPREPOSSESSING and inert, the brain held little interest for ancient doctors who were more interested in the dynamic heart. Early Egyptians, while carefully preserving other organs for eternity, simply pulled the brain out through the nose and discarded it. Ancient Greeks assigned it a larger, though not always central, role. Alcmaeon of Croton, writing around 500 B.C., correctly understood the brain to be the seat of sensation. He traced the optic nerve from the eye to the brain. Aristotle believed the heart was the home of the soul, while the brain cooled the body's circulating heat. The influential Greek physician Galen, born around A.D. 129, believed the heart converted pneuma, the life spirit, into sensation and movement and distributed it around the body through the nerves.

By the time of the Renaissance, some understood the brain to be the center of cognition. For Leonardo da Vinci, as well as for many others, the key structures were the three ventricles inside the brain, which he believed specialized in intellect, sensation, and memory. In 1664, English scientist Thomas Willis published a study on the anatomy of the brain that detailed many structures, including the network of arteries now called the circle of Willis. Willis was the first to attempt a brain map, assigning different functions to different parts of the brain—though for him, the ultimate goal was to understand how the immaterial soul acted on the material body. Not until the late 19th century did researchers begin to understand the role of neurons and the brain's true complexity. ■

A BRAIN ATTACK

STROKES ARE THE THIRD leading cause of death in the United States and the top cause of adult disabilities. They occur because of the intimate interactions between the brain and the cardiovascular system. Also known as a cerebrovascular accident or, more colloquially, as a brain attack, most strokes occur when the blood supply to part of the brain is cut off and, as a result, that brain tissue dies.

TYPES

About 80 percent of all strokes are ischemic strokes: They begin when a blood clot or some other material becomes lodged in a cerebral artery and cuts off the flow of blood to the brain. In thrombotic strokes, another kind of ischemic stroke, the blood clot forms directly in the artery of the brain itself, usually because the blood vessel has been damaged by atherosclerosis. An artery in the brain may also shut down because it has become clogged by fatty deposits.

In embolic strokes, likewise a type of ischemic stroke, the clot travels from elsewhere in the body and lodges in the brain's arteries. Most typically, the clot has formed in a heart that suffers from atrial fibrillation, where the irregular beat and blood flow is not strong enough to prevent clotting.

The remaining 20 percent of strokes are hemorrhagic, occurring when a blood vessel in the brain leaks or bursts outright. Intracerebral hemorrhages happen within the brain tissue, while subarachnoid hemorrhages start in an artery and leak into the fluid-filled space between the arachnoid mater membrane and the brain.

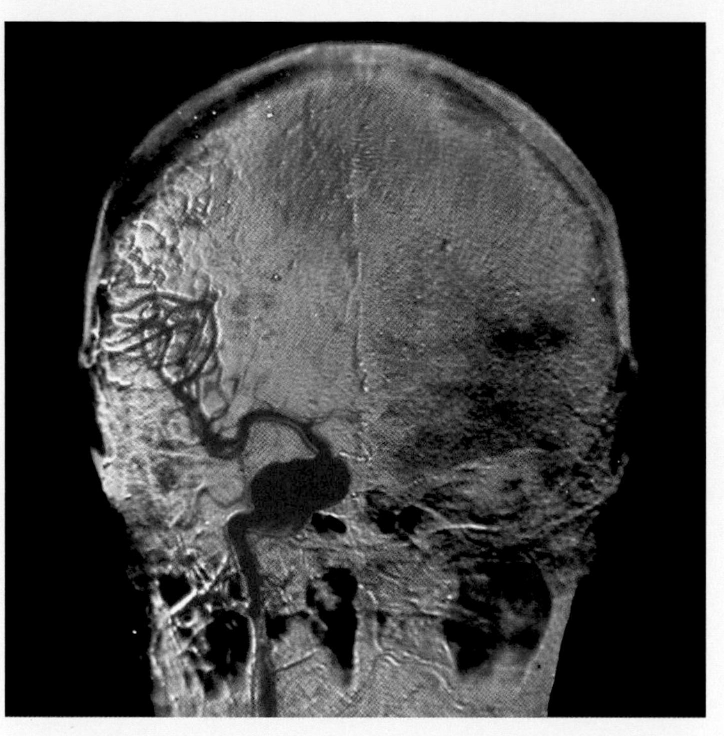

An aneurysm in the carotid artery presages a hemorrhagic stroke.

When this happens, brain cells may die not because they are starved of their blood supply but because the pooling blood puts pressure on the delicate brain tissue. Most hemorrhagic strokes result from blood vessels weakened by high blood pressure or containing thin, ballooning areas called aneurysms. A few also begin with an arteriovenous malformation, tangled webs of thin-walled blood vessels in the brain that are present from birth.

SYMPTOMS

Strokes come on very quickly. Their symptoms typically include one or more of the following signs: sudden weakness or numbness in the face, arm, or leg, usually on one side of the body; sudden difficulty speaking or understanding speech (called aphasia); sudden blurred, double, or lost vision; sudden dizziness, trouble walking, loss of balance or coordination; sudden severe headache with no known cause; and confusion or sudden problems with memory, perception, or spatial orientation.

Sometimes people experience a few of these symptoms fleetingly, feeling them fade in minutes or a few hours. These may be the signs of a transient ischemic attack (TIA), a ministroke, which by definition lasts less than 24 hours. TIAs are probably caused by clots or plaques that temporarily block a cerebral artery but are then cleared or washed away before serious damage can occur. Harmful in themselves, TIAs are also red flags: About a third of people who experience a TIA will later have a major stroke.

TREATMENT

Whether the symptoms are fading or overwhelming, stroke victims need immediate emergency care. Some stroke experts recommend that a companion perform the following quick test, memorable with the mnemonic FAST:

Face: Ask the person to smile. Does one side of the face droop?

Arms: Ask the person to raise both arms. Does one drift downward?

Speech: Ask the person to repeat a simple sentence. Are the words slurred or repeated incorrectly?

Time: If the person shows any of these symptoms, waste no time. Call 911 immediately.

Cardiologists say, "Time is muscle" when dealing with heart attacks. Neurologists say, "Time is brain" in the case of a stroke: The longer it takes to get treatment, the more brain cells will die.

At the hospital, doctors will scan the patient's brain to locate the region of the stroke and find out what kind it is. If a clot is involved and the patient has come to the hospital promptly, physicians can administer a so-called clot-buster drug, properly called a tissue plasminogen activator (t-PA), which may be able to dissolve the clot to open blood vessels. Used within a three hours after the stroke, t-PA can reduce disability by 50 percent. Other surgical treatments can clean out clogged arteries or put a clip at the base of an aneurysm, stopping a hemorrhagic leak.

Exercise and a healthy diet can help prevent many of the risk factors for strokes.

The severity of a stroke depends on the area of the brain damaged and the extent of that damage. The middle cerebral artery is the most common blood vessel involved; it feeds the language and speech areas of the brain, so difficulty in understanding and forming speech is a frequent result of a stroke. Other areas typically affected by strokes include the facial, arm, and leg areas of the left motor cortex, resulting in paralysis of the right side; Wernicke's or Broca's areas, leading to difficulty in forming and understanding speech, as well as reading, writing, or naming objects; facial and arm areas of the sensory cortex, leading to loss of sensation in the right side of the face and right arm; and the optic tract, causing a loss of vision in one-half of the visual field of both eyes. Physical therapy can often retrain a stroke victim in lost skills: Neurons can grow new connections in the brain, and many stroke victims recover substantially.

About 700,000 Americans suffer new or repeat strokes each year, and of that number, over 150,000 die annually. The death rate from stroke is declining, probably due to better control of some risk factors. Some risk factors—including age (over 55), family history, gender (women more than men), and race (African Americans more than Caucasians)—are uncontrollable. But others are treatable; they include hypertension (high blood pressure), high cholesterol, atrial fibrillation (irregular heartbeats), diabetes, smoking, alcoholism, and obesity. With such circumstances, fully 80 percent of strokes experienced today could have been prevented.

interplay of processes in the brain: thoughts, perceptions, emotions, memory, imagination, and will. The human mind is aware of itself and, despite the complexity of its sensations, memories, and emotions, feels itself to be a single, perceptive unit.

Scientists cannot yet map out the origins of consciousness in the human brain. Some researchers study how perceptions, which arrive in tiny fragments, are integrated into single images. Others look into how perceptions move into memory and how memories are recalled. A third avenue of investigation is the study of emotions and how they affect both memory and decision-making.

Scientists are learning, for instance, that emotions are mediated through structures and pathways known as the limbic system, found in the cerebral hemispheres. This system, early to evolve, still controls the fight-or-flight reflex. In humans, it is heavily involved in integrating emotions, physical sensations, and memory.

The little almond-shaped amygdala is a key portion of the limbic system. When stimulated, it produces the sensations of fear; it is also activated when a person recognizes angry or fearful expressions on another person's face. In a frightening situation, the amygdala will trigger the hypothalamus to begin a wave of physical responses, including releasing stress hormones and elevating blood pressure. The amygdala and the nearby hippocampus also play a role in integrating emotion with memories. The limbic system connects to the olfactory bulb, responsible for smell, which may account for strong links between odor, emotion, and memory.

MENTAL FUNCTIONS

Among the most mysterious functions of the brain are its most important tasks: regulating sleep and wakefulness, producing and understanding language, storing memories, and learning. With new brain-imaging techniques, researchers have begun to trace the pathways involved in these functions (see "Breakthrough," below).

Information in the brain flows through networks of neurons that fire together in exact timing when a job must be carried out. The brain is quite plastic—able to reorganize based on new experiences—so strong connections within these networks are reinforced with new synapses and more powerful signals. Weak ones are pruned away.

Learning is a multistage process of acquiring new knowledge through sensory experience and storing that knowledge in memory. Incoming information is held at first in working memory, a kind

+ BRAIN SURGERY +

BRAIN SURGERY has a long history.

+ In Europe and South America as early as 5000 B.C., skulls were drilled open and brain parts were cut away.
+ In 1879 British surgeon William Macewen sucessfully removed a tumor from a young woman's dura mater.
+ In the late 19th century, British neurosurgeon Sir Victor Horsley developed specialized techniques for operating on the brain.
+ In the early 20th century Harvey Cushing, a pioneering American doctor, removed more than 2,000 brain tumors.

BREAKTHROUGH

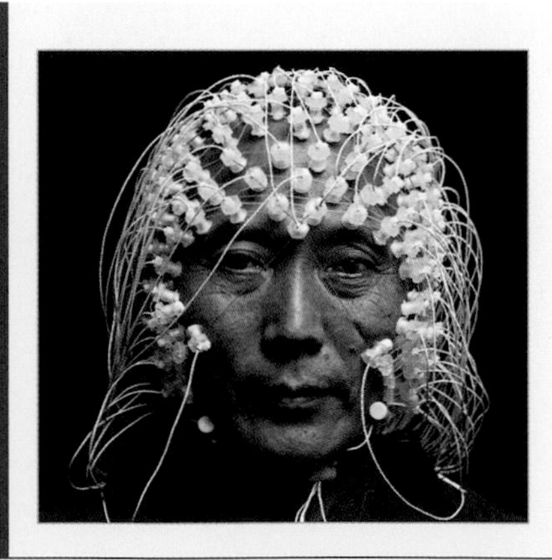

BRAIN MAPPING: The first inside look at brain function was the electroencephalograph (EEG), invented in 1929. The EEG records electrical activity in the brain by means of electrodes on the scalp (left); abnormal brain waves can indicate disorders. Techniques from recent decades include magnetic resonance imaging (MRI), which records how atoms in the brain resonate in a magnetic field. The patient moves through a cylindrical magnet, and the machine produces a series of cross-section images. And magnetoencephalography (MEG), another magnetic imaging technique, can quantify the strength of brain activity in different regions.

PHINEAS GAGE

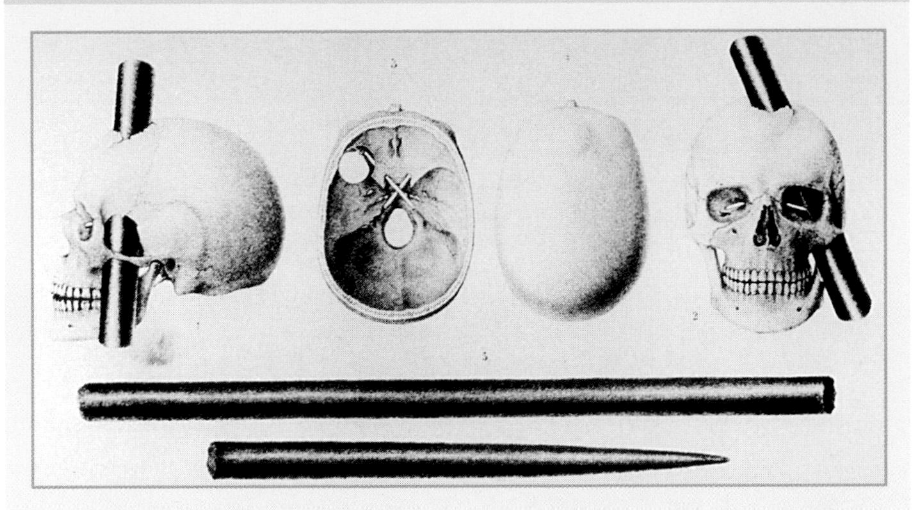

A contemporary lithograph shows the path of the rod through Gage's skull.

IN SEPTEMBER, 1848, while working on a railroad line near Rutland, Vermont, 25-year-old Phineas Gage accidentally set off an explosion that blasted a three-foot-long steel rod competely through his skull.

The rod passed through Gage's cheek and left eye before exiting through the top of his head. Gage fell to the ground—and then, to the astonishment of his coworkers, he stood up again. Driven by oxcart to a nearby hotel, Gage was able to chat while the doctor, John Harlow, inserted his fingers into the holes that the rod has drilled in his skull.

Gage returned to work some months later. Physically fit (although missing his left eye), he was nonetheless a changed man. Previously a well-liked, hardworking foreman, Phineas Gage had turned into a profane, ill-tempered, and indecisive fellow. He was never again able to hold a railroad job, and he died 12 years later after experiencing a series of epileptic seizures.

Dr. Harlow, upon studying Gage's skull, announced his belief that the blasting rod damaged areas of the frontal lobes in the young man's brain. Those lobes, Harlow hypothesized, were responsible for personality—hence the severe personality change experienced by Gage as a result of the accident.

Although Harlow's conclusion was met with some skepticism at the time, the case of Phineas Gage became a famous one in medical circles. It was discussed as part of the larger consideration of whether behavior could be mapped to particular areas of the brain. In 1994, researchers at the University of Iowa used computer imaging to map the trajectory of the blasting rod that penetrated Gage's skull. Their findings agreed with Harlow's: The rod had passed through the ventromedial region on the underside of the frontal lobes—an area we now know to be connected to an individual's personality. ■

of short-term storage in the cells of the prefrontal lobe. Some working memories, perhaps because they are reinforced, then move into long-term memory, a process that seems to be governed by the hippocampus in the limbic system. Patients with damage to the hippocampus cannot form long-term memories.

Scientists are finding that information is broken down into multiple strands that are organized into categories and stored in scattered locations across the cerebral cortex. The memory of your pet, for instance, is not stored in its own bin in the brain, but more likely in a web of associations such as *animal, four legs, small, fur, white, purr, snow, pleasure,* and so on. When recalling the cat, all these locations are activated and send their messages on converging pathways to somehow form the idea, *my cat Snowball.*

Indeed, research now indicates that the brain has at least 20 categories for organizing information, among them plants, animals, body parts, nouns, verbs, proper names, faces, facial expressions, colors, numbers, letters, kinds of sounds, and fruits/vegetables. Studies of people with very localized brain damage show that these categories can have some surprising distinctions: One patient could recognize and name small tools but not large ones; some patients can write words but not numbers; one had difficulty recognizing musical instruments by sight but could name them easily by

PRION THEORY

The carcass of a cow with bovine spongiform encephalopathy burns in an English field.

IN 1982, AMERICAN neurologist Stanley Prusiner (1942–) published an article that set off, as he put it, "a firestorm." Prusiner and colleagues had been studying scrapie, a fatal degenerative disease of the central nervous system of sheep. The disease was similar to a group of rare spongiform encephalopathy diseases in humans, including Creutzfeldt-Jakob disease and kuru, found in the Fore people of New Guinea. In these conditions, brain tissue is gradually destroyed by an unknown process. Prusiner became convinced that the infectious agent was a malformed protein. He named it a proteinaceous infectious particle, or prion.

Prusiner's article caused an uproar. He lost government funding and was denied tenure. Further research would support his theory, however: Prions appear to be abnormal forms of harmless proteins normally found in the brain; they can arise via mutations in the genes encoding the protein or through infection, taken in, for instance, by ingesting infected meat. They can even be inherited. The misshapen proteins apparently induce other benign proteins to fold into the abnormal shape. As they accumulate within neurons, they destroy them, eating holes in the brain tissue.

Prions now appear to be the cause of not only scrapie but also Creutzfeldt-Jakob disease, Gerstmann-Sträussler-Scheinker disease, kuru, and fatal familial insomnia, as well as bovine spongiform encephalopathy—better known as mad cow disease. In 1997, Stanley Prusiner won the Nobel Prize for physiology or medicine for his discovery. ■

sound. Memories of learned physical skills are stored elsewhere from memories of concepts or episodes, which is why patients who can no longer recognize a spouse may still be able to play the piano.

Language, a defining characteristic of the human animal, is dependent on learning and memory. Unlike most functions of the brain, language skills are localized in one hemisphere of the cerebrum, usually the left. Language appears to be broken down into two main components. Speech comprehension involves the prefrontal cortex and Wernicke's area, a region of the temporal and parietal lobes. Speech planning and production occurs in another region, Broca's area, in the frontal lobe. Messages from this region are then passed to a nearby part of the premotor cortex, which controls the larynx and organs of speech.

Far more mysterious even than language or memory is sleep. In fact, even the purpose of sleep is one of the great unknowns of science. We do know that sleep is far more than the absence of wakefulness. It is an active state, and while sleeping an individual passes through several stages, each mediated by various areas of the brain stem and the hypothalamus.

The brain is apparently kept in a state of wakefulness or alertness by nerve impulses sent from the pons and midbrain to the thalamus, which in turn activates the cerebral cortex. It is possible that

SEE ALSO: Chapter Ten, "Producers," PAGE 274

sleep is induced when neurons in the hypothalamus release inhibitory neurotramsitters that turn off arousal systems.

Typically, a person will pass through four stages of increasingly deep sleep, interspersed with a completely different kind of sleep, called REM—rapid eye movement—for a characteristic bodily experience that occurs during this stage. Brain waves during REM sleep indicate increased brain activity, but all large muscle groups are effectively paralyzed during this stage, possibly to prevent the sleeper from getting up and moving around.

Why do we need to sleep at all? Some evidence shows that long-term memories are consolidated during sleep; sleep may also have ties to proper function of the immune system (see "Keeping Healthy," right). But there are still more questions than answers about this fundamental feature of human life.

MENTAL ILLNESS

Mental illnesses such as schizophrenia and depression used to be considered diseases of the mind rather than of the brain. Today the distinction between mind and brain is blurred, as is the line between psychiatry and neurology. Most mental illnesses are now viewed as disorders of the communication systems within the brain, the neurotransmitters that speed signals from neuron to neuron. Many illnesses also have a clear genetic component. Some are linked to abnormal brain structure and physical trauma as well. And the effects of a person's environment, particularly psychological traumas and long-term stresses, are also important factors in mental illness.

Mental illness is common, affecting about 20 percent of the American population. Major depression, bipolar disorder (manic-depression), schizophrenia, and obsessive-compulsive disorder are among the most frequent causes of disability in the workforce.

Schizophrenia is a good example of the complex nature of mental illness. This common disease, which typically shows up during a person's late teens or 20s, is marked by serious disruptions in thinking, perception, and language and in a person's sense of self.

The illness is genetically linked: People with a schizophrenic parent are 12 times more likely to develop the disease, and those with a schizophrenic sibling are 9 times more likely to get it. A person with an affected identical twin has a 46 percent chance of developing the illness. The age of the father plays a part, as well: Men over 50 have three times the risk of fathering a schizophrenic child than those under 25. The disease has also been connected to improper brain development during fetal growth, possibly because of viral infection. The forebrain, the hippocampus, and the amygdala of

KEEPING HEALTHY

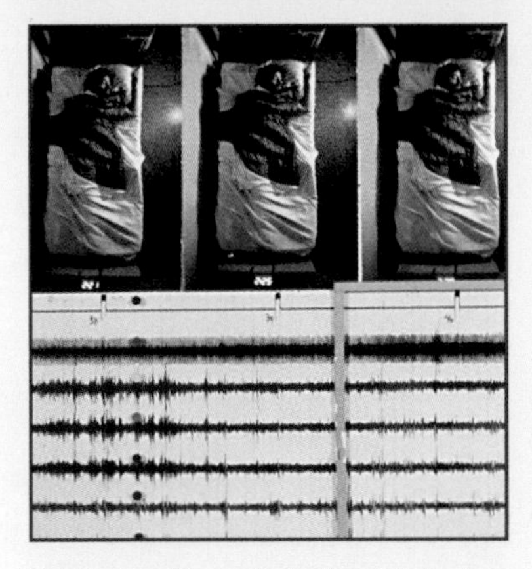

EEG recordings show dreaming sleep.

SCIENTISTS DON'T KNOW why people sleep, but they do know that we can't do without it. Most adults need from seven and a half to eight and a half hours of sleep each night in order to stay healthy and perform their daily functions on an even keel.

Those who go without sleep for several days or more suffer a wide variety of symptoms, ranging from mild to dangerous: irritability, blurred vision, confusion, and visual or tactile hallucinations. People with shortened sleep schedules, down to three of four hours a night, often do poorly on tasks requiring concentration or memory, like taking tests.

Sleep deprivation is also associated with a drop in the number of immune cells in the body. And, of course, the simple fatigue that accumulates with extended sleep loss is a hazard for drivers, factory workers, and the like, all of whom experience a greatly increased risk for accidents if sleep-deprived. ■

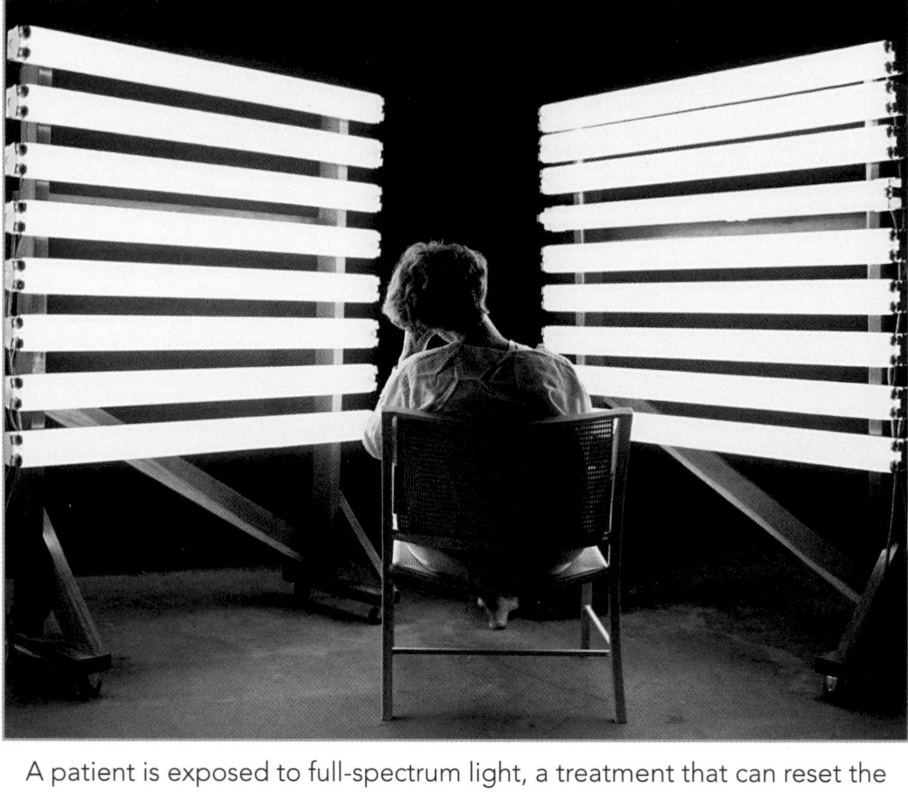
A patient is exposed to full-spectrum light, a treatment that can reset the biological clock and alleviate some forms of depression.

schizophrenics may be smaller than average, while their ventricles—the fluid-filled cavities within the brain—may be larger. And stress is also a major player: A stressful life episode can often trigger schizophrenia.

It is now becoming clear that psychological stress has a powerful effect on the body's hormones, which in turn can reroute the brain's wiring in destructive ways and lead to mental illness. Any external stimulus that threatens the status quo—even just the anticipation of such a threat—can initiate the ancient fight-or-flight reactions within the body.

Under stress, the nervous system signals the endocrine system to release hormones such as epinephrine and cortisol. In the short run, stress hormones pump up the heart, energize the muscles, raise blood pressure, and boost the immune system—all good things if you are facing a charging tiger or fleeing a burning building.

In the long run, however, the hormones released by prolonged stress are damaging, leading to hypertension, hypoglycemia (a form of diabetes), and immune system damage. Stress appears to do harm to the hippocampus, a critical site in the brain for learning and memory. Resarchers have determined that stress is clearly a factor in some mental illnesses. Some people with serious depression, such as those diagnosed with Cushing's syndrome, for instance, secrete excess amounts of the stress hormone cortisol.

Recognizing that most mental illnesses involve an interplay of internal and external factors, doctors typically treat the illnesses with both medication and psychotherapy or behavioral therapy. Most psychiatric medicines attempt to correct possible imbalances in the brain's neurotransmitters. For instance, many antidepressants

WHAT CAN GO WRONG

DEPRESSION is a common and crippling mental disorder that affects one out of every ten Americans, women twice as often as men. Doctors recognize three forms: major depression; a less severe form called dysthymia; and bipolar disorder, a more severe form marked by cycling changes in mood. Major depression is a disabling disease that hamstrings the routines of daily life. People with major depression are 18 times more likely to attempt suicide than people without mental illness. Sufferers typically have some of the following symptoms at least part of the time: persistent moods of sadness, anxiety, or emptiness; feelings of guilt, worthlessness, or helplessness; insomnia, early-morning awakening, or oversleeping; loss of interest or pleasure in hobbies and activities, including sex; restlessness, irritability; decreased energy; difficulty

SEE ALSO: Chapter Ten, "Hormones," PAGE 268

block the reabsorption of serotonin and norephinephrine in the brain, allowing those neurotransmitters to remain in the synaptic space longer. Many antipsychotic drugs are dopamine blockers, since schizophrenia seems to involve an imbalance of the neurotransmitters dopamine, glutamate, and GABA, an amino acid.

Neurotransmitters play a key role in drug addiction, which is considered a brain disorder, though not a mental illness. Drugs such as cocaine, alcohol, and nicotine exert their influences by activating the pleasure circuits in the brain and decreasing feelings of stress or pain. Nicotine, for instance, causes a rush of epinephrine and releases dopamine in the brain, increasing pleasure. Alcohol affects the neurotransmitter GABA and has the effect of calming anxiety. Cocaine also increases dopamine in the brain's reward system, relieving pain and causing euphoria.

So powerful is this reward system that drug users who become addicted will lose control of much of their lives in order to experience it again and again. Researchers are continually challenged to find medications that can pass the blood-brain barrier to lessen the effects of withdrawal from harmful drugs or, even better, block their effects completely.

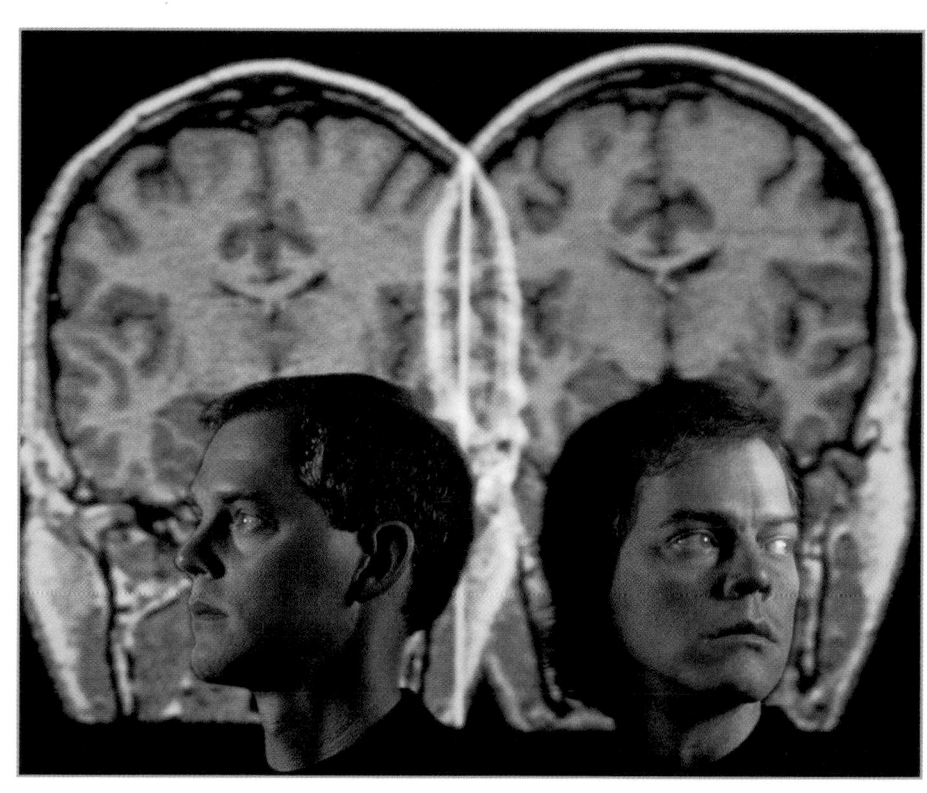

Differences in the brains of identical twins show up in cross-sectional images. The twin on the right has schizophrenia.

+ AUTISM +

AUTISM IS a developmental disorder—more accurately, a wide spectrum of disorders—marked by difficulty with language, poor communication skills, impaired social relationships, and, sometimes, repetitive body movements or obsessions. Autism occurs more commonly in boys. Its causes are unknown and are the object of intense research. Recently, research has linked it to abnormally rapid growth of the brain during an infant's first year.

concentrating or making decisions; appetite changes; sudden changes in weight; thoughts of death or suicide or suicide attempts; and persistent physical symptoms that resist treatment, such as headaches or chronic pain. Like many mental illnesses, depression is probably caused by a combination of factors: psychological, environmental, and biological. It does run in families, although it can occur without a genetic precedent, and it is often triggered by stress or serious illness. Biologists have found that many depressed patients secrete too much cortisol, a stress hormone, and may have abnormal sleep patterns.

Drug treatments for depression typically target the neurotransmitters norepinephrine or serotonin in the brain. A number of current antidepressants are selective serotonin reuptake inhibitors (SSRIs), which block the reabsorption of serotonin in the brain's synapses.

THE SPINAL CORD

CONVEYING THE MESSAGE

THE SPINAL CORD IS the nervous system's central highway connecting the brain to the rest of the body. Information from the body zooms up the cord to the brain, while commands from the brain flash down the cord to muscles and organs. In some cases—such as in the quick reactions of reflexes—signals travel to and from the cord without passing through the brain at all.

The spinal cord and brain together form one continuous entity, the central nervous system. The cord itself extends down from the brain stem and is about 16 to 18 inches long in the average adult. It is held within and protected by the vertebrae, the bones of the spine; the cord runs through the hollow center, called the vetebral foramen, within each stacked vertebra. It is not as long as the spine itself but stops at the first lumbar vertebra, just below the lowest ribs.

The spinal cord is protected by the same three membranes, or meninges, as the brain: the tough, outer dura mater, the weblike arachnoid mater, and the delicate inner pia mater. As in the brain, cerebrospinal fluid fills the space between the arachnoid mater and the pia mater. Between the dura mater and the vertebrae is a padding of fat and connective tissue that gives the cord further cushioning and protection.

The cord, shiny and white, is about three-quarters of an inch thick, but it swells out a bit in two regions. The cervical enlargement,

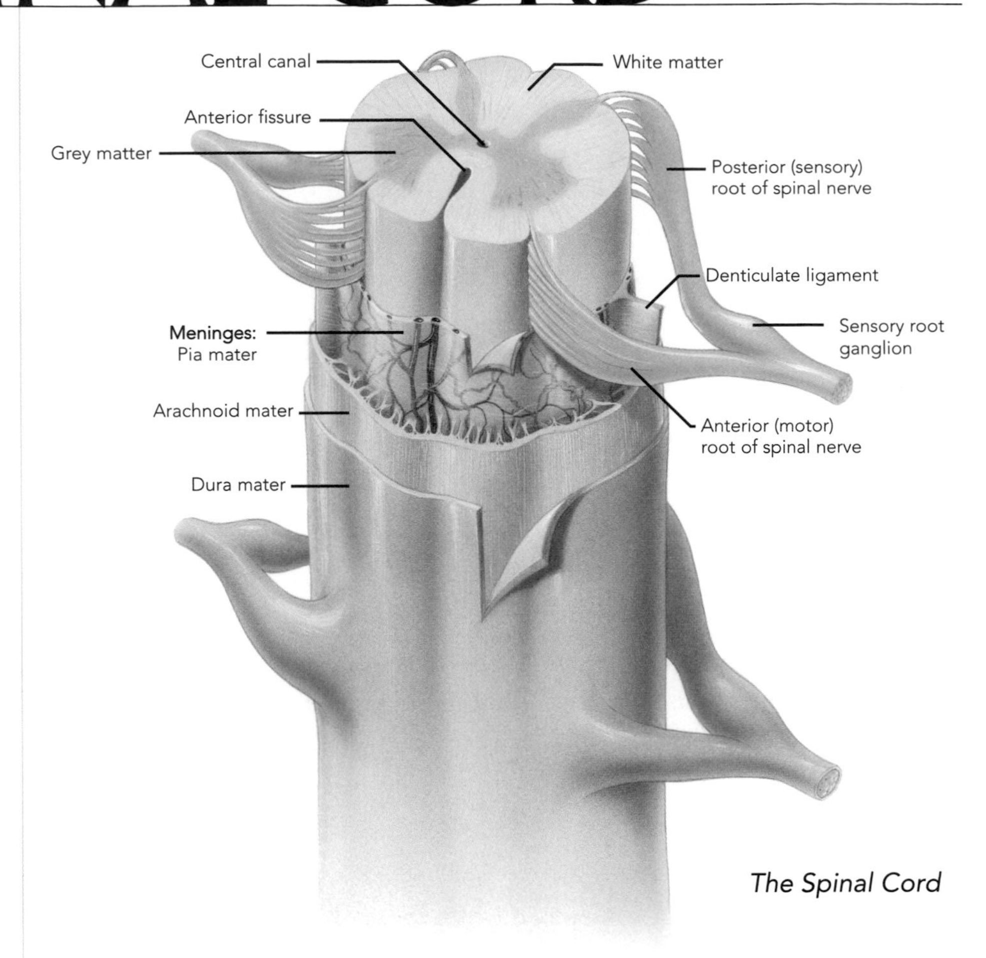

The Spinal Cord

BREAKTHROUGH

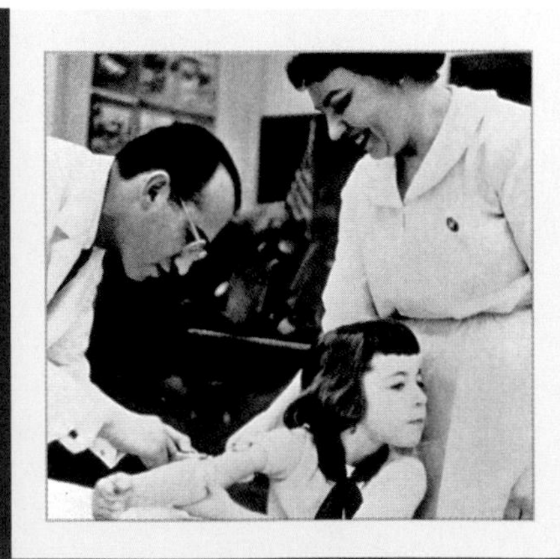

POLIOMYELITIS is an infectious disease of the nervous system spread by viruses. Its most notorious outbreak in modern times, the epidemic of the 1940s and 1950s, was brought to an abrupt end by the introduction of two vaccines: the inactivated, or killed, vaccine developed by Dr. Jonas Salk, left, in 1953, and the live, oral vaccine, developed by Dr. Albert Sabin in the early 1960s. The polio virus enters through the mouth and spreads through the bloodstream, causing flulike symptoms such as fever, muscle pain, and nausea. In its most serious form, it invades the cells of the spinal cord and brain stem, causing paralysis in

SEE ALSO: Chapter Three, "The Spine," PAGE 96

at about shoulder level, is where nerves travel to and from the arms; the lumbar enlargement, in the lower back, is where nerves extend to the legs. In all, 31 pairs of spinal nerves extend from the spinal cord into the rest of the body, each pair emerging from the space between two vertebrae. The lowest pair is part of a fine, hairlike bundle of nerves called the cauda equina ("horse's tail") that drops down from the base of the spinal cord itself.

All of these nerves, part of the peripheral nervous system, connect the brain and spinal cord to glands, muscles, and sensory receptors, some reaching these areas directly but most connecting via other nerves and nerve networks.

The nerve pairs are organized into five groups and named according to the location and level of the vertebral column from which they emerge. Eight pairs of cervical nerves (C1 to C8) extend from the spinal cord in the neck and upper back: Among other things, they connect to the nerves of the skin of the ear, neck, chest, and shoulder and to the muscles of the neck, diaphragm, shoulders, and arms.

SPINAL TAP

A SPINAL TAP, or lumbar puncture, is a method of removing cerebrospinal fluid for testing; it is typically used to diagnose nervous system disorders such as meningitis or multiple sclerosis. It's known as a lumbar puncture because a hollow needle is inserted between the lower lumbar vertebrae, below the end of the spinal cord, to extract the fluid.

Below them on the cord are 12 pairs of thoracic nerves (T1 to T12), which supply the skin of the arms, chest, and abdomen, and the muscles of the ribs, abdomen, and back. Lumbar nerves come in five pairs (L1 to L5) that connect with the nerves of the legs and groin. Below those are five pairs of sacral nerves (S1 to S5) and a single pair of coccygeal nerves (Co1), which mainly serve the buttocks and legs.

Each spinal and cranial nerve can be mapped to the part of the body it supplies, including an area of skin. These nervous system regions are called dermatomes. Doctors can use this knowledge to test for nerve injury; if sensation is lost in a particular area of skin, they can identify the nerve or nerves that may be damaged or severed. Anesthetists use the dermatome map to block pain in particular nerves.

SPINAL NERVES

Each spinal nerve connects to the spinal cord via a two-pronged root, one strand of which consists of sensory axons and the other of motor axons. The cord itself, like the brain, has both gray matter (nerve cell bodies) and white matter (nerve fibers).

Gray matter takes in and integrates information flowing up and down the cord. In cross-section, it forms an H or butterfly shape within the white matter. The upright bars of the H (or wings of the butterfly) are called horns. Those that extend toward the front of the cord contain neurons that control movement of the skeletal muscles. Those that reach toward the back of the cord contain groups of sensory nerves. In some segments of the cord, an additional pair of horns holds the autonomic motor neurons that control glands, cardiac muscle, and smooth muscle.

BREAKTHROUGH

the limbs, throat, or chest. The disease most often infects children. In the 1940s, the National Foundation for Infantile Paralysis sponsored the efforts of Dr. Jonas Salk. Working at the University of Pittsburgh, Salk grew samples of all three virus forms of polio and created a vaccine. By 1961, new cases had dropped by 96 percent. Shortly afterward, Albert Sabin introduced an oral vaccine using weakened live viruses, the type used today. By the 1990s the disease had been eliminated in the Western Hemisphere, although it still remains in Africa and South Asia, where vaccination efforts are still underway today, right.

CHAPTER GLOSSARY

AMYGDALA. Structure in the brain's limbic system that plays an important role in emotional learning.

ARACHNOID MATER. The weblike middle layer of the membranes that protect the brain and spinal cord.

AUTONOMIC NERVOUS SYSTEM. The division of the peripheral nervous system that controls cardiac muscles, smooth muscles, and glands.

AXON. The hairlike extension of a neuron that sends out impulses.

BRAIN STEM. The portion of the brain just above the spinal cord, consisting of the midbrain, pons, and medulla.

CENTRAL NERVOUS SYSTEM. The brain and spinal cord.

CEREBELLUM. Part of the brain behind the medulla and pons; governs coordinated muscle activity.

CEREBRAL CORTEX. Outer layer of the cerebral hemispheres, responsible for conscious experience, thought, and planning.

CEREBRUM. The two cerebral hemispheres that make up most of the brain.

DENDRITE. The branching extension of the neuron cell body that receives electrical signals.

DIENCEPHALON. Part of the forebrain that includes the thalamus, hypothalamus, and epithalamus.

DURA MATER. Outer membrane covering the brain and spinal cord.

FOREBRAIN. The forward portion of the brain that includes the cerebrum and diencephalon.

GANGLIA (sing. ganglion). Groups of nerve cell bodies outside of the central nervous system

HIPPOCAMPUS. Seahorse-shaped structure in the brain's limbic system that is involved in learning, memory, and emotion.

HYPOTHALAMUS. Structure in the brain's diencephalon that monitors the autonomic nervous system.

LIMBIC SYSTEM. Part of the forebrain containing various structures involved in emotions and behavior.

MEDULLA OBLONGATA. The lowest part of the brain stem.

MENINGES. Protective coverings of the brain and spinal cord.

MIDBRAIN. The brain stem between the pons and diencephalon.

MOTOR NEURON. Nerve that carries impulses from the brain and spinal cord to effectors, either muscles or glands.

MYELIN SHEATH. The multilayered fatty covering that insulates most nerve fibers.

NEUROGLIA. Cells of the nervous system that support and protect neurons; also called glial cells.

NEURON. A nerve cell.

NEUROTRANSMITTER. Chemical released by a neuron at a synapse.

PARASYMPATHETIC DIVISION. Subdivision of the autonomic nervous system responsible for overseeing the conservation and restoration of the body's energy.

PERIPHERAL NERVOUS SYSTEM. The portion of the nervous system, consisting of nerves and ganglia, that lies outside the brain.

PIA MATER. The innermost of the brain's three protective coverings.

PLEXUS. A network of nerves.

PONS. The bridgelike part of the brain stem between the medulla and midbrain.

RECEPTOR. Specialized cell or portion of a nerve cell that responds to sensory input and converts it to an electrical signal.

SENSORY NEURON. Nerve cell that carries sensory information into the brain and spinal cord

SOMATIC NERVOUS SYSTEM. The division of the peripheral nervous system that activates skeletal muscles.

SPINAL CORD. The bundle of nervous tissue that runs down the center of the vertebral column, carrying messages to and from the brain.

SYMPATHETIC DIVISION. The subdivision of the autonomic nervous system responsible for overseeing activation of body systems in response to stress.

SYNAPSE. The junction between two neurons or between a neuron and an effector, such as a gland or muscle.

THALAMUS. A structure made of two egg-shaped masses of gray matter in the brain; acts as a relay station for sensory information flowing into the brain.

VENTRICLES. Large interior spaces in the forebrain and brain stem filled with cerebrospinal fluid.

The white matter of the spinal cord is arranged into three columns. Each one consists of bundles, or tracts, of axons—in some cases, over a million of them—with similar origins and destinations. Ascending tracts carry information from sensory receptors up the cord to the brain, while descending tracts convey commands from the brain down motor neurons to muscles and glands.

The names of ascending tracts begin with the prefix *spino-* and end with the region of the brain to which they are connected. Descending tracts begin with the brain region and usually end with the suffix *-spinal.* The anterior spinothalamic tract, for instance, is an ascending tract that carries impulses up the spine to the thalamus. The lateral corticospinal tract brings nerve impulses from the cerebral cortex down the spine.

Different kinds of sensations are transported via different tracts. One set, for instance, carries pain, warmth, itching, tickling, and deep pressure, among other things. Another pair of ascending tracts conveys the sensations of light pressure, vibration, discriminative touch (the awareness of the body part being touched), two-point discrimination (the ability to identify two separate but close points being touched), and proprioception (the awareness of the position of muscles and joints).

The brain integrates this sensory information and may, in response, send commands to the muscles and glands down the descending tracts. Some descending tracts carry impulses that control voluntary movements of skeletal muscles—reaching for an apple, for instance—while others control automatic movements—maintaining posture, for instance.

From the central nervous system—the brain and spinal cord—peripheral nerves branch out through the body.

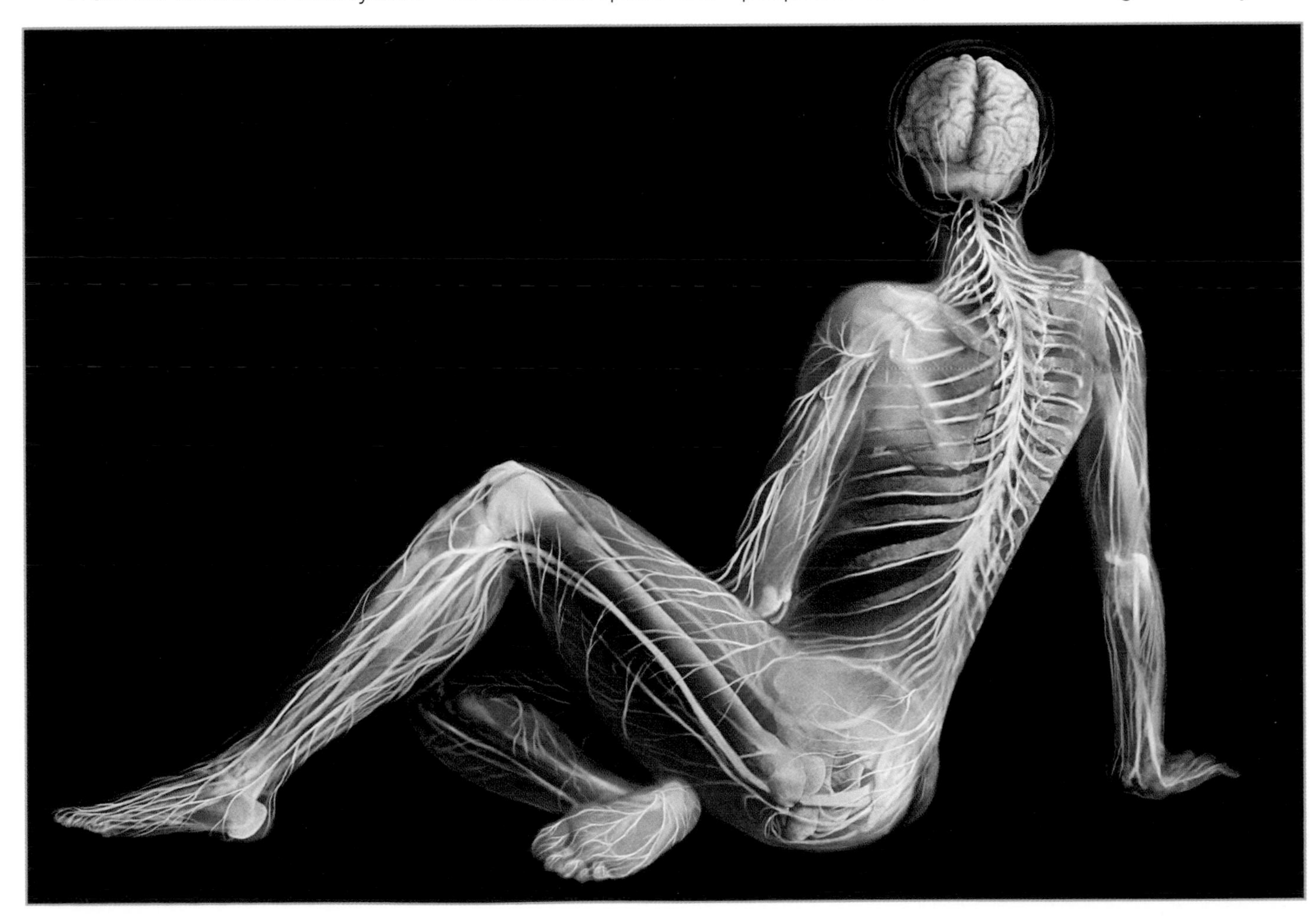

THE NERVES

COMMUNICATING CONNECTIONS

There are millions of nerves that connect the central nervous sytem—the brain and spinal cord—to the rest of the body. Together, they are known as the peripheral nervous system. Broken down into sensory and motor divisions, the peripheral nervous system registers sensations, carries them to the brain, and then takes commands from the brain back to the rest of the body.

Sensation is not the same thing as perception. Sensation is the body's conscious or unconscious awareness of a stimulus, whereas perception is the conscious awareness and interpretation of the stimulus. The brain can receive sensory information about internal body temperature, for instance, without those sensations becoming perceptions.

The sensory division of the peripheral nervous system includes both general and special senses. General senses pull in sensations from the internal organs (visceral senses) and the skin and muscles (somatic senses). Somatic senses include proprioception—position of body parts—pain, temperature, and tactile sensations such as touch, pressure, and vibration. Special senses encompass smell, taste, hearing, vision, and balance.

Sensory receptors come in three basic forms. They can be free nerve endings, simply exposed nerve fibers that register a stimulus. These are found everywhere in the body, but particularly in the skin and connective tissues, and they sense pain, heat and cold, tickles, itching, and some touch. Encapsulated nerve endings, wrapped in a connective tissue capsule, respond to pressure, discriminative touch, and stretching of muscles. More complex are the separate sensory cells found in the eyes or ears. The highest concentrations of sensory receptors are found in the lips, fingertips, and tip of the tongue.

Sensory receptors can also be identified by the kind of sensation in which they specialize. Many respond to tactile sensations only—touch, pressure, vibration, itching, tickling. Some of these receptors are free nerve endings wrapped around

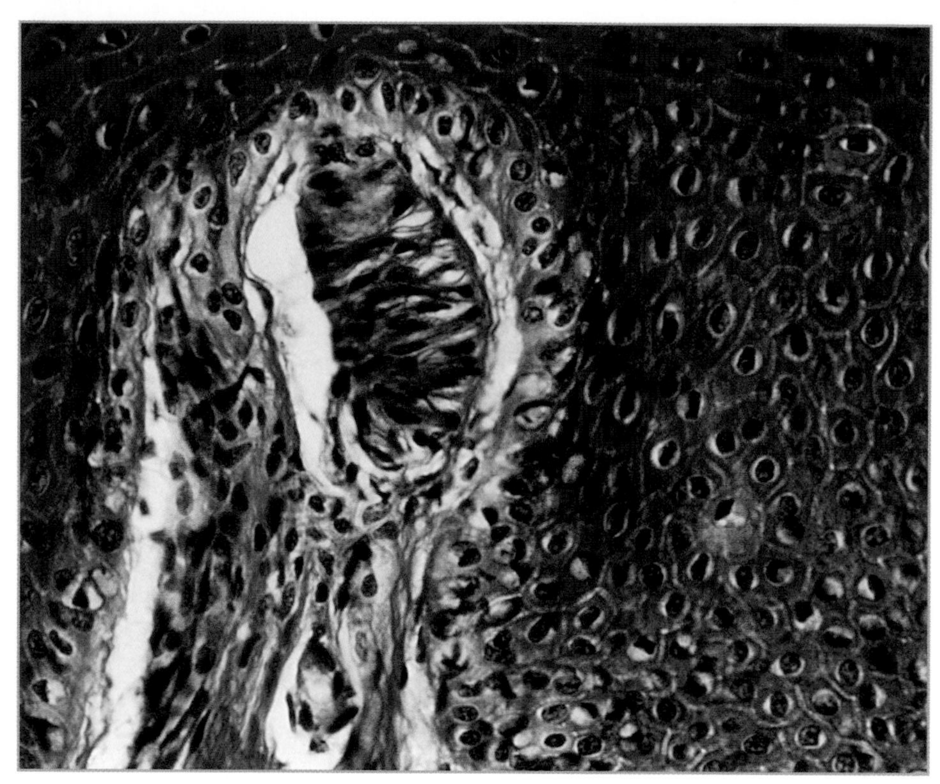

An encapsulated sensory nerve in the skin, called Meissner's corpuscle, is used to covey the sensation of light touches.

HICCUPS

HICCUPS ARE RARELY thought of as a neurological disorder, but they are in fact triggered by an irritation of the phrenic nerve. This nerve, originating high in the spinal cord, controls the motion of the diaphragm. It can send the diaphragm into hiccuping spasms when disturbed. Contrary to popular opinion, standing on your head will not cure hiccups: If this relaxed your diaphragm, you wouldn't be able to breathe in at all.

SEE ALSO: Chapter Two, "The Skin," PAGE 46, Chapter Four, "Skeletal Muscle," PAGE 110 hair follicles, sensing the hairs' movements; others detect slow or rapid vibrations. Sensations of heat and cold are felt by two different kinds of free nerve endings. Free nerve endings known as nociceptors register pain. Found in every tissue except the brain, nociceptors respond to intense heat and cold, unusual stretching, and chemicals released by injury or irritation. Proprioceptors, little-noticed receptors in muscles, tendons, and the inner ear, tell the position of all the parts of the body. When you close your eyes, stretch out your arm, and touch your nose, your are using your proprioceptive senses.

In response to a constant stimulus, many kinds of receptors will adapt by slowing down the impulses they shoot off until the sensation falls below the threshold of conscious perception. An example: After a few minutes in a cold swimming pool, you no longer notice the temperature. Receptors for touch, smell, and pressure adapt quickly, while other sensations fade slowly, if at all, in our perceptions.

Sensations are not perceived consciously until they reach the brain. Impulses from sensory receptors speed along spinal nerves to the spine or along cranial nerves to the brain stem and are passed on from there first to the thalamus and then to the cerebral cortex of the brain. As they enter the brain, the nerves criss-cross, those from the left side of the body entering the right side

LOU GEHRIG

An emotional Lou Gehrig announces his retirement at Yankee Stadium.

AMYOTROPHIC LATERAL sclerosis (ALS), a devastating neurological disease, is commonly known as Lou Gehrig's disease. It gained that name in memory of its most famous sufferer, a popular and short-lived American baseball player.

Gehrig (1903–1941) played first base for the New York Yankees. A powerful hitter, he was nicknamed the "Iron Horse" because of his physical stamina. From 1925 until 1939 he did not miss a game, setting an attendance record that lasted in professional baseball circles until 1995.

In 1938, however, Lou Gehrig began to suffer from cramps and weakness in his muscles. By 1939 he was unable to tie his shoes or sit in a chair. With his batting average down to .143, he took himself out of the lineup, and shortly afterward he received the diagnosis of ALS.

Appearing before a crowd at Yankee Stadium in 1939, Gehrig said, "I may have been given a bad break, but I have an awful lot to live for. With all this, I consider myself the luckiest man on the face of this earth." Gehrig died in 1941.

ALS is a rapidly progressive disease that attacks the motor neurons so that they are unable to send messages to the muscles. Over time, ALS sufferers lose the ability to move or even breathe without assistance. Most die within five years of the onset of symptoms, though a few, such as the physicist Stephen Hawking, may live with the disease for decades.

No cure exists, and the exact cause has not been found. Ten percent of cases are hereditary, and an associated gene has been found, so that genetic screening can be performed in families with a history of the disease. ■

of the cerebrum and vice versa. An area of the parietal lobes known as the primary somatosensory area takes in the sensory information.

Neurologists have mapped the body onto the surface of the brain in this area, finding that parts of the body with many sensory receptors have, not surprisingly, larger areas of the brain devoted to them. Therefore, a substantial portion of the somatosensory area is devoted to fingers, lips, and tongue; the trunk of the body receives relatively little attention. This brain map is not organized like the body, however: Sensations for the thumb register next to those for the eyes, for instance; the genitals, next to the toes. A leading theory about the phantom pain of amputated limbs says that neurons from adjacent areas may be stimulating the part of the cortex devoted to the absent limb—a touch on the face, for instance, might partly transmit to the nearby area for the hand.

VOLUNTARY MOVES

Just in front of the primary somatosensory area of the brain is the primary motor area, where most impulses for voluntary movement begin. Control of complex movements, suppression of unwanted movements, maintenance of muscle tone, and many other subtle balancing acts of the body are coordinated among a variety of brain structures, including the cerebellum, the basal ganglia (nodes of gray matter within each cerebral hemisphere), the thalamus, and motor centers in the brain stem. Commands for movement speed from these areas via the brain stem and spine. Signals travel through cranial nerves (to the face and head) and spinal nerves (to the rest of the body) to reach the muscles. At the end of the journey, the terminals of the motor nerves form junctions with muscle cells, stimulating them to contract.

Although the brain is the executive for most kinds of movements, in some instances sensory impulses and motor commands are integrated in the gray matter of the spinal cord instead. This is the case with spinal reflexes: rapid, involuntary responses to stimuli. Reflexes can be inborn—snatching your hand away from a flame before you're consciously aware of the heat—or learned—stomping on the brake pedal of your car. Pavlov's dogs, conditioned to salivate at the sound of a bell, are a famous example of a learned reflex.

The input and output pathway for a reflex is known as a reflex arc, and it typically passes through three sets of neurons. A sensation ("Heat!") first speeds along sensory neurons to the spinal cord or brain stem. There it swings through an inter-neuron, which sends a signal to one or more motor neurons. A command for movement ("Pull back!") then zips back to the relevant area (in this case, the muscles of the arm).

+ BLINK +

THE AVERAGE person blinks every two to ten seconds. Regular blinking, which keeps the eyes protected and irrigated, is controlled by a blinking center in the caudate nucleus, deep within the brain.

Blinking also occurs as a reflex when the cornea is touched: The impulse flashes along a cranial nerve to the medulla and back. It is the last reflex to disappear under anesthesia.

WHAT CAN GO WRONG

SPINAL CORD INJURY: Severe spinal injuries, like the one suffered by actor Christopher Reeve, left, in 1995, limit the body's ability to move. When the spinal cord is injured or severed, the location of the injury relative to the nerves determines the pain or paralysis. Most spinal cord injuries don't result in a severed cord. A blow to the spine will more typically crush the vertebrae and tear into axons of the spinal cord. If the cord is completely severed, then sensation below the injury will be lost, but an incomplete injury may still allow transmission of messages up and down the spine. Remaining abilities depend on the degree of the damage.

SEE ALSO: Chapter Three, "The Spine," PAGE 96

Reflexes can prompt muscles to stretch, to relax, to flex and withdraw (in response to pain, for example), or, in a more complex transaction, to withdraw while balancing on the other side (as in pulling up your foot from a sharp object while balancing on the other). The classic example is the stretch reflex called the patellar reflex, familiar to anyone who has had a physical exam. When a doctor taps on a patient's patellar tendon under the kneecap while the patient's leg is hanging bent over the edge of a table, the sensation triggers a signal to the spinal cord. Motor commands shooting back from the spinal cord command a muscle spindle in the quadriceps (upper thigh) muscle to contract, extending the leg. At the same time, other signals from the spinal cord tell hamstring muscles to relax, so they don't inhibit the contraction. Tests like this one give doctors a quick indication of the health of the spinal cord and nervous system.

By skipping the trip to the brain, reflexes allow the body to react almost instantly in an emergency. The brain is usually involved at the same time, however, registering the sensation of pain, for instance, or modifying the movement according to the surroundings.

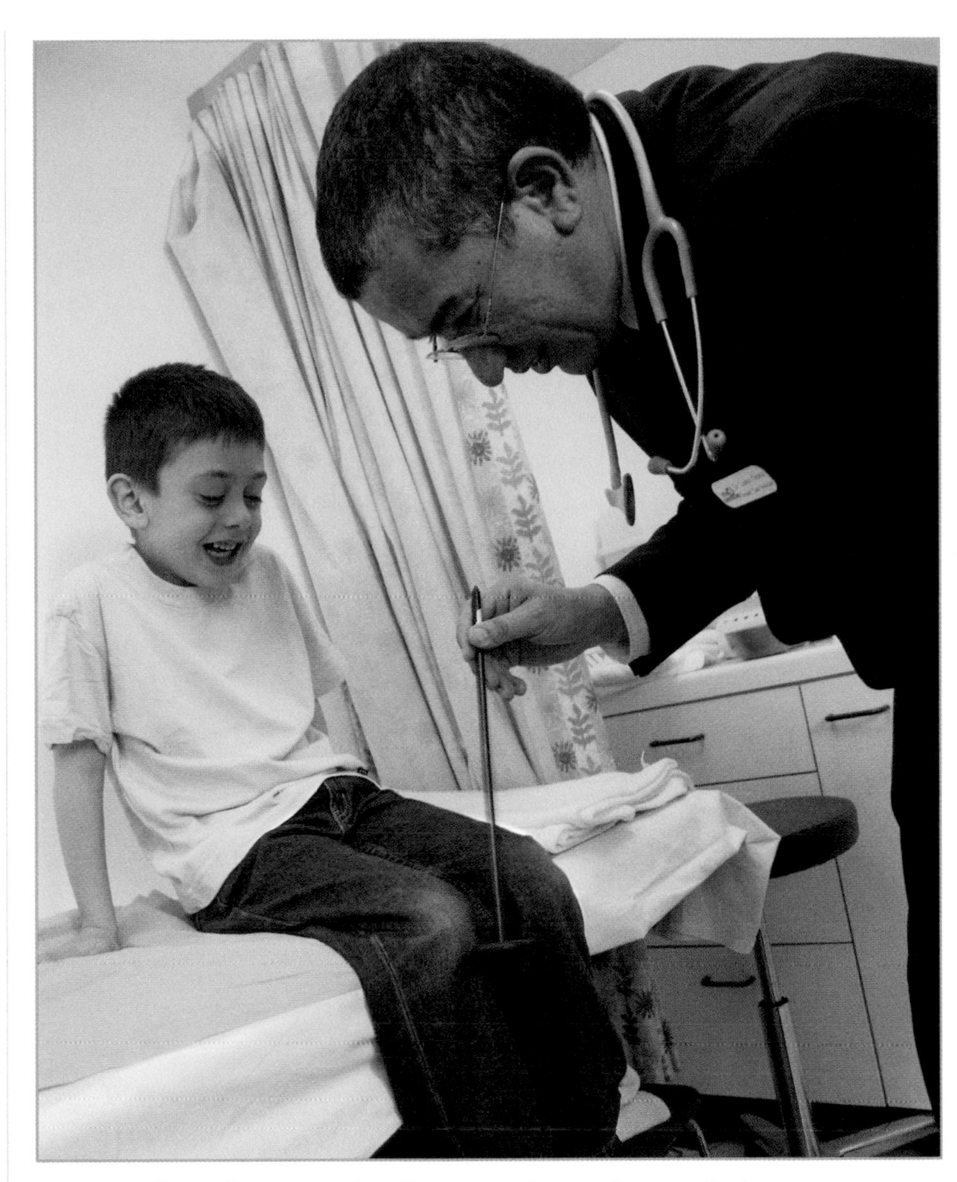

One of the most familiar parts of a routine medical exam, a doctor tests the patellar reflex on a young boy.

AUTONOMIC SYSTEM

In contrast to the somatic nervous system, which involves conscious perceptions and voluntary movements, the autonomic nervous system controls the unnoticed, involuntary workings of the body's interior. Most of the body's internal housekeeping tasks—heart rate, digestion, sweating, opening or constricting blood vessels, dilation or contraction of the pupils of the eye—belong to the autonomic system. More than any other system, it is responsible for maintaining homeostasis: a constant, stable environment within the body.

The input part of this system, composed of its sensory neurons, is similar to that of the somatic system, except for the deeper location of its receptors. These neurons, found within the blood vessels and organs, monitor and send information to the brain about internal conditions such as blood pressure or oxygen levels. Almost all of the sensations reported by these neurons fall under the radar of conscious perception, although certain kinds of pain will break through into awareness.

It is in its output, its motor activities, that the autonomic system is really distinctive. Commanding smooth muscle, cardiac muscle, and some glands, autonomic motor neurons are typically grouped into sympathetic and parasympathetic divisions. Generally speaking, the sympathetic division stimulates and activates the organs in response to stress, while the parasympathetic division inhibits activity and conserves resources at times of rest. Unlike the somatic nerves, which run more or less directly from the spinal column to the muscles, autonomic nerves pass through a junction first: a chain of ganglia. These ganglia, collections of nerve cell bodies, then pass signals on through a second set of neurons to an organ or to several organs, if necessary. Most organs receive signals from both the sympathetic and parasympathetic divisions.

The parasympathetic division of the nervous system is active when the body is at rest, helping it to conserve energy and rebuild resources. It gives priority to digestive activities, decreases heart rate, and constricts blood vessels. It also promotes sexual response, which is restricted during times of stress. Nerves that serve the parasympathetic system emerge either from the brain stem or from the bottom of the spinal cord and pass through ganglia that are found near or within their target organs.

The sympathetic division is more extensive and complex. It is the activating system that responds to stress and readies the body for exercise; it governs the fight-or-flight responses. Among other things, sympathetic nerves cause the adrenal medullae to secrete epinephrine and norepinephrine; they dilate the airways and blood vessels, increase heart rate and the force of heart contractions, promote sweating, dilate pupils, and decrease the motion of the stomach and intestines. Nerves for the sympathetic division extend from the middle of the spinal cord and pass through one of two groups of ganglia. One group, the sympathetic trunk ganglia, runs in two

GOOSE BUMPS

GOOSE BUMPS are visible evidence of the sympathetic nervous system. Cold or fear triggers motor impulses to the nerves controlling the muscles at the base of body hairs. When those muscles contract, they pull the hairs upright and create goose bumps. The reaction, not particularly useful in the sparse hair of human skin, is probably an evolutionary remnant from days when an elevated layer of thick hair would trap heat against the skin.

SYMPATHETIC VS. PARASYMPATHETIC

AFFECTED ORGAN	SYMPATHETIC RESPONSE	PARASYMPATHTIC RESPONSE
Heart	Increased heart rate; stronger atrial and ventricular contractions	Decreased heart rate; decreased atrial contractions
Adrenal glands	Secretion of epinephrine and norepinephrine	No effect
Liver	Glycogen converted into glucose; decreased bile secretions	Glycogen created; increased bile secretion
Iris of eye	Pupil dilation	Pupil constriction
Lungs	Airway dilation	Airway constriction
Stomach and intestines	Decreased motion; contraction of sphincters	Increased motion; relaxation of sphincters
Pancreas	Inhibited secretion of digestive enzymes and insulin	Secretion of digestive enzymes and insulin

SEE ALSO: Chapter Ten, "Hormones," PAGE 268

chains down each side of the spine; nerves extending from this group reach out primarily to organs above the diaphragm. The other group of ganglia, the prevertebral ganglia, run in front of the spine and close to the abdominal arteries. Nerves from these ganglia primarily serve organs below the diaphragm.

Physical activity—even the anticipation of physical activity—will trigger the sympathetic system, but so will a variety of emotions, such as fear, embarrassment, or anger. The hypothalamus, found in the center of the brain, above the brain stem, controls most autonomic functions. It communicates through nerve axons that run from the structure down through the brain stem and spinal cord. The hypothalamus takes in sensory input about the state of internal organs, smell, taste, and blood chemistry.

The hypothalamus is greatly influenced by the nearby amygdala, a little almond-shaped structure within the limbic system that processes emotions such as fear. For instance, a person being approached by a snarling dog will first take in sensory information, the sight and sound of the dog, to be processed in the cerebral cortex. Next the cortex relays those memories and information about the snarling dog to the amygdala, which triggers the emotion of fear and in turn signals the hypothalamus to begin the array of physical phenomena involved in a fight-or-flight response.

Some physiologists identify a third kind of autonomic system: the enteric nervous system. This recognizes the fact that the digestive system is served by an unusually dense, complex network of nerves—as many neurons, in fact, as serve the spinal cord. These nerves control motion and blood flow within the intestines. Although they are influenced by the sympathetic and parasympathetic systems, these nerve networks can also act independently, registering sensations, integrating them, and acting on the intestines without passing through the central nervous sytem. This so-called brain in the gut is probably a reflection of the importance of the digestive system in human evolution.

The sympathetic nervous system gets a workout in a confrontation between leopard and baboon.

THE SENSES

BRINGING INFORMATION TO THE BODY

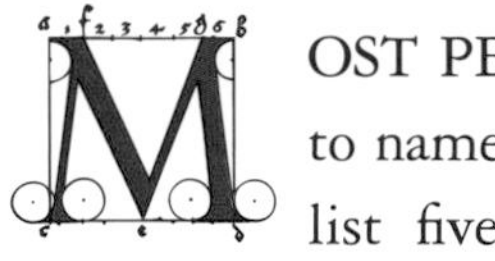

MOST PEOPLE, ASKED to name the senses, will list five: touch, vision, hearing, smell, and taste. But physiologists group them differently. There are the general senses, which include the tactile sensations of touch, pressure, pain, and vibration, picked up by sensory receptors in the skin, muscles, and internal organs. And there are the special senses: vision, smell, taste, hearing, and balance, all of them associated with special receptor cells in the head.

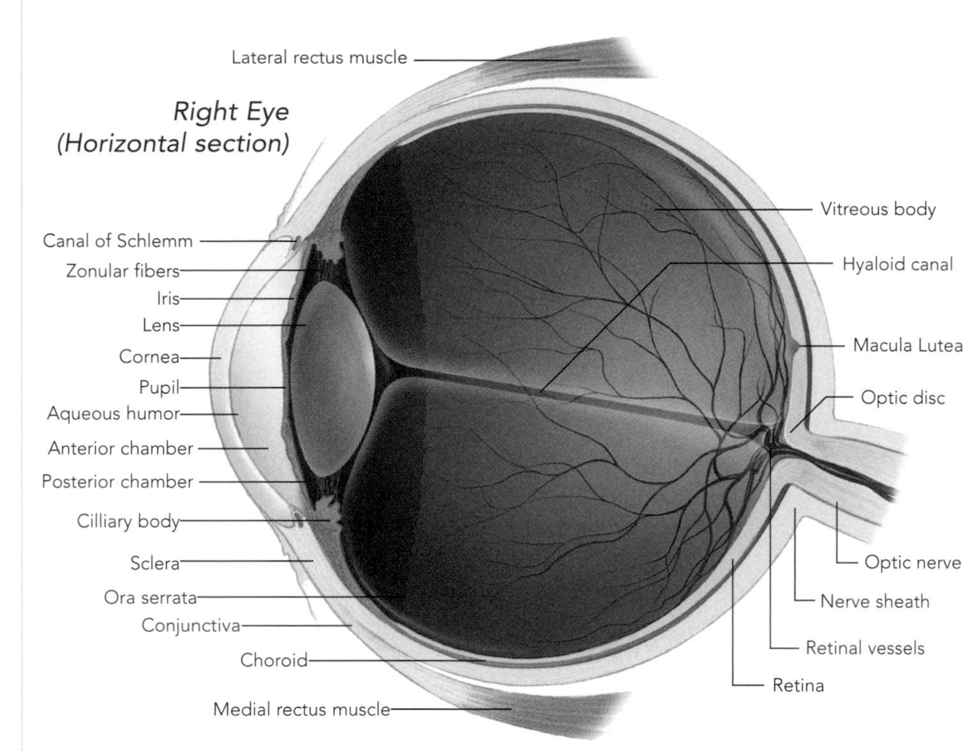

SIGHT

Of all the special senses, vision is probably the most important. The eyes contain about 70 percent of the body's sensory receptors. Vision is the sense most closely associated with our consciousness of ourselves and our surroundings, dominating even our dreams. It is accomplished through a coordination of the delicate structures of the eye and a substantial part of the brain: Nearly half of the cerebral cortex is given over to visual processing.

The adult eye is a sphere about one inch in diameter, with about one-sixth exposed. The rest is sheltered within the eye socket, cushioned by pads of fat and tugged this way and that by six sets of eye muscles. The outermost layer of the eye consists of the sclera, the white of the eye, and, front and center, the transparent area of the cornea, the window that allows light into the eye's interior. The cornea has pain receptors but no blood vessels, which would block the light. Its lack of blood vessels makes it uniquely suitable for transplants, because it does not trigger an immune reaction.

Behind the cornea is a clear fluid, the aqueous humor, which overlays the iris, the colored central part of the eye, and its round opening, the pupil, which dilates and contracts to let light enter. The transparent lens behind the pupil is flexible, changing shape to focus light on the sensory cells in retina.

The interior of the eyeball behind the lens is filled with a clear, jellylike fluid, the vitreous humor. A dark membrane, the choroid, lines most of the eyeball. Its blood vessels feed the organ, while its dark, pigmented cells cut down on reflection within the eye.

Over the choroid is the thin, two-layered retina. The outer layer of the retina, like the choroid, prevents light from scattering within the eye. The inner, neural layer contains 125 million photoreceptors, the light-sensing cells. When light, focused by the lens, hits the retina, it reaches two kinds of photoreceptor cells: rods and cones. Rods, the most numerous, are highly sensitive to dim light but do not provide either color vision

WEBLINK: See the Senses in action at http://nationalgeographic.com/humanbody

or sharp edges. Cones, fewer in number than rods, detect color and sharp details in brighter light.

Signals from the photoreceptors move forward from the retina to an overlying layer of ganglion cells. Axons from the ganglion cells make a sharp turn and run down the back of the eyeball to form the bundle that is the optic nerve. This nerve exits the back of the eyeball at a spot called the optic disc. Because no photoreceptors exist in this area, it is known as the blind spot: Light focusing here cannot be seen, but the eye shifts back and forth so rapidly, we rarely notice this blank space in our vision.

Rods (red) and cones (pink) line the retina in this colorized scanning electron micrograph.

Even if an eye functions perfectly, a person does not see without a working connection to the brain and its vision centers. Optic nerves from each eye reach back into the brain and, at an X-shaped crossing called the optic chiasma, nerve fibers from the innermost halves of each eye cross over to the opposite side of the brain. This means that each hemisphere of the brain receives images from one half of the visual field—for instance, fibers from the left side of the left eye and (inner) left side of the right eye end up in the left hemisphere. People with widespread damage to one hemisphere—from a stroke, for example—may end up seeing only one-half of their visual field, despite having healthy eyes.

Signals moving along two optic tracts, holding the reorganized optic nerve fibers, then reach the thalamus, which in turn passes them on to the primary visual cortex in each of the cerebrum's occipital lobes. There, the visual information is sorted out into such categories as dark and bright edges, object orientation, color, and motion. For the conscious mind to perceive what the eyes have just seen, all these separate categories of perception recombine into an integrated image.

BREAKTHROUGH

LASIK SURGERY: Until recently, most nearsighted and farsighted people corrected by fitting additional lenses over their eyes in the form of either eyeglasses or contact lenses. A new development, surgery to reshape the corneas, began in the early 20th century. This procedure now is dominated by the LASIK technique. LASIK stands for Laser-Assisted In Situ Keratomileusis. During this procedure, eye surgeons first cut a flap in the cornea and then fold it back. Guided by a computer, they then use a special laser to vaporize an inner section of the cornea. The process flattens the cornea into a shape that can focus light accurately on the retina, correcting the misguided path of light rays that actually caused the faulty vision. If all goes well, the patient will be able to see clearly right away, without corrective lenses. The surgery can be used to upgrade vision to 20/15 or better.

TASTE & SMELL

The special senses of taste and smell are quite distinct in the way we perceive them, but they are close partners in the nervous system. Both are chemical senses, registering their sensations through chemoreceptors that detect chemicals dissolved in solution. Taste is heavily dependent on smell, the two senses usually combining to convey flavor, from the salty comfort of chicken soup to the interplay of ripe cherries, vanilla, and oak that might blend to characterize a fine glass of Cabernet.

The sensory receptors for taste are found in thousands of oval structures called taste buds, most of them located on the tongue, but a few also lining the soft palate, pharynx, inner surface of the cheeks, and epiglottis. Taste buds on the tongue sit on top of the little rounded bumps, called fungiform papillae, that give the tongue its rough appearance. Each bud holds sensory receptor cells for taste (called gustatory cells); out of each cell extends a long hairlike microvillus, or tiny projection, bathed in saliva. The variety of chemicals dissolved in saliva stimulate these hairs, which pass the signals down to sensory dendrites at their bases. The mouth is a rough neighborhood, and taste bud cells are frequently destroyed by friction or heat, to be replaced by new cells every seven to ten days.

Wine lovers often inhale deeply before taking a sip. Smell—such as the aroma of fine wine—is an integral component of taste.

A taste sensation is usually a combination of smell and one or more of five primary taste qualities: sweet, sour, salty, bitter, and umami—a fifth quality newly discovered by Japanese researchers, who found that it conveyed the savory taste of steak or tangy cheeses. Organic compounds such as sugars and alcohol trigger the sweet taste. Salty comes from metal ions. Bitter flavors come from aklaloids such as caffeine or strychnine. Sour tastes arise from acids. Umami comes from the amino acid glutamate. Contrary to popular opinion, the tongue has no special areas associated with particular tastes: All kinds of tastes can be detected all over the tongue. The threshold for activating a taste sensation varies. Bitter substances can be detected in minute quantities, while sweet and salty require higher concentrations. This configuration probably evolved because it performs a protective function, since most poisons are perceived as bitter or sour.

Once the sensory dendrites at the base of the gustatory cells have received a stimulus, they pass it on via the cranial nerves to the brain stem, then to the thalamus, and finally to the gustatory cortex in the parietal lobes. As the signals pass through the brain stem, they trigger digestive reflexes. Saliva increases in the mouth, and gastric juices flow into the stomach.

Taste may be limited to five primary qualities, but smell is a

SEE ALSO: Chapter Six, "Respiration," PAGE 156, Chapter Seven, "Ingestion," PAGE 184 different matter. Researchers believe that the body can detect hundreds of primary smells, which combine to create at least 10,000 different odors recognizable to the human brain. These numbers probably pale in comparison to the quantity of odors detectable by many other animals. Smell, like taste, is a chemical sense. After air enters the nose, it flows past a small yellowish patch of epithelium in the roof of the nasal cavity, its size about one inch square. This is the olfactory epithelium, which contains between 10 million and 100 million olfactory receptor cells. Dendrites extending from these cells end in a sprouting mass of little hairlike cilia that spread out in the layer of mucus that covers the olfactory epithelium. These cilia detect the chemicals from the inhaled air that dissolve in the mucus. The olfactory cells are among the few neurons that continue to regenerate; they turn over about every 60 days.

The sense of taste is well developed from birth, but taste preferences do change as people age.

These receptors can be exquisitely sensitive, in some cases detecting an odor from just a few molecules in the air. The foul-smelling chemical methyl mercaptan, for instance, can be detected in minute concentrations, which is why it is added as a warning element to odorless natural gas. Smell receptors adapt quickly, however, and many strong smells sink out of awareness in a few minutes.

Having sensed a particular chemical, the olfactory cells flash signals along their axons through openings in the bone above the nasal cavity, where the axons connect with the olfactory bulbs. These are the ends of the olfactory tracts, which carry the neurons via the thalamus under the frontal lobes to the primary olfactory cortex of the brain. Some nerve fibers also travel from the thalamus to the amygdala and other parts of the limbic system, where the odor becomes associated with emotion and memory. Thus the sensations attached with certain smells often conjure up distinct memories, doubtless an evolutionary remnant from a time when odors carried important messages about danger and desire.

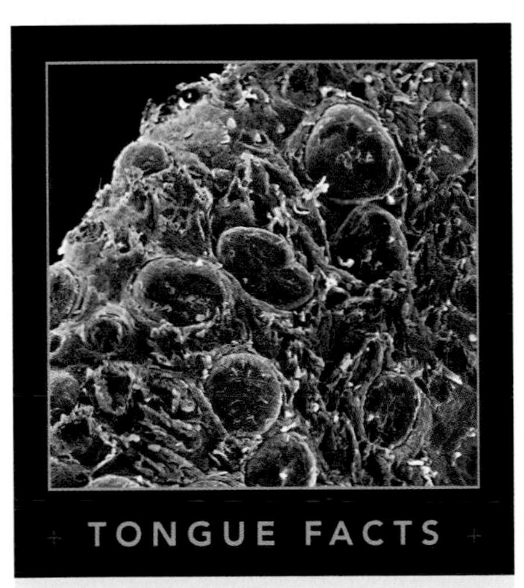

TONGUE FACTS

+ THE TONGUE of a young adult has 10,000 taste buds.
+ Each taste bud contains about 50 sensory cells.
+ The rear of the tongue contains a V-shaped row of extra-large papillae, or receptive cells, each one holding 100 to 300 taste buds, above.
+ Taste aversions are inborn; even a newborn baby will grimace and turn away from a bitter taste.
+ Taste is 80 percent smell.

HEARING

The delicate, curling structures of the ear are home to two disparate senses: hearing and equilibrium, or balance. As different as these senses are, they both rely on the same basic mechanism—liquid sloshing back and forth around receptor cells—to convey an amazing range of sensations to the brain.

The ear itself is usually divided into three parts—outer, middle, and inner—with the inner ear containing the most complex sensory apparatus. The outer ear, the visible part, is made of cartilage covered by skin. Its primary purpose is to funnel sound to the eardrum, or tympanic membrane, a thin partition of connective tissue that divides the outer and middle ears. The middle ear, or tympanic cavity, is a little air-filled cavity in the temporal bone. Within it, connecting the eardrum to the inner ear, are three tiny bones linked by miniature joints: the malleus (hammer), the incus (anvil), and the stapes (stirrup). The stapes is the body's smallest bone, measuring about 0.10 inch long.

The inner ear is far more complicated. Located securely in the temporal bone, behind the eye socket, it is also called the labyrinth because of its maze of spiraling canals. It is, in fact, two labyrinths: an outer bony labyrinth that forms a chamber and encloses an inner membranous labyrinth.

The inner ear's three main areas include a hollow vestibule, holding two little membranous sacs that contribute to equilibrium; three looping semicircular canals, also used for equilibrium; and the coiled, snail-like, fluid-filled cochlea, containing the organ of Corti, the primary organ of hearing. This organ is essentially a sheet of 16,000 hair cells projecting from one membrane, the basilar membrane, and touching another, the tectorial membrane. When vibrations roll through the fluid of the cochlea, the basilar membrane and its attached hair cells bounce against the tectorial membrane, which stimulates the hair cells and generates nerve impulses to the brain.

The pathway of sound waves to the inner ear works like this: First, the outer ear catches airborne sound and directs it to the eardrum. The eardrum vibrates (slowly for low-frequency sounds and rapidly for high-frequency sounds). Next, the eardrum's vibrations shake the attached malleus bone in the middle ear, which passes its vibrations through to the stapes bone. The stapes vibrates against the membrane of the inner ear, shaped like an oval window, which in turn sends pressure waves through the fluid of the cochlea. Then the fluid's pressure waves vibrate the basilar membrane and attached hair cells, which bend against the upper tectorial membrane and generate nerve impulses.

+ EAR FACTS +

+ THE EAR begins to appear 22 days after fertilization in a thickened spot called the otic placode on the embryo's developing head.

+ The human ear can detect frequencies, or pitches, between 20 and 20,000 Hz (waves per second).

+ Humans can detect sounds as quiet as 10 decibels (a mosquito whine).

+ Prolonged exposure to sounds above 90 decibels can permanently damage the hearing.

+ The threshold of pain for hearing begins around 130 decibels. Amplified rock music can exceed 120.

WHAT CAN GO WRONG

TINNITUS IS A ringing, buzzing, roaring, whooshing, or clicking noise that does not have any external cause. It can be almost unnoticeable or loud and maddening. Causes range from earwax to inflammation of the middle or inner ears, long-term noise-related damage (as with the Who guitarist Pete Townshend, (left) high blood pressure, damage to the cochlear nerve leading from the ear, or Ménière's disease, a buildup of excess fluid in the membranous labyrinth. If the cause is related to age or damage, not much can be done; in some other cases, though, the underlying causes can be treated to make the tinnitus disappear.

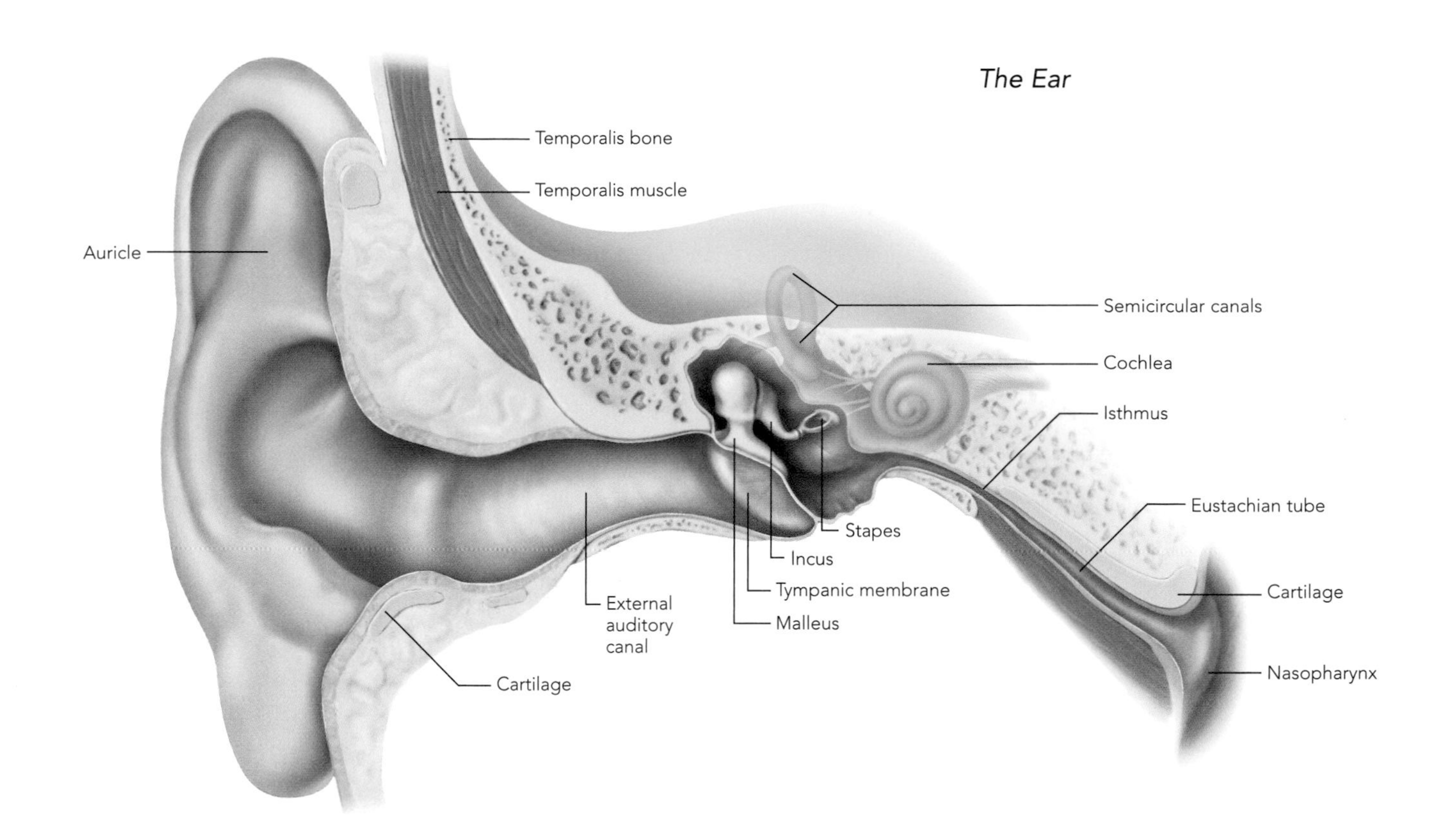

Nerve impulses from the cochlea travel via the cochlear nerve to the cochlear nuclei in the medulla oblongata, then onward to the thalamus, and finally to the primary auditory area of the cerebral cortex. Some of the nerve fibers for hearing cross over to the opposite brain hemisphere, while others do not, so that each hemisphere receives signals from both ears. Brain stem nuclei help to pinpoint a sound's location by comparing the subtle differences in the timing of impulses from the right and left ears.

Equilibrium is the other sense governed within the ears, also through the movement of liquid that passes over hair cells, a process by which signals are sent on to the brain. Two kinds of equilibrium keep our bodies balanced. Static equilibrium monitors posture, the position of the body (particularly the head) relative to gravity, as well as straight-line acceleration, such as the sudden forward movement of a car.

Two little fluid-filled sacs in the inner ear, the utricle and saccule, contain hair cells that detect this kind of motion. Dynamic equilibrium is the sense of rotational movement—spinning about, as in the sensation of an amusement park ride. Hair cells within the semicircular canals pick up this whirling motion. Arranged at right angles to one another, the semicircular canals work together to determine just where the head is moving through three-dimensional space.

Equilibrium is not just a function of the inner ear. It involves other senses as well. Vision, sense receptors in the muscles, and sensory nerves in the skin play a part in maintaining balance. Information about equilibrium needs to be processed quickly, so it travels primarily to the brain stem and cerebellum to produce quick reflex responses in the muscles and eyes. Occasionally the body needs a moment to adjust to equilibrium information, which is why you may stagger and find your eyes jerking back and forth after you have been spinning about.

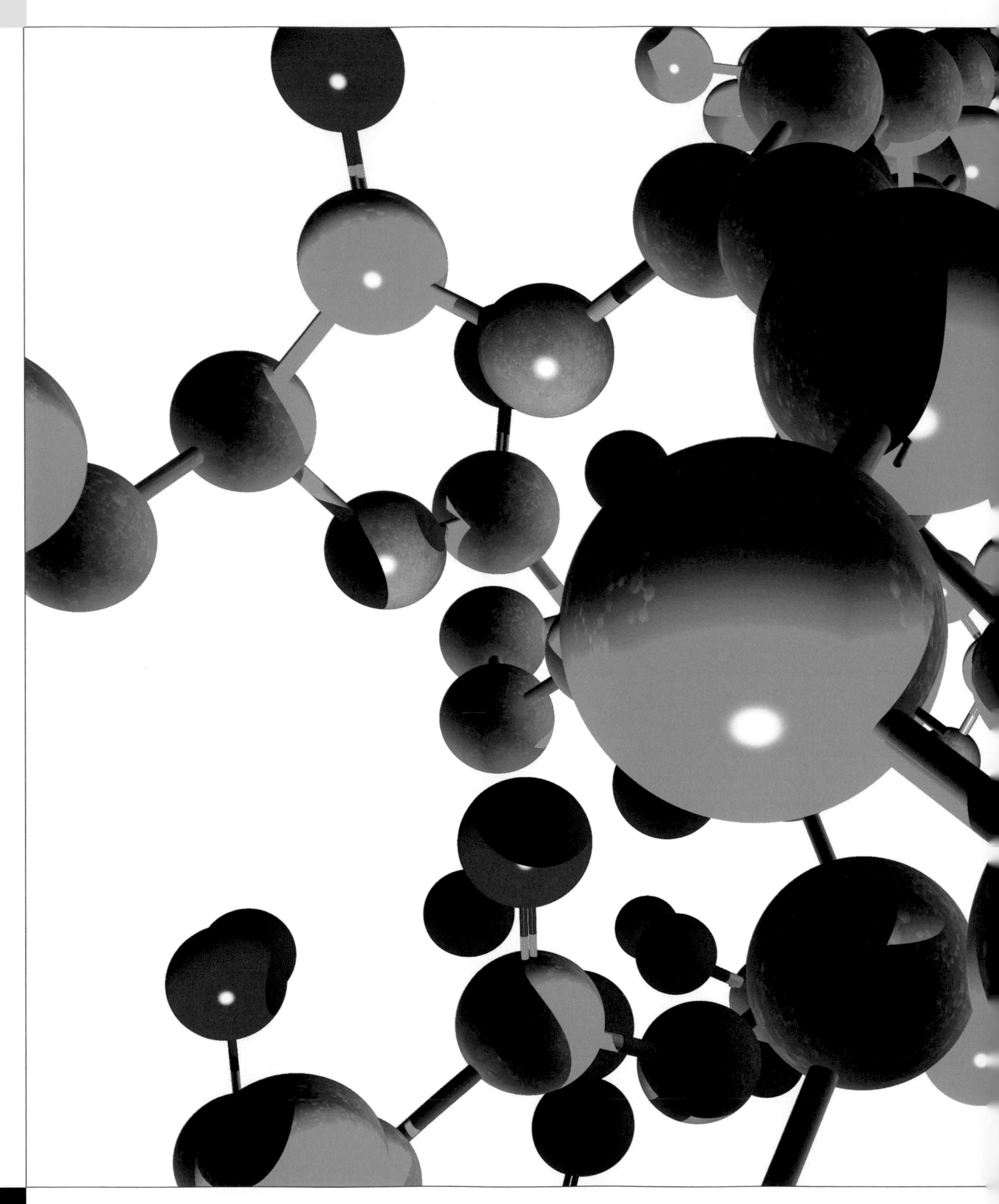

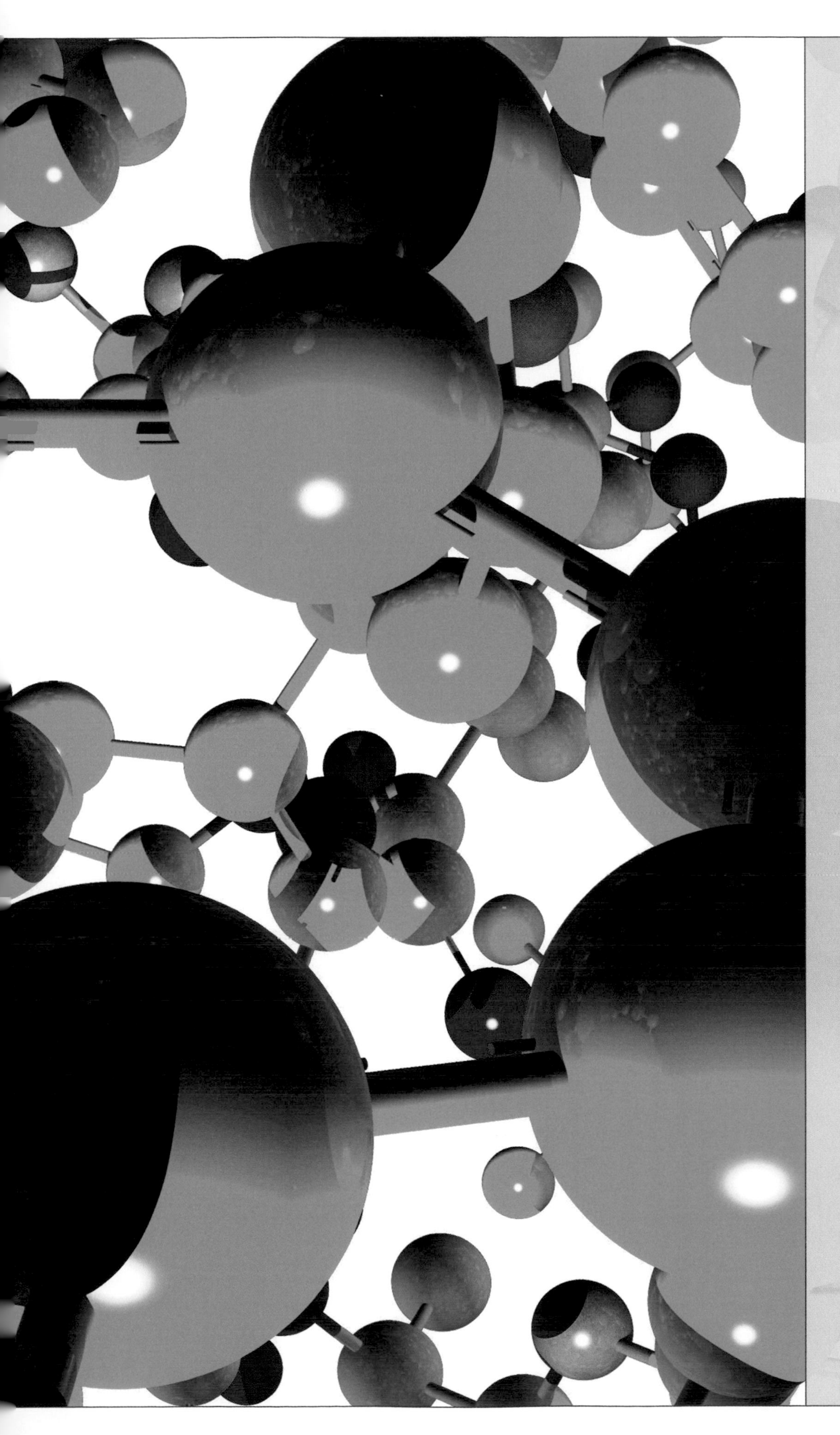

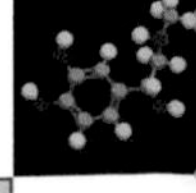

CHAPTER TEN

MESSENGERS

THE ENDOCRINE SYSTEM IS able to communicate and control through chemical messengers that target specific organs throughout the body. Called hormones, these substances influence metabolic activities, or the chemical changes in the tissues of the body. Hormones influence when and how the body develops into adulthood. They dominate the ebb and flow of reproductive processes. They also play important roles in protecting the body from harm, regulating tissue function and metabolism, and maintaining electrolyte, water, and nutrient balances in the blood.

A molecular model of insulin, the hormone that regulates glucose balance in the body

THE SYSTEM

THE UNDERSTATED ENDOCRINE SYSTEM

IF YOU WANTED TO set up a communications network, you could do so in more than one way. You could lay down fiber optic cables and send light signals from one location to another. Or you could devise a system of cell phones and towers that transmit electromagnetic microwaves into the air to be picked up by a base station and relayed to another mobile device.

The human body, too, has multiple ways to communicate, body system to body system. The nervous system works more like the land-based telephone network, with neurons conducting electrical impulses through fibers, from body to brain and back. Alternatively, the endocrine system acts more like a wireless network. Its signals are hormones, chemical in nature. Unlike the brain, the organs and tissues that produce these hormones do not directly connect to their targets.

Instead, specialized endocrine cells release hormones into blood, just as mobile phones transmit microwaves into the air. Many different hormones are circulating through the bloodstream at one time, all searching for specific target cells with special receptors, designed to pick up only specific signals. Only cells that are genetically programmed for a hormone can respond to that hormone.

The Endocrine System

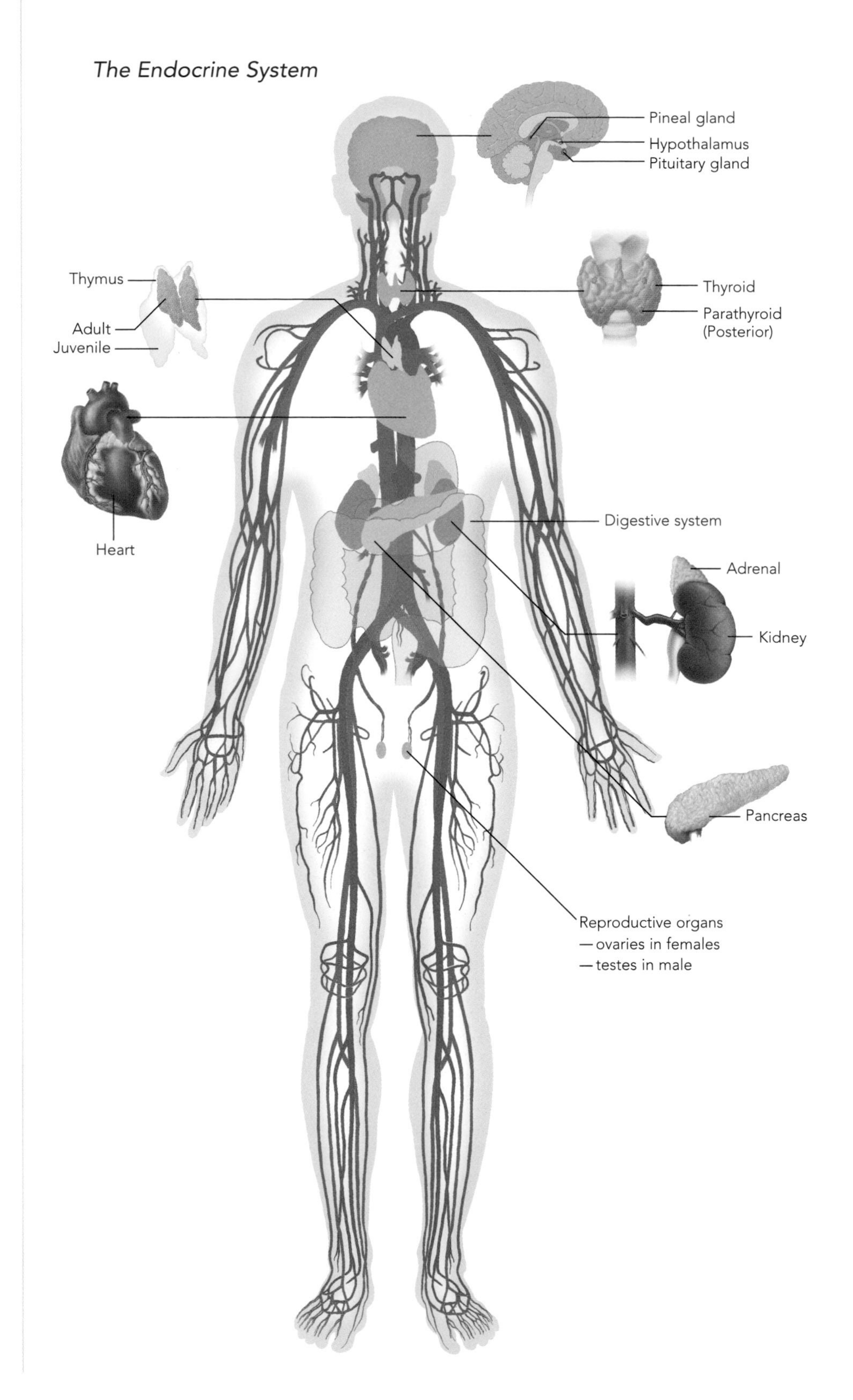

SEE ALSO: Chapter Nine, "The Brain," PAGE 234

In general, the everyday effects of the endocrine system may go unnoticed by the casual observer. The processes and functions regulated by hormones appear seemless and smooth, making it difficult to appreciate the power of the system when everything is running smoothly. But the endocrine system's impact becomes dramatically apparent when the body enters adolescence and begins to go through puberty. Hormones (estrogen in women, testosterone in men) begin to alter the body's appearance, turning girls into women and boys into men. Puberty doesn't occur overnight; it is a series of changes and events that occurs over longer stretches of time, like many processes regulated by the endocrine system.

TIMING

Because of their structural differences, the nervous and endocrine systems also differ in the amount of time it takes for their messages to travel and be received. The time between neuronal impulse transmission and reception is so short, it is often measured in milliseconds. The time between endocrine transmission and reception almost always takes longer. Some reactions, like the fight-or-flight response, can take seconds whereas others, like sleep-regulation controlled by the pineal gland, can last for days or even months.

While responses in the endocrine system—ovulation, for instance—can seem relatively sluggish in comparison, it means that their effects last a long time. Hence the body uses the endocrine system to regulate prolonged processes such as reproduction; maintaining the balance in electrolyte, nutritional, and water concentrations; and fine-tuning cellular metabolism and the flow of energy—all processes that require a steady state of regular rhythms. These body rhythms, guided by waves of chemical messages, represent the major forces at work in the body's endocrine system.

ORGANS & GLANDS

The organs of the endocrine system, like cell phones that do not need a physical base, are small and scattered throughout the body. Hormones in the gastrointestinal system are not even associated with specific glands, nor are endocrine cells in organs such as the lungs and the kidneys. Because their goal is mainly to secrete hormones, the equivalent of organs within the endocrine system are mainly glands. A gland may consist of a single cell or a group of cells that secrete substances into ducts, onto a surface, or directly into the blood.

There are two kinds of glands in the body: exocrine and endocrine. Exocrine glands have ducts, and they produce non-hormonal secretions—like sweat and saliva—that are released in specific places—like the skin or inside the mouth, respectively. Endocrine glands specialize in producing hormones, the messengers that perform the essential functions of this system. On the other hand, endocrine glands lack ducts; their secretions are released into the interstitial fluid and then diffuse into the capillaries and into the bloodstream.

Located throughout the body, the endocrine glands—namely, the pituitary, thyroid, parathyroid, adrenal, and pineal glands—have a webbed, or cordlike, appearance. This anatomical structure allows these organs maximum contact with capillaries, allowing the glands to suffuse their potent

SMALL POWER

+ THE ORGANS of the endocrine system are so small that all the tissue from eight to nine adults would amount to only 2 pounds (about 1 kg) of hormone-secreting tissue.

+ The pituitary gland, which secretes nine hormones, is about the size of a pea.

+ The thyroid, above, is the largest pure endocrine gland in the body.

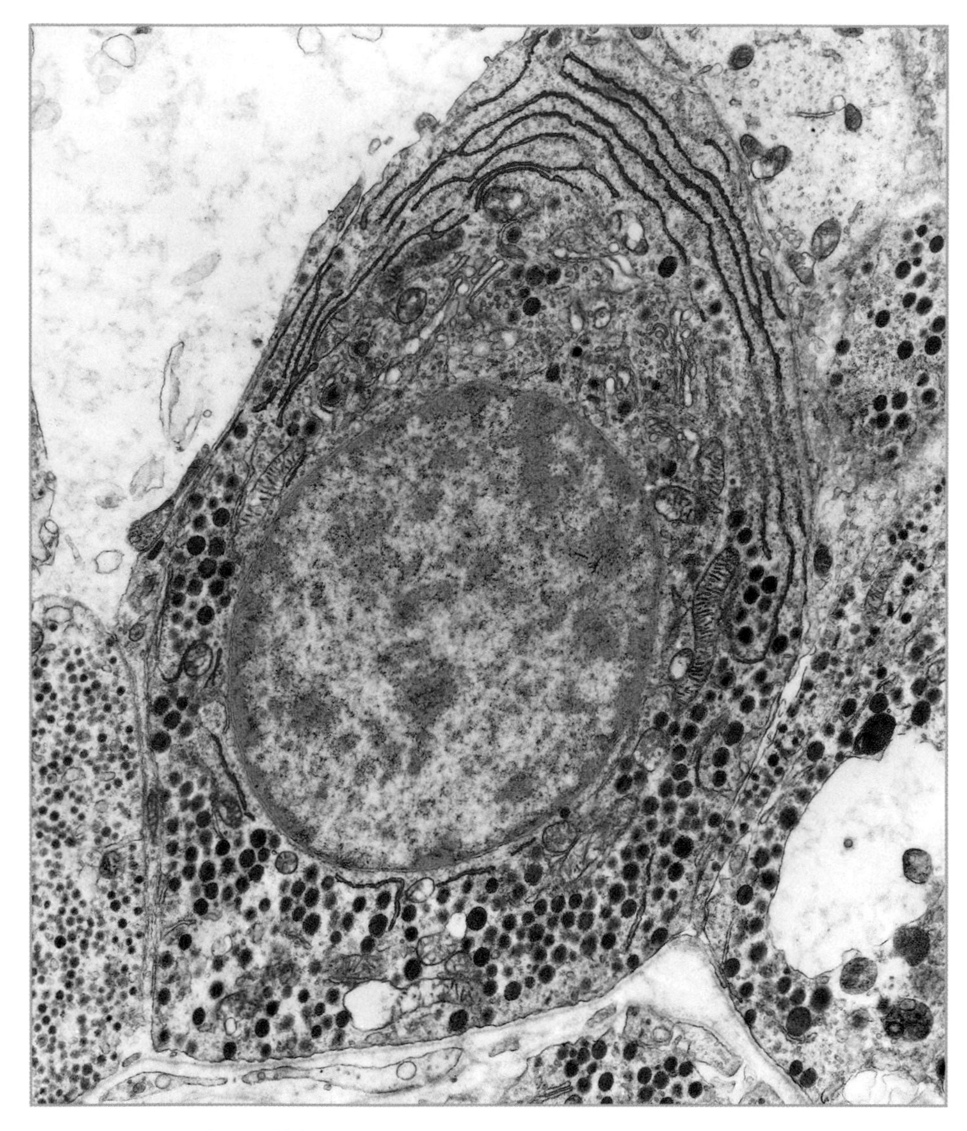

This cell from the pituitary gland, an organ located at the base of the brain, produces growth hormone.

chemicals into surrounding tissue fluid. In most cases, high concentrations of hormones are not necessary to normal function; circulating hormone levels in the blood are generally low.

Other players in the body's endocrine system—such as the pancreas, the hypothalamus, and the gonads (the ovaries in women and the testes in men)—are indeed organs that contain tissue capable of secreting hormones. In the liver, cells make hormones while performing other functions as well.

Meanwhile, in various places throughout the body, isolated clumps of fat cells release a hormone called leptin, which serves the function of regulating fat storage and appetite. Patches of hormone-producing cells line other major organs like the stomach, intestines, kidneys, and heart.

DEVELOPMENT

Because the endocrine glands are found all over the body, they do not develop in a localized fashion when we are in the womb. Instead they develop from widely separated parts of the embryo. The pituitary gland emerges from the ectoderm, about three weeks after fertilization. The thyroid gland develops in the fourth week from the ectoderm. The adrenal glands arise from the mesoderm.

BREAKTHROUGH

THE FIRST HORMONE: In the early 20th century, scientists believed that nervous stimulation controlled digestion. London physiologists Earnest Starling and William Bayliss, left, observed that the pancreas secreted juices in response to food entering the small intestine, but they suspected another player, too. They cut all the nerves to the rats' pancreases and watched with amazement as digestion continued. A chemical secreted from the intestinal lining, not a nerve fiber, caused food breakdown and processing. In 1905, Starling and Bayliss named the substance *secretin*, the first hormone to be discovered.

CHAPTER GLOSSARY

ADRENAL CORTEX. Outer portion of an adrenal gland; produces cortisol.

ADRENAL GLANDS. Pair of glands that controls the stress response mechanism through the secretion of cortisol, epinephrine, and norepinephrine, also known as adrenaline and noradrenaline.

ADRENAL MEDULLA. The inner portion of an adrenal gland. Synthesizes and secretes epinephrine and norepinephrine.

ADRENOCORTICOTROPIC HORMONE (ACTH). Released by the pituitary gland in response to CRF. Targets adrenal glands to boost synthesis of corticosteroids.

CARRIER-PROTEINS. Ferry insoluble hormones to target cells.

CORTICOTROPIN-RELEASING HORMONE (CRH). Stimulatory hormone produced by the hypothalamus.

CORTISOL. Anti-inflammatory and stress hormone produced by the adrenal glands, circulates to the hypothalamus in order to shut down production of CRH and ACTH. Also called hydrocortisone.

CYTOKINE. Protein produced by the lymph system that affects the activity of other cells and is important in controlling inflammatory responses. Interleukins and interferons are cytokines.

ENDOCRINE GLANDS. Ductless glands that secrete hormones directly into the bloodstream or lymph system for a specific physiological purpose or response.

FOLLICLE-STIMULATING HORMONE (FSH). Secreted by the anterior pituitary, FSH stimulates egg development and estrogen secretion in females; initiates sperm production in males.

GLAND. An organ that regulates the secretion or excretion of substances within the body for further use or for elimination.

GONADOTROPIN. Secreted by the pituitary gland; prompts gonadal activity, including the onset of puberty and sexual maturity.

GONADOTROPIN-RELEASING HORMONE (GnRH). Released by the hypothalamus to prompt the pituitary gland to release FSH and LH.

GROWTH HORMONE. Stimulates muscle and bone cells to grow.

HORMONE. Chemical messengers, generally amino acid or steroidal molecules, released into the bloodstream to maintain and regulate individual bodily functions.

HYPOTHALAMUS. Neuroendocrine organ that manages all endocrine functions, working to maintain homeostasis within the body.

LUTEINIZING HORMONE (LH). Secreted by the anterior pituitary. Stimulates ovulation in females, and the testosterone production in males.

MELANOCYTE-STIMULATING HORMONE (MSH). Stimulates production of the pigment melanin in skin and hair and also acts as a neurotransmitter in connection with appetite and sexual arousal.

OVARIES. Female gonads. Produce egg cells, secrete the hormones estrogen and progesterone.

OXYTOCIN. Multipurpose hormone released by the hypothalamus. Stimulates uterine contractions during childbirth; milk production during breastfeeding; orgasm after sex.

PANCREAS. Organ that produces endocrine secretions such as insulin, somatostatin, and glucagon, in order to regulate blood sugar levels.

PARATHYROID GLAND. Regulates increases in calcium levels through the secretion of parathyroid hormone (PTH).

PINEAL GLAND. Secretes melatonin, a hormone considered causally linked to sleep cycles.

PITUITARY GLAND. A neuroendocrine gland with a variety of functions, including regulating the gonads, thyroid, lactation, and water balance.

PROLACTIN (PRL). Protein hormone released by the pituitary gland that triggers milk production in the mother's breasts after birth.

TESTES. Male gonads. Secrete testosterone, create sperm cells.

THYMUS GLAND. Part of the immune system through the development of T lymphocytes.

THYROID GLAND. Regulates metabolism through the secretion of thyroid hormone (TH).

VASOPRESSIN. Produced by the hypothalamus, maintains water balance. Also referred to as antidiuretic hormone (ADH).

HORMONES

CHEMICAL COMMUNICATORS IN THE BODY

ANYONE WHO HAS had the experience of premenstrual syndrome (PMS) or sexual desire can recognize the effects of hormones on the body. The word *hormone* comes from the Greek word meaning "to excite"—and that is exactly what hormones do in the body. They are chemicals excreted at key locations that travel through the bloodstream and stimulate certain cells to cause, control, and monitor the necessary functions that keep us alive, well, and growing.

Hormones trigger many different events in the body. They initiate and control the important processes of growth and life, including reproduction, development, defenses against disease, metabolic balance, and energy regulation. When a cell spews out a hormone, it does so in order to regulate some aspect of body metabolism. The hormone's effect often happens significantly later and in a place far from the body location where it originated.

While some hormones can appear similar in structure, such as the two sex hormones, estrogen and testosterone, hormones are distinctive from each other. They span a broad range of molecular sizes and complexities. Each hormone is as unique as the physiological message it carries.

AMINO ACIDS

Scientists group hormones into two basic categories, one proteinlike and the other fat-based. Proteinlike hormones arise from the building blocks of proteins, called amino acids, hence their name: amino acid hormones. They tend to be the smallest ones, although they do range widely in size. For example, thyroid hormone, which regulates metabolism, is built from parts of two molecules joined together.

Strings of several amino acids join to form hormones called peptides. Longer chains, such as the hormone insulin, are called proteins. All amino acid hormones—

The thyroid gland produces calcitonin, shown here in crystallized form, which regulates calcium levels.

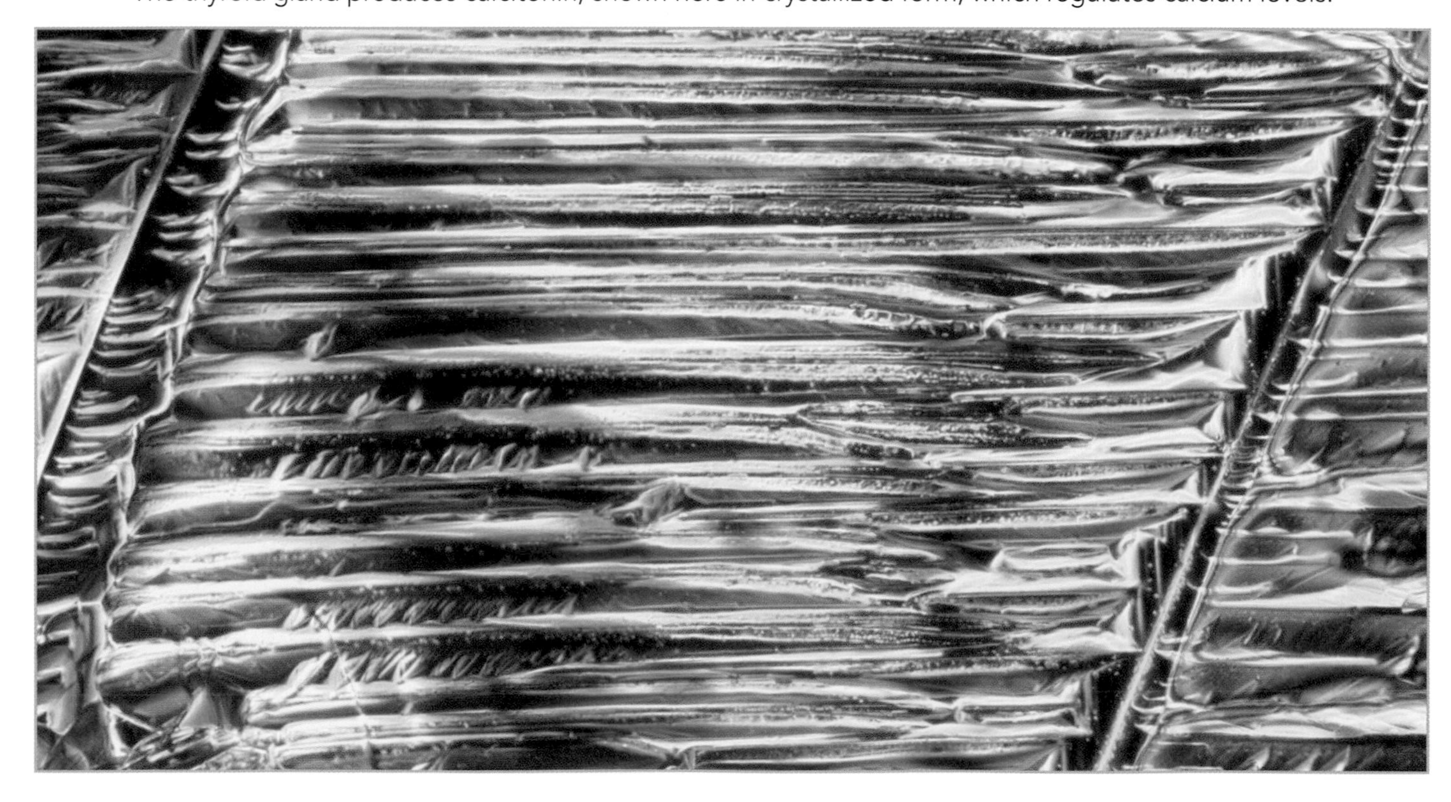

amines, peptides, and proteins—are able to dissolve in water. They are flexible molecules: Some can bend, turn, and twist into a variety of unique shapes. Their specialized structures allows them to perform their very specific functions in target cells.

Unlike the amino acid–based hormones, steroids are fatlike. Examples include the adrenocortical and the gonadal hormones, such as cortisol, estrogen, and testosterone. These steroid hormoncs arc synthesized from a fatty substance called cholesterol.

They are fat-soluble, but they cannot dissolve in water or in the plasma of the bloodstream. These substances are carried through the bloodstream on transport proteins and can slide in and out of oily cellular membranes more easily than some of their water-soluble amino acid counterparts.

"TO EXCITE"

THE WORD *HORMONE* derives from the Greek word *hormon*, "to excite" or "to urge on." Hippocrates used the word to mean "a vital principle." In the early 20th century, Sir Earnest Starling, discoverer of the first hormone, established the modern medical use of the word. As the story goes, he dined with a poet the night before giving a historic speech to the Royal Academy of Physicians on body chemistry. Their dinner conversation resulted in his creation of the new term.

BERTHOLD AND HIS ROOSTERS

Castrated roosters, or capons, do not develop bright red combs like intact males.

IN 1849, German professor Arnold Berthold began chasing roosters. A pioneer in the new field of behavioral endocrinology, Berthold wanted a model for aggressive behavior and how it might be linked to other acts, such as crowing, as well as to details of physical development, such as the growth of a rooster's comb. Berthold hypothesized that because these physical and behavioral traits occurred only in males, they must arise from some exclusively male substance in the body.

He boldly tested his theory by castrating juvenile roosters. True to hypothesis, the unfortunate capons failed to develop proper combs, no longer displayed aggressive behavior, and no longer either crowed or mated. But, good news for the birds, Berthold discovered he could reverse the emasculating effects by implanting normal testes back into the eunuch birds. In fact, after surgery, the remasculated cocksters sprouted male plumage, turned aggressive, crowed, and began chasing hens again with great zeal.

As Berthold reflected on the series of procedures his roosters had undergone, he realized that he must have damaged the nerves connecting the testes to the rest of the body. He reasoned that therefore the nerves could not be the conductor of all these very male characteristics. Instead, Berthold surmised, the testes must secrete an unknown substance into the bloodstream.

Today we know that substance as androgens—sex hormones such as testosterone. Berthold's contemporaries scoffed at his conclusions, but male steroids are a recognized factor in physiology and behavior today. ■

SEE ALSO: Chapter One, "The Cell," PAGE 14, Chapter Twelve, "Body Attacks," PAGE 342

KEEPING HEALTHY

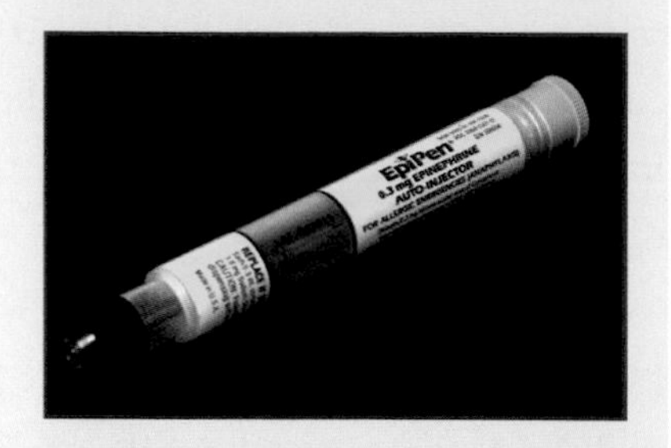

Epinephrine injections can save lives.

ANAPHYLAXIS can be a death sentence. By definition, it is an allergic response that hits multiple systems of the body when it comes into contact with an allergen (like bee venom, certain foods, or medications). Although normally short-term, severe anaphylaxis can become anaphylactic shock, which can lead to respiratory failure, cardiovascular collapse, and, if left untreated, organ failure. Therefore, quick treatment is potentially life-saving.

Out of this emergency, doctors learned that epinephrine, a hormone that naturally occurs as part of the body's stress response, can engender temporary relief of anaphylaxic symptoms. Meridian Medical Technologies packaged the hormone into an ingenious device called an EpiPen, an epinephrine autoinjector. Other pharmaceutical companies have followed suit with their own versions (for example, Twinject, manufactured by Versus Pharmaceuticals). People with severe allergies can carry the pen anywhere to deliver a quick shot. ■

COMMUNICATION

Hormones deliver chemical messages to cells. The cell, in turn, has to be able to receive the message. When a cell answers the signal, its activity is often altered as well. For example, during stress, the hormone epinephrine, produced by the adrenal glands, induces a powerful fight-or-flight response, stimulating the heart and the body's metabolic system. Biochemically, the interaction between a hormone and its receptor means that the hormone binds to smooth muscle cells in blood vessels and unleashes a programmed cascade of chemical reactions that cause the blood vessels of the heart and muscles to open, increasing blood flow to the tissues needed for the action of fighting or taking flight.

But that same hormone, when binding to smooth muscle cells in the stomach, causes their interwoven blood vessels to contract. No one wants blood flowing to the gut when facing a life-or-death situation. Best that it should stay where it's needed, in the muscles, when it's time for a fight. So the question is, how does the same hormone perform two opposing tasks in two different places in the body?

TARGETING

The answer is to be found in the receptors, the receivers that bind to the original hormone signal. Each receptor is linked up to its own biochemical network. That network of molecules, like relay runners, hands off the message, molecule to molecule, and eventually shuttles it to the cell's nucleus, which receives a chemical command to act.

In general, the binding of a hormone to a receptor works like a key in a lock. The shape of both molecules is critical. A cell with the wrong-shaped receptor simply won't bind to an active hormone. That cell won't react, while one with the right receptor will.

Most peptide and protein hormones cannot directly penetrate a

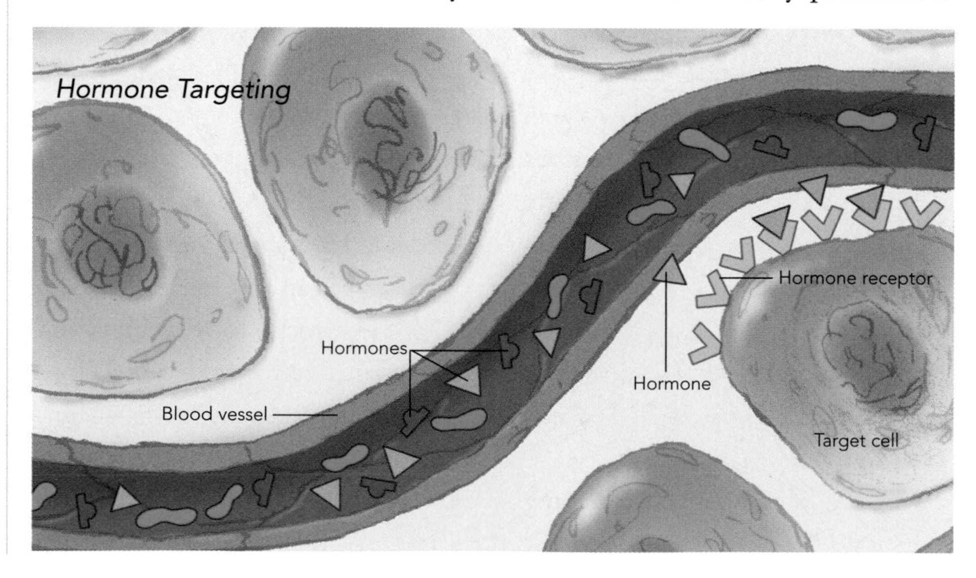

cell. They are water-soluble, while cell membranes are fatty. Therefore these hormones enlist the help of "second messengers," molecules embedded in the target cell.

Generally, a peptide or protein hormone binds to its receptor partner on the cell membrane. In essence, that binding squeezes the receptor and changes its shape. This shape change activates a signal transducer, often called a "G protein." Activation is akin to turning on a molecular switch. A G protein switch initiates the activity of a second messenger, called cyclic AMP (cAMP). This potent molecule, tiny and fast, can move almost anywhere in the cell and trigger cascades of reactions in a cornucopia of body processes.

By using the same messenger, cAMP, to broker so many diverse reactions, the endocrine system is resourceful. It even commissions second messengers to relay to third messengers, in fact. For example, oxytocin, the orgasm-generating hormone, binds to its receptor and activates a second messenger, which then releases calcium as a third messenger. As in the case of orgasm, second and third messengers amplify some reactions. Second messengers also take an ongoing, but aggressive initial reaction (say, the fight-or-flight response) and later, when the stress is gone, dim the reaction so that the body can maintain homeostasis.

Amino acid hormones are not the only players. Adding to the repertoire, steroid and thyroid hormones can diffuse directly into target cells. Testosterone, progesterone, and estrogen all bind to receptors located within the target cells. Activated hormone and receptor cells form a large complex that ambles to the nucleus. There, the complex binds to other proteins, strung along a cell's DNA.

Since DNA is the component of genes, such binding action can turn on a different kind of switch: It can turn on a gene. It will make a copy of its genetic instructions and send that copy back out to the cytoplasm. At that, the cell will make a protein, based on the genetic specifications it has received. That protein, often an enzyme, causes a specific cell action, such as the stress response of cells in reaction to the steroid hormone cortisol.

SMART RECEPTORS

The endocrine system controls a myriad of processes by doling out receptors strategically and placing them in cells where they are needed the most. Most cells of the body bear receptors for thyroid hormones, for example, which stimulate cellular metabolism. All cells need energy to work, and consequently nearly all body cells have thyroid hormone receptors.

Another tactic for control is to raise or lower the levels of a hormone circulating in the blood. Concentration can be a trigger to action. Still another move is to change the number of receptors on a target cell or the strength of hormonal triggers. Depending on circumstances, cells can actually make more receptors in a process called up-regulation. During development, for example, an embryo's primitive heart cells might up-regulate receptors that control new blood vessel growth.

At the same time, the endocrine system can also down-regulate receptors. When people become obese, it has been observed that

VITAMIN D

VITAMIN D GAINED FAME when 20th-century scientists discovered that its deficiency caused bone-crippling diseases. The treatment: Supply vitamin D in the form of cod liver oil. With that, vitamin D gained the status of essential nutrient—something the body needs but cannot synthesize. Scientists discovered that vitamin D is actually a prohormone, the precursor to a hormone. Sunlight on the skin triggers the body to synthesize vitamin D and then converts it into a hormone.

their brain cells will often down-regulate the appetite-dampening hormone called leptin. When food is plenty, and fat stores increase, leptin levels in the bloodstream rise, shutting down appetite. But after too much time at such a high level, the cells of the body tend to become desensitized, and they stop overreacting to a prolonged response. Hence people who are obese may have high leptin levels in their blood, but their bodies have lost the ability to respond appropriately to the appetite-decreasing hormone.

All these control mechanisms, from hormone type to second messenger system to receptor dynamics, compose the diverse showcase of the endocrine system. Control means balance, and that is the job of hormonal flow: maintaining a continual, rhythmic balance through the body's everyday functions.

Powerful attractions are driven by a series of hormones, including testosterone, estrogen, norepinephrine, and oxytocin.

TRANSPORT

Hormones are messengers. They all must travel to reach their targets and convey their messages. But given their widely different functions, they all move at different rates and in reaction to different conditions. The full system of hormonal deployment and transport is elaborate and versatile.

For proteins and other water-soluble hormones, transport is straightforward. These hormones can dissolve in the blood plasma and therefore can float free, unattached to any other molecules. Lipid-soluble hormones pose a different challenge, however, for they do not dissolve in blood plasma. The body has devised a cadre of specific transport proteins, each designed to transport its own steroid or fatty cargo.

Transport proteins also slow the passage of small hormones through the kidney filter, which allows the tiny molecules to stay in the body longer, acting for longer periods of time in the body's various processes. Transport proteins, in essence, sequester and store the hormones, keeping them in ready reserve, available for quick action.

When hormonal action must end, the body has a way to eliminate outdated hormones. The endocrine system has an arsenal of enzymes that chew up unnecessary hormones. The kidneys and liver step in as clean-up crew as well, calling in aging hormones and their breakdown products, targeting both for necessary excretion.

COOPERATION

Hormones are by no means solitary players. Often they work in pairs or relay teams to encourage, build upon, or oppose each other's actions. For example, reproductive hormones spur the development of gonads and other sex-related structures, but reproductive hormones cannot do their job without the action of thyroid hormones. Thyroid hormones allow the reproductive hormones to act, but at the right time, at the advent of puberty.

In another scenario, glucagon, a hormone from the pancreas, acts with epinephrine to prod the liver to release glucose, a sugar, into the bloodstream. The effect of both, acting together, is 150 percent more potent than either hormone acting alone. This effect, a good example of the whole being greater than the sum of its parts, is called synergism. It works to create a big impact in a short time as is needed when the body requires fuel, say, during exercise.

On the other hand, insulin acts to oppose glucagon's effect of raising the body's blood sugar, the measure of glucose in the blood. Called antagonism, this opposition provides the endocrine system with its signature checks in the check-and-balance system.

Hormones are not the only substances that act as checks to the system. Ions, or nutrients, can also rein in hormonal action. For example, calcium in the bloodstream bathes cells of the parathyroid gland. When calcium levels plummet, the parathyroid secretes parathyroid hormone, which works to reverse a low-calcium state in the body.

CONTROL

Hormones can also inhibit their own release in a process known as "negative feedback regulation." In essence, the end product bathes the gland that is producing it, which causes that gland to shut down and halt any further secretion—a feedback process that can happen directly or indirectly.

The endocrine system is stocked full of layers. Glands secrete hormones that target other glands, which in turn produce hormones or other substances to perform a body process. One example of this is the three-layered system involving the hypothalamus, the pituitary gland, and the gonads (the ovaries in women and testes in men). First, the hypothalamus produces a hormone called gonadotropin-releasing hormone. It triggers the pituitary gland to release two other hormones, FSH (follicle-stimulating hormone) and LH (luteinizing hormone). Next, these two chemicals flow into the bloodstream, down toward the gonads. There, in response, the ovaries begin to produce estrogen, and the testes start to produce testosterone. Not only do the sex steroids affect tissues directly; the sex hormone level in the bloodstream also rises, and the hormone flows back to the pituitary, where it halts the production of FSH and LH.

MAJOR HORMONES

HORMONE	ORGAN	TARGET	EFFECT
Growth hormone	Pituitary gland	Liver, Muscle, Bone, Cartilage	Stimulates growth
Thyroid hormone	Thyroid gland	Body cells	Maintains metabolism; regulates blood pressure; promotes normal development of the nervous system, muscles, and skeleton; regulates hydration of skin
Cortisol	Adrenal gland	Body cells	Assists body in resisting stress; mobilizes fats for energy; depresses inflammatory response
Insulin	Pancreas	Body cells	Lowers blood sugar levels
Gastrin	Stomach	Stomach	Secreted in response to food; stimulates production of hydrochloric acid to break down food
Erythropoietin	Kidney	Bone marrow	Stimulates production of red blood cells
Leptin	Adipose tissue	Brain	Suppresses appetite

PRODUCERS

ORGANS OF THE ENDOCRINE SYSTEM

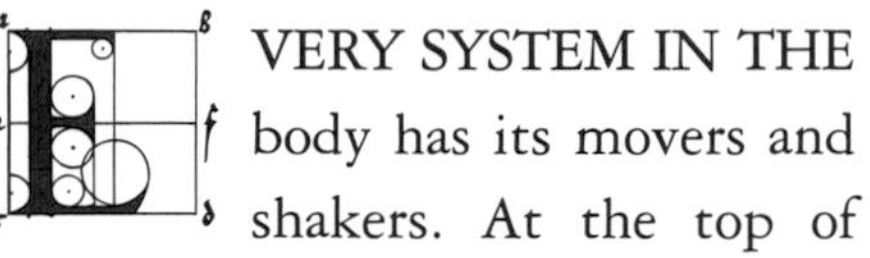

EVERY SYSTEM IN THE body has its movers and shakers. At the top of the chain in the endocrine system sits the hypothalamus, a region of the brain that controls the main glands of the endocrine system: the pituitary, thyroid, parathyroid, adrenal, and thymus.

HYPOTHALAMUS

The hypothalamus is a region of the brain and not an organ on its own. As part of a brain's tissue, it is wired with neurons and has neural functions. But, at the same time, the hypothalamus produces and releases hormones. Therefore it is considered a neuroendocrine organ. The hypothalamus takes its name from its position in the brain, below the portion of the brain called the thalamus.

Deep within, the hypothalamus is embedded at the base of brain stem, ready to perform its role in the body. As leader and guide, the hypothalamus produces a wide variety of hormones. Some stimulate action—such as growth—while others inhibit the action of these same targets.

Why the opposing action? The two hormones illustrate the yin-yang genius of the endocrine system. Once a body response gets going—the growth of a child's bones, for example—there will come a time when that action has to stop—once the child's body reaches adulthood. The hypothalamus is a master of balance, keeping the body in tune and on time. And, as a bonus, the hypothalamus can work in two ways: first, by stimulating target cells directly, as in the case of oxytocin, a hormone that causes the contractions of uterus in childbirth, and second, by activating other endocrine glands, which in turn excrete more hormones that stream toward target cells.

PITUITARY

One of those secondary glands sits right next to the hypothalamus, joined by a stalk of nerves and blood vessels. The pea-shaped pituitary gland is a master in its own right. It is double-lobed, no accident of evolution. The lobe in the back, the posterior pituitary, contains neural tissue. This portion of the gland is therefore dubbed the neurohypophysis. It was born out of brain tissue, which explains its neural nature. This lobe connects directly to the hypothalamus via nerve appendages called axons.

The front-facing lobe has a more glandular role. Called the adenohypophysis, this lobe secretes six potent hormones, each with a widely different function in the body. The adenohypophysis emerges from embryonic oral tissue.

The adenohypophysis over time adheres to the neurohypophysis, separating from its embryonic oral contact. The nervous part of the gland receives chemicals from the hypothalamus, the hormones streaming down the axons for eventual storage in the pituitary gland. The glandular part, while not connected by nerves to the hypothalamus, gets signals

SEE ALSO: Chapter Nine, "The Brain," PAGE 234

WHAT CAN GO WRONG

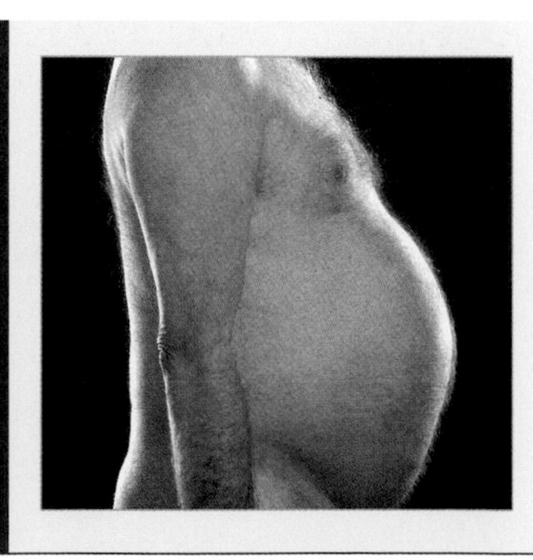

CUSHING'S SYNDROME: When the body produces too many glucocorticoids, the result is characterized by a hump on the back of the neck, a swollen moon-face, and abnormal accumulation of fat around the abdomen, left. Other symptoms include high blood sugar, a drain of muscle and bone protein, and slow wound healing. Caused by high doses of glucocorticoid drugs or by a tumor, the disease unleashes an excess of stress hormones. The body mounts an overzealous response, which can lead to weakened muscles and fractured bones. Stopping the drugs or removing the tumor is the only treatment.

from hypothalamic hormones that travel through the blood vessels that bathe the cells of the lobe.

Both the hypothalamus and the pituitary gland rely on an elaborate system of checks and balances. The hypothalamus begins the job by producing stimulatory hormones such as corticotropin-releasing hormone (CRH), which acts on the pituitary gland to produce and release adrenocorticotropic hormone (ACTH, also called corticotropin).

The ACTH travels to the adrenal glands, which respond by producing cortisol, among other hormones, which acts in inflammatory and stress responses. Circulating cortisol flows back to the hypothalamus and, at the right concentration, shuts down further production of CRH and ACTH.

This loop is called "feedback inhibition." It can affect not only initial hormone production but also the signaling of neurotransmitters in the brain, as well as the manufacture of potent immune-boosting chemicals called cytokines elsewhere in the body.

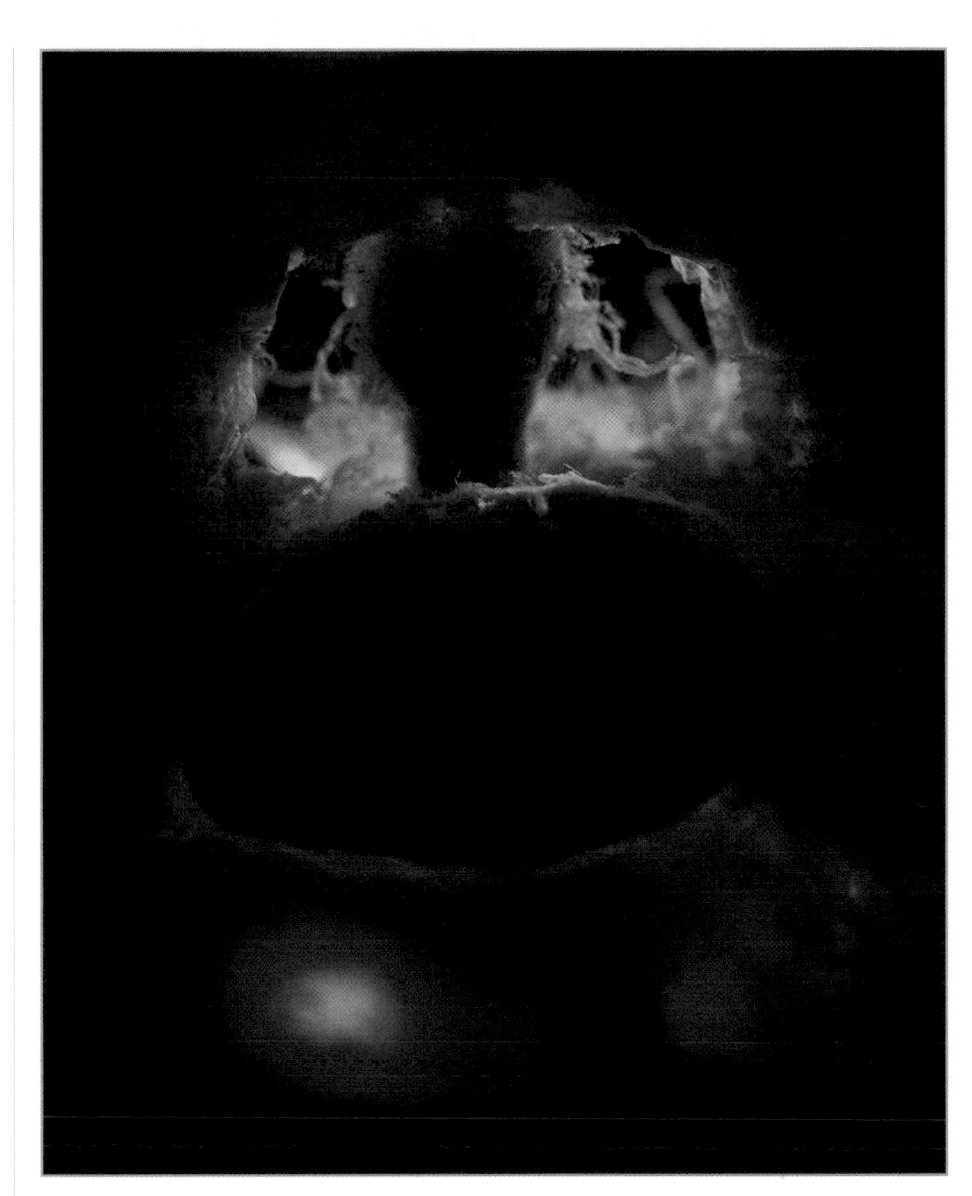

Located in the brain, the pituitary gland is joined to the hypothalamus by a short stalk of nerves and blood vessels.

GROWTH HORMONE

Discussed and debated in the news of late is growth hormone, a molecule that stimulates body cells to grow and reproduce. Bones and muscles reap the most benefit. Cells with receptors specific for the hormone respond by enlarging and proliferating, especially during the period of childhood. Growth hormone is an anabolic or tissue-building hormone.

Besides fashioning bones, this hormone also mobilizes fats from stores in the body and breaks them down into fatty acids for fuel. The action is meant to conserve glucose as an energy source. And, in fact, growth hormone affects gluconeogenesis, the formation of glucose from amino acids, and it can have the effect of adding more sugar to the blood. The effects of too much growth hormone can mimic the symptoms of diabetes mellitus, a condition in which blood sugar levels rise and remain high.

The system needs to be in balance. Too much growth hormone causes gigantism, where a person continues to grow, sometimes up to and beyond eight feet in height. If growth plates of bones are still open and susceptible to growth hormone (GH) action, body proportions remain normal. But if a child has grown and his or her growth plates closed, the result is a condition known as acromegaly, in which a person has enlarged and thickened bones of the hands, feet, and face, especially the jaw.

Too little growth hormone causes no such extreme problems in adults, but it can be associated with loss of muscle and bone mass and unusual accumulation of fat. In children, the lack causes long bones to grow too slowly, resulting in pituitary dwarfism.

Problems with growth hormone deficiencies extend beyond lack of the hormone itself. Growth hormone often exerts its effects by a proxy: insulin-like growth factor 1 (IGF-1). IGF-1 is a protein hormone much like insulin, stimulated by growth hormone. It reaches nearly every cell in the body, and especially cells of muscle, cartilage, bone, liver, kidney, nerves, skin, and lungs.

Approximately 98 percent of IGF-1 is always bound to one of six transport proteins. If any one of these is lacking, though, IGF-1 becomes impotent, resulting in growth failure and certain kinds of dwarfism.

GONADOTROPINS

When it comes to reproduction, the hypothalamus and pituitary work as a winning relay team. The first player is gonadotropin-releasing hormone, produced by the hypothalamus. The hormone nudges the pituitary to secrete two so-called gonadotropins: follicle stimulating hormone (FSH) and luteinizing hormone (LH). These, in turn, regulate the function of the ovaries and testes. Depending on the body's gender, FSH works to stir either egg or sperm production. The hormone also coaxes the maturation of eggs and follicles, the cells surrounding a developing egg, in the ovaries of females.

LH works on the ovaries or the testes and prods them to produce their own hormones, estrogen or testosterone. LH action is special because the pituitary sends it out in pulses. Although no one knows exactly why, researchers have discovered that the staccato rhythm of LH production is crucial to proper ovulation and the synthesis of ovarian hormones.

Of note, prepubertal boys and girls have virtually no gonadotropins in their bloodstreams. Only when puberty dawns does the pituitary animate and begin the reproductive relay. Later, hormones such as testosterone and estrogen feed back to the hypothalamus to suppress FSH and LH release. In this way reproduction, the most crucial system to the survival of the species, remains in balance.

The hypothalamus and pituitary gland (both shown in green) sit squarely in the middle of this scan of the human brain.

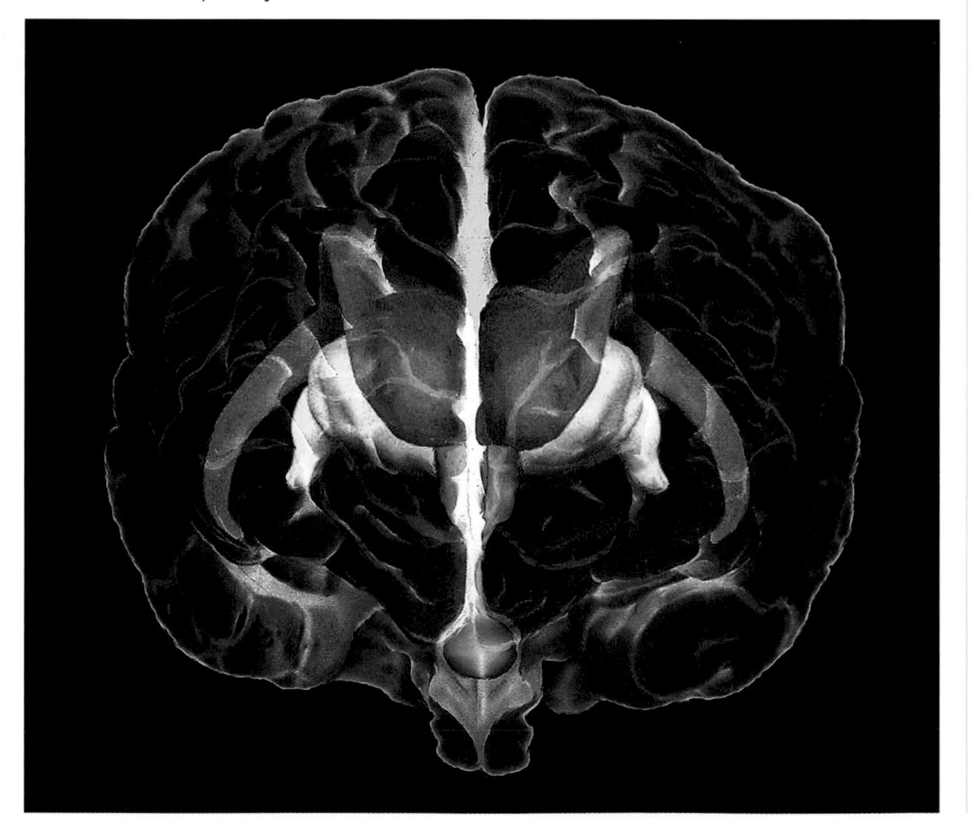

PROLACTIN

This protein hormone is known as the trigger for milk production

SEE ALSO: Chapter Twelve, "Male System," PAGE 294, "Female System," PAGE 300 in the breasts after birth. Despite its brief cameo role in any single woman's lifetime, it is crucial to survival of the species as a whole.

The hypothalamus makes a releasing factor to stimulate prolactin production via the pituitary. Oddly, scientists still do not know this releasing factor's true identity. They suspect that the neurotransmitter serotonin causes prolactin synthesis and release. On the other hand, scientists do know the opponent of this releasing factor: dopamine. This inhibitory hormone prevents prolactin secretion.

Male bodies produce a prolactin-inhibitory hormone in slightly larger amounts than female, although the differences diminish once a woman has had children. The level of its production rises and falls with the tide of monthly estrogen fluctuations in a woman's bloodstream. Low estrogen means

HORMONE FACTS

+ GROWTH HORMONE promotes growth of muscle and bone. It also controls protein, fat, and carbohydrate metabolism.
+ Thyroid-stimulating hormone affects metabolism rates.
+ Adrenocorticotropic hormone influences the stress response.
+ Prolactin causes the mammary glands to produce milk.
+ Antidiuretic hormone controls conservation of body water.
+ Oxytocin stimulates uterine contractions and orgasm.

WORLD'S TALLEST MAN

At age 13, Robert Wadlow stood over 7 feet tall and wore size 25 shoes.

ROBERT WADLOW went down—or up—in history as the world's tallest man, reaching a height of 8 feet, 11.1 inches. A tumor in his pituitary gland caused it to produce far too much growth hormone.

Born in 1918, Wadlow grew at a normal rate up to the age of 4. Once his first growth spurt started, it never abated. He stood 6 feet tall at the age of 8. By 17, he had surpassed 8 feet. His feet reached size 30, and his hands measured more than a foot long. Scientists believe that Wadlow was still growing when he died.

His height caused medical complications. He could not walk without leg braces; he lost sensation in his feet. He may have suffered from heart problems. He died in 1940, at the age of 22, from an infected blister caused by a faulty leg brace. ■

high prolactin-inhibitory hormone release; high estrogen means high prolactin-releasing hormone levels. Women experience the ebb and flow of estrogen in their bodies as breast tenderness and abdominal swelling before their menstrual periods. Prolactin levels are surging at this time, but only briefly.

With pregnancy, however, everything changes. Prolactin levels rise dramatically at the end of pregnancy, as a woman's body begins to produce milk to feed the coming newborn. Suckling triggers prolactin release further, along with encouraging ongoing milk production.

Pituitary gland tumors can bring on results that include the secretion of too much prolactin. Other symptoms include inappropriate lactation, lack of menses, infertility in females, and erectile dysfunction in males.

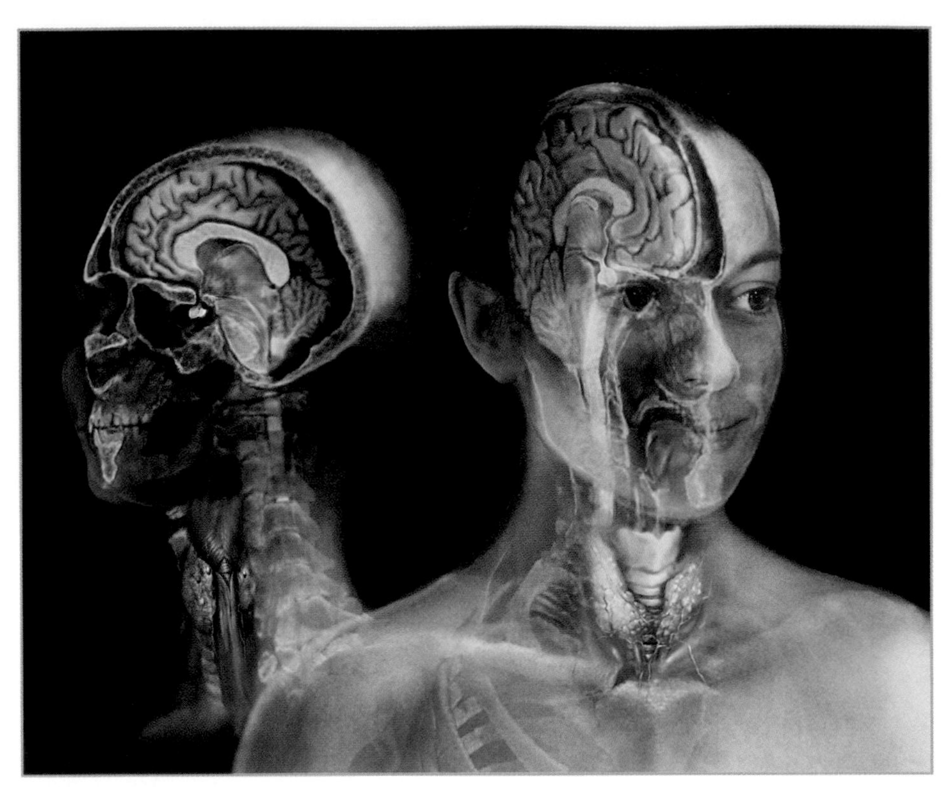

Electronic scanning created this 3-D depiction highlighting the hypothalamus, pituitary gland, thyroid gland, and parathyroid glands.

OXYTOCIN

Of all the hormones in the body, none more than oxytocin can be dubbed the "love and cuddle" hormone. It is a substance associated with explosive spasming. During childbirth, oxytocin mediates the contractions of the uterus that push the baby out of the mother's body. During breastfeeding, oxytocin targets special cells surrounding the mammary glands that squeeze out milk, commonly termed the "letdown reflex" experienced by nursing mothers. For both men and women, oxytocin stimulates sexual arousal and orgasm.

The hormone's production begins with neural triggering of the hypothalamus. That connection in the brain also explains the link of each of these processes —childbirth, nursing and sexual climax—to intense emotions. The hypothalamus is intricately linked to the limbic system of the brain, the site of pleasure, pain, and rage. Thus, oxytocin is a wonderful kind

BREAKTHROUGH

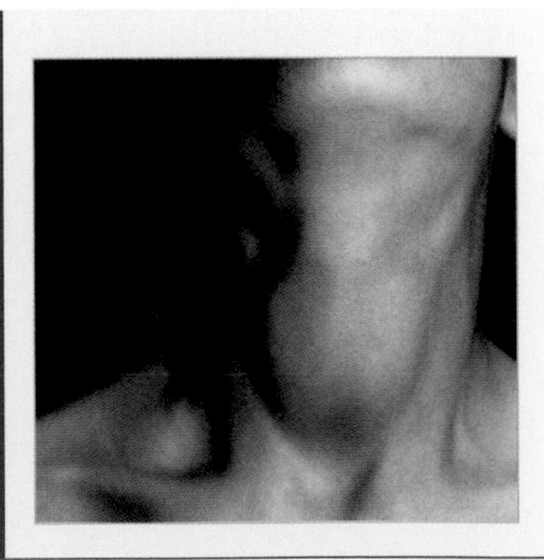

GOITER, A DISEASE known by abnormal swelling in the neck, left, drew attention during World War I when soldiers drafted from the Great Lakes region started showing up for active duty with its symptoms. Doctors knew that goiter arose from enlarged thyroid glands, but they did not understand why a cluster of cases would come from one geographic region. Eventually investigators linked hypothyroidism to a lack of iodine in the diet. The iodine content of Midwestern soil was low. Saltwater fish, another natural source of iodine, was not in ready supply. So these men lacked the iodine needed to

SEE ALSO: Chapter Eight, "The Kidney," PAGE 212

of Cupid inside the body, stirring feelings of satisfaction, of nurturing, and of affection as these body processes unfold.

ANTIDIURETICS

Diuresis means urine production, hence the antidiuretic hormone (ADH) stops the process of adding water to urine. Produced directly by the hypothalamus, this hormone tempers the water balance in the body, preventing wide swings like dehydration to bloating. A high level of ADH causes blood vessels to contract—a characteristic that gives the hormone its other name: vasopressin.

The keys to this system are the osmoreceptors in the hypothalamus: structures that can sense the level of dilution or concentration in a body fluid. The receptors' job is to monitor the level of electrolytes (solutes) and water in the blood. If solutes, such as salt, are at levels that are too high—after excessive sweating, for example—the osmoreceptors send off impulses to the hypothalamus to trigger the synthesis and release of ADH.

The pituitary stores and releases the antidiuretic hormone into the bloodstream, sending it toward its target, the kidney tubules. They remove water from the urine as it is forming and return it to the blood. On the other hand, if the solute concentration drops, the osmoreceptors stop firing, effectively ending ADH release.

Many people have experienced ADH action firsthand after a night of drinking. Alcohol inhibits ADH secretion, generating lots of urine output. The next day, the body experiences dehydration and the dry mouth of a hangover.

DISCOVERY

IN 1880, SWEDISH medical student Ivar Sandstrom discovered the parathyroid gland—but to little fanfare. He displayed data from extensive dissections in several species and reported the gland's anatomical position, but no one believed that a mere student could find a heretofore unknown organ. Sandstrom even predicted that the gland had an endocrine function, but only later did the discovery gain acceptance.

Drugs called diuretics actually encourage this same process of ADH inhibition. They cause water to flush from the body. This can be a necessary step in the treatment of hypertension and congestive heart failure, conditions in which excess water accumulates in the tissue and must be eliminated.

Individuals with eating disorders often take diuretics in a misguided attempt at weight loss. The practice is dangerous because it often ends in dehydration, electrolyte imbalance, and sometimes even death.

ADRENAL HORMONES

The adrenocorticotropic hormone (ACTH) is a polypeptide hormone that targets the adrenal gland to boost the synthesis of corticosteroids (glucocorticoids and sex steroids). One glucocorticosteroid called cortisol has been fashioned into a drug, cortisone, used in the treatment of rheumatoid arthritis because of its anti-inflammatory effects (see "Hormone Manufacturer," page 281).

The entire process begins in the hypothalamus with the produc-

BREAKTHROUGH

keep their thyroids functioning properly. That breakthrough in understanding led to a new treatment for the prevention of goiter: iodide supplements. Only about 150 mg of iodide keeps the thyroid functioning properly, but people worldwide often lack even that little. More than 8 million people worldwide suffer from complications arising from iodide deficiency. In regions of South America, Southeast and Southern Asia, and Africa, where malnutrition is prevalent, the condition is seen quite often. In the United States, goiter is generally avoided because iodide is customarily added to commercial table salt, right.

tion of corticotropin-releasing hormone (CRH). The pituitary also secretes a related molecule called proopiomelanocortin (POMC), a prohormone, in that it is a precursor cut up by enzymes in the body into several other hormones. Among them are two opiates, enkephalin and beta endorphin, known for their feel-good effects. Both target the μ-opioid receptor in the brain, the same receptor that responds to chemicals derived from opium, such as morphine and codeine. All these substances provide relief from pain; the drugs, though, bring the risk of addiction along with pain relief.

One other progeny of POMC is melanocyte-stimulating hormone (MSH), which stimulates special cells in the skin and hair to produce and release the pigment melanin. Coincidentally, MSH also acts as a neurotransmitter in the brain with profound effects on appetite and sexual arousal. That dual role helped in POMC's discovery, since mice defective in it grew orange hair and obese bodies.

THYROID & PARATHYROID

At the front of the throat, underneath the voice box, sits one of the largest and most important glands in the endocrine system: the thyroid gland. It controls the essential process of metabolism, by which the body converts food into energy. Therefore this 10- to 20-gram, butterfly-shaped gland affects every cell and most systems of the body, essential in processes ranging from the development of the fetal and neonatal brain to the regulation of many functions in the adult body: temperature, protein synthesis, growth, heart rate, fertility, and central nervous system matters.

Smaller glands known as parathyroid glands attached to the thyroid lobes help to control blood calcium and phosphate levels. Follicles within the thyroid gland encircle a protein-rich material called colloid, which absorbs and stores iodine ions from the body.

Iodine is critical to the process of synthesizing the hormones thyroxine (T_4) and triiodothyronine (T_3). In fact, it is so important to proper thyroid functioning, the body stores 25 percent of its iodine supplies in the follicles of the thyroid gland (see "Breakthough," page 278).

Thyroid hormones rely on an intricate system of control that adjusts itself continually to maintain balance. Deficient production of thyroid hormones can lead to infertility, chronic fatigue, and, in severe cases, mental retardation in infancy or coma in adulthood. Other symptoms of thyroid deficiency include an intolerance for cold, muscle pain and weakness, dry skin, a slow heart rate, and excessive sleepiness. Overproduction of these hormones can lead to

In the abdomen, the pancreas butts up against the kidneys, where the adrenal glands are perched.

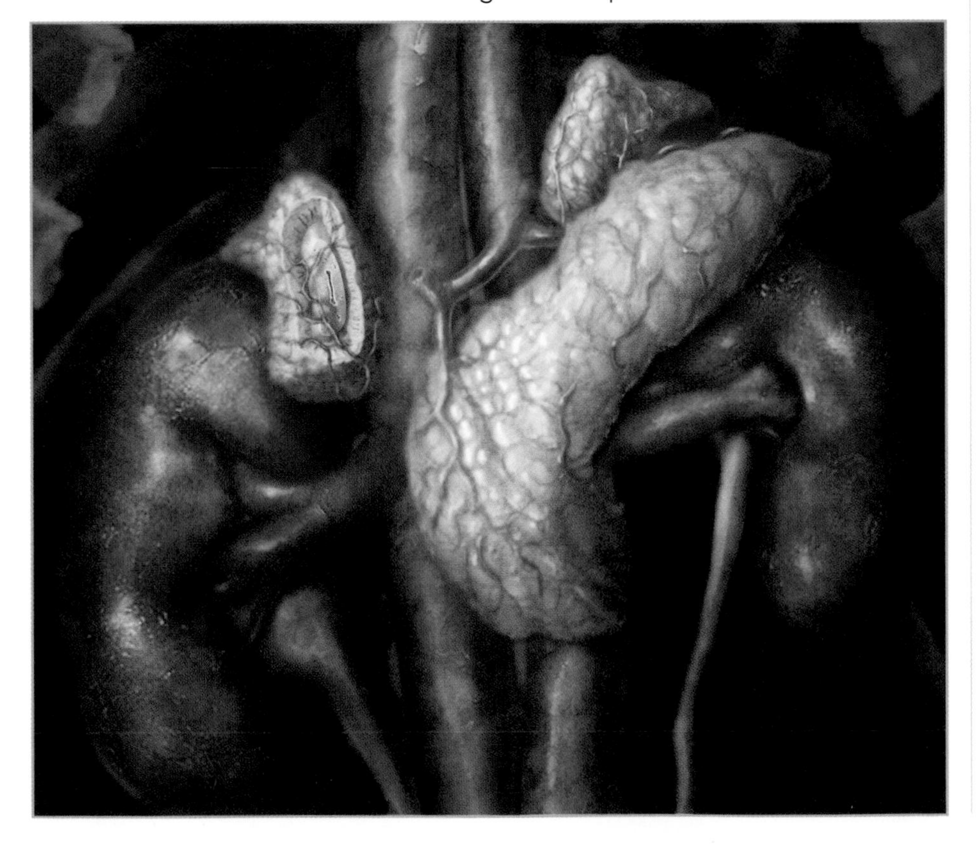

nervousness and anxiety, insomnia, and increased heart rate.

Thyroid hormones begin their journey at their site of synthesis, the epithelial cells that line the outer edge of the follicles. Thyroid hormones then enter the bloodstream, searching for their target cells. Carrier proteins ferry the potent hormones through the body via a complex vein and artery system to target cells within the body and brain, where carriers release their hormonal cargo and target cells take it in.

Once inside a cell, these hormones bind to intercellular receptors, proteins that act as signaling towers that relay messages to the cell's DNA. These messages trigger or inhibit the expression of particular genes.

While this system works well for delivering hormones to most cells of the body, the hormones used by cells of the brain face an additional challenge. They must cross the formidable blood-brain barrier, a wall of cells that acts as gatekeeper and restricts the passage of unknown or harmful substances into the brain.

One other important thyroid regulator is the hormone calcitonin, which stimulates the movement of calcium from the blood to the bones and helps prevent osteoporosis. Calcitonin also inhibits the activity of osteoclasts, cells that otherwise would deprive the bone of calcium and transfer it back into the blood. Calcitonin prevents the absorption of calcium

HORMONE MANUFACTURER

Percy Julian's method of synthesizing cortisone made the hormone widely available.

BEFORE THE TURN of the 20th century, doctors used cortisone, a vital adrenal hormone, to treat a variety of inflammatory ailments. It was an effective treatment, but hard to come by. Pioneering drug producers had to extract natural cortisone from the adrenal glands of oxen. Thus it was incredibly expensive, not to mention unpleasant, to formulate.

Then, in the 1950s, a breakthrough came. Percy Julian, a chemist with the Glidden Company in Chicago, started with soybean oil and synthesized a compound with cortisone-like properties. Julian had already successfully synthesized male and female hormones from soybeans. His success meant that he created a synthetic cortisone substitute cheap enough to help millions.

Percy Julian's life tells the story of a pioneering spirit. He was an African American born in 1899 in Birmingham, Alabama, growing up during the difficult Jim Crow era, when African Americans in the South rarely received an education beyond the eighth grade. But Julian, and his brothers and sisters, continued on to graduate degrees.

Percy Julian became the second African American to receive a doctoral degree and the first African-American chemist to be inducted into the National Academy of Sciences. He is generally considered a pioneer in the synthesis of drugs from plants. His work laid a foundation for the advancement of several other medical and chemical advancements, including the birth control pill. ■

by the tubules of the kidney as well, which causes more calcium to be excreted in the urine.

The parathyroid glands secrete an opposing hormone, parathyroid hormone (PTH), when blood calcium levels fall too low. PTH stimulates the removal of calcium from the bone and raises the blood calcium levels. It also increases the activity of osteoclasts and triggers absorption of calcium by the kidney tubules, which suck the calcium back into the blood. PTH also indirectly affects calcium absorption in the small intestine, introducing more calcium into the blood. The system works in yin-yang balance to maintain daily homeostasis.

ADRENALINE

A PERSON WHO seeks dangerous thrills earns the title "adrenaline junkie," but the adrenaline part of that colloquialism is a misnomer. The rush is actually the complex interplay of several hormones. Adrenaline and noradrenaline are released from the adrenal medulla, while the adrenal cortex secretes cortisol, aldosterone, and corticosterone.

ADRENAL GLANDS

Resting atop the kidneys, two wedge-shaped glands, the adrenals, control the stress response, one of the body's life-or-death reactions. When faced with danger, hunger, or any other threat to survival, the adrenal glands respond, sending the body into its mode of fight-or-flight response.

Anatomically, the adrenal system seems small, almost inconsequential. The adrenal glands are dwarfed by the kidney, yet they interact with it in one of the most beautiful orchestrations of two body systems. The adrenals also coordinate with the nervous system, which allows quick reactions to mental and physical stress.

The adrenal glands contain two parts, the cortex and the medulla. The cortex, composed of three distinct layers of cells arranged in concentric layers, produces a critical stress hormone called cortisol. The inner medulla, or the bull's eye, synthesizes and secretes epinephrine and norepinephrine, more commonly known as adrenaline and noradrenaline. These hormones work together and engender a repertoire of responses that keep the body in balance.

Cholesterol, the foundation for several adrenal hormones, grouped as "steroid hormones," is critical to adrenal functioning. It can be absorbed from the diet or synthesized within the body from precursor metabolites, small molecules produced during the metabolism of food. Once formed, cholesterol begins an odyssey toward its eventual targets: the middle and inner rings of cortex cells.

As the journey begins, cholesterol travels with transport proteins that shuttle it across the cell membrane. Because cholesterol cannot cross from blood to cell interior on its own, a single defect in the transport protein can block or entirely shut down the function of the adrenal cortex.

Inside the cell, cholesterol must cross through another barrier: the membrane of the mitochondria, tiny organelles that function as power plants and provide the energy needed for all cell processes. In the mitochondria, various enzymes convert cholesterol into a variety of essential hormones, including glucocorticoids.

WHAT CAN GO WRONG

IN 1855, THE BRITISH PHYSICIAN Thomas Addison witnessed a strange, debilitating syndrome. Patients who presented chronic fatigue, nausea, and muscle weakness also showed, even more oddly, changed skin pigmentation. For five years Addison worked diligently, describing the disease. Then, tragically, Addison committed suicide, the culmination of years of chronic depression. His legacy did live on, however. Researchers named the illness after the pioneer who first identified it and learned that Addison's disease stems from a failure of the adrenal glands to produce enough of the stress hormone cortisol. In some

SEE ALSO: Chapter Eight, "The Kidney," PAGE 212

Skydiving induces thrills and excitement, emotions enhanced by hormones from the adrenal glands.

Of these steroid hormones, cortisol is the most renowned.

Cortisol clearly provides a demonstration of how the body can closely integrate the nervous and the endocrine systems. When either mental or physical stress occurs, the body alerts the hypothalamus, located in the brain. In response, the hypothalamus secretes a corticotropin-releasing hormone, which in turn stimulates the release of the adrenocorticotropic hormone (ACTH) from the anterior pituitary gland. ACTH travels through the bloodstream to the adrenal cortex, where it binds to surface receptors and exerts an influence on its cells. In response, they unleash a stream of cortisol.

Every cell in the body has a glucocorticoid receptor, so every cell in the body has the capability of reacting to stress, as signaled by cortisol. Like cholesterol, cortisol needs a transport protein to taxi the steroid hormone to the receptors of its target cells. Inside those cells, special molecules transport the hormone into the nucleus, where it signals

cases, the body also fails to produce the right balance of the hormones aldosterone and epinephrine. This failure leads to an array of symptoms, from low blood pressure and tremors to fatigue, dehydration, and a craving for salty foods. If left untreated, the disease can be fatal, but having the disease does not mean one cannot have an active life: President John F. Kennedy, opposite, had Addison's but managed to keep his condition a secret.

There are two varieties of the disease. The most common involves genetic variations combined with unidentified environmental factors, which leads to an autoimmune attack on the adrenal glands. A rarer scenario, affecting only men, is a single but powerful defect in one gene found on the X chromosome. Treatment involves replacing the cortisol, and, if necessary, aldosterone and epinephrine. Severe episodes require intravenous cortisol along with glucose and salts.

INSULIN DISCOVERY

Carbon, hydrogen, oxygen, nitrogen, and sulfur atoms make up the insulin molecule.

FOR CENTURIES, diabetes was a death sentence. People with the disease tragically wasted away, while physicians could only manage their symptoms. Doctors did know the disease arose from sugar imbalances and that the pancreas was involved. In fact, in the first two decades of the 20th century, researchers lowered blood sugar in test animals by giving them extracts from the pancreas—effective, but impure and toxic.

In 1921 Frederick G. Banting, a young orthopedic surgeon from Ontario, Canada, began to study the so-called islets of Langerhans, found in the pancreas. Researchers believed that these cell clusters secreted a mysterious substance that could regulate sugar metabolism. In a laboratory at the University of Toronto, Banting worked with a student assistant, Charles Best, to study the physiology of dogs that had been rendered diabetic by having their pancreases removed. Banting needed to find a way to isolate the substance made in the islets of Langerhans before the pancreatic juices had time to digest it. He found that by surgically tying off the ducts from the pancreas, he could leave the islets and their hormonal products intact.

He and Best succeeded in isolating an extract from the pancreas, a substance that reduced the high blood sugar in the diabetic dogs. Over time the procedure was modified, resulting in extracts that successfully treated humans with diabetes. In 1958 Banting received the Nobel Prize for discovering insulin. ■

the expression of a wide variety of genes. These genes catalog information that, when expressed through molecules such as RNA and proteins, stimulates the breakdown of fat for energy, the conservation of glucose through a process called gluconeogenesis, and the release of protein building blocks called amino acids from muscle tissue. All of these processes coordinate to regulate the proper levels of blood sugar, or glucose, and the metabolism of carbohydrates.

In addition to carbohydrate metabolism, glucocorticoids can block inflammation or suppress immune system functioning. Researchers have exploited this property by formulating medications derived from cortisol, also known as hydrocortisone, to treat inflammatory diseases, such as arthritis, as well as autoimmune disorders, such as lupus, in which an overzealous immune system attacks the body's own tissue.

While these responses take place in the inner or middle cortex,

+ PANCREAS FACTS +

+ A SMALL, FLAT organ, the pancreas is about 5–6 inches long.
+ The pancreas contains both exocrine and endocrine cells. Called acini, the exocrine cells make up 99 percent of the organ.
+ Groups of endocrine cells are clustered together in islets of Langerhans, named for the German physician who first described them in 1868.

cells in the outermost layer conduct yet another symphony of hormonal synthesis and body regulation. This layer synthesizes and secretes mineralocorticoids, hormones that regulate electrolytes—ions that carry an electrical charge and can act as conduits for molecules such as water, moving in and out of cells, thus regulating the flow of fluids in the body.

The most important mineralocorticoid is aldosterone. It travels through the bloodstream and targets the cells of the kidneys. Its action stimulates the absorption of water and sodium ions and the excretion of potassium ions.

The kidneys constantly filter the body's total blood supply. By regulating the concentration of the sodium and potassium ions, they can control blood pressure and volume. Attesting to this system's importance, a body devoid of mineralocorticoids cannot properly balance fluids in the circulation. The kidneys malfunction, the heart fails and the body eventually dies.

When blood pressure or blood volume drops, special cells in the kidney release a substance called renin into the blood. Renin interacts with plasma protein, ultimately forming angiotensin II, which stimulates the release of aldosterone and helps to raise blood pressure.

Finally, the adrenal medulla, the most interior part of the adrenal glands, synthesizes and secretes adrenaline and noradrenaline, the "fight or flight" hormones. Unlike the steroids, such as cortisol or aldosterone, adrenaline and noradrenaline are protein-based hormones. The body constructs them both from the amino acid tyrosine.

Cells of the medulla store the two hormones, ready and waiting for the time when the brain signals that it recognizes danger. Within milliseconds, brain cells release the neurotransmitter acetylcholine. In response, the medulla deploys a flood of adrenaline and, to a lesser extent, noradrenaline, which travel to receptors on the surface of their target cells, including those of the heart, lungs, and intestines.

As with other hormonal relay systems, triggered receptors pass their hormonal message to the DNA in the nucleus. Genes turn on and alter their expression patterns. The body experiences this change in a number of ways. The heartbeat speeds up; the bronchial tubes dilate, or enlarge, allowing more oxygen intake; the pupils dilate, absorbing more light; and fat and glycogen begin to break down, speeding metabolism and supplying a rush of energy.

The adrenaline cascade also coordinates an overall give-and-take in the body, giving to organs in need while taking from others deemed temporarily inconsequential. For example, during a rush of adrenaline, the blood vessels in the leg muscles expand while secretion of gastrointestinal juices in the digestive system halts. Adrenaline is often used to treat anaphylaxis, a severe allergic reaction. Because of its bronchodilator effects, the same hormone is used to treat asthma.

In an islet of Langerhans, the red alpha cells (left) emit glucagon while the green, yellow, and brown beta cells (right) secrete insulin.

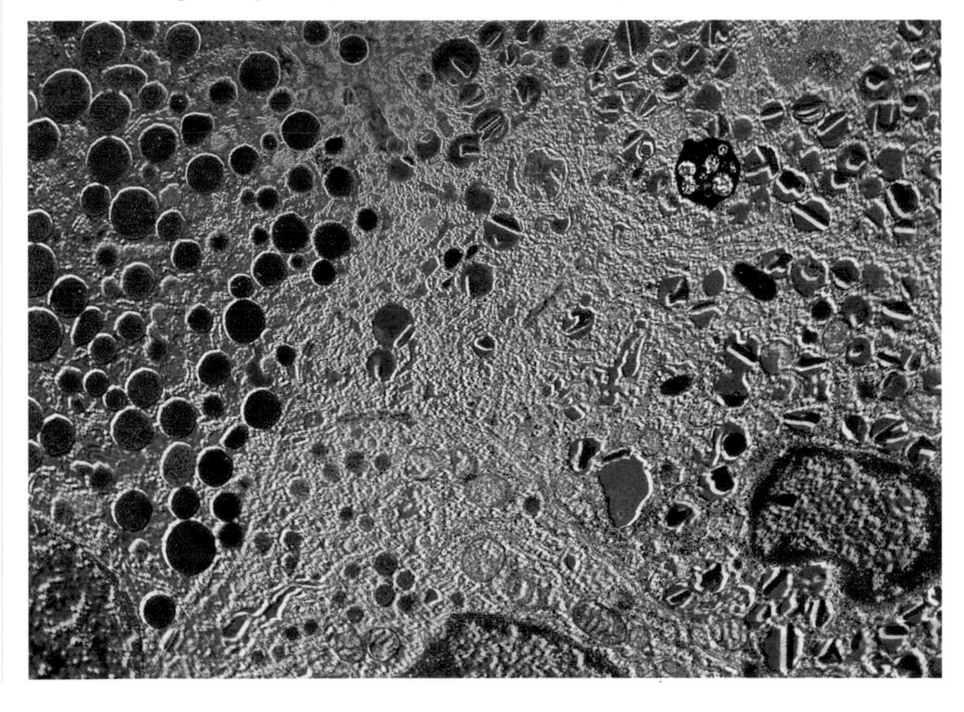

LIVING WITH DIABETES

DIABETES, ALSO CALLED diabetes mellitus, occurs when the body experiences a chronic deficiency in either the secretion or the action of insulin, a hormone made by the pancreas. Those who suffer from this condition must manage chronic problems with blood sugar regulation that grow more severe over time.

TYPES 1 & 2

There are two major types of diabetes. Type 1 diabetes shows up primarily in children and young adults. Often genetic factors play a role in an individual's predisposition for this type of diabetes. Type 1 diabetes often involves damage to cells in the pancreas, almost always the result of an autoimmune attack, that is, a situation in which the body's own cells militate against its proper functioning.

When the pancreas fails to make insulin, the body experiences abnormally high glucose levels. With Type 2 diabetes, enough insulin is initially made, but cells in the body do not respond appropriately, either due to a lack of insulin receptors or to an improper functioning of the receptors. This condition is known as insulin resistance.

Doctors used to think of Type 2 diabetes as an adult disease. People do tend to develop the condition predominantly in midlife or later. With the recent upsurge in childhood obesity—cases of it have doubled in the last 30 years—younger and younger patients are now being diagnosed with Type 2 diabetes. Today, nearly 21 million Americans live with diabetes, while 41 million more are prediabetic, meaning their blood sugar levels are significantly above normal levels, putting them at risk.

In 1946, these diabetic English children learned to give themselves insulin injections to manage their own care.

SYMPTOMS

Both types of diabetes share basic features. Facing a lack of insulin, the body strives to keep blood sugar levels high all the time. Despite the effort, the sugars—essential fuel for the body—cannot get into cells without the hormone insulin, so cells in a diabetic body start to starve for glucose. In response, the body reacts by making still more sugar, breaking down glycogen and using amino acids from protein to make glucose in a process called gluconeogenesis. The blood sugar level soars even higher, spilling over into a person's urine. This is why diabetes can often be detected in the results of a urine test.

As the problem worsens, sugar cannot be used as cellular fuel. In some cases, especially among Type 1 diabetics, the body breaks down more fat. Ketones, the product made by the body during the breakdown of fat, flood the bloodstream. When they accumulate in the blood, the blood's pH drops, acid levels rise, and the body falls into a dire state called ketoacidosis because the blood has become far too acidic. Symptoms including nausea, vomiting, and rapid breathing can ensue, leading potentially to coma or even death.

Diabetes may show itself in symptoms that include excessive urine output, thirst, and hunger. As glucose spills into urine, it forces water to follow. This process often causes such extremes of urination that dehydration can occur. A person who has no insulin function, in essence, has a body that believes it has not eaten. This person is starving in the land of plenty. Before the identification of insulin in the early 20th century, undetected diabetics lost weight precipitously. Today, physicians are sensitive to the signs, and in the developed countries, at least, diabetes rarely goes undiagnosed.

Obstetricians test pregnant women routinely for gestational diabetes, which arises for the first time in a woman during pregnancy. While this condition usually abates after giving birth, it leaves the mother more likely to develop diabetes later in life. It is also unhealthy for the developing fetus, equally sensitive to high glucose levels in the blood.

In times past, women who already suffered from diabetes were discouraged from getting pregnant. Today, with available blood sugar monitoring and management techniques and a fuller understanding of diabetes in general, diabetic mothers are simply advised to remain vigilant in watching blood sugar levels during pregnancy.

TREATMENTS

Before the discovery of insulin in 1921, doctors could do little for diabetes patients but manage their symptoms. But in the 21st century, those suffering from the disease are often able to live normal, active lives. Type 1 diabetes, caused by a lack of insulin in the body, can be treated with insulin supplements, often given as injections up to four times daily.

In the continuing search for a diabetes cure, cellular array technology is used to analyze the genome of blood cells.

Even so, some people with the disease develop long-term circulatory and nervous system problems. These can be minimized by constant monitoring and management of an individual's blood sugar level. Eating right can also reduce complications. A diet high in fat and high blood cholesterol levels can cause further problems, including heart disease, stroke, kidney shutdown, and blindness.

Type 2 diabetes, in which there is insulin resistance and therefore too much insulin in the body, requires a more complicated treatment. Doctors prescribe a combination of many different categories of antidiabetic drugs. Some drugs have been designed to reduce the diabetic body's resistance to insulin; others have been engineered to increase the sensitivity of hormone receptors, so that the body's natural insulin is more likely to be put to use.

At the same time, physicians encourage a program of weight loss, exercise, and healthy diet in addition to a drug regimen in the treatment of Type 2 diabetes. Studies show that Type 2 diabetes can be significantly delayed, or even prevented, through the loss of 5 to 10 percent of body weight in persons at high risk. Overall, physicians recommend regular blood sugar screening to assess whether a person is at risk for Type 2 diabetes. Medical researchers are hoping to find methods of measuring blood sugar that are easier and do not require drawing blood.

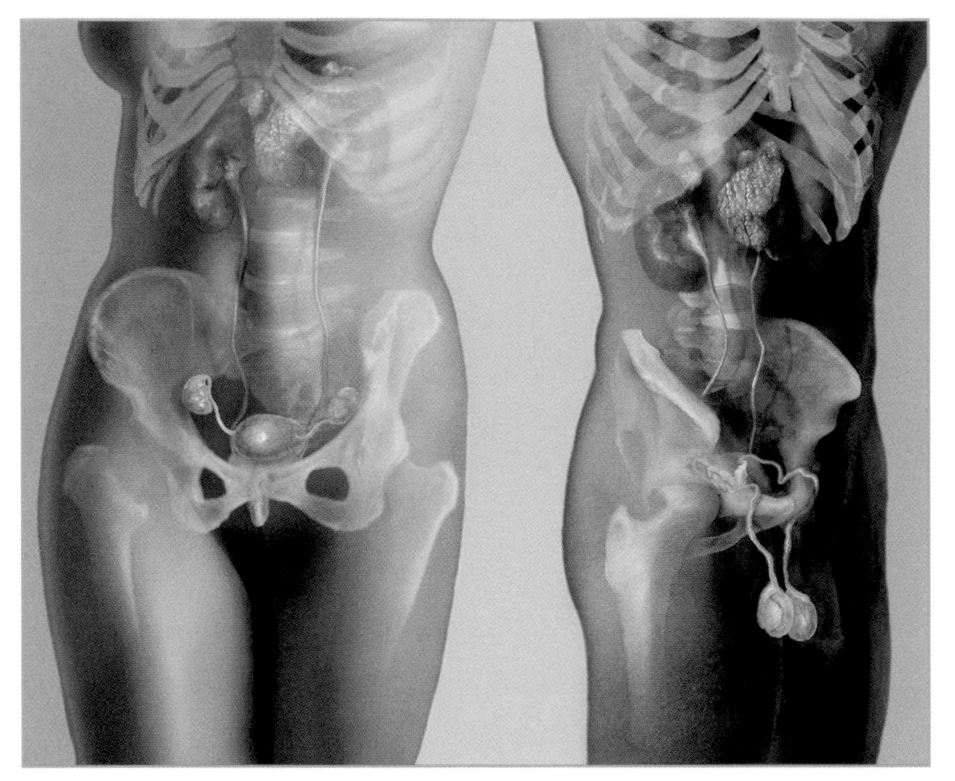

The gonads create reproductive cells and are different in men and women. Women's gonads are ovaries, left, and men's are the testes, right.

THE PANCREAS

Some would think that the stomach is the organ of hunger. But actually the pancreas, tucked underneath the stomach, wins the title of hunger regulator, managing its important feat with the help of blood sugar monitoring. The organ itself resembles a tapered cylinder, divided into three parts: a head, a body, and a tail. This anatomy, however, has nothing to do with distinctions in function.

Instead, function arises from two types of tissue: endocrine tissue, which secretes hormones, and exocrine tissue, which secretes digestive juices. These two types of tissue work next to and with each other. The endocrine tissue grows in small circular clusters, known as the islets of Langerhans; the exocrine tissue surrounds those clusters. Within the islets, three types of cell secrete essential hormones.

Alpha cells secrete the protein hormone glucagon. Low blood sugar, also known as hypoglycemia, triggers the production of this hormone. When blood sugar levels plummet—many hours after a meal or after exercise, for example—alpha cells go to work. They excrete glucagon, which travels to special receptors on liver cells and spurs the process of glycolysis, breaking down the molecule glycogen into glucose. Newly made glucose flows from liver cells into the bloodstream. The body now has a fuel source to continue functioning. If the sugar produced from glycolysis fails to meet body needs, however, glucagon has another trick up its sleeve. The hormone also stimulates amino acids to convert to glucose, a process called gluconeogenesis, which involves the production of a steady stream of glucose.

Meanwhile, back in the islets of the pancreas, beta cells, bathed in sugar-laden blood plasma, monitor glucose levels. When too high, the cells go to work, secreting insulin, which reverses the effects of glucagon; insulin lowers high blood sugar by binding to receptors on target cells and allowing glucose to enter cells.

Again, the liver is the workhorse. It converts glucose from the blood into glycogen, a strategy to store glucose for future needs. Insulin is such a versatile hormone that it also prevents the breakdown of fat, promotes the synthesis of fat, and inhibits the degradation of protein during its conversion into energy. The combination is breathtaking. Blood sugar drops, and the system regains its balance.

Insulin is such an important molecule, all mammals produce it. The insulin molecule is very similar, no matter whether it is found in voles or dogs or humans. Scientists have exploited this phenomenon, extracting insulin from a large mammal such as a cow and using the bovine insulin thus derived to treat insulin deficiencies in people, including those suffering from diabetes.

SEE ALSO: Chapter Seven, "Absorption," PAGE 194

While the body is experiencing the balancing act between insulin and glucagon, yet a third type of pancreatic cell, the delta cell, is synthesizing and secreting a molecule called somatostatin. Somatostatin inhibits the secretion of insulin and glucagon as well as other hormones. It is the body's attempt to impose yet another layer of control on a system that extracts, stores, and uses fuel critical for life.

THE GONADS

The endocrine system underlies the very perpetuation of the species. As the body makes eggs and sperm cells, and the two unite to form a new being, the endocrine system is there, guiding and shifting components within the complex reproductive systems.

The main endocrine organs involved in reproduction are the gonads: testes in males and ovaries in females. These organs create the gametes, or reproductive cells (egg cells in women, sperm cells in men). The body is ingenious in how it uses the same hormonal players to control development in both sexes. The key is adding variations, depending on sex.

In both males and females, the whole process begins with the hypothalamus, which pulses out gonadotropin-releasing hormone (GnRH). That jars the pituitary gland to produce follicle-stimulating hormone (FSH) and luteinizing hormone (LH).

STEROIDS

Baseball player Jose Conseco chronicled his use of steroids in his 2005 book Juiced.

ATHLETES MAKE headlines these days not just for their record performances but for the steroids they are using. Variations of the sex hormone testosterone, these anabolic steroids increase bone and muscle mass during development, especially puberty. Drug companies first brought steroids to market in the 1950s as a treatment for anemia and muscle-wasting diseases. Athletes soon caught wind of the drugs and by the 1960s, body builders and other elite athletes were taking megadoses, hoping to better their performances and masculinize their bodies.

No one can estimate how many people are taking steroids today. Football linemen, baseball outfielders, and Olympic athletes have admitted to taking the drugs. Combined with a program of heavy resistance training, the drugs do seem to increase isometric strength and body weight.

But doctors and trainers still debate whether athletic performance on steroids is actually better and, if so, how long the effects can last. There is also a risk of serious side effects: acne and hair loss, shrinking testes, infertility, liver damage, liver cancer, heart disease, and, at the worst, death. For females, side effects include smaller breasts, a larger clitoris, excess body hair, and thinning of scalp hair.

Today's latest steroid is androstenedione, sold over the counter as a nutritional booster. A precursor to testosterone, this substance has the same potential as other steroids but is also prescribed at times to prevent bone and muscle loss during aging. ■

In males, FSH and LH target the testes in what biologists refer to as the "brain-testicular axis." This allows a male, after puberty, to develop all the sexual characteristics that make him male, including the ability to manufacture sperm. The male endocrine system then adds in a protein called inhibin, produced by cells in the testes, as well as the steroid hormone testosterone, which both feed back to dampen GnRH production. In essence, testosterone keeps the male reproductive system in check. But because it is an anabolic steroid, it also builds up the muscles and controls behavioral traits such as aggression.

Meanwhile, in the female, tiny pulses of FSH and LH stimulate growth, maturation, and estrogen secretion from the follicles in the ovaries. Over the course of 14 days, estrogen levels climb to a threshold that feeds back to amplify a surge of LH and FSH from the pituitary. This hormonal gush stimulates the developing egg to ovulate. The ruptured egg sac becomes a structure called the corpus luteum, which causes the production of progesterone. The uterus thickens, and the pituitary gland receives signals that cause it to shut down its production of LH.

If fertilization does not occur, hormonal sustenance of the uterine lining dries up. Menstruation results. At the same time, the brain senses lowered LH levels. These trigger the hypothalamus to begin pulsing out GnRH once again, the beginning of another reproductive and hormonal cycle.

PINEAL GLAND & THYMUS

Hanging like a pinecone within the brain, the pineal gland secretes melatonin, a hormone that affects sleep cycles. The actual cells that do the job are called pinealocytes. Developing in compact clusters, the cells are surrounded by a bed of calcium salts, which show distinctly in a brain x-ray.

Melatonin honors the nocturnal cycle, waxing at midnight and waning at noon. The hormone's production is linked to the visual pathway of the brain. Light and dark flip this hormone off and on, a process at the heart of melatonin's function as a biological clock.

The hormone works by targeting a patch of tissue deep in the hypothalamus called the suprachiasmatic nucleus. It is stocked with melatonin receptors. Their exposure to bright light resets the clock and connects the night and day cycle in with other rhythmic body processes, such as body temperature, sleep and waking, and—in mammals others than humans—seasonal mating.

Another tiny but crucial endocrine structure is the thymus, located deep in the sternum area of the chest cavity. The thymus makes a family of peptide hormones thought important in the development of immune responses. Early on in human life, the thymus is relatively large. It diminishes in size in adults.

OTHER PRODUCERS

Other organs have cameo parts in the performance of the endocrine system. The gastrointestinal tract is well stocked with special cells that release amino acid and peptide hormones such as secretin and cholecystokinin (CCK). CCK is an

BREAKTHROUGH

BODY FAT CONTAINS a hormone that tracks fat (or lipid) stores and should keep them in perfect balance with energy output. That lipometer is leptin, a protein hormone that has profound effects on appetite. In 1994, endocrinologist Jeffrey Friedman, left, at Rockefeller University discovered the hormone in a strange breed of mice. The rodents were obese, three times the weight of normal mice. Friedman's team determined that the animals got this way because they carried mutations in their so-called obesity genes, which encode the protein leptin. These obese mice lacked the substance.

SEE ALSO: Chapter Twelve, "Male System," PAGE 294, "Female System," PAGE 300 appetite-regulating peptide that pours out in response to food. It triggers the gall bladder to release bile and opens the sphincter entering into the small intestine.

Enteroendocrine cells, as they are called, act a lot like neurons, earning their moniker as "the brain in the gut." Some peptide hormones of the stomach and intestines, such as serotonin and histamine, are identical to neurotransmitters in the brain and act locally on tissues nearby. There are many unique gastrointestinal peptides as well. This wealth of hormones allows precise and local regulation of hunger, satiety, and energy balance—along with a connection to brain sensations of appetite and fullness.

PINEAL GLAND

+ Doctors used to think the pineal was vestigial, with no use to humans.
+ Julius Axelrod, above, linked the pineal gland to its role in the sleep cycle and melatonin production.
+ The pineal begins to produce melatonin in humans at age three months. Production peaks at puberty.
+ Melatonin secretion may be suppressed by light. It may also figure in seasonal affective disorder (SAD).

The top of the heart, the atria, contains special cardiac muscle cells that secrete atrial natriuretic peptide, which targets the adrenal cortex, inhibits aldosterone, and signals the kidneys to increase the salt level in the urine. Why? Removing salt and water from the bloodstream reduces blood volume, pressure, and sodium concentration. The process gives the heart a rest when the body is at rest.

Cells in the kidneys make the glycoprotein called erythropoietin. This hormone's function is to stimulate bone marrow and step up the manufacture of red blood cells. Not surprisingly, drugs related to erythropoietin are used to treat anemia.

When summer arrives, people start thinking of the beach and poolside. It's the season to pay attention to skin in its relation to the endocrine system.

Skin cells produce cholecalciferol, an inactive form of vitamin D. When cells are exposed to ultraviolet light—during an afternoon outdoors, for example—the latent hormone enters the bloodstream through capillaries buried in the dermis. The liver takes the next step, modifying the prohormone, and sends it to the kidneys for ultimate activation. The final product, vitamin D or calcitriol, works as part of a transport system and helps the body take in calcium, a mineral critical for bone strength, from ingested food.

One last endocrine player is adipose tissue, or fat cells. These energy packets take in glucose and lipids during a meal and, if the body does not immediately need the fuel, convert them to fat. As this occurs, fat cells churn out leptin, which binds to its receptors on neurons in the brain. This complex circuitry controls appetite, temperature, and reproductive prowess.

BREAKTHROUGH

Friedman's group and others traced the site of hormone production not to a gland but to fat cells themselves. For clues to what the hormone does, Friedman needed only to look at the mice. Missing the hormone, they ate as if in a state of perpetual starvation, unable to stay warm or grow normally. Now researchers know that leptin levels increase in the bloodstream as fat deposits grow. The hormone signals the brain to stop eating, increase activity levels, and thus raise body heat. Ironically, the bodies of people who are obese actually make too much leptin. Overweight and obese individuals develop leptin resistance, probably due to downregulation of leptin receptors or to malfunctioning of receptor response. Since the discovery of the hormone, researchers have identified a host of other molecules that regulate appetite and fat storage.

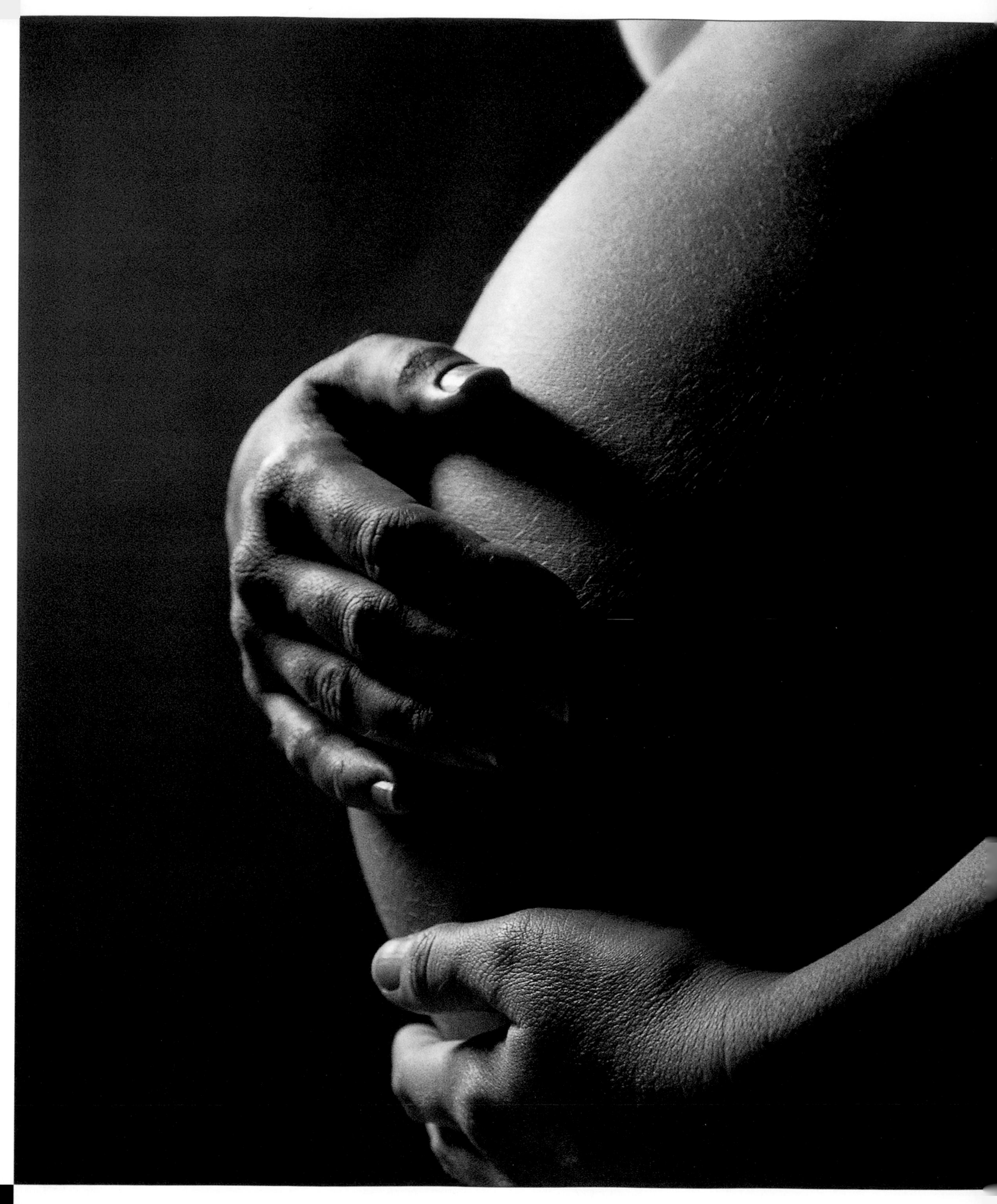

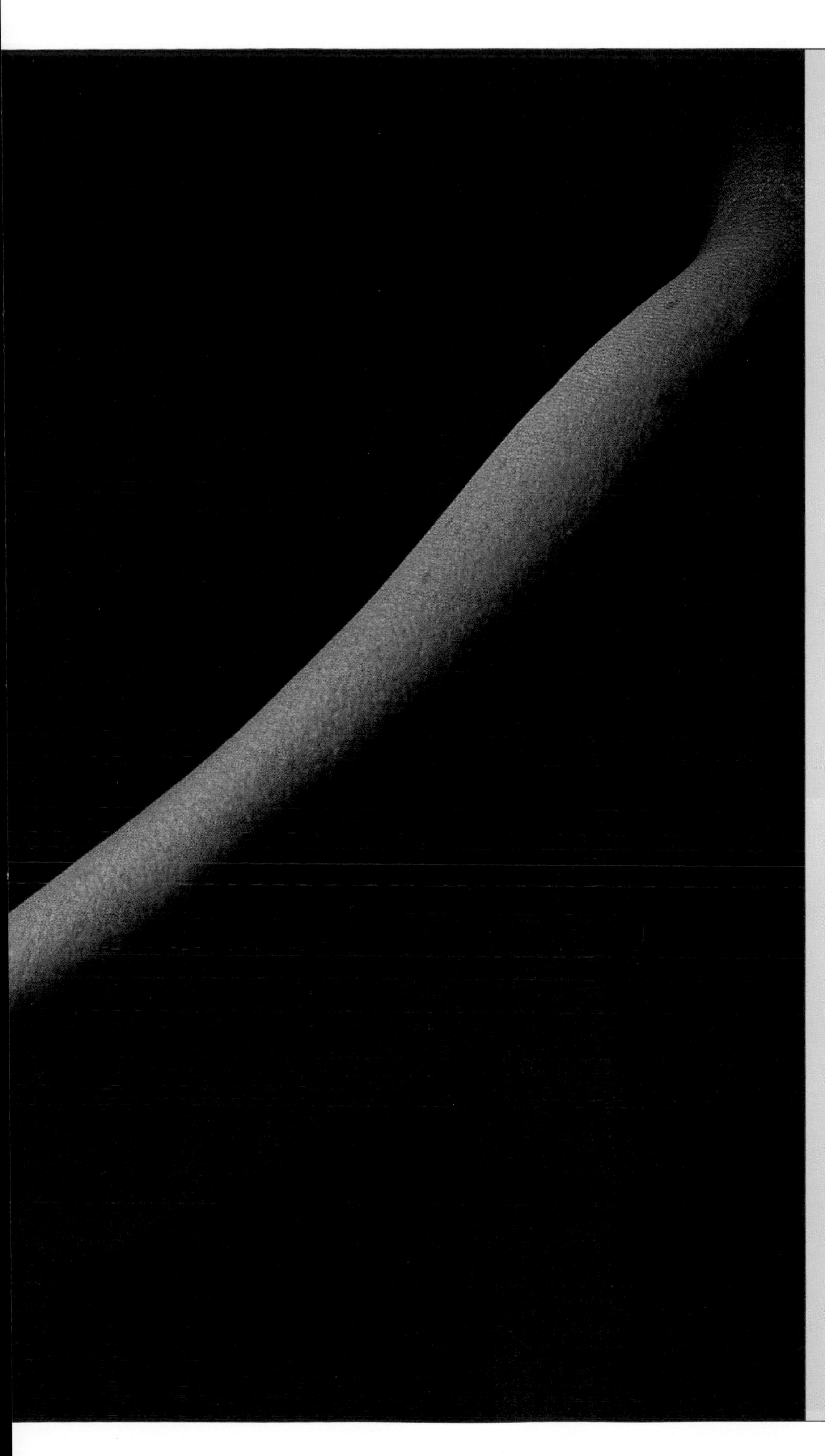

CHAPTER ELEVEN

REPRODUCTION

LIFE BEGETS LIFE. THE HUMAN reproductive system, which drives the entire process of life-giving, is therefore essential to human existence. Superficially, the reproductive system is a series of tubes, ducts, glands, and organs. Underlying it all is a delicate chemical dance, a pulsation of hormones and neurochemicals within men and women. Male and female systems are complementary. Each performs specialized tasks with the common goal of joining the sperm and the egg into a single unit, which grows into a new life within its mother's womb.

An expectant mother's body swells as the new life within develops.

MALE SYSTEM

THE MASCULINE BODY AND ITS FUNCTIONS

IN REPRODUCTION, both the man and the woman have important parts to play. The man's main biological function is to supply sperm, the cells that combine with the woman's egg when conception occurs. For that reason, the male reproductive anatomy is relatively simple and centered on the production, storage, and delivery of sperm. The operation of those systems, however, involves a complex interaction with glands and hormones that begins even before birth.

The Male Reproductive System

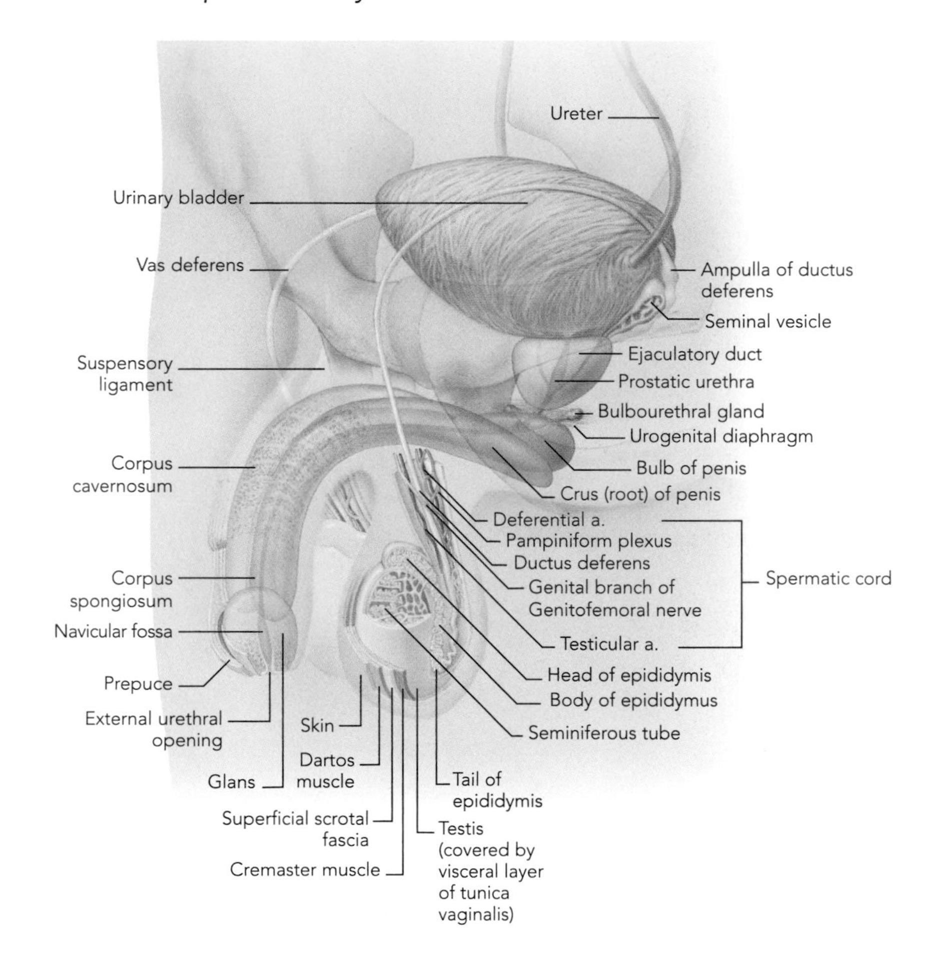

MALE ANATOMY

The testes, or testicles, are two plumlike glands suspended within the scrotum, a skin sac that hangs outside the body. The testes produce and cultivate spermatozoa, immature sperm cells that contain a man's entire genetic heritage. Reproductive glands and vesicles create fluids such as semen that lubricate, feed, and propel potent sperm cells on their journey toward new life.

The rest of the male reproductive anatomy is a delivery system, including the epididymis, coiled and nestled against each testis; the vas deferens, joining the epididymis to the ejaculatory duct; and the urethra, part of the urinary system and also the sperm's exit path. The anatomic pinnacle of this system is the penis, which delivers sperm cells to the female reproductive tract. The penis consists of the root, the shaft, and an enlarged tip, called the glans penis. It is cuffed by loose skin called the foreskin, which many cultures remove surgically in a procedure called circumcision (see "Circumcision," opposite).

The endocrine and nervous systems produce hormones that jump-start and run the male reproductive engine. Meanwhile, neurotransmitters imbue a sense of gender and sexual drive. All of

+ PRIAPISM +

+ NAMED FOR THE Greek god of fertility, Priapus, priapism is a condition of prolonged erection.
+ It can be caused by many factors, including the use and abuse of erectile dysfunction therapies and injury to the spinal cord.
+ To treat the often-painful condition, physicians commonly aspirate blood from the erectile tissue.

SEE ALSO: Chapter Eight, "Urinary Tract," PAGE 222, Chapter Ten, "Hormones," PAGE 268 these systems interact to create male identity.

TESTOSTERONE

Long before birth, male fetuses "masculinize" within the womb. The process hinges on the male sex hormone testosterone, manufactured by cells of the testes. Prenatal testosterone bathes the male fetus's brain and primes it for later development. This exposure is so critical that if a male fetus lacks the correct hormonal environment, he can grapple with infertility later in life (see "What Can Go Wrong," page 297).

Testosterone works alongside three other regulatory hormones: gonadotropin-releasing hormone (GnRH), which pulses from a brain structure called the hypothalamus; and the gonadotropins, follicle-stimulating hormone (FSH) and luteinizing hormone (LH), both produced by the pituitary gland, a bulbous structure suspended from the hypothalamus. The four hormones pulse and work as a team, one hormone triggering the next in what biologists call the "brain-testicular axis."

This interaction in a male infant produces high levels of GnRH, FSH, and LH from birth until he reaches six months of age. In fact, just after birth an infant's hormonal state equates to that of a midpubertal boy. Within months, however, a restraining system develops. It involves a protein called inhibin, produced by cells

CIRCUMCISION

Italian artist Bartolomeo Veneto painted The Circumcision *in the 1500s.*

CIRCUMCISION, the practice of removing the skin covering the head of the penis, reaches back to antiquity. The ancient Egyptians surgically removed males' foreskins, as evidenced by tomb artwork from the 6th Dynasty (2325–2175 B.C.). In the fifth and sixth centuries B.C., Semitic peoples also embraced the practice and made it part of their culture. The foreskin may enhance erogenous sensation by protecting the highly sensitive penile tip; its removal possibly symbolized the sacrifice of "sinful" human enjoyment for holiness' sake.

Conversely, the ancient Greeks and Romans valued sensual pleasure. They avoided circumcision, and after the conquests of Alexander the Great in the fourth century B.C., the practice plummeted in popularity. The religions of Judaism and Islam held to the practice, however, and shepherded it into contemporary times. In the 19th century, physicians in the United States and Great Britain linked circumcision to hygiene and disease control. The practice fell out of favor in Britain after World War II but remains widespread in the U.S. (for both religious and secular reasons) despite opposition from those who question the surgery's benefits and the ethics behind it.

However, recent studies have found that circumcision does provide measurable health benefits. Because it causes keratinization, a hardening of the skin surface, circumcision reduces the likelihood of contracting sexually transmitted bacterial infections. Circumcision also helps protect against penile cancer and removes Langerhans' cells, the specific target of the HIV virus. In recent studies in South Africa, Uganda, and Kenya involving more than 10,000 men, researchers found that if men were circumcised, they contracted HIV at nearly half the rate of uncircumcised men. ■

of the testes, as well as testosterone. These substances act like a molecular brake upon GnRH production.

Puberty is simply a release of this biochemical brake. Scientists speculate that the trigger is a tiny molecule secreted by the pituitary gland. Once the putative trigger begins to circulate through the body, the reproductive system, on hold for years, begins to rev up. Testosterone stimulates sperm production. The body develops sexual characteristics, including the sprouting of pubic, underarm, and facial hair. The voice deepens. The skeletal muscles increase in size and mass. From the perspective of biology, a boy becomes a man.

Spermatogenesis

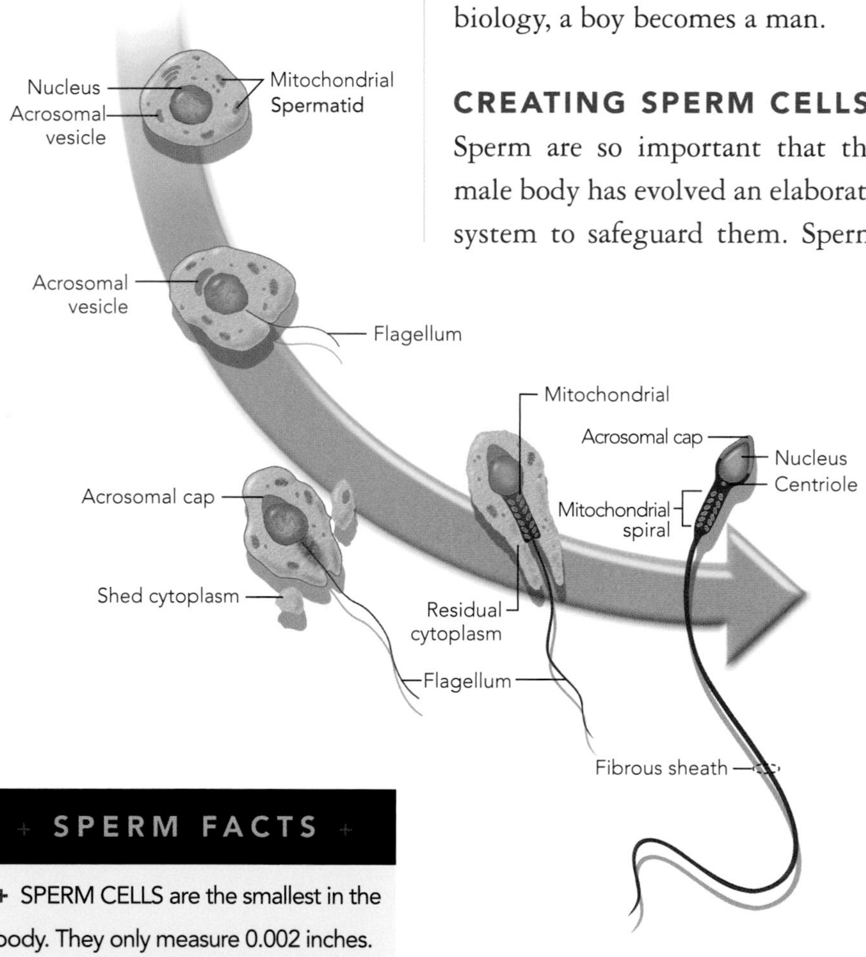

CREATING SPERM CELLS

Sperm are so important that the male body has evolved an elaborate system to safeguard them. Sperm thrive only if incubated three degrees below normal body temperature. Hence, the scrotum, which holds the testes and their vulnerable sex cells, hangs below the too-warm abdomen. The scrotum also has built-in structures to control this positioning. When the body is cold, for example, the cremaster muscles—the muscles by which the testes are suspended—contract to pull the testes up toward the belly. The dartos muscle, woven within the tissue of the scrotum, wrinkles the skin, reducing the surface area available for testicular heat loss. At other times, the cremaster muscles loosen, the dartos slackens, and a massive blood vessel system, comprising almost two thirds of the scrotum, cools the blood as it circulates around the testes.

Mature sperm are streamlined cellular rockets that shuttle genetic material to their target, an egg. Each sperm possesses a tail, which is connected to the cell body by a robust molecular engine. This allows the sperm, upon launch through ejaculation, to swim relatively great distances through the female reproductive tract. Sperm also possess special "heads" that can penetrate the formidable capsule that protects the egg.

But sperm do not start out as accomplished voyagers. They begin as immature spermatogonia: tailless, impotent, nonmotile stem cells buried deep within the testes. From a boy's infancy to puberty, spermatogonia divide to reproduce but remain immature. Then, at about age 14, a boy enters puberty. From a sperm cell's perspective, puberty means his spermatogonia are also ready to mature.

All living organisms carry the genetic instructions for life in the

+ SPERM FACTS +

+ SPERM CELLS are the smallest in the body. They only measure 0.002 inches.
+ The male body creates 300–400 million sperm cells per day.
+ Formation of mature sperm cells takes 64–72 days.
+ A milliliter of semen contains 50–130 million sperm cells.
+ Sperm cells live only 48 hours after ejaculation.

SEE ALSO: Chapter One, "Inheritance," PAGE 24

form of DNA. In humans, DNA is carried on 46 chromosomes in the nucleus of every cell, 23 from the mother and 23 from the father. During fertilization, the parents combine their genetic material into one embryo. It is the sperm, however, that carries the X or Y chromosome that, when added to the egg's X chromosome, determines whether the child will be a girl (XX) or a boy (XY).

But before that union can happen, sex cells must halve their chromosome number to 23. This reduction occurs in a process called meiosis, or "lessening." During meiosis, immature spermatogonia, now called spermatocytes, cycle through two precisely orchestrated cellular divisions to become four daughter cells, each with half as many chromosomes as a typical human cell.

At this stage, the male sex cells are called spermatids. Even though they possess the right chromosomal number for fertilization, they are still nonmotile and far too bulky to make a lengthy journey. A streamlining process must occur, in which the spermatids shed excess cell contents, grow tails, and compact DNA into their bulletlike "heads."

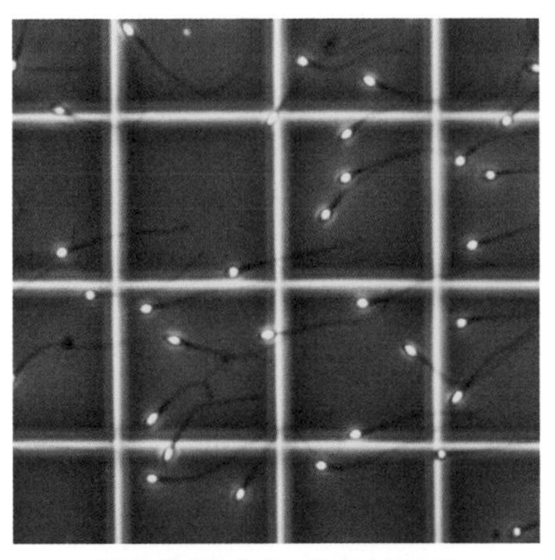

NORMAL COUNT
About 113 million sperm per milliliter of ejaculate populate a normal sperm count.

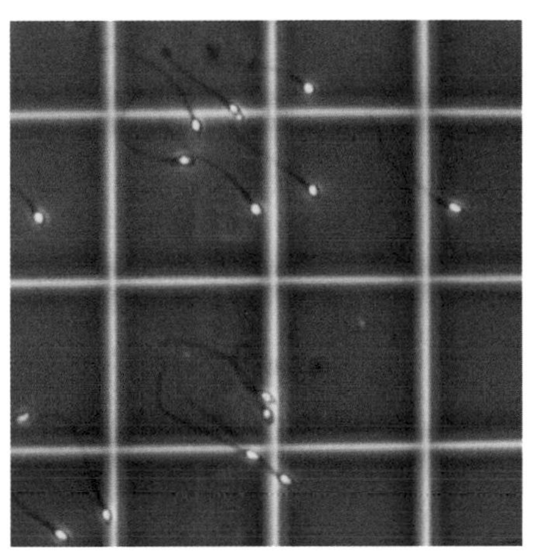

LOW COUNT
Only 60–70 million sperm per milliliter of ejaculate are found in low sperm counts.

The human body has evolved an anatomical pipeline through which a sperm cell must travel as it develops. The entrance is one of four tightly coiled seminiferous tubules that make up the bulk of the testis. Most of the cells lining these tubules are sperm-forming cells in various stages of development. The seminiferous tubules function as virtual sperm factories, providing supporting cells that bathe, feed, and anchor the developing sex cells.

Spermatids on their way toward maturity enter the epididymis, which hugs the outside of the testis. Because these spermatids do not yet have tails, muscle contraction and pressure from testicular fluid propel them from the tubules into the epididymis. In essence, the epididymis is 20 feet of tightly wound tubing. Developing spermatids have 20 days to travel its length while accomplishing the many tasks of development, such as learning how to swim on their own. If sperm cells remain within the epididymis too long, they are consumed by neighboring cells and do not complete the journey.

WHAT CAN GO WRONG

MALE INFERTILITY often goes undetected because there are few outward symptoms. Most men only become aware of it after trying to become fathers. Recent studies have shown that the male factor is part of the problem for 30–40 percent of infertile couples. Infertility has a variety of causes, from congenital defects and blockages to environmental and lifestyle factors. Obesity, alcohol and drug abuse, exposure to lead, and malnutrition all affect a man's reproductive health. To diagnose infertility, doctors use a medical history and a physical exam. Blood tests can detect possible hormonal imbalances. Semen analysis reveals sperm appearance, number, and motility. Surgery can correct a common cause, a varicocele or varicose vein around the spermatic cord. If hormonal disorders are the problem, hormonal stimulation may increase sperm production.

During ejaculation, smooth muscles within the epididymis contract to expel sperm into the next duct of the system, the vas deferens. This duct squeezes the sperm forward and out the urethra.

In a sterilization procedure known as a vasectomy, surgeons cut the vas deferens and tie off the end. The body continues to make sperm even though the sperm cells cannot exit the system. The body will eventually begin to recognize these as "foreign" cells and produce antibodies to consume them.

SEMEN CREATION

But sperm cells do not make the next step alone. They must combine with other substances to create semen, a life-giving elixir that is a mixture of DNA-carrying sperm and the fluids that support their ability to achieve fertilization. Sperm mingle with a small amount of fluid in the vas deferens, but most of the fluid in semen is secreted by the prostate gland and fingerlike structures called the seminal vesicles that are perched on the backside of the bladder.

The seminal vesicles secrete seminal fluid and, together with the vas deferens, form the ejaculatory ducts. The thick, yellowish seminal fluid contains sugar for fuel; vitamins to regulate cellular actions; enzymes that help speed sperm velocity; and regulatory molecules (called prostaglandins) that will thin the mucus that guards the entrance to a woman's cervix, the gateway to the womb. As the sperm exit the vas deferens, they mix with seminal fluid in the seminal vesicles. During ejaculation, the semen enters the first stretch of the urethra.

There, semen flows by the prostate, a walnut-size gland that encircles the neck of the urethra. The prostate pipes in yet another secretion—a milky, slightly acidic fluid that contains citrate, enzymes, and prostate-specific antigen (PSA). These molecules will protect and activate the sperm for optimal performance as they exit the urethra on their way toward their ultimate destination—a union with an egg.

A VITAL LIFE FORCE

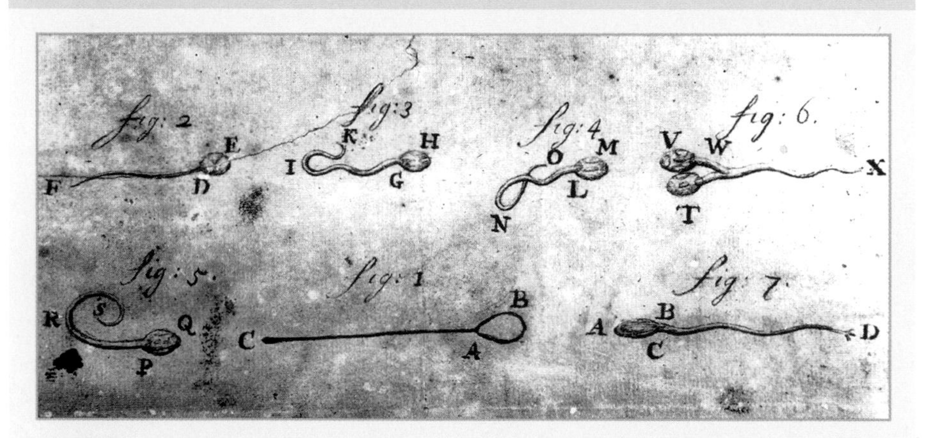

Anton van Leeuwenhoek made these drawings after seeing semen under a microscope.

THE GREEK philosopher Aristotle (384–322 B.C.) believed that semen contained a vital life force, a combination of distilled brain fluid and fundamental elements. It was this life force, he wrote, that sparked fetal development from the material supplied by the woman, the menstrual blood. Centuries later, British physician William Harvey (1578–1657) built on these views. He championed the idea of embryonic development whereby an undeveloped egg met with semen, whose union created a "formative virtue" that powered life and guided its development. Later generations favored preformation, the idea that a miniature, fully formed human preexisted and enlarged within the womb. This view gained traction in 1677, when the Dutch scientist Anton van Leeuwenhoek (1632–1723) put semen under a microscope and observed "little animals," the single-celled spermatozoa. Spermists theorized that these cells contained little people. Dutch physicist Nicolas Hartsoeker (1656–1725), in his now famous drawing of a homunculus, imagined what such a "little man," huddled within the sperm, might look like. Eventually, the preformationist theory collapsed, as critics noted that each little human would have to contain another little human, the next generation, within itself, like Russian dolls, ad infinitum. ■

SEE ALSO: Chapter Eight, "Urinary Tract," PAGE 222

CHAPTER GLOSSARY

CERVIX. The lower neck of the uterus, which extends into the vagina or birth canal.

CLITORIS. A small, sensitive structure, richly innervated and composed of erectile tissue; located on the exterior female genitalia.

CORPUS LUTEUM. A ruptured ovarian follicle that secretes progesterone and estrogen after ovulation has occurred.

EJACULATION. A smooth-muscle contraction that expels semen from the male reproductive tract through the urethra during male orgasm.

ENDOMETRIUM. The inner uterine membrane; shed monthly and regenerated if no pregnancy occurs.

EPIDIDYMIS. A curled portion of the male duct system in which sperm mature. Emerges from the testes and connects to the vas deferens.

ESTROGEN. Female sex hormone.

FALLOPIAN TUBES. The tubes through which the egg travels from the ovary to the uterus.

FERTILIZATION. The joining of the sperm and egg nuclei.

FOLLICLE. An ovarian structure that houses a developing egg and contains layers of follicle cells.

GAMETE. A male or female reproductive cell; the sperm or egg.

GENITALIA. The internal and external reproductive organs.

MEIOSIS. Cell-division process that reduces the number of chromosomes by half and results in the formation of four cells.

MENOPAUSE. A period of life when, due to changes in hormonal secretions and cycles, female ovulation and menstruation cease.

MENSTRUAL CYCLE. A series of uterine changes, including ovulation and bleeding, which occur cyclically in response to the fluctuating blood levels of ovarian hormones.

OOCYTE. Immature female sex cell, also known as ovum or egg. Contains a woman's genetic material.

OOGENESIS. Process of egg production in the female

OVARY. The chamber for egg production in the female reproductive system; the female gonad.

OVULATION. The monthly release of an immature egg from the ovary.

PENIS. A male organ used in reproduction and urination. Consists of a connecting root, a shaft, and an enlarged tip.

PLACENTA. A temporary organ formed from maternal tissues during pregnancy to nourish the fetus. Delivers nutrients, removes wastes, and produces pregnancy hormones.

PREGNANCY. The 280-day time period during which a woman carries a developing baby in the womb.

PROGESTERONE. A female hormone that prepares the uterus for a fertilized egg.

PROSTATE. A chestnut-size gland that encircles the male urethra and is responsible for a part of the process of semen production.

PUBERTY. The period of life in which the levels of gonadal hormones, estrogen and testosterone, rise. This increase causes the reproductive organs to fully develop and gain function and any secondary sex characteristics to appear.

SCROTUM. The external sac found in the male, hanging below the abdomen. It houses the testes and helps regulate their temperature.

SEMEN. Milky, slightly acidic fluid containing sperm and various other secretions that sustain the sperm cells in the female reproductive tract.

SPERM. The male gamete, which holds a father's genetic material.

SPERMATOGENESIS. The process of sperm formation.

TESTES. The male gonad, which produces and cultivates spermatozoa, immature sperm cells.

TESTOSTERONE. The male sex hormone, produced by the testes.

UTERUS. Organ within the female reproductive system in which the fertilized egg is received and in which the embryo will develop after conception.

VAGINA. The canal in the female system extends from the cervix to the exterior of the body; also referred to as the birth canal.

VAS DEFERENS. In the male system, the duct that propels sperm from the epididymis to the urethra.

VULVA. The external female genitalia that surrounds and protects the vaginal opening.

FEMALE SYSTEM

THE FEMININE BODY AND ITS FUNCTIONS

BEFORE THE 18TH century, females were viewed by physicians as incomplete males. A superficial knowledge of anatomy helped to make the case. Early physicians saw that females had a womb, with a shape on top not unlike a scrotum but with the testes, represented in the woman by the ovaries, lying outside it.

The womb opened into a tubular canal, something like a penis turned inside out. Thus, the notion of sexual difference put forth by the second-century Greek physician Galen prevailed for many years: Females must be males who, because of a lack of "vital heat," retained their reproductive structures inside their bodies, while those of true males emerged outside.

HYSTERIA

THE TERM HYSTERIA comes from *hystera*, ancient Greek for uterus. As early as the sixth century B.C., Greek, Roman, and Egyptian physicians believed that womb could detach and move around a woman's body. The wandering womb put pressure on her internal organs, causing emotional distress and physical illness. Pleasant fragrances were used to treat it.

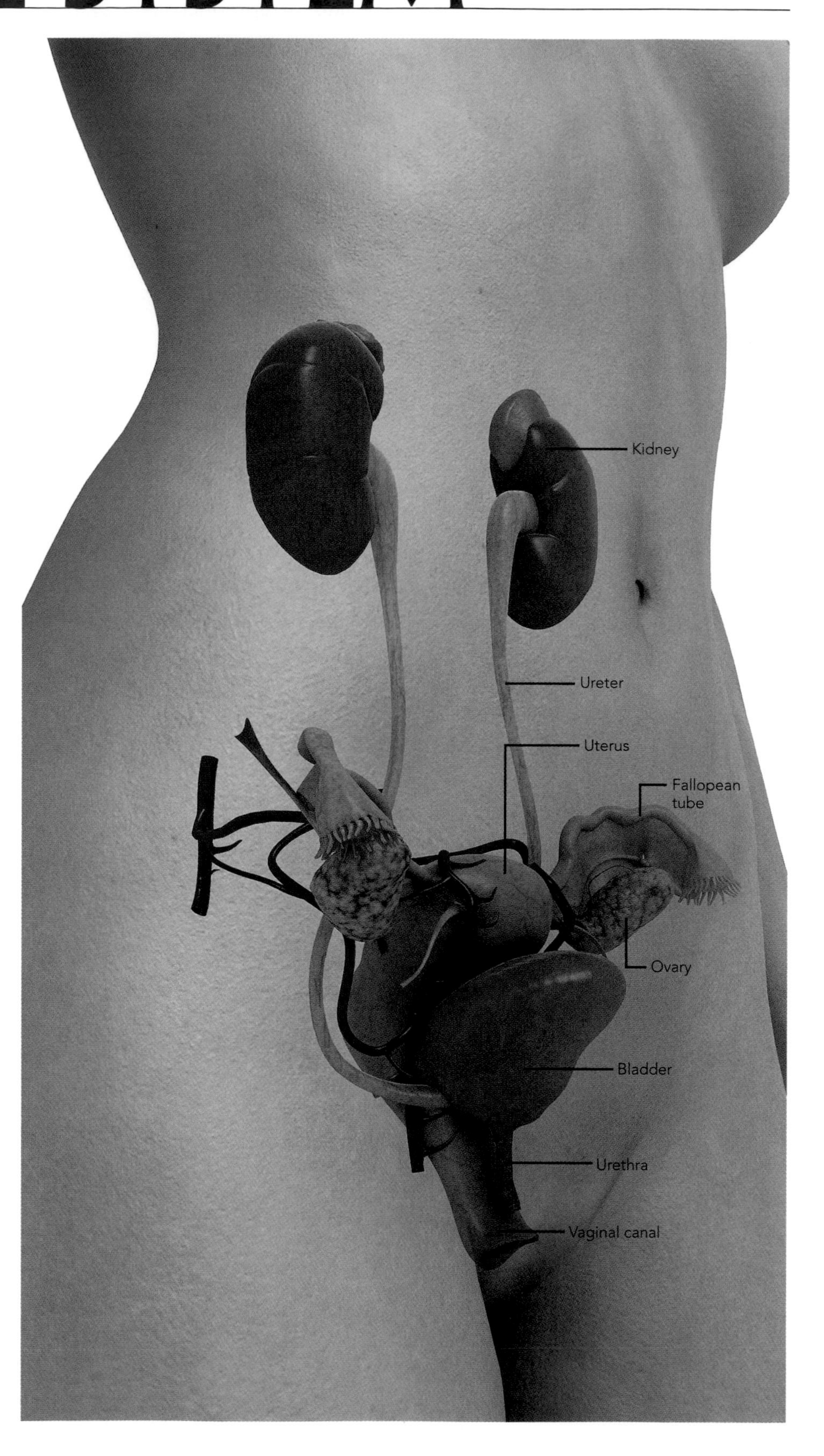

The organs of the female reproductive system are nestled alongside the urinary tract.

SEE ALSO: Chapter Eight, "Urinary Tract," PAGE 222

In fact, modern medicine has come to understand that the female reproductive system is actually more complex than its male counterpart. Females, after all, must create and store eggs for fertilization, provide an environment where those eggs can develop and thrive, and house and nurture a growing fetus.

FEMALE ANATOMY

Given this dual context, the ovaries command central focus. The shape of an almond and twice the size, each ovary is both an egg production factory and a hormonal powerhouse. The ovaries grow the eggs, or ova, and prime their development. The ovaries also make and dispatch cycles of sex hormones that steer the development of both mother and baby.

The female reproductive system has a second seat of command: the womb, or uterus. This inverted-pear-shaped structure is a muscular organ interwoven with an extensive blood-vessel system to feed menstrual flow. The top of the uterus reaches toward the ovaries, and their egg supply, with arms called oviducts, or fallopian tubes. The bottom of the uterus curves into a neck called the cervix. The cervical canal leads to a three-inch vagina, or birth canal. These structures are anything but an incomplete male tract. They are the centerpiece of "femaleness," the creator and container of a growing baby, and a locus of sexual pleasure.

Despite differences, there is a parallel between the male and female reproductive systems. Modern scientists have studied development of the external genitalia and found that many male and female structures arise from the same type of embryonic tissue.

In females, the area containing the external genitalia is called the vulva. It includes the mons pubis, the mound of fatty tissue above the pubic bone that becomes covered in hair after puberty and that contains sebaceous glands that secrete sexual attractants called pheromones. At the lower end of the mons pubis are the labia majora ("large lips"), which protect the other external genitalia from injury and guard the internal genitalia from infection; they are considered the female counterpart of the scrotum. They lie outside the labia minora, the inner lips that surround the vaginal opening. These "small lips," which swell with blood during sexual excitement and become sensitive to touch, are analogous to the ventral penis.

Between the labia minora, the projecting clitoris is composed of erectile tissue, richly innervated, and capable of engorgement during tactile stimulation, which can produce orgasm. Thus, the clitoris is penislike. In fact, it is hooded by a skin fold called a prepuce, which is much like the male foreskin.

Unlike in the male, however, in the female the clitoris and surrounding structures are separate from the urinary tract. The urethra

KEEPING HEALTHY

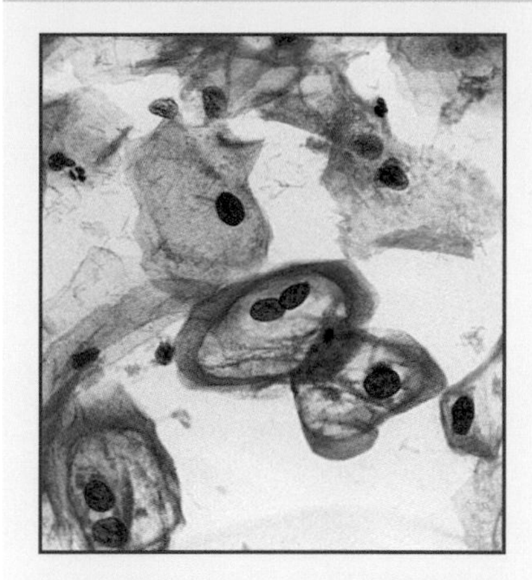

Cervical cells are infected with HPV.

CELLS OF THE cervix constantly reproduce to replace those sloughed off by menstruation or sexual activity. Cervical cancer, the second most common cancer in women, occurs when cervical cells begin reproducing uncontrollably. Recent studies suggest that more than 90 percent of cervical cancer cases are caused by the human papillomavirus (HPV), a sexually transmitted virus that infects the cells and mucous membranes of the cervix. In 2006, the Food and Drug Administration approved the first HPV vaccine, Gardasil. While celebrated as a medical breakthrough, the vaccine is not foolproof. There are 120 known strains of HPV. The vaccine targets strains that cause about 70 percent of all cervical cancers and two other HPV strains that cause most forms of genital warts. It is effective for a minimum of four and a half years in females unexposed to those HPV strains. ■

embeds itself in the front wall of the vagina, allowing the latter to evolve into a structure capable of great strength and flexibility. The vagina is stocked with muscles that stimulate the penis. Lubricants, secreted by Bartholin's glands alongside the vaginal opening, and elasticity allow the vagina to withstand friction during intercourse.

Women and men now celebrate and debate their similarities and differences. As the debates continue, their reproductive anatomy remains what it has always been: analogous but unique.

PUBERTY

For four years after the onset of puberty, a girl will experience hormonal changes that cause development of breast buds first, then pubic and underarm hair, and an overall growth spurt. Breasts are a combination of connective tissue, fat, and mammary glands. The latter are present in both males and females but only functional in women. Mammary glands are actually modified sweat glands. Each connects to a duct which opens into the nipple. Each is stimulated to produce milk during pregnancy by the hormones progesterone and prolactin.

The hormone estrogen triggers fat deposition. Adolescent girls experience this as an increase in the size of breasts and hips. The hormone also lightens a girl's pelvis in anticipation of holding a growing fetus. Her bones experience increased calcium uptake to sustain density of her skeleton. Her body goes through an estrogen-triggered growth spurt around the age of 13. Estrogen maintains a low total cholesterol and high HDL. All these changes coalesce during adolescence to allow a girl to reach womanhood, in biological terms.

Her developing system must learn how to time itself with the appropriate hormonal levels. When she achieves this, she experiences her first menstrual period, called menarche. It heralds a 28-day cyclic pattern of hormonal interactions that will take about three more years to become regular. Producing a mature egg, called ovulation, does not usually occur with menstruation for approximately the first year to 18 months.

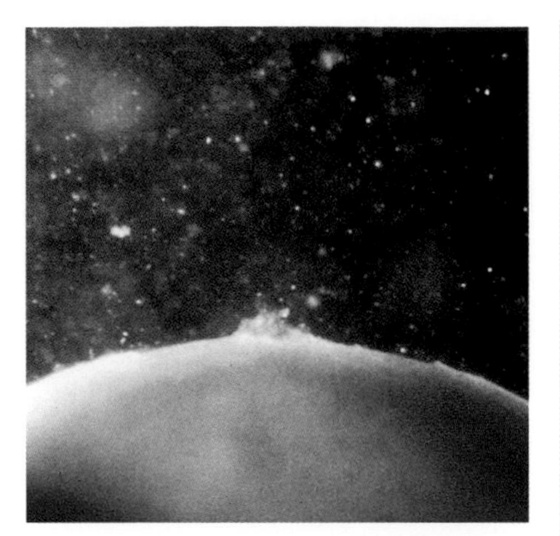

OVULATION BEGINS
In the primary stages of ovulation, the ovarian follicle bursts open.

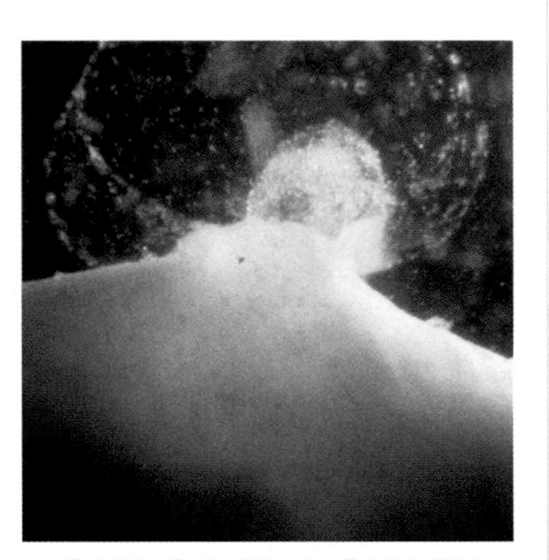

OUT OF THE OVARY
A ripe ovum begins to emerge from the ovary wall.

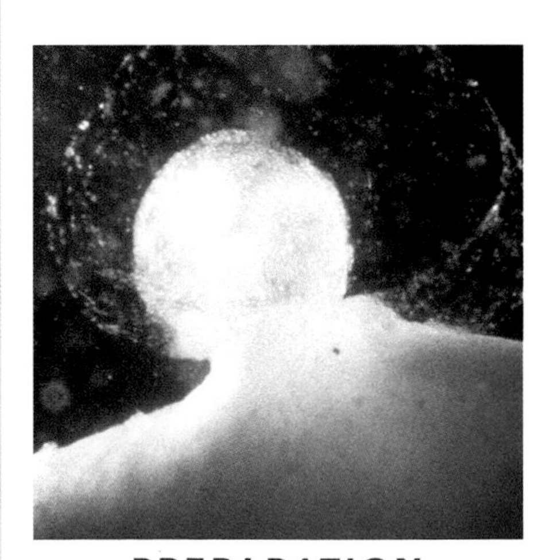

PREPARATION
Smaller, protective cells encircle the ovum before its journey.

THE EGG

When a girl is born, her complete potential egg supply is born with her. While still in the womb, she creates about seven million egg cells, but most of these will die. At birth, about two million remain, dormant and primordial.

By puberty, only about 400,000 are left, many times more than she will need during her life. Yet only a small number of these will mature into eggs, with one egg typically released during each menstrual cycle. She has until menopause, when her menses cease—which typically occurs between ages 45 and 55—to execute her chances at perpetuating her genes.

SEE ALSO: Chapter One, "Inheritance," PAGE 24, Chapter Ten, "Hormones," PAGE 268

As with developing sperm cells, immature ova must go through a development process to become fertile. The female counterpart to spermatogenesis is oogenesis. Before birth, the immature egg cells, called oogonia, proliferate by mitosis (cell division) and then, through partial meiosis, differentiate into primary oocytes. These infant eggs, in developmental terms, have two sets of chromosomes, not the single set needed to form half of fertilization's egg-sperm genetic partnership. The primary oocytes remain in this state of arrested development until the girl reaches puberty. After puberty begins, a young woman's ovary will pick only one primary oocyte to complete the first phase of meiosis.

In females, this process differs in one key aspect from meiosis in male spermatogenesis. The cell division of oocytes produces two daughter cells that are unequal in size. One, called the secondary oocyte, gets all the cytoplasm and organelles of the mother cell. The other, called the polar body, gets little more than genetic material. The female body engages in this scheme because the potential egg will need nutrients for its seven-day journey to the uterus. The polar body will eventually degrade and disappear.

The secondary oocyte then starts phase two of meiosis. Again, the cell halts before completion. The oocyte is hedging its bets, waiting for ovulation and a sperm cell. Only if a sperm cell penetrates will the developing egg, now called the ovum, quickly finish meiosis. The process spits out a second polar body while the ovum keeps all the nutrients. Researchers still do not consider this ovum fertilized. Fertilization occurs only when the nuclei of ovum and sperm unite.

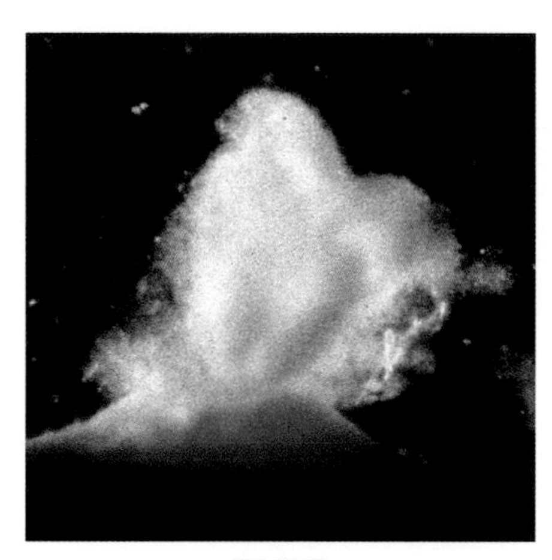

EXIT
A plume of follicular fluids follows the ovum's departure.

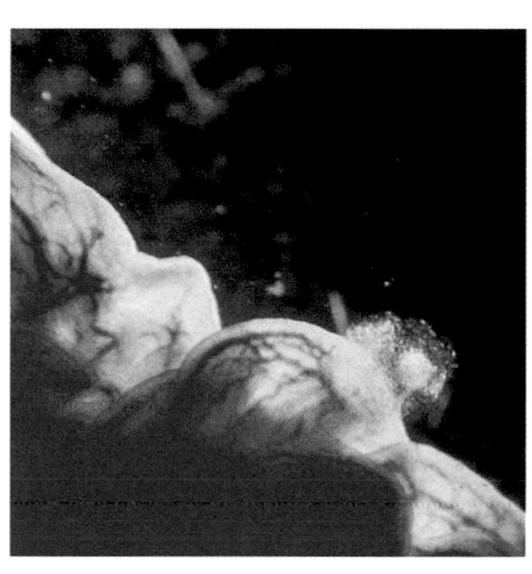

JOURNEY BEGINS
The ovum drifts towards one of the fallopian tubes.

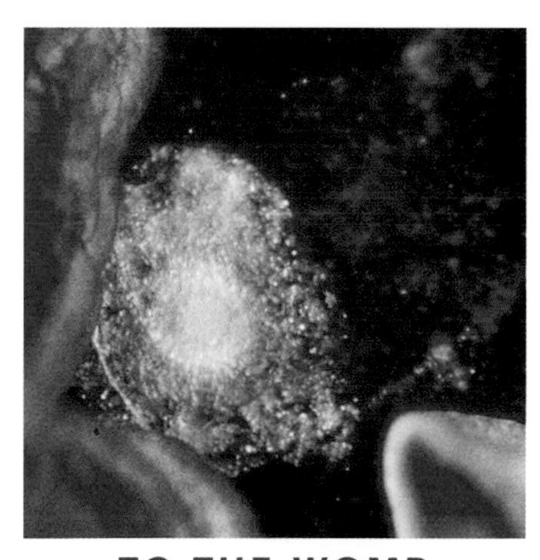

TO THE WOMB
Smaller escort cells surround the ovum as it moves into the tube.

BACKUP PLAN

Fewer than 500 secondary oocytes are released during a woman's lifetime. One might ask, "Why such an inefficient system—seven million oogonia cut down to 500 ova?" Perhaps the reproductive system has evolved a strong backup plan that includes producing a superabundance of egg cells, and then timing their release with precision.

"Backup" also applies to other parts of the egg-making system. For example, follicle cells cover, coax, and cater to developing oocytes throughout their lifespans. Follicle cells begin as a single layer encasing primary oocytes. Later, the cell covering divides into two layers around secondary oocytes. The follicle cells do more than protect. They also secrete a hormonal precursor to estrogen. This hormone guides the development of the immature egg inside its follicular shell, as well as timing and triggering other aspects of the female menstrual cycle.

As the day of ovulation draws nearer, the follicle cells grow and

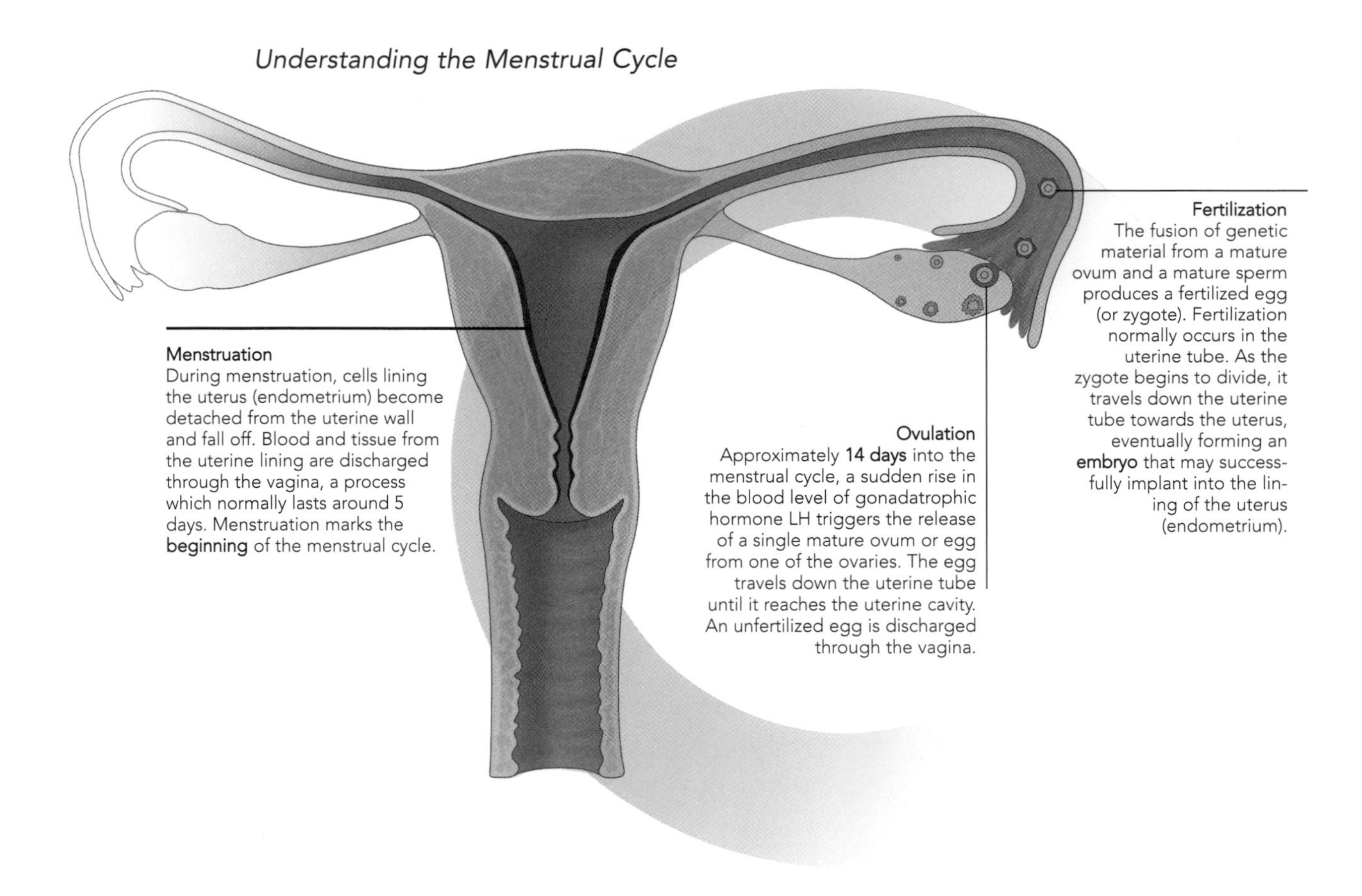

shape themselves into a fluid-filled cavity around the egg. On about day 14 of a woman's menstrual cycle, one dominant, full-size follicle bulges like a boil from the outside of the ovary. The neck of the follicle stretches into a stalk and erupts from the ovary surface, releasing the oocyte into the body cavity near the oviduct that will usher it toward the uterus.

Many women experience this eruption a pinching or cramping sensation in the abdomen, sometimes called "mittelschmerz." The follicle then releases its secondary oocyte. The rest of the follicles that had been developing in parallel simply degenerate, allowing a woman to ovulate only one eventual egg per cycle.

Meanwhile, the ruptured follicle collapses and fills with clotted blood. The transformed follicle then fashions itself into a corpus luteum, a working endocrine gland. For the next 10 days, it secretes progesterone and some estrogen as the oocyte travels down the oviduct toward the uterus. If a sperm fertilizes the egg, the corpus luteum will continue its hormonal mission until the placenta takes over hormone-producing duties. Otherwise, the corpus luteum degenerates, triggering a woman's body to begin again the cycle of egg development and release.

+ EGG FACTS +

+ THE EGG is the largest cell in the human body.
+ The size of a pinhead, it is the only human cell that can be seen with the naked eye.
+ The release of eggs does not alternate between the two ovaries but seems to be random.
+ After ovulation, an egg is available to be fertilized for 12 to 48 hours.

THE MENSTRUAL CYCLE

While the developing oocytes wait,

SEE ALSO: Chapter Ten, "Hormones," PAGE 268

formed and potent in their follicular nests, two other anatomical structures stand ready at the base of the brain to trigger the next phase. The hypothalamus, the hormonal conductor, and its target, the pituitary gland, will direct a waltz of hormones to the ovary.

Together, the hypothalamus, pituitary, and ovary form a trio that plays a wonderful song of fertility from puberty to menopause. The hormonal notes are identical to those in male fertility: gonadotropin-releasing hormone (GnRH) pulses from the hypothalamus to the pituitary gland; follicle stimulating hormone (FSH) and luteinizing hormone (LH) ebb and flow from the pituitary.

The ultimate hormonal target differs, however. While FSH and LH stimulate the testes in males, the hormones animate the ovaries in females. And the response, a 28-day menstrual cycle, is much different from the process of sperm development for ejaculation.

In females, tiny pulses of FSH and LH stimulate growth, maturation, and estrogen secretion from the follicles in the ovaries. Over the course of 14 days, estrogen levels climb to a threshold that feeds back to induce LH and FSH release from the pituitary. This hormonal release stimulates the dominant primary oocytes to begin maturing into a secondary oocyte that bursts forth in a five-minute ovulation.

The ruptured follicle then curtails its production of estrogen, as it metamorphosizes into a corpus luteum. Luteinizing hormone, which prompts this transformation, then prods the corpus luteum to produce progesterone as well as estrogen. These two hormones flow toward the uterus, which, in response to rising estrogen, has been building up a blood-enriched lining, called the endometrium.

The uterine lining thickens from day 6 to day 14 of a woman's monthly cycle. Within it endometrial cells grow receptors for the hormone progesterone, which will maintain the lining, helping it to secrete mucosa and nutrients. These would bathe, cushion, and sustain a fertilized egg, if it arrived, until it implanted. Progesterone, along with rising estrogen, also feeds back to the pituitary gland to shut down its production of LH.

If fertilization does not occur, progesterone will lose its target, the thickened lining of the endometrium. LH coming from the pituitary begins to dwindle. The hormonal supply from the corpus luteum decreases as well. Devoid of hormonal sustenance, blood vessels that feed the uterine lining kink and spasm. Cells within self-digest. The uterine blood vessels constrict and relax. Blood pours into weakened capillary beds and they fragment. All but the deepest layer of the endometrium sloughs off over the next three to five days. A woman experiences this process as menstruation, as the lining sheds. Some women experience symptoms of cramping and back pain as the uterus strongly contracts to expel its unused lining.

Meanwhile, the brain senses lowered LH levels. They trigger the hypothalamus to once again begin pulsing GnRH—the beginning of another cycle and another chance to pass on her genes. And so it repeats, month to month.

BREAKTHROUGH

IN THE 1950S, birth control activist Margaret Sanger asked researcher Gregory Pincus to develop a pill to prevent pregnancy. Pincus teamed with Dr. John Rock, and their research led to the the Food and Drug Administration's 1957 approval of Enovid for the treatment of menstrual disorders. Word spread of Enovid's contraceptive abilities; by 1959 over half a million women were taking it. In 1960, the FDA approved Enovid's use as birth control, giving women unprecedented control over their fertility. When *Time* put "The Pill" on its cover in 1967, right, more than 12.5 million women worldwide were on "The Pill."

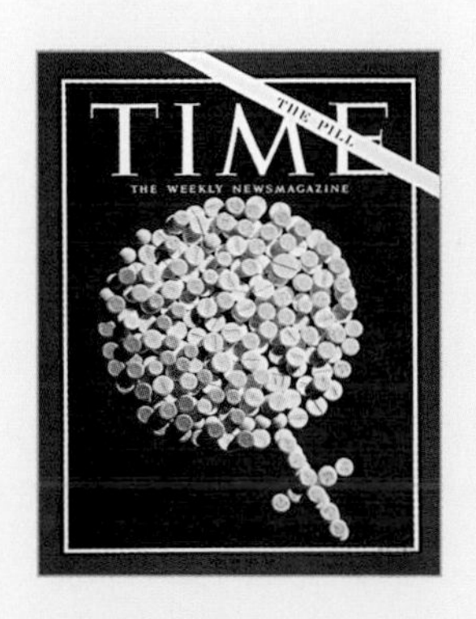

CONCEPTION

HOW BABIES ARE MADE

NATURE HAS GIVEN humans a virile incentive to reproduce: the feeling of sexual arousal. During sexual excitement, a host of stimuli bombard the parasympathetic system, a division of the autonomic nervous system that, as one of its functions, influences sexual stimulation. Erotic touch, sounds, smells, or visual cues all can activate parasympathetic neurons. So, too, does the thought of a sexual encounter. (As therapists often note, the mind is the most powerful sexual organ of all.)

AROUSAL

For males, arousal centers on blood flow. A mesh of arteries supplies the erectile tissue in the penis. Normally, they are constricted, and the penis stays flaccid. When a man becomes sexually excited, nerve cells promote release of a gas called nitric oxide. It prompts penile arteries to dilate, flooding the spongy tissue with blood. The result is an erection: the penis extends outward and grows large and stiff so that it may penetrate the vaginal canal to deposit sperm inside.

Women experience arousal much as men do, but sensations are not limited to the genital area. The clitoris does stiffen, and the vagina lubricates. The breasts engorge, and the nipples become erect.

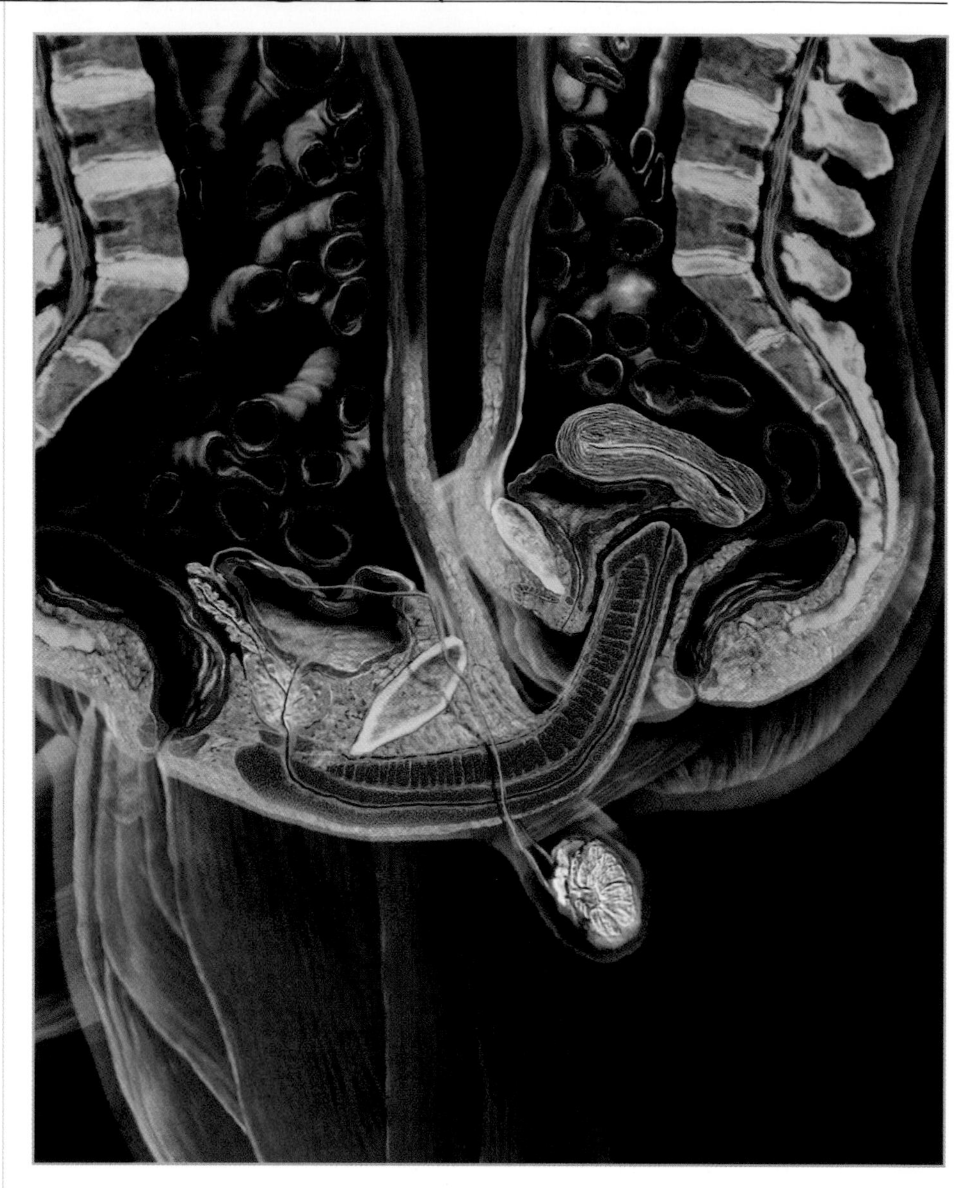

Three-dimensional scanning allows an internal view of the mechanics of sexual intercourse.

As arousal continues, impulses build until they unleash a massive discharge of nerve impulses over sympathetic nerves. The result is called an orgasm, a climactic burst of pleasure accompanied by a rise in blood pressure, an increase in heartbeat, and powerful muscular contractions throughout the body.

In men, the exploding, squeezing sensation of orgasm involves ejaculation, the propulsion of semen from the male duct system at tremendous speed. This high-velocity force during intercourse is paramount. As the penis spasms inside the vagina, it must shoot potent sperm long distances

SEE ALSO: Chapter Ten, "Hormones," PAGE 268

toward an ovulated egg. With female orgasm, women experience sweeping muscle tension and rhythmic contraction of the pelvic floor muscles.

Orgasm in men is followed by a refractory period, in which they cannot produce another erection for minutes to hours. By contrast, women may experience multiple orgasms in a single sexual experience. And women, unlike men, do not need an orgasm to conceive.

JOINING TOGETHER

Fertilization, the next step in the union of the reproductive systems, involves a complexity that approaches the miraculous. The journey proceeds on two fronts. Sperm ejaculated into the vaginal canal swim toward the egg, which is viable only 12 to 24 hours after ovulation. The sperm themselves have only 24 to 72 hours to complete their reproductive jobs.

And so the millions of sperm launch and propel through a gauntlet of acid in the vagina and

UNIQUE SYSTEM

THE REPRODUCTIVE system differs from all others in the body because

+ It is dormant for the first 10–15 years of a person's life. The reproductive system animates only after puberty occurs.
+ In order to fufill its primary function, it depends on interaction with the complementary system of another person. It cannot act alone.

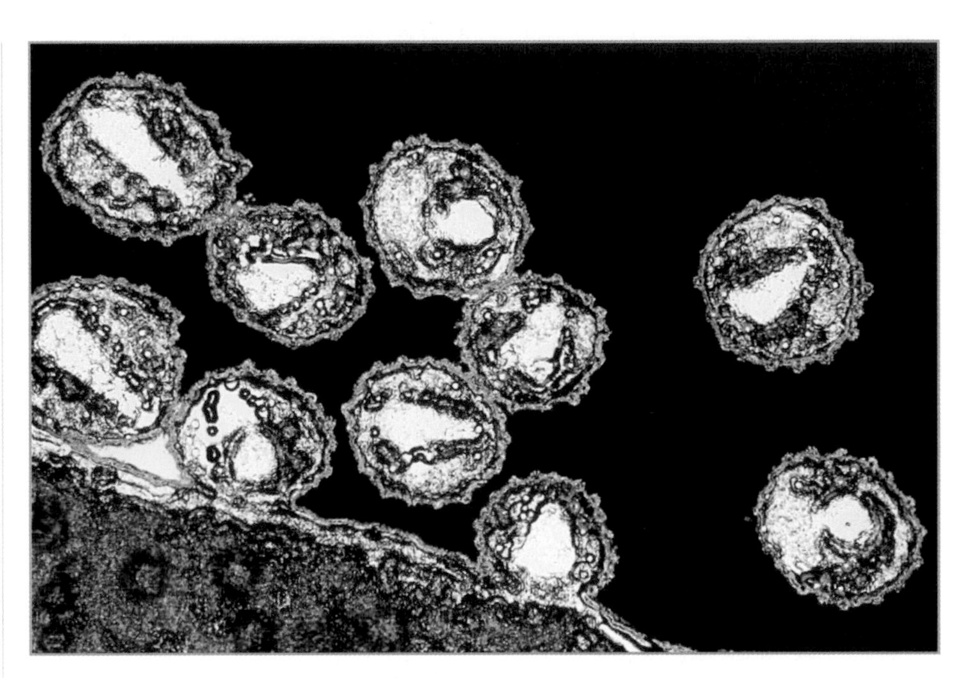

An electron microscope allows a close-up view of the human immunodeficiency virus (HIV), the cause of Acquired Immune Deficiency Syndrome (AIDS).

SEXUALLY TRANSMITTED DISEASES

More than 12 million people in the U.S., a quarter adolescents, acquire sexually transmitted diseases (STDs) every year. Historically, the most notorious were gonorrhea and syphilis, both caused by bacteria and now treatable by penicillin and related antibiotics. More recently chlamydia, genital herpes, and AIDS have taken center stage.

	DESCRIPTION
CHLAMYDIA	Chlamydia is the most common STD in the U.S. Afflicting 4-5 million people annually, chlamydia is an obscure and silent epidemic. Part of the reason is that 80 percent of infected women and half of infected men show no symptoms. Chlamydia is easily detected by a body fluid or urine test and treatable by the antibiotic tetracycline.
GENITAL HERPES	Most commonly caused by the virus herpes simplex type 2, herpes infections are marked by sudden breakouts of painful sores on the reproductive organs. In pregnant women, the virus can cause severe birth defects. The anti-viral drug acyclovir can speed healing and reduce flare-ups.
AIDS	The most fatal is AIDS (Acquired Immune Deficiency Syndrome), caused by human immunodeficiency virus (HIV). More than 900,000 Americans have contracted AIDS since 1981, and nearly a million are currently infected with HIV. There is no cure, but researchers have discovered cocktails of drugs that may slow the spread of HIV in the body.

THE KINSEY REPORTS

A trio of women read a 1953 review on the Kinsey Report on sexuality.

NO SCIENTIFIC discussion of human reproduction can ignore sexuality. The two infuse and drive each other. No one believed this more than Alfred Charles Kinsey, whose infamous Kinsey Reports spawned the study of human sexuality, now called sexology. Born in 1894, Kinsey began his career as a biologist. He was fascinated by gall wasps, tiny insects with a medley of mating practices that led him to wonder just how much human sexual practices varied.

In 1938, Kinsey explored that idea in a groundbreaking lecture at Indiana University, Bloomington. He decried the "widespread ignorance of sexual structure and physiology" and called the scientific community to action. With thousands of interviews, personal experimentation, and a devoted following of graduate students, Kinsey composed *Sexual Behavior in the Human Male* (1948) and *Sexual Behavior in the Human Female* (1953), popularly called the Kinsey Reports.

The reports were bestsellers, but Kinsey's subject, methods, and conclusions drew heated attack that continues today. One of the most controversial was his theory that sexual orientation exists as a spectrum. Kinsey formulated a scale from 0 to 6, where 0 is exclusively heterosexual and 6 is exclusively homosexual. Most people, he found, were somewhere in the middle. Kinsey died in 1956, 17 years before the American Psychiatric Association, influenced by his work, removed homosexuality from its list of mental illnesses. ■

mucus in the cervix. Those conditions kill sperm, as does spill-out from the vagina after intercourse. But estrogen (secreted after ovulation from the empty follicle, the corpus luteum) thins the cervical mucus, allowing a few thousand sperm to reach the uterus.

Still, the sperm are readying themselves to penetrate the egg ahead of them. Swimming through the cervical mucus, the uterus, and the uterine cavity, the tadpole-like sperm begin to shed proteins buried in their cellular membranes, which exposes the tips of the sperm heads, called acrosomes. They hold powerful enzymes that, if launched too soon, would disintegrate the lining of ducts and tubes of the male and female tracts.

When the sperm reach the egg, they cannot immediately penetrate it. The oocyte has three protective coverings: a cellular membrane on the inside, a tough zona pellucida in the middle, and, on the outside, a "radiating crown" of cells.

And so the sperms' enzymes, mainly hyaluronidase, have stayed dormant throughout development. Only when the sperm reach and cling to an oocyte's zona pellucida do they let loose with an explosion of acrosomal enzymes. These eat away at the fortified zona pellucida. Many sperm will spill out all their enzymes and die as they break down the layer. Finally, one sperm is able to bind to a pocket within a receptor on the surface of the secondary oocyte.

FERTILIZATION

In a two-part process, the sperm works it way into the egg's membrane, and the egg assists by internalizing the sperm. The two unite, triggering three biochemical gates to shut, in order to prevent more sperm from entering. If two sperm get in, the oocyte will die.

But this oocyte still is not fertilized. The sperm must travel to the center of the oocyte. There a second polar body emerges, leaving an ovum nuclei. Both it and the sperm head expand, forming male and female pronuclei. Working together, they send out cables called mitotic spindles that connect male to female genetic material. The nuclear membranes break open. Chromosomes spill out. They unite on the mitotic spindle, like towels pinned onto a clothesline. With that, the cell, now called a zygote, is truly fertilized, ready to divide into a new life—a preembryo.

The zygote must then make a week-long journey down the fallopean tube. It is assisted by cilia, waving hairlike structures that line the walls. Thirty-six hours after fertilization, the zygote cleaves into two daughter cells, called blastomeres. The daughter cells diminish in size as compared with their mother. This size reduction is important because the growing zygote, now referred to as a preembryo, needs cells with a high ratio of surface area to volume, in order to more easily absorb nutrients and excrete wastes through their membranes.

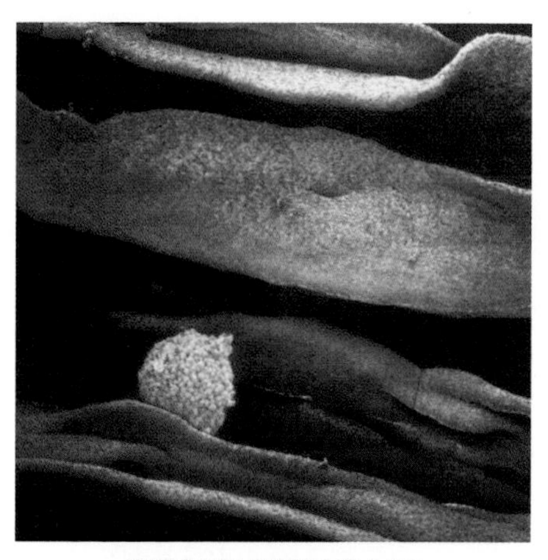

RENDEZVOUS
An egg travels in the fallopian tube, where sperm will swim to join it.

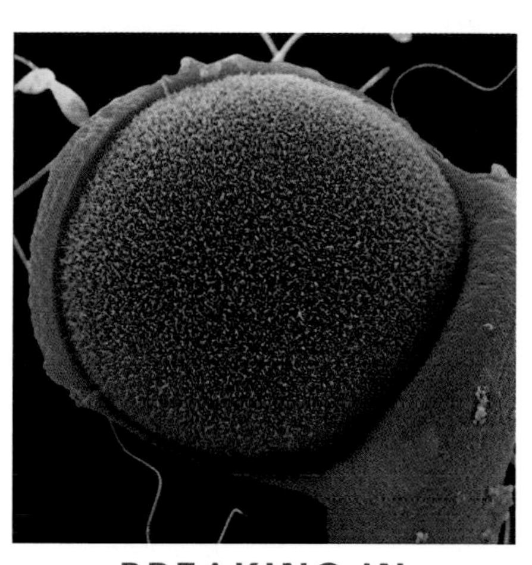

BREAKING IN
Sperm attack the zona pellucida, a membrane encasing the egg.

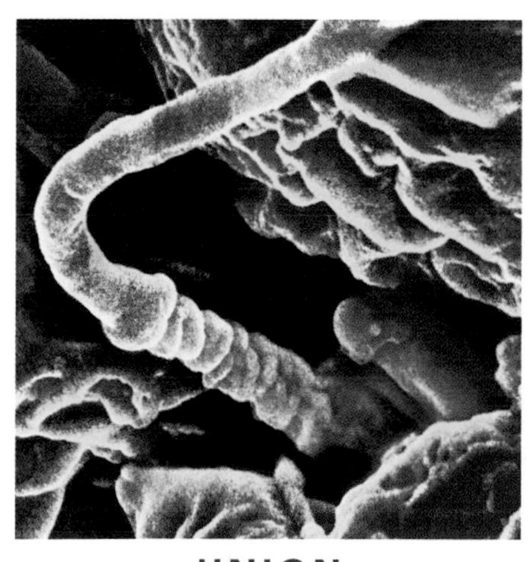

UNION
A sperm plunges into the egg's plasma membrane.

In 72 hours, after more cell division, the dynamic ball of 16 cells is called a morula. Day five after fertilization, it proliferates to a 100-celled structure, the protective zona pellucida breaks down, and a blastocyst forms. This is a fluid-filled cavity composed of cells called trophoblasts, the future placenta, and an inner cell mass that will grow into the actual embryo.

IMPLANTATION

Meanwhile, hormones secreted by the corpus luteum have primed the uterus with a thickened endometrium. About a week after ovulation, large protrusions with sticky protein surfaces emerge.

After exiting the fallopian tube and entering the uterus, the preembryo looks for a docking station. If the chemical environment is ripe, the blastocyst sticks to the endometrial wall and implants high up in the uterus. If the vascular wall is not ready, the blastocyst floats freely a bit longer and attempts it again later, adhering to a lower position in the uterus.

Implantation is a process that is chemical as much as a physical. Once anchored, trophoblast cells of the blastocyst secrete digestive enzymes and growth factors. They inflame the endometrium in order to weaken it enough to burrow inside. Fourteen days after fertilization, endometrial cells cover the precious embryo with a cellular blanket of protection and nutrition in which it can grow.

OUTSIDE THE WOMB

WHILE RESEARCHERS have dissected fertilization into a series of perfectly timed steps, the process still holds many mysteries. Couples trying to have a baby but who do not conceive within a year are considered to have a "fertility problem." Given the emotional costs of infertility and the desire to explore the yet-unsolved puzzles surrounding conception, researchers have long proposed means to help women conceive. In vitro fertilization (IVF) refers to techniques in which clinicians extract eggs and fertilize them outside a woman's uterus.

BEGINNINGS

The idea of assisted fertilization became the buzz in the 1930s when an anonymous editorial in the prestigious *New England Journal of Medicine* suggested that the practice, if successful, would be "a boon for the barren woman with closed tubes!" The article triggered an explosion of controversy, as the public grappled with the prospect of doctors' "playing God." The archbishop of Canterbury in 1948 even recommended banning the practice.

Still, researchers persisted, coming close in Australia in 1973 with the first IVF pregnancy, which did not come to term. Meanwhile, in England, Cambridge University scientist Robert Edwards and Oldham gynecologist Patrick Steptoe had begun working together on IVF. Steptoe had pioneered the technique called laparoscopy, by which a fiber optic light could be inserted into the abdomen to inspect blockages in fallopian tubes. Using this same technique, he developed a method of extracting eggs from the ovaries for fertilization by means of a suction needle in the laparoscope.

Edwards had traveled to Baltimore's Johns Hopkins University in the summer of 1965, where he worked on perfecting the techniques for fertilizing a human egg outside the womb. While there, he was helped by gynecological surgeon Howard Jones and his wife, Georgeanna Jones, a reproductive endocrinologist, who together later pioneered IVF in the United States.

Over the years, the British team faced hostility and skepticism, as critics claimed IVF would produce abnormal babies. Finally, ten years after Steptoe and Edwards first fertilized an egg in a petri dish, they announced the birth of Louise Joy Brown, the world's first "test tube baby," born July 25, 1978. The healthy baby weighed 5 pounds 12 ounces. The birth led to an uproar over whether this was a one-time event and, if not, what meaning this conception method had in terms of morality and safety.

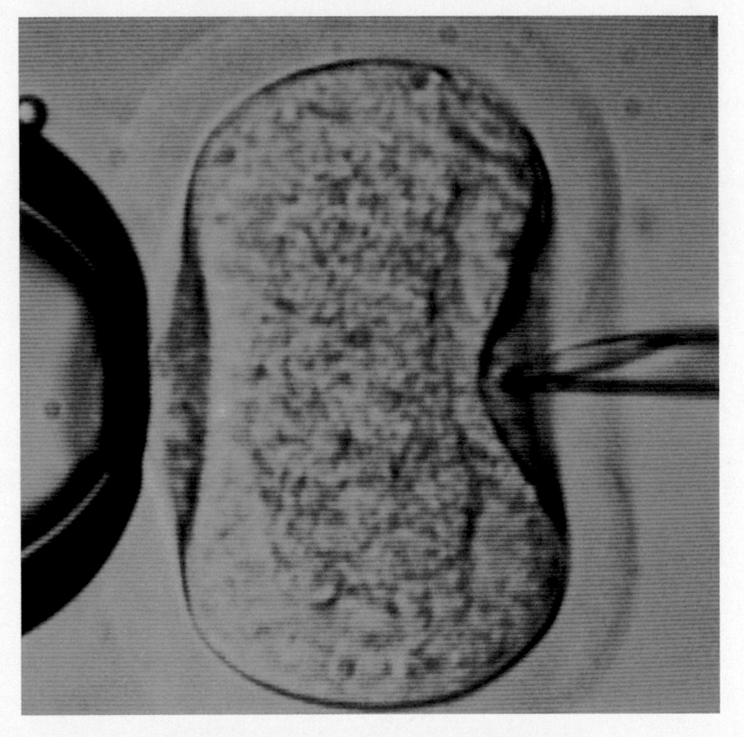

With high magnification, it's possible to see how a needle (at right) injects sperm directly into a human egg.

OH BABY!

Despite these concerns, other in-vitro babies soon followed, including the first success in the United States. Elizabeth Jordan Carr was born December 28, 1981, in Norfolk, Virginia, at the Jones Institute for Reproductive Medicine at Eastern Virginia Medical School, the fertility clinic founded in 1980 by Howard and Georgeanna Jones. Eventually, nearly 3,000 children would be born through the efforts of the Jones Institute.

In 1984 French researchers developed a new technique for retrieving eggs. Steptoe's laparoscopic procedure, though minor surgery, still required the patient to be anaesthesized and hospitalized. In the French method a long aspiration needle is inserted through the wall of the vagina into the ovary, where it extracts fluid containing eggs. The procedure, which is guided by ultrasound, can be done in an outpatient setting. It is simpler, cheaper, and reduces risk to the patient; for those reasons it can increase success because more patients are willing to repeat the procedure multiple times.

One-year-old Louise Brown, the first "test tube" baby, smiles as she appears on television in 1979.

NEW TREATMENTS

Today, IVF has evolved into a popular practice, a chance at parenthood for the estimated 7 percent of U.S. couples who are infertile and others like them worldwide. Today, couples may also try assisted reproductive technologies (ART), fertility treatments in which both eggs and sperm are involved. In general, ART procedures begin as gynecologists surgically remove eggs from a woman's ovaries, sometimes after hormonal treatment. Technicians combine her eggs with the father's sperm in a culture dish for about 18 hours. If sperm counts are low, technicians can even inject a single activated sperm directly into the egg in a technique called intracytoplasmic sperm injection, or ICSI. After about 48 hours, when the preembryos reach the six- to eight-cell stage, a doctor returns several to the woman's uterus, where—if successful—they will implant and develop. While the success rates for ART vary from clinic to clinic, according to the Centers for Disease Control and Prevention, fertility experts performed a reported 122,872 ART cycles in 2003, leading to the birth of 48,756 infants, including some twins and triplets. That same year, Louise Brown celebrated her 25th birthday. Some 5,000 test-tube babies were invited to the party, hosted by Robert Edwards, who had helped pioneer the IVF techniqes that brought Louise into the world.

CONTINUING DEBATES

Researchers continue to push the frontiers of assisted fertilization, sometimes provoking new controversies. The American Society for Reproductive Medicine, among others, has drafted policies to regulate procedures. Still, ethical and religious dilemmas abound. Most recently, debate has erupted about pregnancy after menopause, the use of genetic diagnosis before implantation to screen embryos, and the possibility that embryos might be created to be stem cell sources.

Fears that the children born of these scientific techniques would be abnormal has long since been laid to rest. Louise Brown's parents went on to have another daughter, Natalie, also by means of IVF. In 1999, Natalie became the first test-tube baby to give birth. Then, in December 2006, Louise Brown, the world's first test-tube baby, bore a son, Cameron Mullinder, conceived naturally.

PREGNANCY

A BODY CHANGES AS A NEW LIFE GROWS WITHIN

As a pregnancy develops, it also changes a mother-to-be's body. Hormones wax and wane, resetting her body chemistry in profound ways. Some women will experience numerous symptoms as a result; others, realtively few as the baby grows over the next nine months.

EARLY STAGES

Pregnancy is divided into three-month stages called trimesters. In the earliest stages of the first trimester, a common symptom is "morning sickness." Despite the name, bouts of nausea and vomiting can occur at any time of the day. In rare cases, nausea is constant, and vomiting occurs several times a day. The condition, called hyperemesis gravidarum (HG), typically subsides by the 20th week of pregnancy but may require brief hospitalization to correct fluid and metabolic imbalances.

Surging estrogen and progesterone cause blood vessels to sprout throughout a woman's reproductive organs, giving her genitals a purplish cast and increasing their sensitivity. Blood supply to the breasts increases, sometimes appearing as a network of bluish lines. The breasts swell and become more tender as the number of milk glands increase.

Beyond hormonal changes, the first trimester holds many surprises. A growing embryo, measuring about three inches in length and weighing about half an ounce by the end of 12 weeks, fills the pelvic cavity. Weight gain is slight, only about a pound per month. The

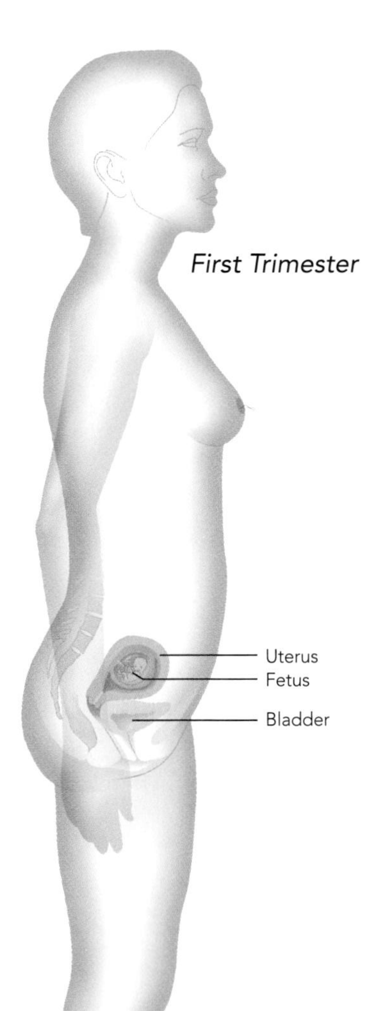

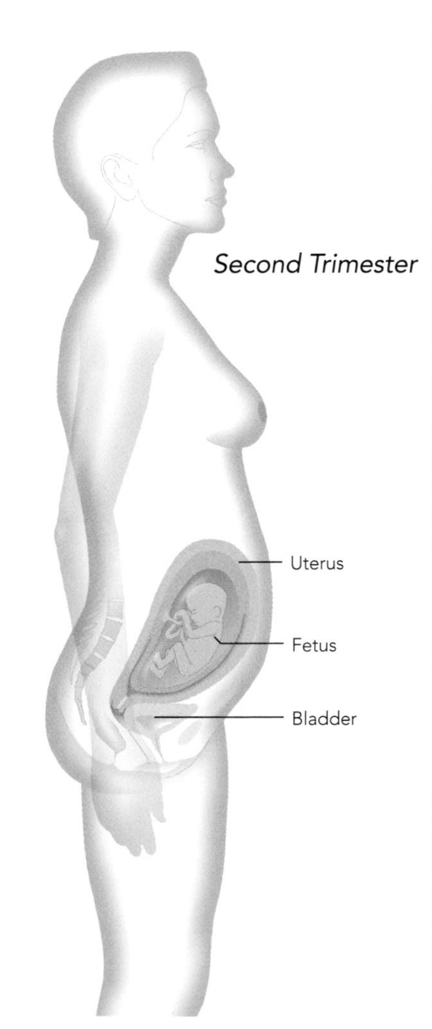

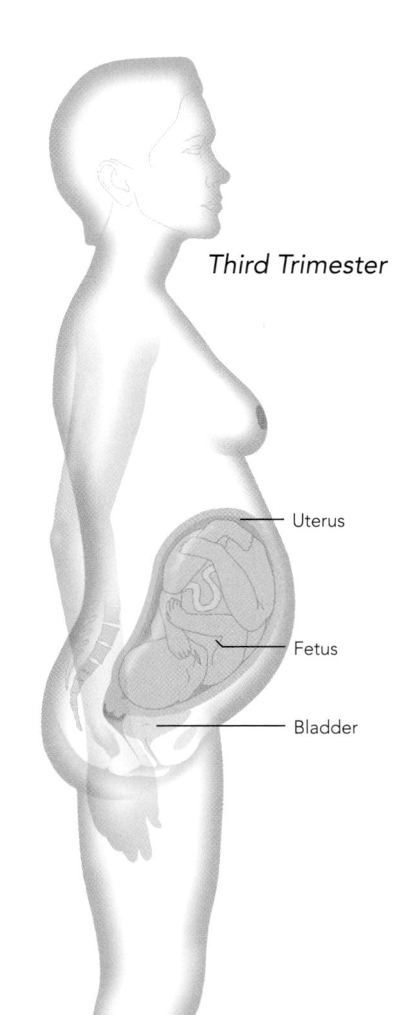

SEE ALSO: Chapter One, "Development," PAGE 34

abdomen may protrude slightly by the end of the trimester. Other symptoms include backaches and more frequent urination as the embryo, in its fluid-filled amniotic sac, begins to press upon the bladder. The heartbeat speeds up as it pumps increasingly more blood to the uterus. Feelings of fatigue are typical as the body adjusts.

PREGNANCY TEST

+ AN EGYPTIAN PAPYRUS from 1350 B.C. describes a pregnancy test in which a woman would urinate on wheat and barley seeds. If the seeds grew, she was pregnant.
+ In the early 20th century, a woman's urine was injected into rabbits, rats, and frogs to detect human chorionic gonadotrophin (hCG), a hormone produced only by a developing embryo.

LATER TRIMESTERS

During the second trimester, the placenta provides the embryo with food, energy, and hormones through two arteries in the umbilical cord while removing waste through a single vein. It also takes over the role of secreting estrogen and progesterone, a task formerly performed by the corpus luteum and the trophoblast cells of the embryo, buried in the uterine lining. (If this transfer fails, the embryo will be naturally aborted. One-third of pregnancies end after the first trimester because of failures in implantation or placental development.)

In the second trimester symptoms of nausea and fatigue fade, and hormone levels begin to stabilize. The mother's metabolism begins to process more fatty acids and less sugar (glucose) to preserve as energy for the baby. In some women, this change can cause gestational diabetes mellitus. The condition normally ends at childbirth, but about half of the women in whom it occurs will develop Type 2 diabetes later in life. During this trimester, weight gain increases to a rate of about a pound per week. Maximizing nutritional value to provide for the growing fetus is vital, as is keeping weight gain within healthy limits.

During this second trimester, the mother feels the baby's first movements. The abdomen swells as the uterus expands upward, toward the diaphragm, and down, toward the bladder. The breasts continue to activate more milk glands, and the areolas, the pigmented areas aroung the nipples, may darken and enlarge.

In the third trimester, the continued growth of the baby causes increased pressure on the digestive and other internal organs. The developing baby needs more nutrition than ever, causing the mother's total volume of body water to rise. Blood volume increases 25 to 40 percent to carry more nutrients where they are needed; the ankles, fingers, and face may swell as a result. Pressure from the swollen uterus pushing on the pelvic blood vessels can cause varicose veins. In the last couple of weeks of pregnancy, estrogen levels have climaxed, causing irregular ripplings of the uterine muscles—called

WHAT CAN GO WRONG

ECTOPIC PREGNANCY occurs when an embryo (shown in the yellow area) implants outside the uterus. Most ectopic pregancies occur in the fallopian tube carrying the embryo to the uterus. Blockages to the tubes, birth defects, or a tubal ligation may cause an ectopic pregnancy. Supplemental estrogen and progesterone also can slow a fertilized egg in the tube enough to cause ectopic pregnancy. Since a placenta cannot develop and an embryo cannot grow in a tube, half of ectopic pregnancies end naturally, without serious damage or future infertility; others require surgery to prevent rupture of the tube.

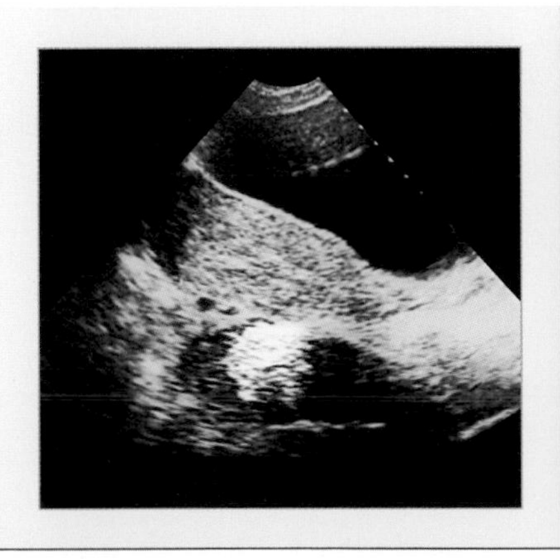

IGNAZ SEMMELWEIS

A portrait of Ignaz Semmelweis as he appeared in the 1800s.

A GREAT OBSTETRICAL breakthrough was a relatively simple procedure: hand washing, first championed by Ignaz Philipp Semmelweis. Born in Hungary in 1818, Semmelweis practiced obstetrics when a form of blood poisoning called puerperal fever (also called childbed fever) killed many new mothers and their offspring. The disease tended to strike the hardest among the lower socioeconomic classes, many of whom gave birth in hospitals where the fever ran rampant.

As head of Vienna General Hospital's First Obstetrical Clinic, Semmelweis noted that his wards' mortality rates were triple those of the wards staffed by midwives. Yet the delivery techniques of the two did not differ. In 1847, Semmelweis discovered why after a colleague punctured his finger during the autopsy of a puerperal fever victim and died of infection. Pathology results showed that the woman and the physician died from similar causes, leading Semmelweis to postulate that an infectious particle had passed between them. Physicians could transport pathogens after conducting postmortem examinations to women in childbirth. By contrast, midwives did not perform autopsies, so they were not exposed.

Semmelweis mandated that physicians in his ward scrub their hands with a wash of chlorinated lime to kill the agent. The hand washing policy reduced mortality rates from 18 percent to less than 2 percent. In 1861, Semmelweis published his findings, showing the decline in death rates after instituting hand washing. Most doctors, however, did not adopt his practices until 1878, when Louis Pasteur published his theory of germs. ■

Braxton Hicks contractions, or false labor. Estrogen also ripens the endometrium, causing it to sprout receptors for the potent pregnancy hormone oxytocin, which will play a large role in childbirth.

BIRTH

Signs of labor may be a series of fluttering, false contractions, as a rupture of the amniotic sac (or "water breaking"), or as full labor. As it begins, fetal cells excrete oxytocin, which causes the uterus to begin contracting. Placental cells make prostaglandins, while the mother's hypothalamus churns out more oxytocin. The result is rhythmic, expulsive contractions, which squeeze the upper uterus at first in intervals about 15 to 30 minutes apart.

The contractions gain force and propagate downward, reinforced by muscles of the lower uterus. The cervix thins and softens, prompting the birth canal to open. The cervix dilates, and as the baby moves into position the mother

C-SECTION

WAS THE CESAREAN section named for Julius Caesar's surgical delivery from the womb? It's unlikely. In Roman times, cesareans were typically performed if a mother was dead or dying, and Caesar's mother lived to see him to adulthood. Another possibility: It comes from the Latin *caesones*, the term for male infants born by a postmortem cesarean.

SEE ALSO: Chapter One, "Development," PAGE 34

feels an intense urge to bear down. Lasting from six to twelve hours, this first phase of labor is generally the longest.

Eventually the cervix fully dilates to about ten centimeters. With the mother's pushing, the uterine muscles compress until the largest part of the baby's head reaches the vulva, an event called "crowning." The baby's head is delivered, followed, much more easily, by the neck, shoulders, and body. After the baby is delivered, the umbilical cord is cut. In the last act of birth, the placenta is shed and expelled.

Before and after the birth, hormonal interactions prepare the mother's body for lactation, or milk production by the mammary glands. Hormones, including lactogen, trigger the hypothalamus to excrete prolactin-releasing hormone (PRH), which stimulates the pituitary to make prolactin, the milk-triggering hormone.

This process takes about three days, during which the mammary glands produce colostrum, a low-fat, vitamin-stocked substance that, like milk, helps sustain a nursing baby and prevent infections. Hormones continue to exert influence in breastfeeding. When an infant latches onto its mother's nipple, touch sensors there trigger the production of PRH and prolactin. Milk supplies accumulate in the sinuses of the mammary glands. Then oxytocin causes the "let-down reflex" in which milk is ejected from the glands and the mother feels a sense of well-being, in parallel with her baby's comfort at being fed. Oxytocin also stimulates contractions of the uterus and that organ's return to its normal size. During this phase, mother and baby form a strong emotional bond. And so the species lives on.

Enjoying a quiet nap, this baby girl embodies the end result of the reproductive process.

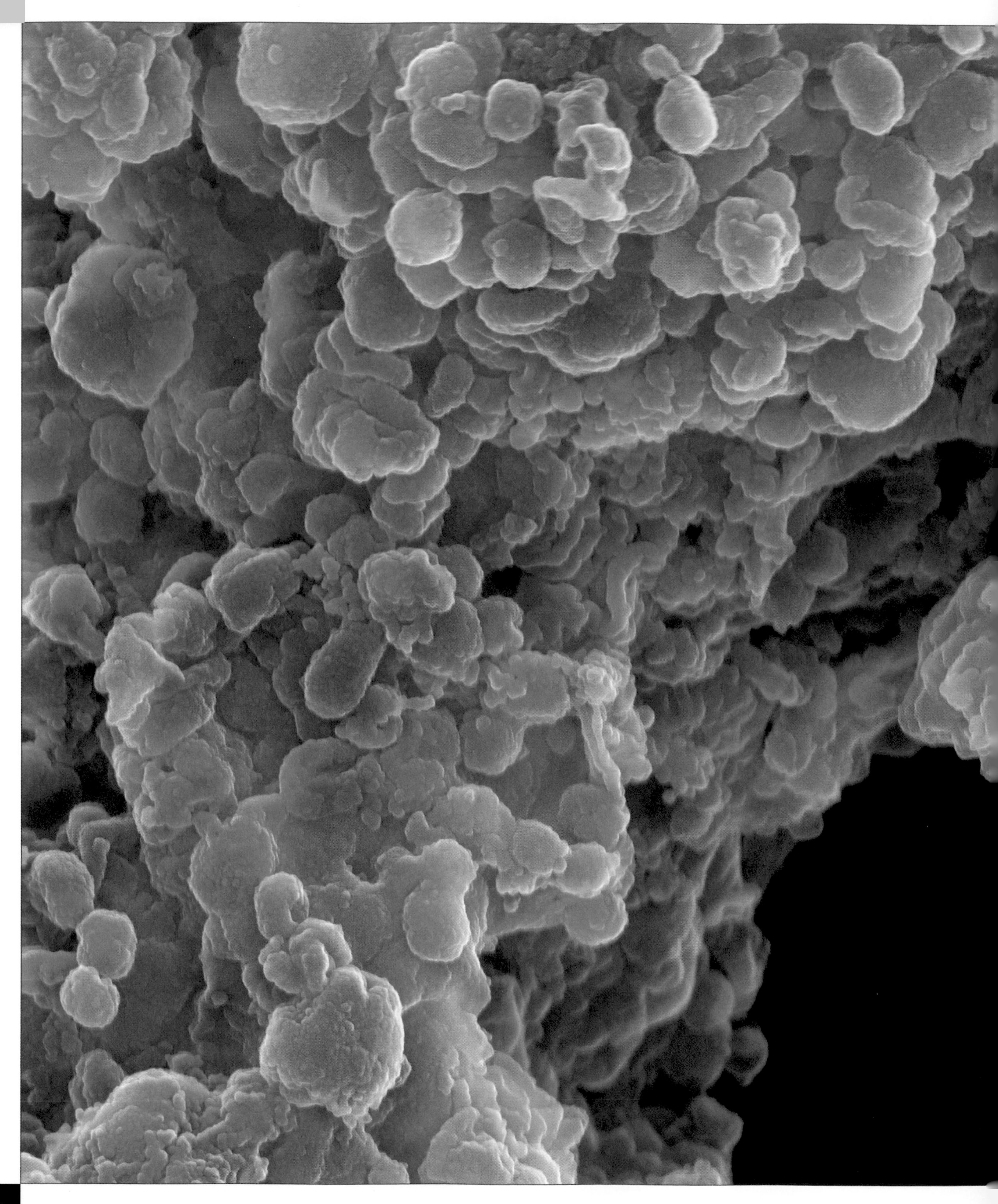

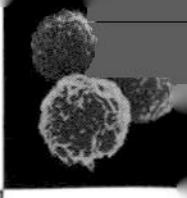

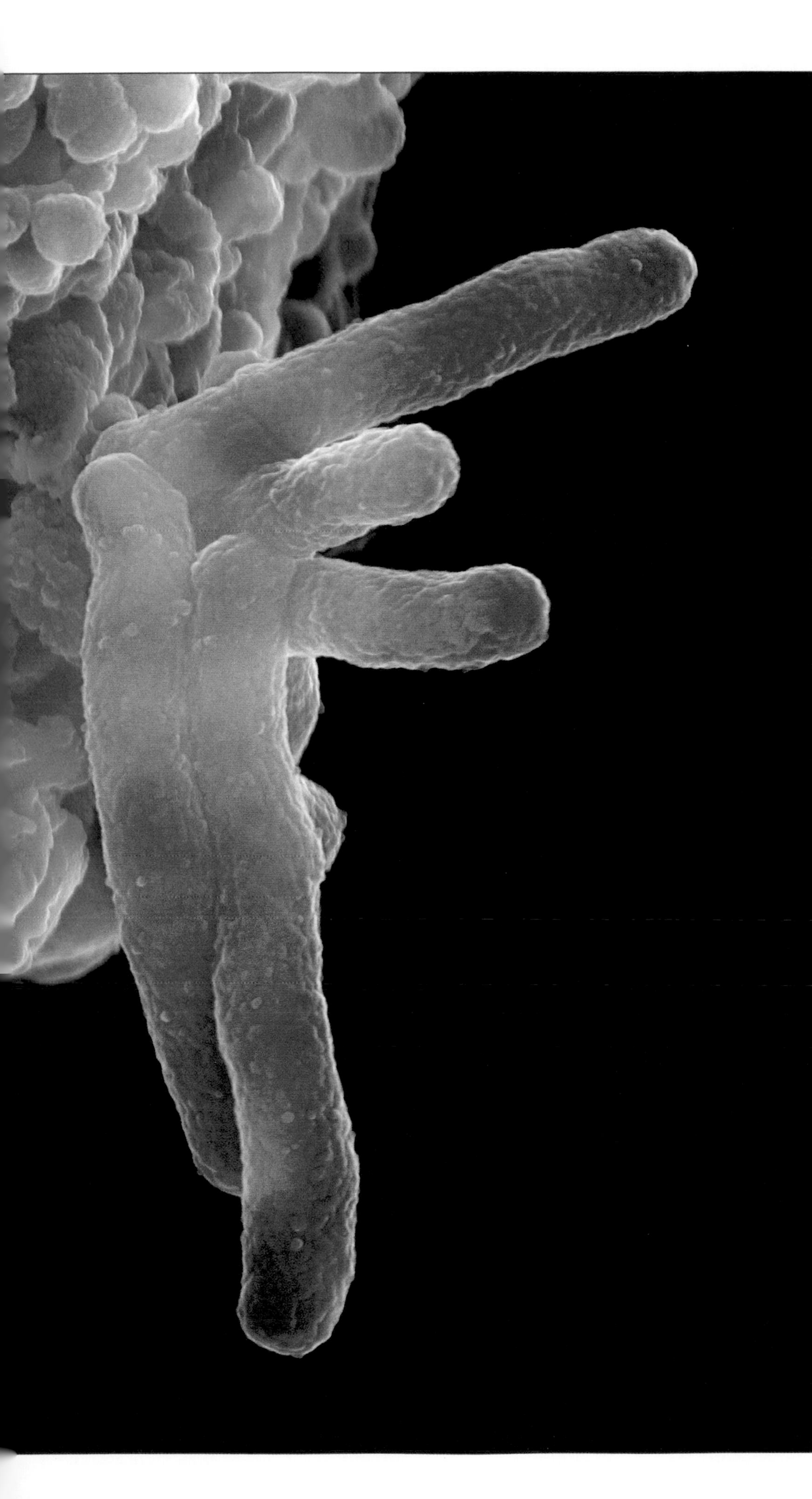

CHAPTER TWELVE

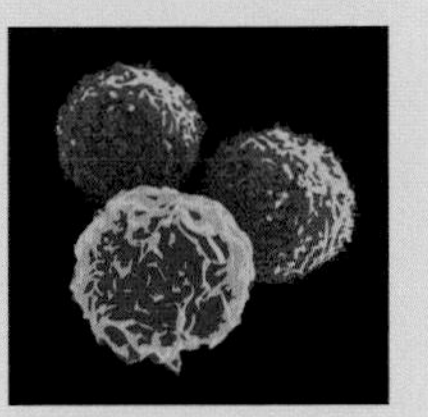

PROTECTION

EVERY DAY, THE BODY IS attacked by many potential invaders. The invaders, microorganisms called pathogens, come in the form of viruses, bacteria, parasites, and fungi. The body's multitiered defense is known as the immune system. One branch, the innate system, provides prompt emergency response to incoming pathogens; the other, the adaptive defense system, targets each pathogen in a specialized way. This strong defense keeps everything running smoothly by keeping out the invaders.

A protective macrophage (blue) prepares to attack an advancing tuberculosis bacterium (purple).

IMMUNE SYSTEM

KEEPING THE BODY HEALTHY

UNLIKE SOME OF THE body's other systems (such as the digestive and circulatory systems), the immune system is a functional system that operates throughout the human body, its work carried out by the body's trillions of immune cells and specialized molecules.

The first line of defense is the skin. Our body's covering of tissue protects us from all kinds of dangers large and small. One obvious invasion of this first defense is a bump, bruise, burn, or cut; nerves in the skin register pain and tell us to avoid the rock or the hot stove that caused the problem. A bump resulting in a bruise means that blood vessels have broken beneath the skin, and the escaped blood has formed little clots called hematomas. A cut allows microorganisms to breach the first wall—the skin—and trigger internal defenses, which we will examine in greater detail later in the chapter.

Injuries are generally the least of the immune system's problems. Armies of germs—viruses, bacteria, parasites, and fungi—make daily sorties against our body, whether we are injured or not. Through the mouth, nose, eyes, ears, genitals, and skin pores, these saboteurs infiltrate the body. Since skin can't seal out every possible opening (or else we could not breathe or eat), the first line of defense has

SEE ALSO: Chapter Two, "The Skin," PAGE 46, Chapter Five, "The Blood," PAGE 132

IMMUNOLOGY

A plague epidemic took the lives of more than 100,000 in 1720 Marseille, France.

IT HAS LONG BEEN known that the body has natural defenses and that it can, to some degree, heal itself when infected. Immunology, the science of how these defenses work on a cellular level, has developed only within the past one hundred years. It was known during the Revolutionary War that soldiers who injected themselves with a dose of smallpox had a better chance of not contracting the disease. Then in 1796 British physician Edward Jenner took the risk factor out of vaccination by creating one of the first successful vaccines for smallpox (see "Discovering Immunization," page 339). He did not know that smallpox was caused by a virus or the mechanisms by which the body's immune system could be primed by a small amount of it.

Not until the mid-1800s did the understanding that microorganisms caused infection and disease become accepted. Discoveries then began piling up. French chemist Louis Pasteur developed several groundbreaking vaccines for farm animals, including ones against rabies and anthrax in cattle. By the end of the century, scientists had discovered phagocytes, antibodies, and the fact that vaccines work by stimulating the production of antibodies.

During the first half of the 20th century, researchers began unlocking the mysteries of how antibodies respond to specific antigens. In the mid-1960s scientists described the workings of B cells and T cells, both adaptive immune cells.

Much research continues into the present time, with progress being made recently in understanding the genetic controls of the immune system and how each of the immune system's millions of different antibodies are produced. ■

a backup—the mucosal lining. Pathogens entering the gastrointestinal tract, genitourinary tract, and lungs are generally stopped by the sticky, acid-secreting mucous membranes lining the organs.

A second line of defense comes into play against those persistent invaders not blocked by the skin and mucosal lining. This second line is composed of cells, which include a broad category called phagocytes, or engulfing cells-whose basic job is to cat thc invaders. Among the phagocytes are white blood cells called neutrophils and eosinophils, which help initiate inflammation. Another broad category of generalized cells, the natural killer cells, are unleashed immediately against viruses, cancer cells, and other foreign cells. In addition to these immune cells, numerous chemical compounds respond to infection and injury, move in to destroy pathogens, and begin repairing tissue. While these cells and chemicals are at work, the body often makes another generalized response to foreign invasion—it heats up, or produces a fever.

The body's third line of defense—the adaptive defense system—is often slower to react, because the body must determine what kind of attack is underway and organize this final, more specific response. When the first two lines of defense fail, specialized warriors then go into action against the most perfidious of pathogens, whether they be microbes from the outside or internal cells in revolt such as cancer cells. These elite fighting units are not just lying around in wait; instead, they are trained on the job—that is, they are created in response to a virus, for instance, that the body has not seen before. The adaptive defense system is thus an improvised response to a specific invader.

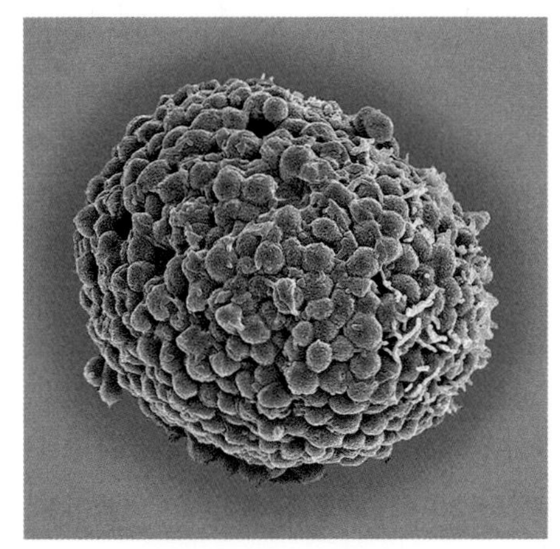

MAST CELL
This kind of white blood cell is activated by allergic reactions and physical injuries.

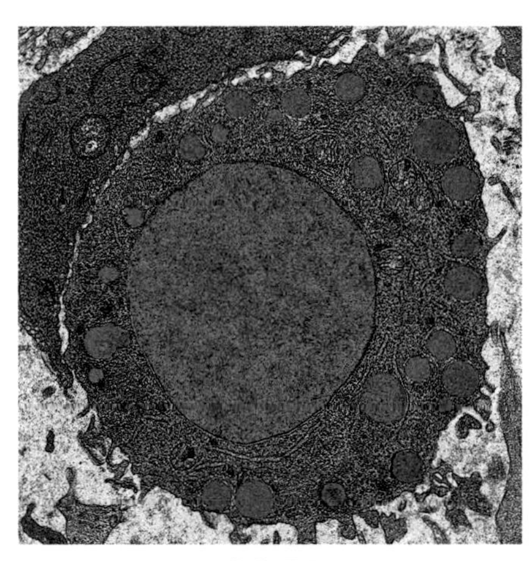

BASOPHIL
Basophils are the smallest and least common white blood cell at play in the immune system.

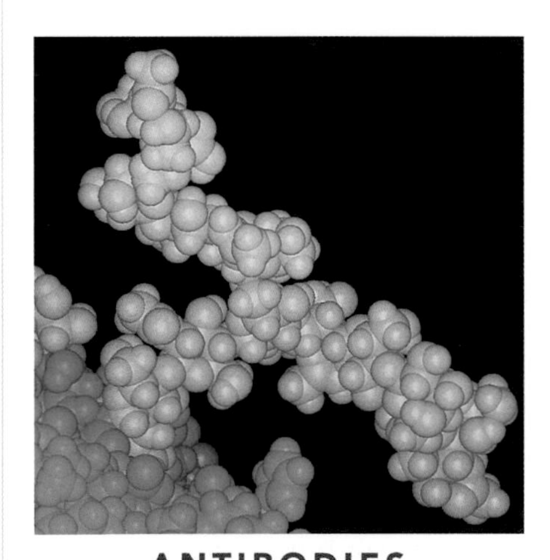

ANTIBODIES
These specialized proteins combat specific microorganisms.

Chief among the immune cells of the adaptive defense system are white blood cells called lymphocytes that come from bone marrow and thymus and circulate through the body. The body also produces antigen-presenting cells that swallow pathogens or chemical constituents of the pathogens called antigens and cause lymphocytes to start reacting against the invaders. Two remarkable qualities of the adaptive defense system are that it works systemically—that is, once activated in one part of the body, it functions throughout; and it memorizes the antigens so that the next time they come along the body hits back quicker and harder.

REACTIONS

The nonspecific, or innate (natural), reaction requires no previous exposure to the antigen being encountered. Something potentially harmful enters the body, and its defenses go up. The nonspecific immune system can kill or limit the spread of pathogens, but it does not increase its efficacy against any specific invader—it doesn't learn to recognize one particular antigen as more harmful than any other. It simply does the best job it can and leaves the rest up to the body's specific immune response.

The specific immune system is sometimes called the adaptive, or acquired, immune system, because unlike the innate system it does learn to distinguish between particular pathogens and antigens. People afflicted with the chicken pox virus are sick for about a week; after this they are usually immune to the disease for life. Why? Specific antibodies and T lymphocytes recognize the virus upon re-exposure and eliminate it before it causes any harm. Mumps, measles, and most other infectious diseases can also be memorized by the body's immune system so that when the pathogens appear, the body quickly starts making the appropriate antibodies; each kind of antibody is a specific key that works to unlock and block only that specific pathogen's antigens. Immunizations for these diseases are a refined shortcut: They train the body to recognize specific antigens without your having to get sick.

Adaptive immunity can be divided into two basic kinds—humoral and cell-mediated. The humoral immune response is carried out by antibodies. Also known as immunoglobulins, these antibodies are molecules of protein that derive from B lymphocytes and their progeny, plasma cells. Antibodies circulate in the blood and other bodily fluids—hence the name "humoral," from the ancient concept of body humors. The cell-mediated immune response is an immunity granted specifically by activated T lymphocytes, which destroy (or lyse) infected cells, cancerous cells or foreign cells in transplanted organs.

INNATE
Bumps, bruises, and swelling are all signs that the innate defenses are at work.

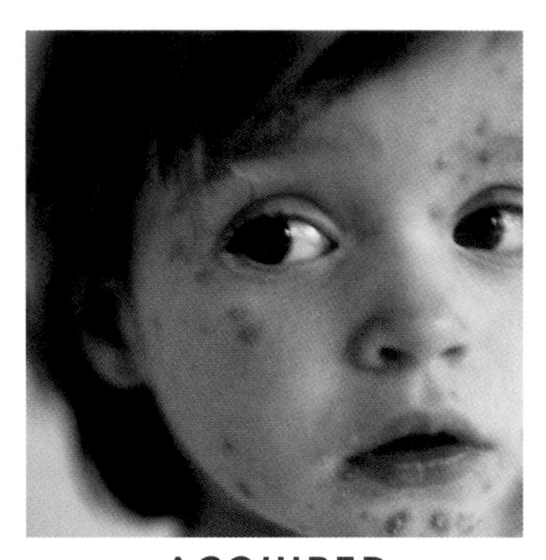

ACQUIRED
The body's adaptive defenses can learn to target pathogens, like chickenpox, after exposure.

+ COMPARISONS +

+ THE INNATE reponse is broad-based; the adaptive response targets specific pathogens.
+ The innate response is found in all organisms; the adaptive response is found only in jawed vertebrates.
+ The innate response has no "memory;" the adaptive response memorizes individual pathogens and increases its effectiveness on re-exposure.

There are a number of other organs and tissues involved in the body's immune response. Bone marrow and thymus generate the important white blood cells known as lymphocytes. They migrate to the body's lymph nodes, which are found in the neck, armpits, groin, abdomen, and at the base of the lungs.

The body's largest lymph organ is the spleen, a five-inch-long organ located on the left side beneath the diaphragm and behind the stomach, tucked safely underneath the rib cage. It produces antibodies, is the site of activation of T lymphocytes, destroys aging and damaged blood cells, and detects and destroys certain pathogens.

The thymus, located between the top of the lungs at the base of the neck, is a small organ involved in developing the T-cell part of the immune system. Another aggregate of lymphoid tissue, the tonsils are situated in the throat. In addition to these lymphatic masses, the body has a network of lymphatic vessels that transport lymph—a protein-containing fluid. Lymph is a fluid, while lymphocytes are white blood cells found in both blood and lymphatic tissues.

CHAPTER GLOSSARY

ALLERGY. Hypersensitivity or overly aggressive immune response to a perceived invader, or antigen.

ANTIBODY. Protein molecule produced by a plasma cell (progeny of B lymphocytes) in response to a specific antigen; also called an immunoglobulin.

ANTIGEN. A substance (chemical constituent of a microbe or other chemical material) recognized as a threat and thus provoking an immune response.

ANTIGEN-PRESENTING CELL. Cell that can be fixed or migratory; presents antigens to T cells in an immune response.

AUTOIMMUNITY. An immune response where the body attacks healthy body tissues.

B CELL. Lymphocyte that can develop into an antibody-producing plasma cell or a memory cell.

BASOPHIL. The smallest and least common type of white blood cell. Releases histamines and other inflammation mediators.

CELL-MEDIATED IMMUNITY. Wing of the adaptive immune system in which activated T cells and other effector cells destroy microbes and other antigens.

HISTAMINE. Substance released by the immune system that dilates blood vessels.

HUMORAL IMMUNITY. Wing of the adaptive immune system in which immunity is provided by antibodies.

IMMUNIZATION. The conferring of immunity, whether actively (through administration of a vaccine) or passively (via the placenta, breast milk, or through the administration of of an "antiserum").

IMMUNODEFICIENCY. Deficient function of the immune system.

INFLAMMATION. Nonspecific defensive response to injury. Indicated by redness, increased temperature, swelling, and pain.

LYMPH NODE. One of many bean-shaped structures situated along the lymph system.

LYMPHOCYTE. Type of white blood cell in the blood, lymph, and immune organs involved in both cell- and antibody-mediated immune responses.

MACROPHAGE. A large phagocyte derived from a monocyte; it phagocytizes bacteria and other debris, and it acts as an antigen-presenter for T and B cells.

MAST CELLS. Immune cells that detect foreign substances in the body and activate local inflammatory responses to them.

MHC PROTEINS. Major histocompatibility complex proteins that occur on all cells in the body in a configuration unique to each individual person.

NATURAL KILLER (NK) CELL. Defensive cells of the body's innate immune system; lyse (or destroy) pathogens, cancer cells, and transplanted foreign cells before activation of the adaptive defenses.

NEUTROPHIL. Most common kind of white blood cell, the neutrophil phagocytizes and destroys bacteria and other pathogens.

NONSPECIFIC IMMUNE SYSTEM. A generalized immune response to a pathogen, requiring no previous encounter. Also called "innate defense system."

PHAGOCYTE. White blood cell that engulfs foreign matter and cellular debris.

SPECIFIC IMMUNE SYSTEM. The specialized wing of the immune system, which targets specific microorganisms and other antigens. Also called "adaptive defense system," or "acquired defense system."

SPLEEN. Largest of the lymphoid organs; houses B and T cells, phagocytizes aging and damaged blood cells, and filters out and destroys certain pathogens.

T CELL. Lymphocyte that controls major functions of the adaptive immune system and mediates cell-mediated immune responses as helper, cytotoxic, suppressor, or memory T cells.

THYMUS. Organ in the chest at the base of the neck in which T cells develop an ability to recognize self antigens.

VACCINE. Preparation of killed microorganisms, living attenuated organisms, or fully living organisms that is administered to produce or artificially increase immuinity to a specific disease.

INNATE DEFENSES

A BROAD RESPONSE

EVERY PERSON FROM birth is protected against a wide variety of pathogens and foreign substances by his or her innate defense mechanisms. The skin and mucosae form a first line of defense. A second line of defense, also part of the innate system, is composed of phagocytes and other destroying cells. Other nonspecific defenses of the second line include antimicrobial proteins, inflammation, and fever. Both these lines of defense—sometimes called external and internal—are broad-based, relatively unsophisticated ramparts against bodily invasion. Although in a sense they are less powerful than the body's adaptive immune system, they are able to keep out a large number of potentially harmful agents and thus save the adaptive system for only the most troublesome of pathogens.

Powerful magnification zooms in on cells from the lining of the nose. Inside the nasal cavity, they secrete mucus to trap foreign particles.

BARRIERS

Let's take a closer look at how the skin keeps out invaders. Composed of tough squamous epithelial cells, the epidermis or outside layer of the skin forms a strongly protective sheath against microbes and environmental toxins, while keeping vital moisture and nutrients in. A keratinized epithelial membrane, the skin, like hair and nails, is heavily endowed with the protein keratin, which is impervious to weak acids and bases, as well as bacterial enzymes. Sebaceous glands in the skin produce an oily substance called sebum, which films the skin; the unsaturated fatty acids in sebum make the skin partially water-resistant and restrict the growth of some bacteria and fungi.

Fatty acids and other skin secretions such as lactic acid are acidic enough to prevent bacterial proliferation. Sweat not only helps keep the body cool, but also washes the skin and contains the enzyme lysozyme, which breaks down the cell walls of some bacteria. Regular shedding of dead skin cells helps cleanse the body of bacteria and other pathogens. These invaders get no farther than the surface of the skin, unless they can find an entry point, as through a cut, burn, or body orifice, for example.

Body cavities opening to the outside—the respiratory, digestive, urinary, and reproductive tracts—are not lined with skin, but with mucous membranes. These mucosal layers secrete mucus, a viscous fluid that lubricates and moistens the body cavity lining. Since it is sticky, it traps a large number of germs. The respiratory tract is further protected by hairs in the nose that screen out microorganisms, dust, and many other airborne

SEE ALSO: Chapter Two, "The Skin," PAGE 46, Chapter Six, "Respiration," PAGE 156 pollutants. Those invaders that make it past the nose face a gauntlet of tiny cilia in the cells of the respiratory tract. These microscopic hairlike projections continually sweep mucus-trapped dust and microbes up to the throat where they can be coughed or sneezed out. Another trap the body secretes is earwax, which prevents microbes, insects, and other foreign bodies from invading the ears.

The mouth is protected by secretions of saliva, which wash away bacteria from the teeth and oral cavity, preventing them from colonizing and gaining a foothold. Similarly, tears cleanse the eyes, whether you cry or not—blinking does the job. Both saliva and lacrimal fluid (tears) contain the bacteria-destroying enzyme lysozyme.

Likewise, the stomach protects itself by secreting concentrated hydrochloric acid and protein-digesting enzymes, which are toxic to many harmful microbes. While these agents are at work, a larger, mechanical process often takes over in cases of stomach bugs or food poisoning. Reverse peristalsis, or vomiting, physically rids the body

A RIVER OF TEARS

A CLEAR SALINE FLUID, tears are secreted by lacrimal glands located above each eyeball and spread over the eye when the eyelid blinks. Tears drain via the lacrimal canal into the nose. They flow as a reflex response to irritants, bright lights, spicy foods, and heightened emotions. The act of blinking is both voluntary and reflexive. In addition to washing away foreign substances, tears fight infection with lysozyme, an enzyme that dissolves the coating of many bacteria.

EXTERNAL DEFENSES

DEFENSE	METHODS
Skin	Physical barrier against microbes; secretes perspiration and sebum, which contain anti-bacterial chemicals; keratin in skin cells guards against acids, bases, and bacterial enzymes
Mucous membranes	Though not as impervious as skin, still prevent entry of various microbes
Mucus	Secreted by mucous membranes; traps pathogens and foreign particles in respiratory, digestive, and urogenital tract
Nasal hairs	Coated with mucus; help trap dust and other foreign substances
Cilia	Lining the respiratory tract, sweep out microbes and dust
Tears	Dilute and wash out eye irritants; contain lysozyme
Saliva	Cleanses teeth and mouth of microbes; contains lysozyme
Urine	Flowing action clears urethra of microbes
Vaginal secretions	Acidic fluid inhibits bacterial and fungal growth
Gastric juice	Contains hydrochloric acid and enzymes to kill pathogens in the stomach
Vomiting and defecation	Expel pathogens from gastrointestinal tract

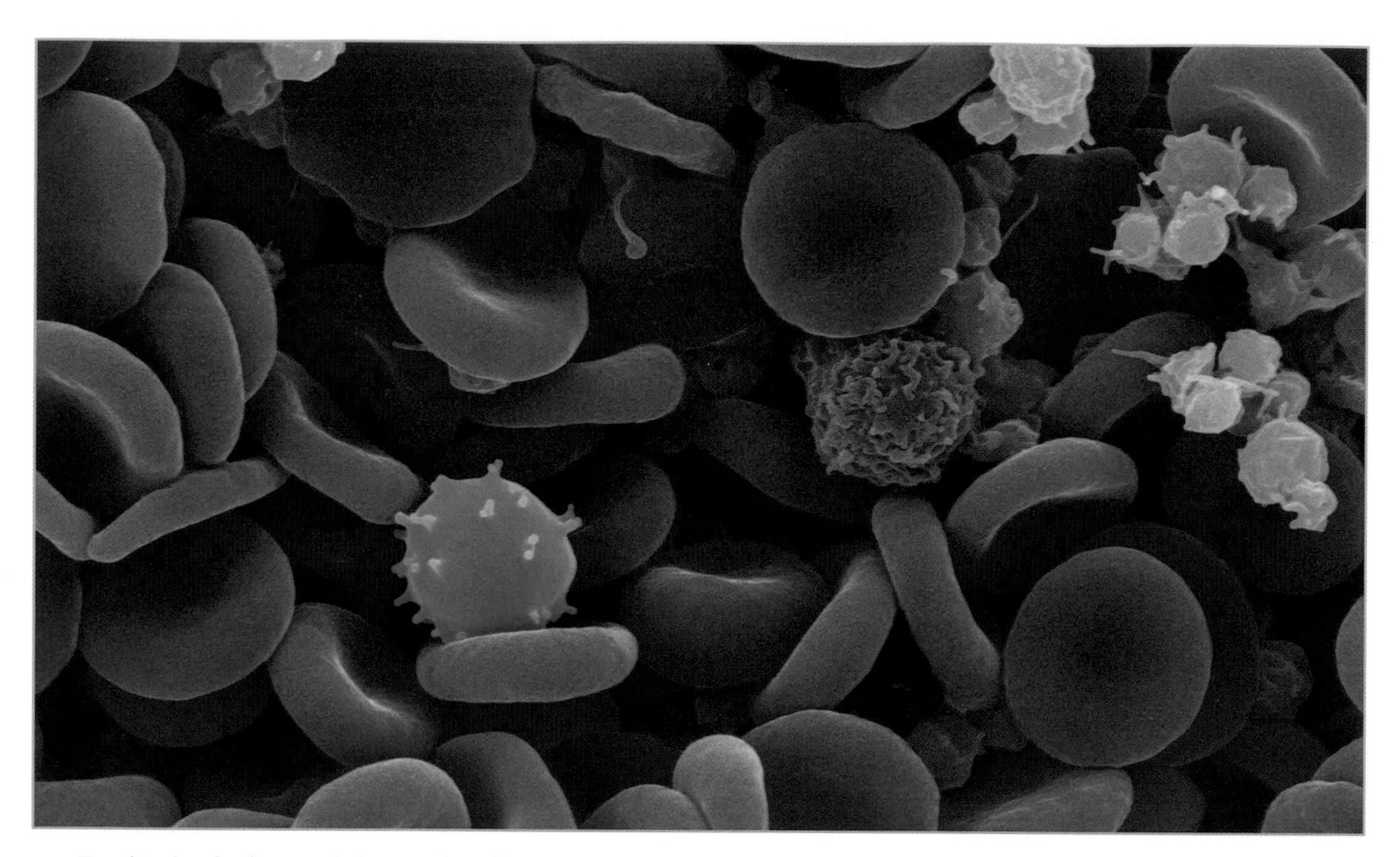

Two kinds of infection-fighting white blood cells (green, purple) work alongside platelets (brown) in the blood.

of pathogens, as does diarrhea or normal defecation. Urination has a cleansing function in addition to its obvious job of ridding the body of waste—it scours the urethra, expelling any bacteria attempting to lodge there. Likewise, secretions from the vagina make for a cleansing tide, and the acidity of vaginal secretions is yet another shield against bacterial growth.

ALL ABOUT PUS

A FEW DAYS AFTER the body's response to an infected wound, an aggregate of dead neutrophils, tissue debris, and live or dead microbes form in the wound. This creamy yellow fluid is called pus. Normally the pus either oozes out onto the skin or drains into an internal cavity. When it is sealed into a pocket of collagen fibers, it becomes an abscess that may need to be surgically drained.

INTERNAL DEFENSES

The pathogens that get past the skin and mucosae encounter a second line of defense, the internal innate defense. This system includes cells, antimicrobial proteins, inflammation, and fever. Here we will examine the two kinds of cells of the innate system involved: phagocytes and natural killer cells (or NK cells).

The phagocytes are cells that eat microbes and other particles. The main phagocytes are macrophages, the big guns of the innate immune system. Literally "big eaters," macrophages develop from monocytes, white blood cells that exit the bloodstream and grow into macrophages in the body's tissues. Wandering macrophages move about in search of infections, where they feast on microbes and debris. Fixed macrophages live in such organs as the liver and the brain and stand ready to swallow up invaders. Each of these fixed macrophages has a name—Kupffer cells in the liver, microglia in the nervous system, histiocytes in the skin, tissue macrophages in the spleen, and so on.

After macrophages, the most important kind of phagocytes are

SEE ALSO: Chapter Five, "The Blood," PAGE 132

neutrophils, which rank as the most abundant type of white blood cell. Another kind of white blood cell, the eosinophil is not a heavy eater as phagocytes go, but it is important in killing worms and other parasitic invaders.

Phagocytes kill their prey in a five-step process called phagocytosis. In the first step, chemotaxis, phagocytes sniff out their prey, lured by chemical signals from the offending microbes, damaged tissues, or white blood cells. Then, in adherence, the phagocyte attaches itself to the microbe or other particle. This can only happen if the phagocyte can read the surface structure of the pathogen. A pathogen with a complicated structure or protective capsule, such as the pneumococcus bacterium, is not easy for phagocytes to latch onto. During ingestion, the third phase, the phagocyte surrounds the microbe by extending projections called pseudopods that engulf the microbe in a sac called a phagosome.

The fourth step is digestion. In this key stage, organelles called lysosomes within the phagocyte's cytoplasm join with the phagosome and induce metabolic production of hydrogen peroxide and other highly toxic oxidants, produce a bleach-like chemical with chloride ions, and release lysozyme and other digestive proteins destructive to pathogens. Lysozyme is an enzyme that breaks down the invader's cell walls. Other digestive enzymes within the lysosome act on proteins, carbohydrates, lipids, and nucleic acids.

In some cases, such as with the tuberculosis bacillus, the pathogen inhibits the joining of the lysosomes with phagosomes and simply multiplies within the phagocyte. This is when the cavalry shows up—cells of the adaptive immune system come and stimulate the release of more powerful chemical mediators that release cell-killing free radicals such as nitric oxide. In the final step in phagocytosis—the killing—the enzymes and oxidants take the microbe apart; residual bodies of indigestible material leave the cell through exocytosis. Neutrophils are such thorough killers that they take themselves out in the process.

The other family of cells involved in the innate immune response are the natural killer cells, which compose some 5 to 15 percent of all lymphocytes. Like the phagocytes, they also reside in the spleen, lymph nodes, and red bone marrow. These helpful assassins can terminate a number of infectious microbes and cancer cells, zeroing in on cells that have inadequate or abnormal plasma membrane proteins called major histocompatibility complex (MHC) antigens. We will take a closer look at MHC molecules later; for now suffice it to say that each person has a unique MHC signature, allowing for "self-recognition." Foreign cells without proper identification are thus subject to destruction by killer cells.

Part of a small cadre of large granular lymphocytes, natural killers are indiscriminate, or nonspecific, in their killing, earning them the adjective "natural." Instead of engulfing microbes as do phagocytes, natural killers shoot their targets full of holes: They attack the plasma membrane of a target microbe with chemicals called perforins, which make the microbe

BREAKTHROUGH

ANTISEPTICS: Before the introduction of antisepsis, up to 50 percent of surgery patients died from infection. Knowing French chemist Louis Pasteur had proved that disease was spread by microorganisms, English surgeon Sir Joseph Lister began using carbolic acid as a wound dressing in 1865. The acid, a sewer cleaner, reduced infections so dramatically that in four years mortality in his ward dropped from 45 to 15 percent. Realizing that germs were not just airborne, he used antiseptics to sterilize surgeon's hands and instruments. For his pioneering work, he was made a baron, the first doctor elevated to peerage.

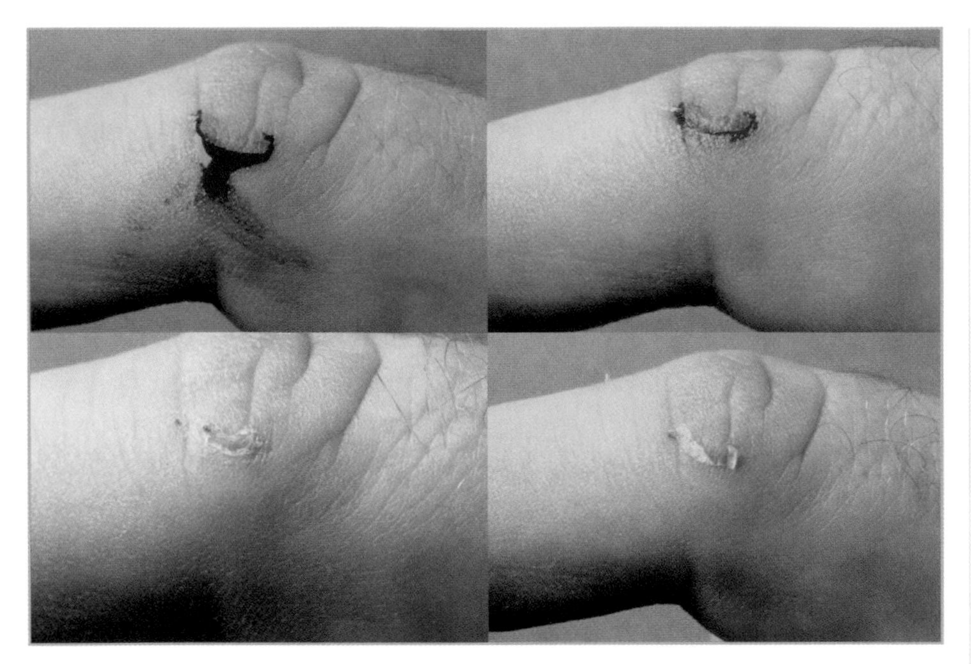

A minor cut on a finger invokes the inflammatory response, which begins the healing process.

leak to death. They can also kill by releasing into target cells molecules that lead to apoptosis—programmed cell death in which the membrane stays intact but the nucleus and cytoplasm shrink, the DNA breaks apart, and the mitochondria quit functioning.

INFLAMMATION

When you injure yourself—say you sprain your ankle, or cut or burn your hand—what happens? The affected area reacts with an inflammatory response, just the same as it does when invaded by microbes. Inflammation is yet another weapon in the body's innate defense arsenal, and there is more to it than meets the eye.

Any tissue damage, no matter what the cause, triggers the body's inflammatory response. Physical trauma, heat or chemical burns, infection by bacteria, parasites, viruses, and fungi—all these elicit inflammation. The four signs that signal inflammation are pain, redness, heat, and swelling. Loss of function is sometimes considered a fifth sign—the immobilized area (a joint, for instance) is forced to rest and thus can heal more quickly. Inflammation is not simply a reaction. It serves three main purposes: It rids the injured area of harmful pathogens and cell debris; it prevents pathogens and toxins from spreading; and it prepares the area for tissue repair.

Inflammation occurs in three stages—vasodilation and increased permeability of blood vessels, emigration of phagocytes, and tissue repair. The first thing that happens at the site of an injury is vasodilation of arterioles: Vessels dilate so that they can carry more blood to the injured area. At the same time, capillaries become more permeable, allowing antibodies, oxygen, nutrients, and clotting proteins to swarm the inflamed area. The extra flow of blood helps wash out microbes and dead cells; it's the body's first way of cleaning itself. Think of this stage as the emergency medical team roaring down the highway to the accident site.

What causes vasodilation and increased permeability? How do the vessels know when they need to dilate and become leaky? Several chemicals are released after an injury, stimulating vasodilation. Let's take a closer look at the major ones. When an injury occurs, mast cells, platelets, and white blood cells sound an "alarm" by releasing histamine, causing vasodilation and leaky blood vessel walls. In addition, kinins stimulate chemotaxis (the attraction of leukocytes to microbes), and they make neutrophils release lysosomal enzymes, which help produce even more kinins. Kinins also induce pain.

+ WARNING SIGNS +

+ PAIN: from neuron injury, toxic chemicals from microbes, vasodilating chemicals, and edema
+ Heat: from increased blood flow in affected area
+ Redness: also from increased blood flow in affected area
+ Swelling: from greater permeability of local blood vessels

SEE ALSO: Chapter Two, "Healing," PAGE 54, Chapter Five, "The Blood," PAGE 132

Another group of chemicals, the prostaglandins heighten the effects of histamine and kinins, and they produce free radicals, which themselves can trigger the inflammatory response. Prostaglandins may also prompt the second stage of inflammation—emigration of phagocytes from the blood into interstitial fluid. Leukotrienes stimulate increased vascular permeability, and they attract phagocytes and help them adhere to pathogens. Complement protein, a group of about 20 plasma proteins, enhance phagocytosis, lyse microorganisms, stimulate the release of histamine, and attract neutrophils.

Besides increasing the flow of blood and associated antibodies, vasodilation and increased vascular permeability cause three of the four symptoms of inflammation—heat, redness, and swelling. The localized increase of blood and metabolic activity raises the temperature at the injury site and accounts for the redness (erythema) beneath or within the epidermis. The swelling (edema) occurs when the more permeable capillaries leak fluid into the interstitial spaces and surrounding tissues. The fourth symptom, pain, results from damaged nerve endings and the toxic chemicals of microbes. Kinins can produce inflammatory pain, and prostaglandins can intensify and prolong it. Pressure from swelling also induces pain.

About an hour after the inflammatory response has begun, phagocytes arrive and start emigrating from the blood vessels into the damaged tissues. This emigration process is caused by chemotaxis. Once out of the vessels, neutrophils (a kind of phagocyte) go to work eating up microbes. The bone marrow keeps a steady supply of neutrophils coming, increasing their numbers in the bloodstream four to five times during times of bacterial and certain fungal infections and with severe tissue injury. The neutrophils are followed shortly by monocytes, which within 8 to 12 hours develop into macrophages, the serious microbe-devouring professionals. After a few days the phagocytes die off, and they and the dead tissue cells clump together to form pus (see "All About Pus," page 324). Blood-clotting agents have in the meantime been leaking from the capillaries and weaving a mesh of gel-like fibrin threads, which prevents the spread of microbes. This mesh, along with the work of the busy phagocytes gobbling up microbes and debris, sets the stage for tissue repair.

While inflammation is a localized innate response, a more general response is fever. During a microbe invasion, leukocytes and macrophages release pyrogens, chemicals that reset the body's thermostat above the normal temperature of 98.6°F (37°C). A moderate fever helps inhibit microbe multiplication, but the mechanism for this is still not well known. By speeding up metabolic activity, fever can also promote healing.

KEEPING HEALTHY

Ice packs can reduce swelling and pain.

SERIOUS ATHLETES and weekend warriors know the danger of sprained joints and strained muscles. While inflammation occurs when you pull or tear a muscle, you can speed up healing by following the first-aid procedure of R.I.C.E: rest, ice, compression, and elevation. Rest to prevent further injury and allow the injured area to begin healing. Apply ice packs for up to 20 minutes to reduce swelling, pain, and muscle spasms. Compress by lightly wrapping the affected area with an elastic bandage. Finally, elevate the injury above the heart to further reduce swelling. Swelling helps by diluting toxic substances, but since muscle strains do not introduce foreign microbes, swelling can be counterproductive by slowing down the healing process, causing pain and spasms. If you suffer a serious sprain or any injury that causes pain for more than 48 hours, see your physician or visit the emergency room. ■

ADAPTATION

ACQUIRING SPECIALIZED DEFENSES

GREEKS IN THE FIFTH century B.C. noted that people who survived the plague never suffered from it again. Somehow their bodies remembered the disease and were resistant to it in the future. What the Greeks observed was the body's acquired or adaptive immune system in action. Unlike the broad attack of the innate defense system, the adaptive system attacks pathogens with specific, made-to-order antibodies and cells. The phagocytes of the innate system don't discriminate between harmful agents; the lymphocytes (white blood cells) and their derivative antibodies do. In fact, the antibodies and lymphocytes of the adaptive system can recognize millions of different pathogens, some not even found in nature, and they can distinguish infected, cancerous, or foreign cells in an organ from normal cells of the same type.

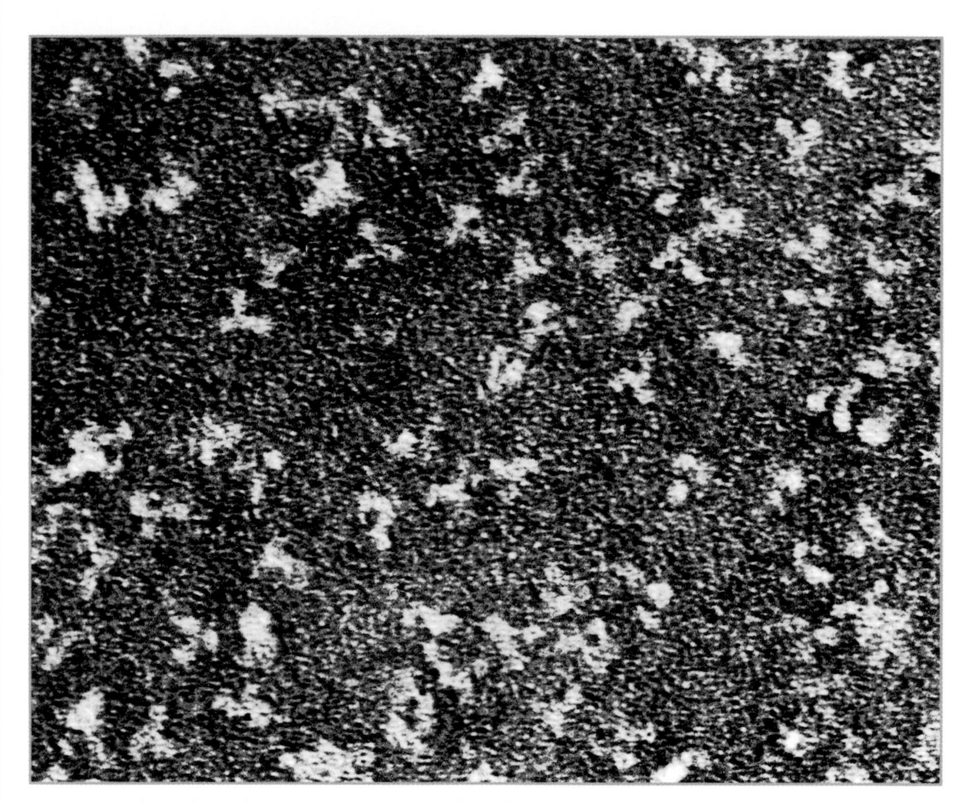

Found in the plasma, human antibodies (yellow) help defend the body against infection by binding to foreign proteins.

Each antibody and lymphocyte recognizes only one kind of antigen (a substance that provokes an immune system response). The receptors on the surface of each lymphocyte recognize the specific chemical structure of the antigen and fit the antigen like a key sliding into a lock. The body must be primed by an introduction to these antigens. Then the old adage applies: That which does not kill you makes you stronger. The adaptive immune system manufactures the necessary antibodies and cells so that upon re-exposure you can fight back and win. Thus, unlike the innate system, the adaptive system is a highly specialized fighting force with a "memory."

From stem cells within the bone marrow, some one million lymphocytes are created every second. At any given time, billions of lymphocytes are moving through your bloodstream and tissues, each of them seeking out the antigen that fits its receptor. Since there are millions of possible antigens, the body may hold only a few of each type of lymphocyte programmed to attack each type of antigen. Upon finding a match, the lymphocyte rapidly multiplies. Immunity in the

DISEASE VECTORS

BROADLY SPEAKING, a disease vector is a vehicle or pathway through which an infectious disease spreads; specifically, it is an organism that transmits a pathogen. Malaria is spread through the mosquito vector—mosquitoes can carry the malaria parasite and transmit it to the blood of a human host via a bite. Similarly, lice carry the typhus parasite. Control of these diseases is a matter of eliminating the middle man—the vector.

SEE ALSO: Chapter Three, "The Bones," PAGE 76, Chapter Five, "The Blood," PAGE 132 adaptive system is granted systemwide, throughout the body, instead of just at the site of infection.

The two kinds of adaptive immunity are humoral and cell-mediated. Humoral immunity is that provided by antibodies, which circulate in bodily fluids or "humors," especially blood and lymph. These antibodies, also called immunoglobulins, are protein molecules produced by B lymphocytes and their progeny, plasma cells. Antibodies seek out bacteria, viruses, and other pathogens, deactivate them, and tag them for destruction by phagocytes. In cell-mediated immunity, the lymphocytes themselves go after cells infected with viruses, cancerous cells, or foreign cells; after that they either lyse the cells or summon macrophages or other killers.

FIGHTING CELLS

The three kinds of cells of the adaptive immune system are the T and B lymphocytes and the antigen-presenting cells. T lymphocytes, or T cells, are the engines of the cell-mediated wing of the adaptive immune system. They comprise 65 to 85 percent of all bloodborne lymphocytes. Unlike cells of the humoral wing, T cells produce no antibodies but directly destroy microbes. All blood cells, lymphocytes included, come from stem cells in red bone marrow. All lymphocytes start life practically identical; as they mature, they differentiate into T and B cells. The

CONQUERING YELLOW FEVER

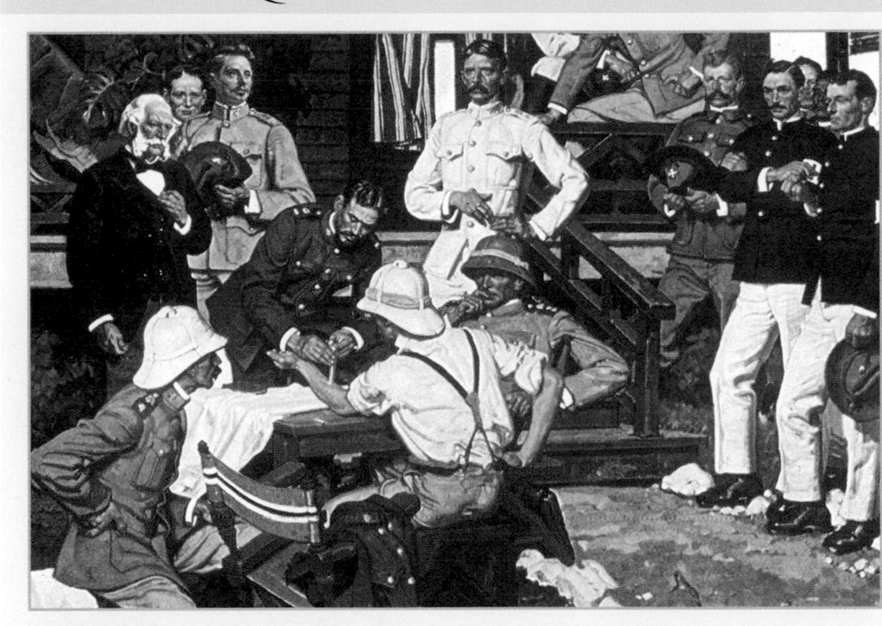

Carlos Finlay, left, and Walter Reed, center, observe a yellow fever inoculation.

ONCE CONSIDERED a major plague of the New World, yellow fever is now confined mostly to western and central Africa and tropical regions of the Americas. Named for its ability to cause jaundice, yellow fever generally induces flulike symptoms; severe cases may lead to liver or kidney failure, abnormal bleeding, and convulsions. Many yellow fever victims die. Those who recover have a lifetime immunity, thanks to their adaptive immune systems. Though there is no cure, an immunization provides protection for 10 years. Before immunization, yellow fever epidemics broke out in places as far from the tropics as New England, Spain, and France.

It was in 1881 that Cuban epidemiologist Carlos Finlay postulated that the vector for yellow fever transmission was a mosquito now known as *Aedes aegypti*.

U.S. Army surgeon and pathologist Walter Reed followed up on Finlay's work by showing that the disease passes from person to person by mosquitoes carrying the infectious agent. Control of the *Aedes* mosquito led to virtual elimination of the disease from Cuba; similar methods led to successes in Brazil and Panama.

Scientists from the Rockefeller Institute for Medical Research in New York proved in 1927 that yellow fever is caused by a virus, isolated strains of which led to the production of a vaccine. Further study in Africa and South America demonstrated that the disease could be passed first to animals such as monkeys and then to humans, via infected mosquitoes. Altough the disease has been drastically reduced, still about 30,000 people die every year in places where yellow fever is endemic. ■

earliest T cells multiply in the thymus (hence the name T, for thymus-derived), and then mature over a two- to three-day span. T cells that poorly recognize antigens do not survive, nor do those which strongly bind to self-antigens (surface proteins of body cells). The latter process, negative selection, prevents the immune system from attacking the body.

The B cells (from "bone-marrow derived") mature in the bone marrow. Comprising the humoral wing of the adaptive immune system, B cells produce antibodies—protein molecules also called immunoglobulins—that provide defense against extracellular microbes and other foreign substances.

After the T and B cells become immunocompetent they have one of 10,000 to 100,000 possible chemical receptors on the surface of each of their cells. The receptors of one cell are capable of binding to only one specific antigen. For instance, the hepatitis A virus and each of the more than 200 kinds of viruses that cause common colds are recognized by only one receptor combination on a lymphocyte.

Though the body makes the necessary lymphocyte force to fight a particular antigen, scientists believe that the body cannot create a new immune cell receptor upon meeting an antigen. Rather, the body is genetically programmed to produce all the possible receptors it might need. Thus genes, not antigens, dictate what your body can recognize and resist. Many of the kinds of lymphocytes you make will never be needed in the environment you inhabit; they simply stand on reserve for life.

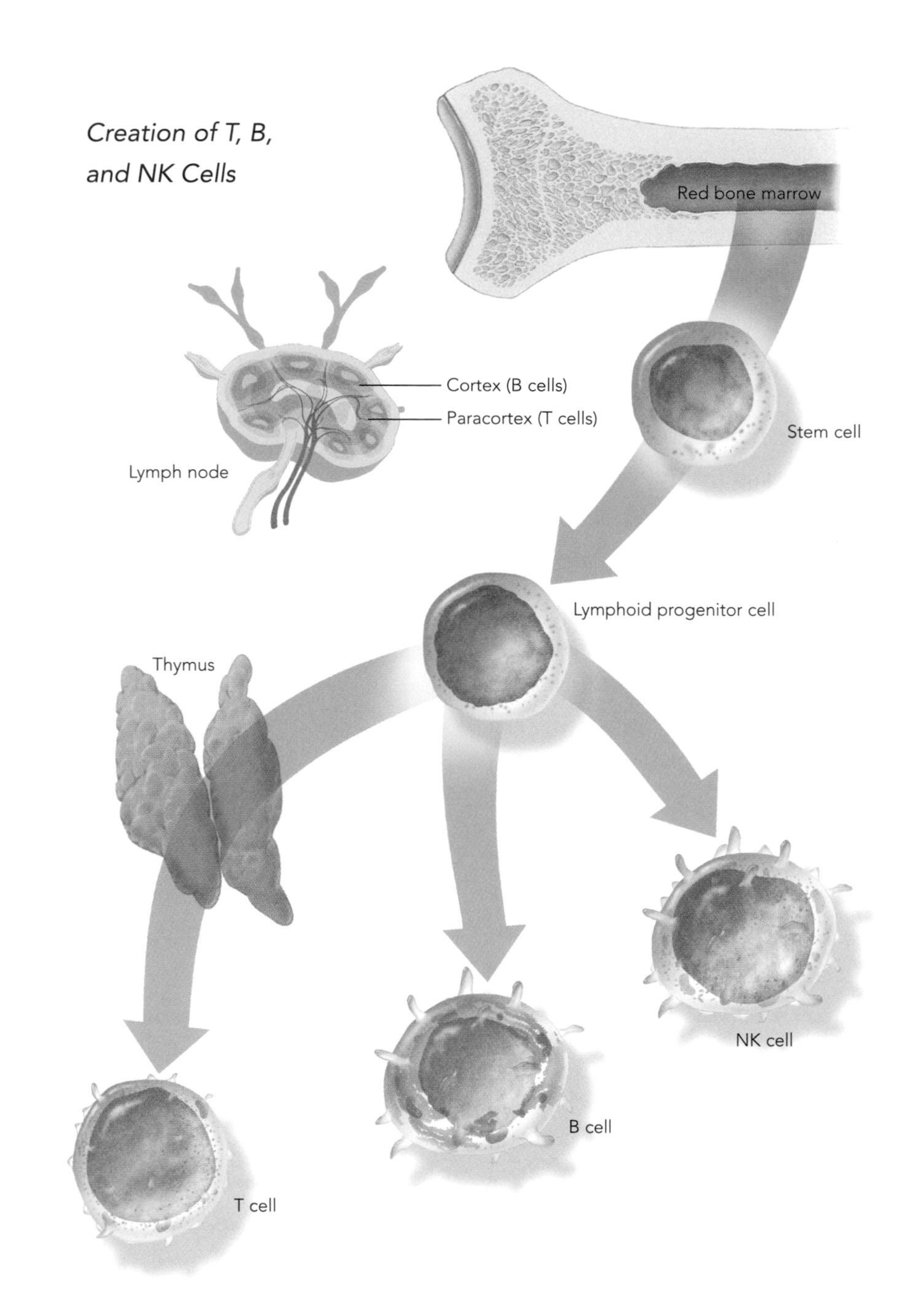

AGING CELLS

BEFORE BIRTH, the human body produces huge numbers of T cells and B cells. But as we age, the production slows down, and the thymus begins to atrophy. It is unclear why the immune system breaks down as we age, but because it does we become more susceptible to diseases such as cancer. Even though the thymus diminishes, T cells are replaced by cell division outside the thymus and thus cell-mediated immunity continues.

SEE ALSO: Chapter Ten, "Producers," PAGE 274

Upon reaching immunocompetence, T and B cells move to the lymph nodes, spleen, and other secondary lymphoid organs. At this point, the T and B cells are not fully mature because they have not matched with antigens. In the secondary lymphoid organs they can bind with antigens and become activated, fully functional T and B cells.

The antigen-presenting cells (APCs) swallow microbes and other antigens and then display a piece of them like signal flags to T cells. Thus they present the antigen to the lymphocyte. The major APCs are dendritic cells, macrophages, and activated B cells. B cells can be either antibody producers or antigen presenters. It is also important to understand that APCs don't simply signal T cells; they help activate them. In turn, T cells then crank up the maturation and mobilization of APCs.

Lymphocytes circulate throughout the body and thus increase the likelihood they will encounter antigens. Once they are needed in a specific area, homing signals call them into the appropriate tissues. Some lymphocytes hang out continuously in lymph nodes, waiting for lymph capillaries called lymphatics to bring pathogens by. Either strategy—circulating or standing in one key place—works well for encounters. Bloodborne antigens are often encountered in the spleen; invaders of the mouth and nose are often met by lymphocytes and APCs in the tonsils.

THE BOY IN THE BUBBLE

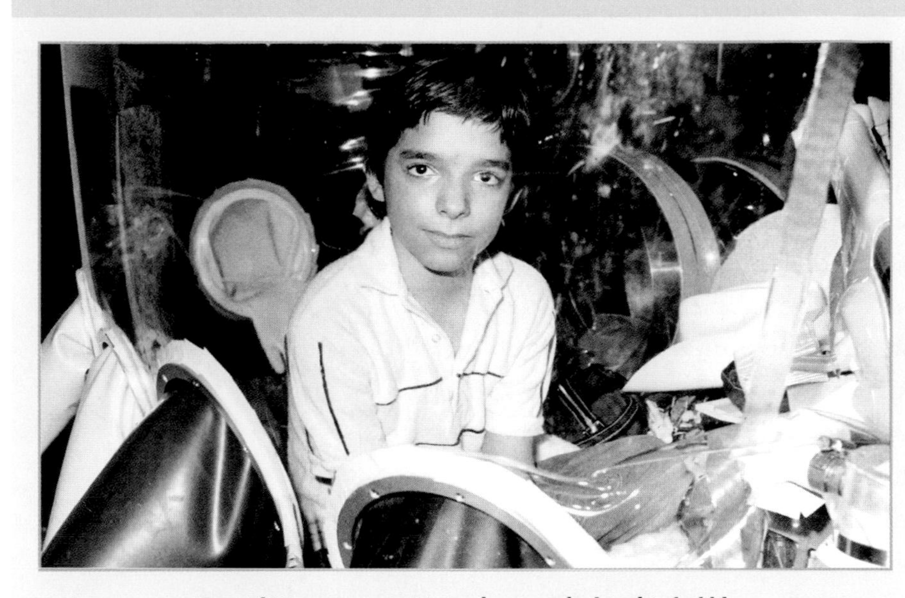

David Vetter at age 11 is photographed in his bubble.

WHATEVER KNOWLEDGE of the immune system scientists gained from the famous "Boy in the Bubble," the most important lesson may have been that efforts to prolong life can have tragic consequences. David Vetter was born in Houston in 1971 with severe combined immune deficiency syndrome (SCIDS), a genetic defect that results in a deficit of B and T cells. This deficit leaves children vulnerable to all germs. A brother had died of the same disease. David's parents knew there was a good chance another child could have SCIDS.

Within 20 seconds of his birth, David was placed inside a sterile plastic isolator. Until the last two weeks of his life, he would never be free of his bubble. Great effort was taken to sterilize everything: Bottles, food, diapers, and clothes were put into chambers filled with ethylene oxide gas at 140°F for four hours and then aerated for up to a week. Heavy rubber gloves in the walls of the bubble allowed David's parents and doctors to handle him. Research grants funded the project while the Vetters hoped for a cure.

In the meantime, David became something of a celebrity. Not always cooperative with the media, he had to be coaxed into exploring an 11-foot-long addition to his 6-by-12-by-4.5-foot bubble. As he grew older, feelings of anger and isolation often caused him to act disruptively. He had vivid nightmares about germs and was afraid to use a specially made space suit. Desperate for a solution, doctors attempted to transplant bone marrow from his sister when David was 12 years old. But a few months after the operation he developed such severe diarrhea, vomiting, and fever he had to be placed in intensive care outside the bubble. He died 15 days later of Burkitt's lymphoma. ■

ADAPTIVE RESPONSES

After maturing in the bone marrow or thymus, lymphocytes travel to all parts of the body via the bloodstream. Certain sites, though, have a special strategic importance. Within the skin and internal mucosa, there are lymphocytes lying in wait for any microbes that attempt to invade. Then the lymph tissue itself houses vast numbers of lymphocytes. Within the tissues of lymph organs, including lymph nodes, lymphocytes multiply and stand ready for activation or transport to other sites.

Lymph nodes filter lymph as it travels back to the bloodstream. Located in the neck, under the arms, in the groin, and within the chest and abdominal cavities, these half-inch-long, bean-shaped organs lie along the network of lymphatic vessels. They keep microorganisms from entering the blood and help activate the immune system by spotting antigens. Lymph nodes can become swollen and tender when they detect invaders (like bacteria or viruses) and become activated. Once the nodes have cleaned the lymph, the fluid is carried through the lymph vessels, a one-way system that returns fluid to the heart; up to three quarts a day of interstitial fluid is thus returned to the bloodstream.

The tonsils are lymph organs positioned around the tongue and throat to filter out microbes from food, water, and air. The tonsils and adenoids form a ring of lymphatic tissue. The tonsils in effect "invite" infection by trapping bacteria, but for this risk the body gains a memory of harmful antigens.

The largest lymphoid organ is the fist-size spleen. As with the lymph nodes, the spleen acts as a site for lymphocyte proliferation and immune surveillance and response. Here also, old and damaged blood cells are destroyed. The spleen saves the iron from destroyed red blood cells for recycling in the bone marrow, stores blood platelets, and produces blood cells in fetuses. If the spleen ruptures (from infection or injury) or is removed surgically, the bone marrow and liver pick up some of its duties, though a person with no spleen is more vulnerable to some bacterial infections.

A humoral immune response begins with an antigen challenge. In the lymphatic tissue—usually the spleen or lymph nodes—the invading pathogen or other antigen encounters an immunocompetent but inactivated B lymphocyte. The antigen is detected by antibodies found on the surface of B lymphocytes. Antigen-presenting cells similarly ingest the pathogen or antigen, process it, and present the processed antigen to helper T lymphocytes programmed to recognize the antigen using a customized surface receptor. At this point the antigen-bearing B lymphocytes and antigen-presenting cells induce the helper T lymphocytes to produce signaling chemicals, called cytokines, that stimulate the antigen-bearing B cells to proliferate and evolve into antibody-secreting plasma cells, each capable of producing 2,000 antibody molecules per second. The antibodies seek out antigens and tag them for destruction. Within five days the plasma cells die.

Instead of evolving into plasma cells, some of the antigen-specific B cells become memory

SEE ALSO: Chapter Three, "The Bones," PAGE 76, Chapter Five, "The Blood," PAGE 132

WHAT CAN GO WRONG

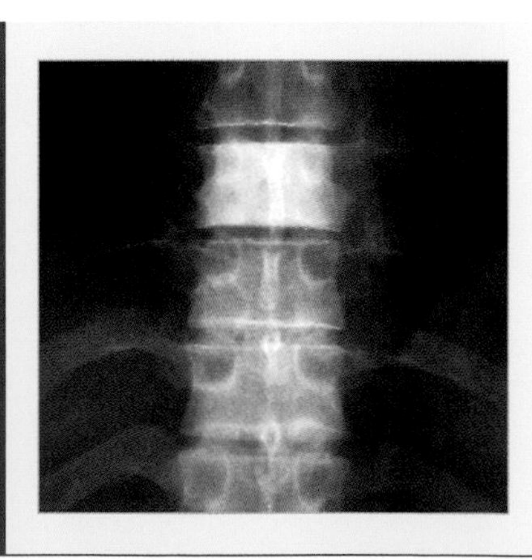

HODGKIN'S LYMPHOMA was first described in 1832 by English physician Thomas Hodgkin. Of the several types of lymphoma (cancers of the lymphatic organs) the two main kinds are Hodgkin's and non-Hodgkin's (which includes all the other types). In both, immune cells become cancerous and divide rapidly in a lymph node or other sites. The tumors grow from B or T lymphocytes. The disease may spread to other nodes and tissues, including bone, like the ivory vertebrae at left. Radiation therapy and/or chemotherapy are used to treat some lymphomas with a cure rate of about 75 percent.

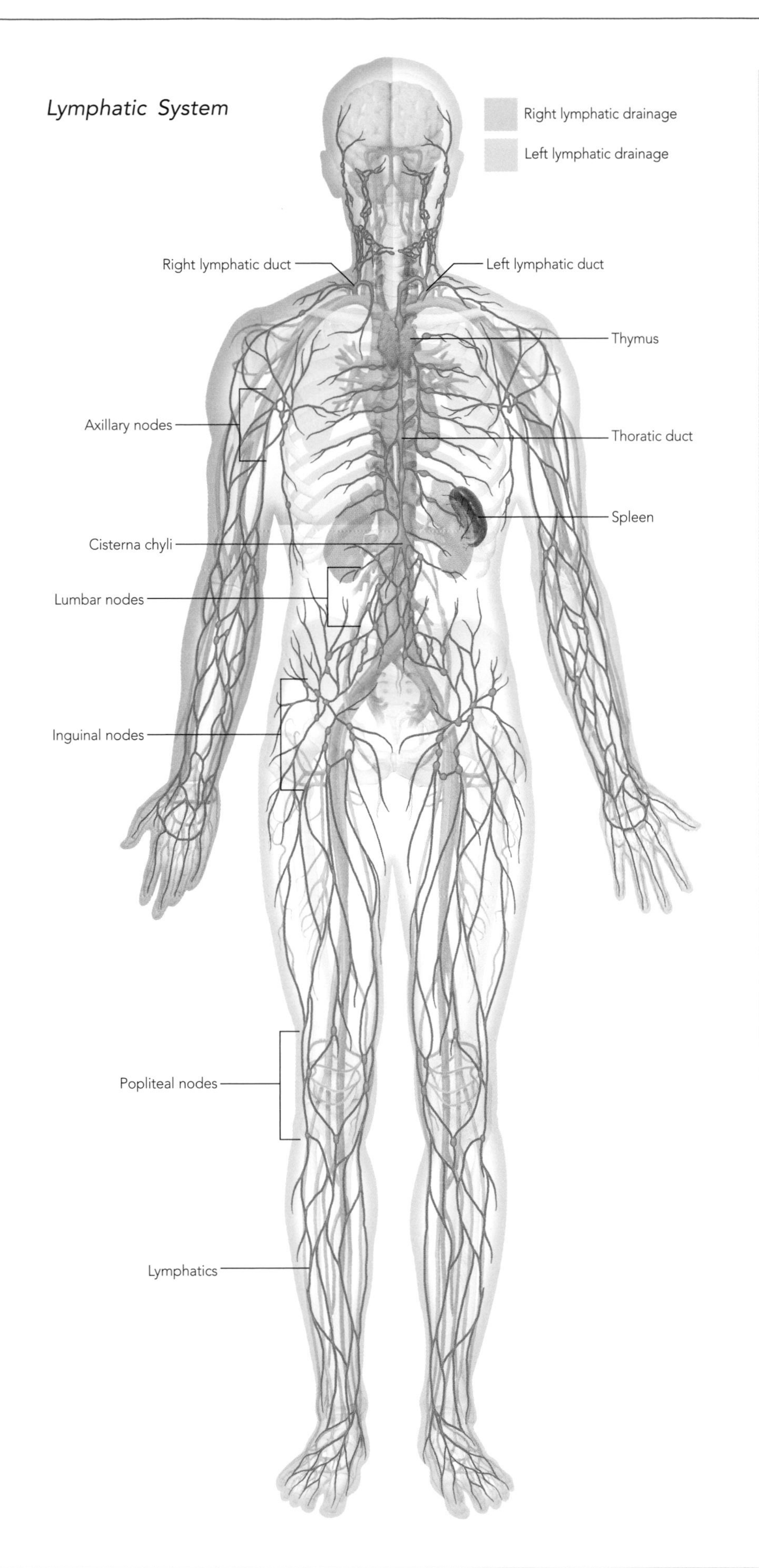

cells. These important players can launch a humoral immune attack upon recognizing an antigen in the future. The memory cells thus provide immunological memory. While a primary humoral response can take three to six days to build up enough antibodies to begin fighting an invasion, a secondary response—one that relies on immunological memory—is much quicker and stronger. Within hours an army of plasma cells is mobilized; in two or three days the antibody level is higher than it gets at the ten-day peak of a primary response. Antibody levels then remain elevated in the blood for weeks or months.

CELL MEDIATION

The antibodies of the humoral immune system provide only a simple, limited defense against pathogens. What they leave unfinished,

THE SPLEEN

LARGEST OF THE lymphoid organs, the fist-size spleen lies on the left side of the abdominal cavity beneath the diaphragm. The spleen serves as a base for lymphocyte proliferation, as well as immune surveillance and response. In addition, it helps clean the blood of worn out and damaged red blood cells. At the same time, the spleen saves the iron from destroyed red blood cells for future use by the bone marrow; it stores blood platelets; and it produces blood cells in fetuses.

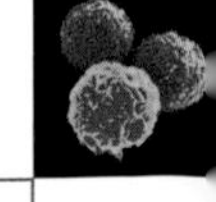

the cell-mediated division picks up. Whereas antibodies only deactivate antigens or tag them for destruction, the activated T cells (along with cells from the innate system) actually do the killing. T cells may also take aim on body cells, either cancerous or foreign cells in a transplanted organ. Since T cells cannot distinguish free-floating antigens, they rely on antigen-presenting cells to process antigens first and then present them to the T cells to begin the cell-mediated response.

Several kinds of T lymphocytes make up the cell-mediated wing of the adaptive immune system. The helper T cells are crucial links in the immune response. Once it is activated by the antigen on an antigen-presenting cell, a helper T cell then begins to stimulate the production of more T cells and B cells. The helper T cells are the directors of the cellular immune response, a job that they perform by both direct contact with other immune cells and by the release of chemicals called lymphokines or cytokines, proteins that enhance the immune and inflammatory responses. Lymphokines mobilize a formidable army of cells into action against the antigens—more T cells, B cells, macrophages, and other white blood cells heed the call and muster in overwhelming numbers to the battlefield.

Cytotoxic T cells, or killer T cells, are the only T cells that attack and directly destroy other cells. Once activated by an antigen, a cytotoxic T cell is recruited and pumped up by a helper T cell. It then begins stalking the blood and lymph systems and lymphatic organs, hunting down cells with the antigen it is programmed to destroy. These are usually virally infected cells, cancer cells, or cells from transplanted organs. They may also hunt down tissue cells infected by certain bacteria and parasites. In this way they stop the invasion of microbes that are hidden from antibody detection within host cells.

To kill their prey, the cytotoxic T cells first have to dock onto the target cell. All body cells have a surface antigen ID unique for each person (except for identical twins). Called the major histocompatibility complex (MHC) antigen, this ID flashes its "self" identity. An infected or cancerous cell flashes a fake ID or an inadequate self ID. This fake or inadequate ID is what the cytotoxic T cell binds to. The mechanism by which it kills the bad cell is not completely understood, but in many cases the T cell releases a chemical called perforin into the target cell. Perforin lyses the target cell by creating pores in its membrane, allowing cytoplasm to leak and causing the cell to die.

Once the intense battle against the antigens is drawing to a close, something has to declare victory

+ VIRUSES +

NEITHER PLANT or animal, a virus is something between a simple microorganism and a complex molecule. As parasites, viruses cannot reproduce or live outside a host cell. Viruses consist of nucleic acid (either DNA or RNA) wrapped in a protein sheath. They are pieces of bad genetic code in a protective protein wrapper, causing disease by entering a host cell, taking over the cell's machinery, replicating, destroying the host cell, and spreading to other cells.

WHAT CAN GO WRONG

INFECTIOUS MONONUCLEOSIS, often referred to as "kissing disease," is a form of infection by the Epstein-Barr virus (EBV). Most people are infected with EBV when they are young children and have symptoms of a bad cold or mild flu. Infectious mononucleosis occurs when EBV infection is postponed until later in life. The illness mainly strikes people from 10 to 35 years old and is passed primarily through the exchange of saliva, but it can also be caused by exposure via coughing and sneezing or sharing food utensils or a glass. Fatigue and severe sore throat are the most common symptoms of the disease; enlarged lymph

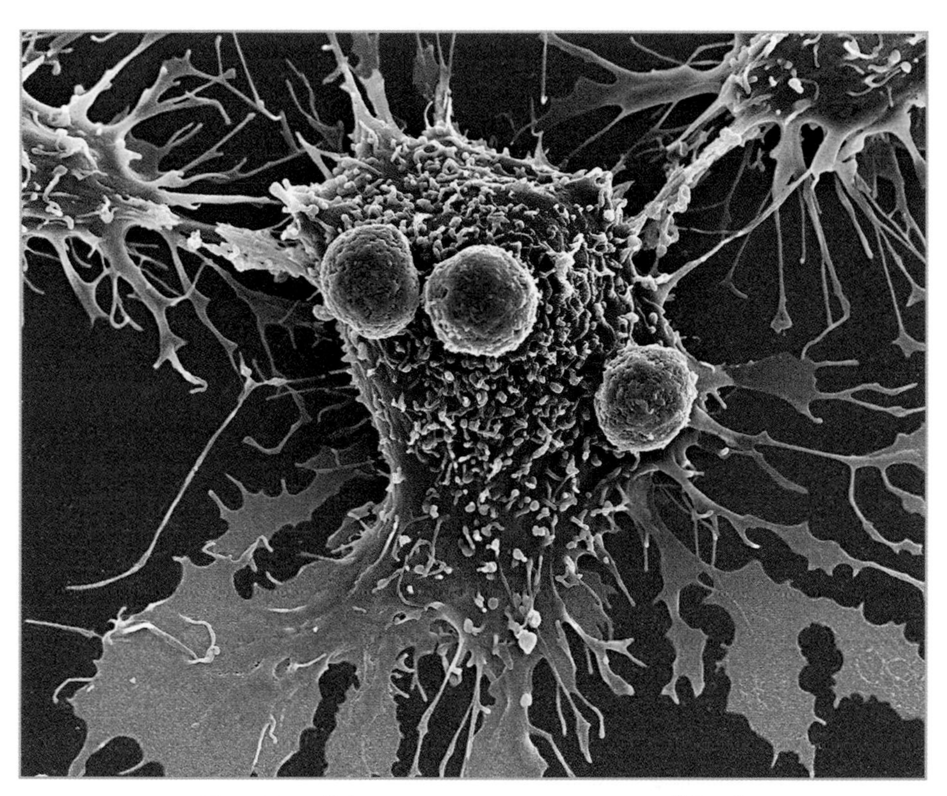

After identifying antigens on a cancer cell (red), three T lymphocyte cells (purple) bind to it.

and call off the winning force. That's where the suppressor T cells come in. The suppressors slow down B-cell and T-cell activity. Without these specialized T cells, the immune system would keep turning out more and more lymphocytes, like a Hydra that cannot stop itself. It is likely that suppressor T cells play an important role in prohibiting the autoimmune response—the production of lymphocytes that attack a person's own healthy tissues.

As with B cells, there are also memory T cells. Clones of T cells remaining after a cell-mediated immune response carry immunological memory and are thus termed memory T cells. At any given time there will be thousands of memory T cells available the next time the same pathogen invades the body. These cells will then react swiftly and efficiently to halt the invasion, generally before the infection has had time to produce symptoms.

In summary, the various kinds of T cells work with each other and with other immune cells to launch attacks on invading pathogens. With helper T cells directing the battle, the cytotoxic T cells kill cells hosting viruses and other microorganisms, as well as cancerous or transplanted foreign cells. The helpers also enlist B cells, and together they stimulate the body's innate defenses against invaders that are hidden away in infected cells or free-floating in blood and tissue spaces. The suppressor T cells call the action off when the antigen threat is neutralized. Meanwhile, memory T cells are spawned as cell clones, ready for the next attack of the same antigen.

IMMUNIZATION

When your B cells have encountered and acted against antigens, your immune system is showing active humoral immunity. This immunity is exhibited whether

nodes and spleen, fever, headache, and muscle soreness are also typical. The occurrence of mononucleosis is three times higher in females than in males. From the saliva, the virus moves through the mucosa in the upper respiratory tract and into the bloodstream, where it infects and multiplies in B lymphocytes. The attack on the B cells alerts and activates T cells, which turn on the cell-mediated and humoral arms of the immune system. An elevated white-blood-cell count ensues with the highly activated T cells in the blood. These white blood cells with a one-lobed nucleus, called mononuclear leukocytes, give rise to the name mononucleosis. Treatment includes bedrest and the administration of plenty of fluids. Patients are usually back to their normal routines within two to three weeks, though the fatigue can last for months. The virus remains in the body for life.

DEFENSE DESTROYER

One of today's most serious diseases, acquired immune deficiency syndrome (AIDS) targets the body's immune system and cripples its ability to defend itself from infection. First reported in 1981, the disease is thought to have originated in Africa, possibly first in monkeys.

Since then, the disease has spread around the world, through sexual contact, sharing of contaminated needles by intravenous drug users, and, before effective blood-screening methods were developed, through transfusion of contaminated blood and blood products. First recognized when young homosexual men reported strange cases of pneumonia and skin cancer, the disease is now primarily spread through heterosexual contact.

Currently, more than 40 million people worldwide have HIV/AIDS. Some 5 million new cases arise annually. Since 1981 more than 20 million have died from AIDS, and 3 million more die each year. About 70 percent of all AIDS cases are in sub-Saharan Africa. In the United States, upward of 1 million people are infected with the virus that causes AIDS.

CAUSES

HIV (human immunodeficiency virus) is the virus that causes AIDS. It is transmitted through blood and body fluids, primarily semen and vaginal fluid. During unprotected sex contaminated fluids make contact with mucous membranes lining the vagina, penis, rectum, or mouth, but in order to enter the bloodstream, the virus has to gain access through an abrasion or cut in the membrane. The virus can also be passed along by a contaminated needle or syringe, such as those shared by intravenous drug users. Before routine and effective blood screening for HIV was initiated in the mid-1980s, people receiving transfusions were at risk. The virus can also be spread through the placenta or breast milk of a mother. It is not spread by coughing or shaking hands, and although saliva, sweat, and tears can carry the virus in low levels, it is not believed that transmission occurs through them.

The mold Aspergillus causes the fungal lung infection aspergillosis, a common ailment in AIDS patients.

ATTACKS

Once inside the body, HIV attacks the immune system's helper T cells, also known as CD4+ cells because of the surface protein CD4. The helper T cells orchestrate the body's entire immune response to pathogens, activating B cells, macrophages, and other T cells. HIV binds to the CD4 molecule on the surface of the helper T cell, which allows the virus entry into the cell's cytoplasm. Once inside the cell, the virus sheds its protective protein coat and hijacks the cell's engines to make a DNA copy of its own ribonucleic acid (RNA), which is then incorporated into the host cell's DNA. From then on, every time the cell divides, it adds a copy of the viral DNA. Furthermore, the virus can force the host to make millions of copies of viral RNA, as well as protein coats. In effect, one virus can enslave a cell to make an army of viruses. In a devious twist, the production of HIV is often imperfect. Thus, mutations crop up frequently as the virus replicates, allowing it to evolve rapidly and keep ahead of antiviral immune defenses, drugs, and potential vaccines.

It takes about a day for an infected T cell to die. The immune system responds by helping to kill infected cells, but many of the virus particles move into other T cells. The body continues to produce more T cells, but it is a losing battle: Over time more are lost than replaced.

The first stage of HIV infection generally results in a short flu-like illness. Symptoms include fever, fatigue, swollen lymph nodes, rash, sore throat, and joint pain. At this point the levels of HIV in the blood are still so low that routine blood tests do not detect the virus. But within a few weeks, the immune system has created enough antibodies that a test can reveal HIV-positive status. For several months to years the HIV virus will then continue to replicate, and the afflicted person will show no symptoms.

After the helper T cells decline to about 20 percent of normal levels, opportunistic infections begin occurring that the body cannot fight. The patient now has AIDS. Among the more common infections are a lung disease caused by the parasite *Pneumocystis carinii*, tuberculosis, herpes simplex, and toxoplasmosis (which affects the brain). Additionally, cancers such as lymphoma and Kaposi's sarcoma (causing skin and other lesions) are frequent, as is dementia. One or more of these infections or cancers is likely to kill the patient.

Covering the National Mall in Washington, D.C., the AIDS Memorial Quilt commemorates the lives lost to the disease.

TREATMENTS

There is no cure or vaccine for HIV infection. But there are drugs that can slow its progress, and researchers continue to develop new treatments. Some drugs, including AZT (zidovudine), inhibit the action of the reverse transcriptase enzyme that enables the viral RNA to be copied. Another group of drugs interferes with the enzyme protease, which helps make protein coatings for the HIV viruses.

Since 1996, many physicians have been treating HIV cases with a "triple therapy" called HAART (highly active antiretroviral therapy), which calls for a combination of two or more reverse transcriptase inhibitors and one protease inhibitor. This triple cocktail has met with success in drastically reducing the number of viral particles in an infected person. The downside is that it cannot completely eradicate the virus, and some people cannot withstand its toxicity. Also, at a cost of more than $10,000 per year, the therapy is often beyond the reach of many people in the world who need it.

Education about the dangers of unprotected sex has often proved to be the best hedge against AIDS. In Uganda, awareness campaigns begin in schools as early as kindergarten, with songs and posters aimed at getting children to realize their ability to stop the AIDS epidemic. In Senegal, prostitutes must be licensed and routinely tested for HIV. Other West African campaigns warn men during soccer matches. National campaigns in other countries still face an uphill climb in battling superstitions about the disease.

the infection is naturally acquired through a microbe invasion or artificially acquired through a vaccination. In both cases, antibodies are produced against a specific antigen; helper T cells then organize an attack on the antigen. Vaccination with dead or weak pathogens, or more recently with genetically engineered fragments of microbial antigens, simply saves us the suffering that we might otherwise go through to achieve active humoral immunity. With immunity from natural infection, memory B and T cells are created during the primary response—the first exposure to the pathogen. While the primary response may take days to build up to its maximum strength, the secondary response generally takes only a few hours. A secondary response is any re-exposure to the pathogen after the primary response. In this case, the humoral and cell-mediated arms of the immune system are already primed by the existence of thousands of memory B and T cells. When they recognize the same old antigen, they go into immediate action, literally dividing and conquering—that is, they multiply and differentiate into plasma cells (antibody-producing B cell progeny), cytotoxic T cells, and so on.

Secondary responses are quicker and more potent: The antibodies produced on re-exposure to a pathogen can be hundreds of times greater than during the primary response. These antibodies exhibit more affinity with the antigens—they bind more readily and tightly to quickly overcome them.

Memory cells can remain in a person for decades. Of course, the original ones die, but their progeny maintain a steady population, a vigilant unit of trained guards. Often with a secondary infection with the same pathogen the response is so fast and overwhelming that you show no symptoms of the infection.

Vaccinations rely on the body's immunological memory. A small amount of killed or attenuated microbes, or portions of microbes'

Routine vaccinations are an important—but often least favorite—part of any childhood trip to the doctor.

FIRST VACCINES

1798 Smallpox	1945 Influenza
1885 Plague	1955 Polio
1917 Cholera	1963 Measles
1917 Typhoid	1967 Mumps
1923 Diphtheria	1969 Rubella
1927 Tuberculosis	1982 Hepatitis B
1927 Tetanus	1995 Hepatitis A

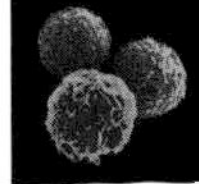

antigens, are introduced into your body. The vaccination is enough to activate your B cells and T cells and produce memory cells. The body can then handle natural infections without ill effects. Booster shots are often given after a period of several years to reinforce the immunological memory gained by vaccination.

Early efforts at immunization were often desperate attempts to stop epidemics, such as smallpox. Crude preparations of live attenuated microbes or microbe components closely mimic a real infection and thus lead to long-lasting protection. They may cause mild symptoms; immunocompromised patients may even suffer severe infections. In recent years, recombinant DNA technology, or genetic engineering, has helped refine vaccinations by producing purer forms of microbial antigens.

PASSIVE IMMUNIZATION

Like active immunity, passive humoral immunity can also occur in two ways. First it can be acquired naturally, when a mother's antibodies pass through the placenta to a fetus and in breast milk to her baby. Or it can be acquired artificially by injection with an "antiserum" containing gamma globulin. Such borrowed protection is short-lived. Antibodies are simply introduced; since there is no antigen challenge, there is also no active immunological response, and once the passively injected antibodies are lost, no new ones are produced.

THE FIRST VACCINATION

A 1796 illustration shows Edward Jenner, center, giving the first vaccination.

THE IDEA OF injecting a disease into a healthy person as a defense against that very disease took a leap of faith. Smallpox was such a deadly scourge that people were willing to take the risk. Caused by the *Variola major* virus, smallpox first appeared in Mesopotamia some 7,000 years ago. Pustules on the mummy of Egyptian Pharaoh Ramses V resemble those of smallpox victims. By the 16th century, smallpox had likely surpassed plague as the most dreaded pestilence in Europe. Highly contagious, smallpox could devastate entire communities. Victims suffered severe headaches, fever, and red rashes that left disfiguring scars; up to 40 percent of victims died.

Centuries before a vaccine was developed, the Chinese inoculated people by blowing bits of smallpox scabs into the nose of a healthy person, or by injecting them with pus from an infection. That technique was also practiced in Europe and America in the 18th century; the 2 percent risk of death from the inoculation was better than dying during an outbreak.

Then in 1796 English physician Edward Jenner took a calculated risk by purposely infecting a boy with material from skin lesions on the hand of a milkmaid with cowpox, a relatively innocuous disease caused by a near relative of the virus that causes smallpox. He had observed that dairy workers infected with cowpox seemed immune to smallpox. After the boy recovered from a mild infection of cowpox, Jenner inoculated him with smallpox, and the boy had no ill effects. Based on this discovery, Jenner coind the term *vaccine* from the Latin word *vacca,* which means cow. The 20th century saw a steady advancement in the field of immunization. In 1980 smallpox was declared eradicated. ■

Travelers to foreign countries often get a shot of gamma globulin to boost their immunity to a variety of infectious diseases. Other antiserums are used for emergency treatment of rabies, snake bites, tetanus, and botulism.

Let's take a look at two serious diseases, diphtheria and tetanus. A vaccine for diphtheria came out in 1923, one for tetanus in 1927. Most children in the United States today receive a series of five immunizations for these two diseases, in combination with a pertussis (whooping cough) vaccine: The DPT vaccine is administered at 2, 4, 6, and 18 months and before entering school. Then, the general recommendation is that people receive boosters for diphtheria and tetanus every 10 years. Clearly the adaptive immune system takes a lot of priming before it is fully immune to these two diseases.

Vaccines, of course, are active immunizations. But what happens if you have not been vaccinated or are not up on your boosters? Passive immunization can come to the rescue. If you step on a dirty nail, there is a chance that tetanus bacteria from the ground could get deep enough into the wound that, deprived of oxygen, they could germinate and produce the tetanospasmin toxin. As its name suggests, this toxin causes muscle spasms, particularly in the neck and jaw. What starts as a minor stiffness could, if left untreated, develop into actual lockjaw and painful convulsions within a week or two. If the convulsions begin early, the bacteria have propagated rapidly, and your chances for recovery are not good.

For any possibly contaminated wound, you should consult your doctor. In the meantime, clean the wound as thoroughly as you can. If you have had a recent tetanus shot, you are probably in the clear. If not, you will probably be given a booster. If you have had no tetanus immunizations, your doctor may give you a shot of tetanus immune globulin along with a standard shot of tetanus vaccine. This immune globulin is made from blood that contains antibodies for tetanus. A form of passive immunization, the antiserum is not the same as that of your body producing its own antibodies, and thus it provides protection for only a few weeks. But that is usually enough to fight off the bacteria and get you past the dangerous stage.

Another bacterial disease, diphtheria is now very uncommon in the United States. The bacteria

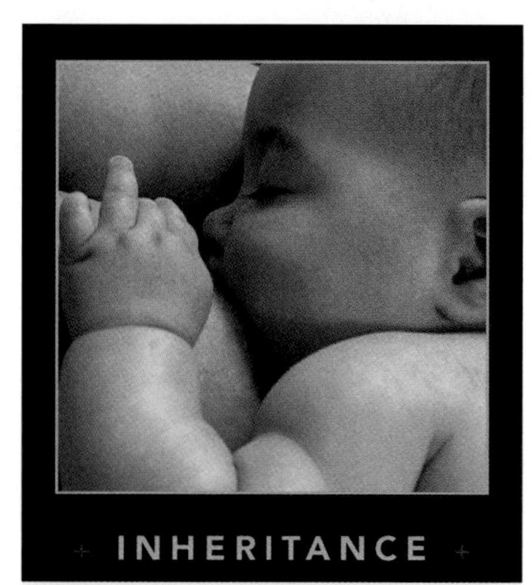

INHERITANCE

A FORM OF passive immunization is the immunity conferred by a mother on her fetus. The placenta screens maternal lymphocytes from the uterus, but the mother's antibodies travel across the placenta and into the fetal bloodstream, giving the fetus protection against the diseases the mother is protected against. The protection lasts a few weeks; then the baby's own immune system takes over. A similar boost comes from mother's milk.

BREAKTHROUGH

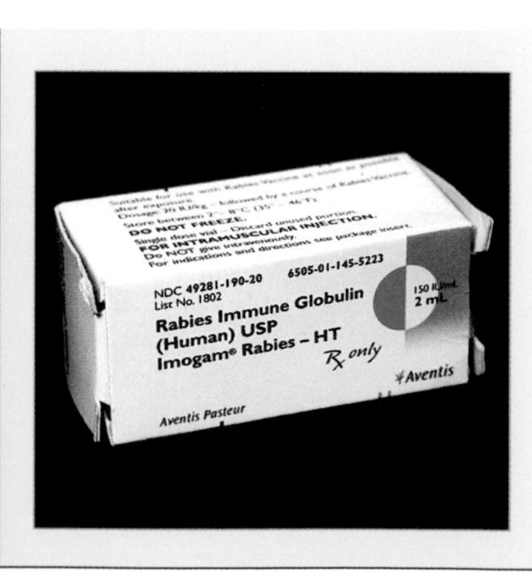

RABIES VACCINE: After Jenner's landmark smallpox vaccine in 1798, another 87 years would elapse before the breakthrough in rabies immunizations, left. The scientist was Louis Pasteur, founder of microbiology, born in eastern France in 1822. From early on, he began applying his skills as a chemist to problems in real-life situations, including the wine, beer, and silk industries. Using his microscope, Pasteur discovered bacteria that were spoiling wine; mildly heating it killed the bacteria. Named in his honor, the process of pasteurization of milk and other perishable liquids soon became a standard precaution.

usually are inhaled in airborne water particles, coughed or sneezed out by someone infected with or carrying the disease. The bacteria incubate for up to a week. Then you develop a mild sore throat, fever, and malaise. After this, a thick gray membrane coats the throat and tonsils, and your breathing can become labored. More serious symptoms include vomiting, heart infection, and paralysis of the respiratory muscles. Death may even occur. Those with proper vaccinations will probably develop no symptoms. Without vaccinations you must rely on passive immunization by a diphtheria antitoxin. This emergency measure usually destroys the infection. During a diphtheria outbreak in Nome, Alaska, in 1925, antitoxin serum was delivered in a heroic down-to-the-wire dogsled relay that took five and half days.

A white blood cell (red) engulfs a tuberculosis vaccine (blue). The bacteria in the vaccine will prime the immune system without causing sickness.

A more deadly disease, rabies (from the Latin for "madness") is caused by a virus; untreated, it is almost always fatal. The rabies virus is usually transmitted through the saliva of an animal during a bite. The virus generally incubates for three to seven weeks before symptoms appear, but they can appear as early as ten days or as late as two years later. The symptoms begin with tingling around the bite area; the skin becomes very sensitive, and the patient begins producing excessive saliva. Swallowing becomes difficult. Periods of intense anxiety and agitation ensue, followed by uncontrollable muscle spasms, convulsions, paralysis, and death. Since the disease is so dire, bites from wild animals are often treated as though for rabies. A combination of passive antibody and a vaccine is given in five separate doses, which can be very painful. The first rabies vaccine was invented in 1885 (see "Breakthrough," below).

BREAKTHROUGH

After years of groundbreaking work with various animal diseases, Pasteur, right, turned his attention to the deadly disease rabies. During experiments with dogs, he found that by using dried tissue from infected animals he was able to isolate weakened viruses. In 1885, he used a solution of these viruses to save the life of a boy, nine years old, who had been bitten by a rabid dog. The boy developed no serious symptoms. With donations from around the globe, the Pasteur Institute was founded in 1888. It evolved into one of the world's outstanding biological research institutions. Pasteur directed it until his death in 1895.

BODY ATTACKS

ALLERGIES AND AUTOIMMUNITY

As wonderful and nearly flawless as the immune system usually is, in some people it can be too sensitive. Pollen, animal dander, or other substances that are not intrinsically harmful to the human body are "read" by some immune systems as harmful antigens. The result is an allergic reaction, which might include sneezing, tissue inflammation, mucus secretion, and more serious symptoms. In short, the body reacts against the antigen, this time called an allergen, the substance it perceives as a foreign invader. Allergies or hypersensitivities—though usually not life threatening—may cause tissue damage and thus need to be monitored and controlled.

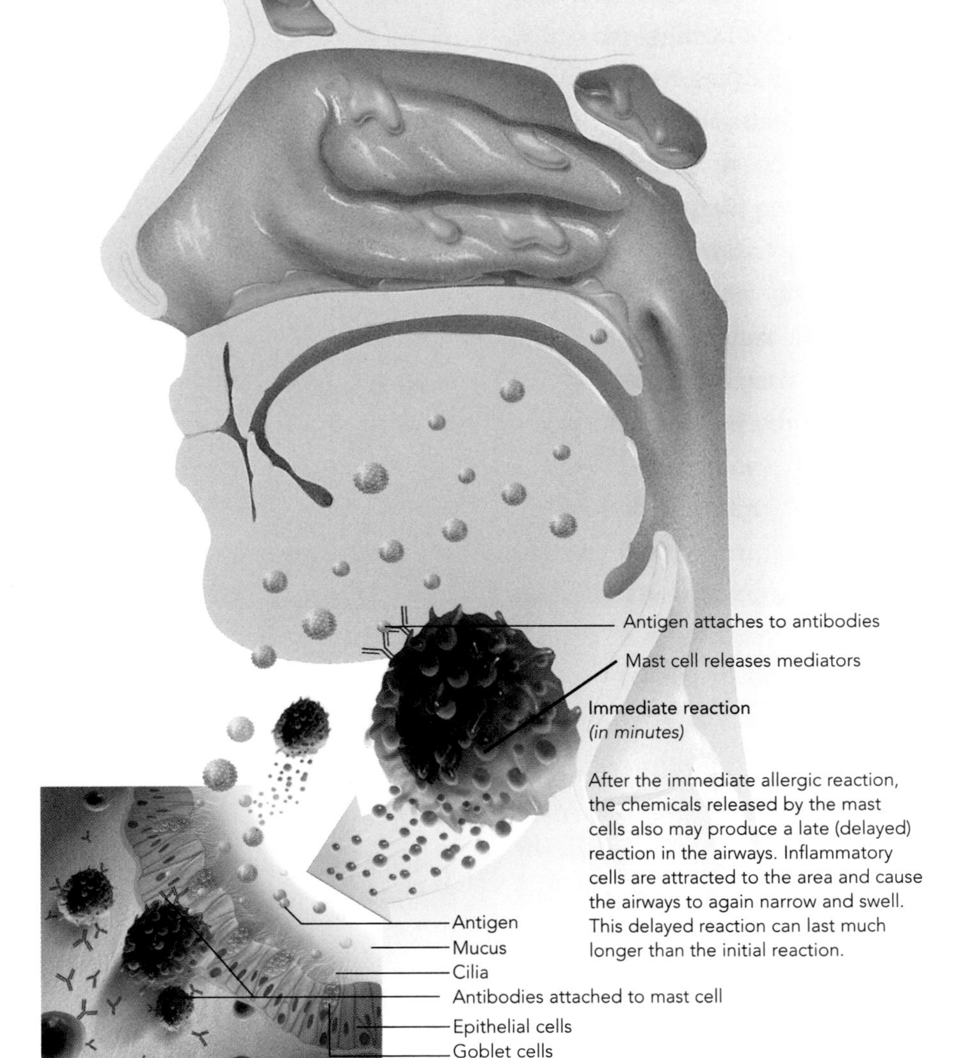

ALLERGIC REACTIONS

There are four main kinds of allergic reactions: anaphylactic, or immediate; cytotoxic; immune complex-related; and delayed. The first three are antibody-mediated, the last is a cell-mediated response.

In anaphylactic ("against protection") hypersensitivities, the reaction begins within seconds of the allergen introduction and lasts for about a half hour. During the first encounter with an allergen, antigen-presenting cells eat the innocuous "invader" and present its pieces to T cells. The T cells then secrete chemical mediators or cytokines that tell B cells to grow into antibody-secreting plasma cells. Special antibodies called IgE bind to mast cells within tissues and to white blood cells known as basophils. The immune system is now primed and ready.

At all subsequent exposures to the same allergen, the antibodies bind with the allergen, and the mast cells and basophils release chemicals: histamine, prostaglandins, leukotrienes, and kinins. All of these combine to cause an

SEE ALSO: Chapter Six, "Breathing," PAGE 168, Chapter Ten, "Hormones," PAGE 268 inflammatory response: vasodilation, increased capillary permeability, bronchial muscle contraction within the lungs, and excessive mucus secretion. The affected person may have itching, watery eyes, a runny nose, and troubled breathing. Hay fever, or allergic rhinitis, is one of the most common examples: Millions of people are somewhat allergic to pollen and suffer allergic rhinitis (hay or fever have nothing to do with it). If the allergen is ingested rather than inhaled, the person may experience cramping, vomiting, or diarrhea.

The anaphylactic reaction occasionally is bodywide, instead of just local. Usually this occurs when the allergen gets into the bloodstream right away and travels throughout the body. Some bee stings, spider bites, and injections such as penicillin may act this way, causing anaphylactic shock. Since the reaction occurs all over the body, the results can be quite serious: Breathing may be impaired by constriction of the bronchioles in the lungs and swelling of the larynx; widespread vasodilation can lead to hypotensive shock; and death can occur in minutes. The hormone epinephrine, a vasoconstrictor, is often used as a lifesaving antihistamine during episodes of anaphylactic shock.

Some immediate hypersensitivity reactions require no previous sensitization to the allergen. The tendency to have localized immediate reactions to allergens, called atopy, is hereditary. About 10 percent of the U.S. population has such reactions.

Cytotoxic, or type II, allergic reactions take up to three hours to develop; the effects may last for several more hours. Again, antibodies are reacting against antigens, but this time on the surface of cells. These antigens may be located on micro-organisms, red blood cells, lymphocytes, platelets, or transplanted tissue cells. When the antibodies have attached to the antigens, the plasma proteins called complement are activated, and cells bearing the antigen are destroyed by lysis. These reactions happen when incompatible transfused blood is rejected by the body.

Immune-complex, type III, reactions occur when soluble antigens and antibodies form insoluable complexes. The antibody-antigen complexes can persist deep within walls of blood vessels and other tis-

KEEPING HEALTHY

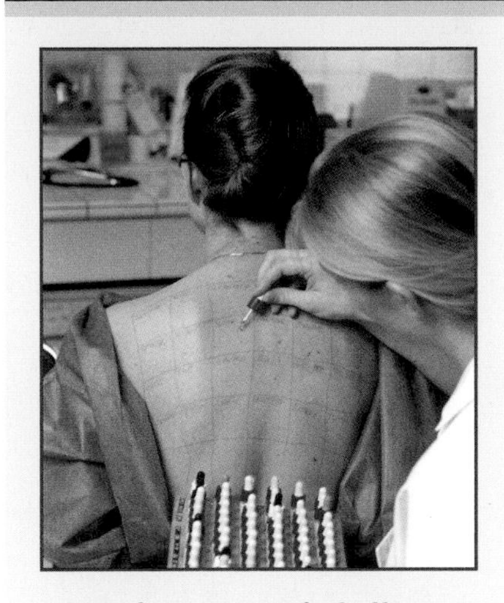

A prick test exposes food allergies.

ALTHOUGH MANY people believe themselves allergic to certain foods, fewer than 2 percent of Americans are affected. An allergic reaction to a food causes the immune system to produce histamines and other chemicals in response to the allergens. Food sensitivities, such as lactose intolerance, sometimes result in symptoms similar to allergic reactions but are not caused by an antibody response, and there is no release of histamine. Children are more likely than adults to have food allergies, but they often grow out of them. About 90 percent of food allergies are caused by cow's milk, peanuts, egg whites, wheat, or soybeans. Some people may be allergic to shellfish, berries, tree nuts, or corn. Most food allergies are type I, or immediate, reactions, occurring within seconds of eating. Symptoms include swollen lips, tongue, and throat; nausea and vomiting; hives; and nasal congestion. It can be a life-threatening emergency. ■

ALLERGY

THE TERM "ALLERGY" was coined in the early 1900s from the Greek *allos* meaning "other," by Austrian physician Clemens Baron von Pirquet. In 1905 he observed that patients receiving a second injection of smallpox vaccine or horse serum had a quicker reaction than that produced by the first. He theorized that primed antibodies were reacting to antigens in the serum. His findings led to a skin test for tuberculosis: A bump indicated a positive test.

sues where they can cause inflammation and eventual tissue damage. Rheumatoid arthritis and some types of glomerulonephritis are examples of type III allergic reactions.

Finally, cell-mediated, or type IV, allergic reactions are also known as delayed hypersensitivity reactions. Symptoms appear 12 to 72 hours after exposure. Antigen-presenting cells (such as those in the skin) bind to the allergen and carry it to lymph nodes, where they present it to T cells. The T cells then multiply, and some of them return to the allergen's point of entry. They activate macrophages, tumor necrosis factor, and other agents that then create an inflammatory response.

Allergic contact dermatitis is one of the most commonly encountered type IV reactions, caused by exposure to poison ivy, some cosmetics, heavy metals, and other seemingly innocuous materials. When skin brushes against poison ivy, the plant resin urushiol works into the skin and binds with self-proteins. Then these are bound by antigen-presenting cells located there and attacked as invaders by the immune system. Within a couple of days an itchy rash develops.

Physicians treated lupus patient Katherine Hammons with stem cells from her bone marrow. Here she comforts another patient.

AUTOIMMUNITY

In some people the immune system can work against itself. Instead of recognizing a "self" cell as friendly, it attacks as if the cell were an invader, producing antibodies and cytotoxic T cells that destroy normal body tissues. Called autoimmunity, this condition affects 5 percent of adults in North America and Europe, with women having twice the incidence of men.

Normally, B and T cells learn to recognize one's own MHC (major histocompatibility complex) proteins, a personal ID attached to all body cells. The immune system's ability to recognize and ignore its own cells is called self-tolerance. Early in life, immune cells that do not recognize your MHC proteins

SEE ALSO: Chapter Two, "Healing," PAGE 54, Chapter Ten, "Hormones," PAGE 268

WHAT CAN GO WRONG

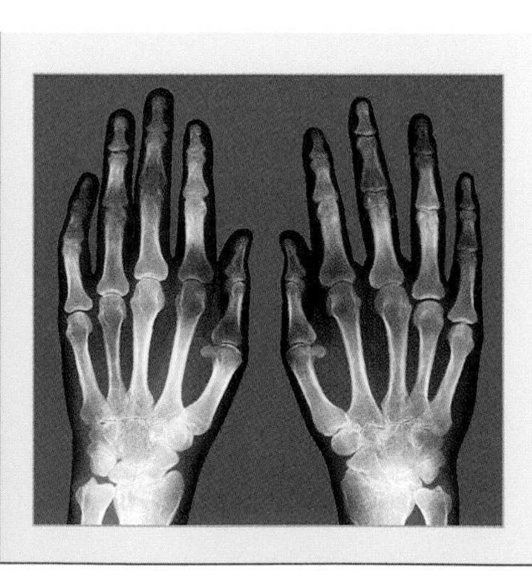

RHEUMATOID ARTHRITIS is the most debilitating form of arthritis. It is an autoimmune disease that usually begins in people aged 40 to 50. Though the initiating autoimmune mechanism is poorly understood, some researchers believe that molecules in the joints so closely resemble those of a microbe or some other antigen encountered by the patient that the immune system attacks the joints almost as innocent bystanders. Genetic predisposition appears to play a large role in activating the immune system against the body's joints. More than 1 percent of Americans suffer from the disease. Rheumatoid arthritis begins with are eliminated or deactivated. Sometimes this safeguard doesn't work as intended. Environmental cues or faulty genes produce clones of self-reactive B and T cells that become activated and fuel antibody- or cell-mediated immune responses against "self" cells.

Besides renegade immune cells, autoimmunity can be caused by the appearance of new self-antigens, or self-proteins not previously exposed to the immune system as a result of a genetic mutation or a chemical alteration of existing self-antigens, possibly during infections. There are also foreign antigens that so resemble self-antigens that the self-antigens become targets. In rheumatic fever, for example, antibodies attacking streptococcus bacteria cross-react with heart antigens and damage heart muscles, heart valves, and joints. Another example of an autoimmune disease, multiple sclerosis (MS) causes degeneration of tissues in the brain or spinal cord and can lead to partial or complete paralysis. Cytotoxic T cells become activated and destroy the myelin sheaths around nerve axons.

In the case of Type 1 (juvenile) diabetes mellitus, the T cells attack pancreatic beta cells, which produce insulin. Without insulin, carbohydrates cannot be normally processed. There is a period of several years in which the T cells weed out beta cells, possibly because their proteins resemble those of a virus the immune system is primed to attack. After this period symptoms may appear, usually before the person is 15 years old. Symptoms include increased thirst and urination, weight loss, fatigue, and ketoacidosis (increased ketones in the blood). Today some 750,000 Americans have type 1 diabetes.

LUPUS

NAMED FOR ITS characteristic skin lesions that resemble the bite of a wolf, lupus is an autoimmune disorder. Its most common form, systemic lupus erythematosus, is more prevalent in women than men, affecting people aged 15 to 50; the hormone estrogen may play a role in its development. Symptoms include redness or rash on the cheeks, sensitivity to light, inflammation of the lining of the heart or lungs or joints, damage to glomeruli of the kidneys, and the production of anti-DNA antibodies.

Autoantibodies can cause other autoimmune diseases. In Graves' disease, they mimic a thyroid-stimulating hormone that leads to hyperthyroidism. Autoantibodies may block receptors for the neurotransmitter acetylcholine; nerves thus cannot properly communicate with weakened skeletal muscles, resulting in the disease myasthenia gravis. Some autoimmune diseases stem from wrongly inactivated helper T cells or from the excess output of the antiviral protein gamma interferon. Other autoimmune diseases include lupus (see "Lupus," left), rheumatoid arthritis (see "What Can Go Wrong," below), glomerulonephritis, pernicious and certain types of hemolytic anemia, Addison's disease (adrenal gland failure), Hashimoto's thyroiditis, and ulcerative colitis.

Treatments for autoimmune diseases seek to suppress parts of the immune response without interfering with the whole system. Injections of genetically altered antibodies help some MS patients.

WHAT CAN GO WRONG

inflammation of the joints in the feet and hands, opposite, affecting the left and right at the same time. It then progresses to the wrists, knees, hips, and shoulders, thickening the synovial membranes and causing scarring and permanent damage to the joint capsules and cartilage. During inflammation, lymphocytes and other white blood cells pool in the joints and cause swelling. The thickened synovial membrane forms a pannus, abnormal tissue that adheres to the cartilage. Scar tissue fuses bones across the joint, rendering the joint useless. Pain and stiffness in a joint is followed by visible swelling and degrees of immobility. Fatigue, muscle weakness, and weight loss often accompany the disease. Although there is no cure, anti-inflammatory medications such as aspirin and ibuprofen help with pain relief. Some sufferers require a stronger anti-inflammatory such as prednisone or other corticosteroid.

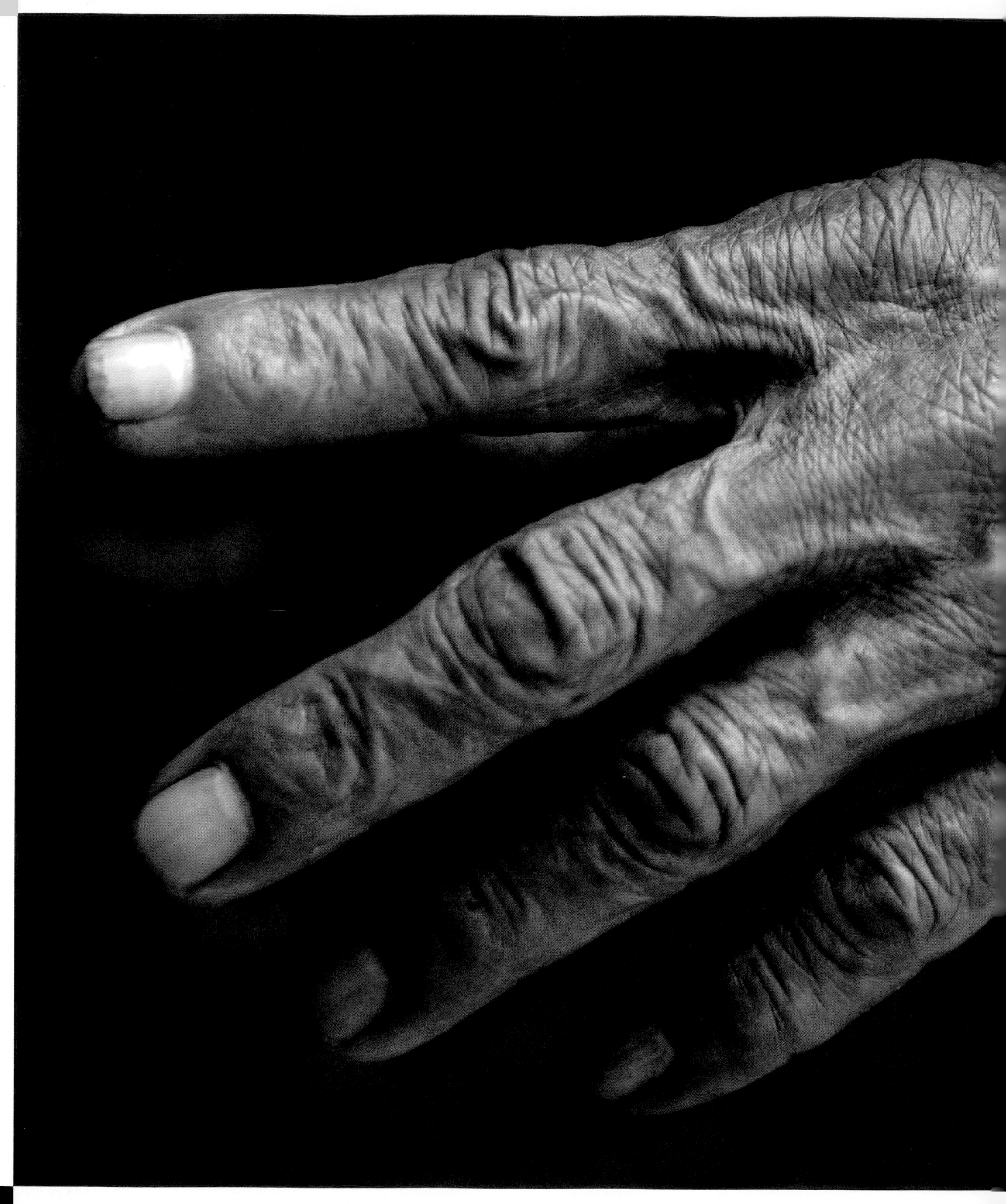

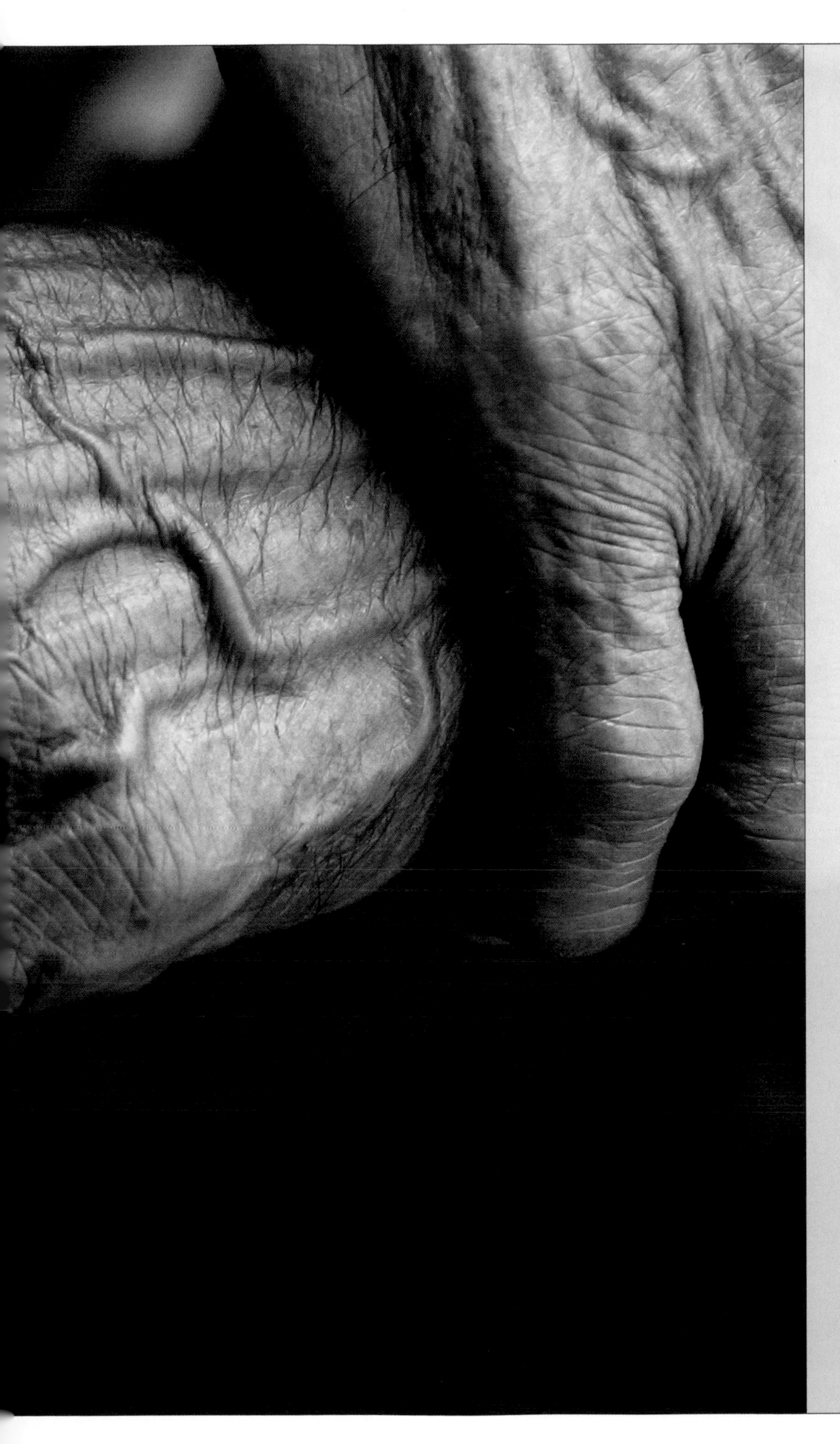

CHAPTER THIRTEEN

AGING

THE HUMAN BODY CAN LIVE longer than ever. Thanks to advances in medicine and a better understanding of the body, people are enjoying longer, more active lives. Increased knowledge about the effects of time on the body has led to better long-term strategies to maintaining good health. The decisions made every day will determine—to a large extent—the quality of life we may enjoy down the road.

The hands of an Indonesian man show the effects of age.

MATURITY

WHY THE BODY AGES

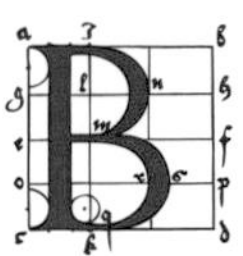

EGINNING AT BIRTH, aging is an ongoing process of growth and maturation. It starts in infancy and encompasses the healthy development of a child into a young adult as well as the journey from young adulthood into maturity. When most think of the grimmer aspects of aging, they are actually contemplating senescence: the process by which the body gradually breaks down and becomes unable to function properly, leading to death.

Even in the same family, the aging process can differently manifest itself, as with this pair of brothers.

Even this form of aging needs some clarification. To most of us, senescence is synonymous with disease and impairment: hearing loss, arthritis, osteoporosis, heart disease, dementia, and so forth. And it's true that old age is typically marked by increasing illness and disability. But these are still disorders, not examples of normal body functioning at any age. Usual aging may include these disorders, but normal aging does not.

Aging—senescence—is a set of cellular changes that occurs to the body over time in adulthood. As the body ages, its cells work less efficiently. Eventually cells stop dividing and die. As a result, tissues shrink and organs don't function as well as they once did. Almost all of the body's systems are affected by these changes: the senses, the digestive organs, the cardiovascular system, the immune system, the bones, and the muscles. Interestingly, the central nervous system—the brain and spinal cord—is among those least affected by age. In most of these body systems, the decline in function is not drastic. Only in situations of stress or disease does the older body lose its ability to work well.

Aging ranks with sleep as one of the fundamental mysteries of human biology. What causes the body to slow down, its cells to stop dividing, its organs to fall prey to increasing illness and disability? What biological processes dictate this decline? And what might be the evolutionary advantage in aging? No one has a definitive answer to these questions yet, although many theories have been propounded. The theories can be grouped into the gradual-damage-over-time camp and the genetic programming camp.

The first group of theories holds that the body ages because of wear

+ SUPERCENTARIANS +

MORE PEOPLE are living past 100 now, but there are a few people who made it well past the century mark.

+ Thomas Peters, Netherlands: 111 (1745–1857). First on record
+ Delina Filkins, USA: 113 (1815–1928)
+ Sarah Knauss, USA: 119 (1880–1999)
+ Jeanne Calment, France: 122 (1875–1997)
+ Thomas (1786–1893) and Elizabeth (1786–1891) Morgan, Wales: 106 and 105. Oldest married couple

SEE ALSO: Chapter One, "The Cell," PAGE 14

and tear that accumulates in the tissues over the years: Waste products build up in cells, backup systems fail, repair mechanisms gradually break down, and the body simply wears out, like an old car. Researchers have found that damaged proteins and destructive molecules called free radicals accumulate in cells as a result of metabolism over time; the cells' DNA also picks up increasing damage and mutations.

The second group of theories says that aging is driven by our genes—by an internal, molecular clock set to a particular timetable for each species. Support for this theory comes from animal studies: Scientists have been able to alter just one gene in some species of worms, mice, and other animals to increase their life spans.

As for the evolutionary issue, biologists point out that the benefits of natural selection greatly decline after reproductive age. In humans, natural selection will favor individuals who remain strong and healthy through the child-rearing years, so that their children survive to spread their genes. Therefore, evolution favors genes that are beneficial early in life rather than those that are beneficial later. The body does not have limitless resources and energy: It must make trade-offs. The more resources it expends on reproduction, the fewer it has to repair other body systems.

Yet even these commonsense theories have counterexamples. Some species of plants and animals

JEANNE CALMENT

Jeanne Calment celebrated her 117th birthday in 1992.

BORN ON FEBRUARY 21, 1875, in Arles, France, Jeanne Calment lived through the invention and growing dependence on electric lighting and power, the development of the telephone, the rise of the automobile, and the discovery and widespread use of antibiotics. She survived World War I and World War II. She saw the rise and fall of the Soviet Union. She was born into an age when the telegraph was new and exciting, and she lived well into the age of computers and the Internet.

As a child, Jeanne Calment met Vincent Van Gogh. The Impressionist artist came to buy art supplies at her father's store. He was "dirty, badly dressed, and disagreeable," Calment remembered later in life.

Marrying in 1896, Calment far outlived her husband, who died in 1942. In fact, she outlived two more generations. Her one and only daughter died in 1934, and her one and only grandson died in 1963.

Calment engaged in many activities that experts say can lead to a longer life. She stayed physically and mentally active well into old age. At 85, she took up the sport of fencing. And when she turned 100, she was still regularly riding a bicycle.

In any number of ways, Jeanne Calment defied the odds. She smoked up through the last two years of her long life. When asked to what she attributed her longevity, she mentioned her diet rich in olive oil, wine—and laughter.

At the age of 121, Jeanne Calment decided it was time to share some of her secrets to longevity. She released a compact disc of reminiscences, entitled *Time's Mistress*. She made sure that the recording appealed to generations coming after her, including hip-hop music in the background. On August 4, 1997, Jeanne Calment passed away. At 122 years and 164 days old, she was the oldest person on record in world history. ■

IN SEARCH OF THE FOUNTAIN OF YOUTH

The fountain of youth rejuvenates the old in a 15th-century Italian painting.

THE QUEST FOR immortality—or, even better, eternal youth—is as old as human civilization. One of the great works of early literature, the *Epic of Gilgamesh* (written before 2000 B.C.), depicts the efforts of its hero to avoid death by obtaining a magical plant. A snake carries off the plant before the hero can test it, and with that, "Gilgamesh sat down, weeping, his tears streaming over the side of his nose."

Ancient Chinese alchemists also sought the elixir of immortality, devising solutions of liquid gold, mercury, or arsenic, which may have contributed to the deaths of numerous Chinese emperors. Rumors of a fountain of youth on the island of Bimini intrigued Spanish explorer Juan Ponce de León in 1513, although it's unclear if he ever really looked for it.

More reasonable quests for healthy longevity, as opposed to eternal life, led to sensible and not so sensible advice from early scientists and philosophers. In the 13th century Roger Bacon advocated exercise, a controlled diet, good hygiene, and inhaling the breath of a young virgin. Wealthy Europeans in the early 20th century even tried extracts from the testicles of apes.

As a higher percentage of the population began to live into old age, the fields of gerontology (the study of old age) and geriatrics (medical care of older people) truly began. British surgeon Marjory Warren is considered a founder of the field of geriatrics. After a study of older patients in a county hospital, she recommended making the treatment of the elderly a medical specialty, with its own training and special units in hospitals. Warren set the tone for modern practice, which focuses on overall care of the aging and on improving quality of life rather than simply treating diseases as they arise. ■

appear not to age at all. One species of sponge may live as long as 15,000 years, while a bristlecone pine can survive for 4,000 years without aging. Some rockfishes live over 200 years, many turtles and tortoises beyond 100. What keeps them going—and what might keep us going—is one of biology's hottest areas of research.

LIFE SPANS

How long can a human being live? Is longevity increasing? Without records from prehistory, it's hard to know. Skeletons of Cro-Magnon humans indicate that some lived to at least 60. In recent centuries, a few people in each generation have lived past 110, and in the 20th century a remarkable Frenchwoman, Jeanne Calment, made it to 122 (see "Jeanne Calment," page 349). Judging by these examples, the healthiest people on Earth can apparently make it to an upper age limit of 125 or so.

What has changed, and continues to vary across populations, is the average—not the ideal—human life expectancy. In ancient Rome, a 40-year-old could expect to live to age 63. By the Middle Ages, even in the years before the Great Plague, life expectancy had dropped: A 40-year-old European would live, on average, to 56.

In the United States in 1900, an average 40-year-old white male would live another 27 years, to 67. By 2004, a 40-year-old European man would live to 78, a 40-year-old woman to 82. By the late 20th century, improved nutritional practices and sanitation methods, the advent of antibiotics, and other key advances of modern medicine had reduced infant mortality and childhood death to lower levels in industrialized countries. These successful advances meant that, for the first time, average life expectancy at birth and at maturity were almost the same.

Health care for the old has improved as well, and the numbers of older people in the overall population have grown. In the 20th century, the population over 65 increased tenfold, while the overall population only increased threefold. In 2000, the percentage of the population over 65 was 12.4 percent; in 2040, with the baby boom cohort among the very old, 20 percent of the population will be older than 65 and 19 percent of that group will be over 85. At that point, there will be as many people over 80 in our society as under 5. The number of 100-year-olds—centenarians—will likewise increase in this century, with the Census Bureau projecting that 1 in every 26 baby boomers will reach the century mark.

This recent increase in life expectancy is due largely to the great strides made in treating heart disease and stroke. Between 1980 and 2002, the mortality rate for ischemic heart disease dropped by 50 percent in people from 65 to 74. Mortality for stroke decreased by 45 percent in 65 to 84 year olds. As more people were surviving heart disease, though, the number of deaths from cancer rose slightly. As healthy as they have become, the elderly still suffer in large numbers from chronic diseases, particularly hypertension (affecting more than 50 percent of those over 65), arthritis, hearing problems, coronary artery disease, and various cancers.

No one has yet found the fountain of youth, or even the pond of guaranteed healthy old age, but doctors have identified common-sense practices that keep the elderly living longer and healthier lives. Among them are healthful eating, getting 30 minutes or more of exercise most days of the week, quitting smoking, limiting alcohol intake, preventing injuries, keeping up with vaccines (such as tetanus, influenza, and pneumonia), and screening for diabetes, cancer, and depression.

LIFE EXPECTANCIES

ON AVERAGE, WOMEN live longer than men. Why is unclear, but men's riskier behavior and higher mortality rates from cancer may account for some of the difference. The gap in life expectancy at birth is seen consistently around the world.

	Men	Women
Japan	78	85
Iceland	79	83
Australia	78	83
Sweden	78	82
France	76	83
United States	75	80

BREAKTHROUGH

CALORIE RESTRICTION: In 1935 Cornell nutritionist Dr. Clive McCay found that mice fed 30 percent fewer calories than usual lived 40 percent longer, with fewer diseases. Today, the findings of an ongoing experiment at the University of Wisconsin-Madison reflect the same trends. Studying the long-term effects of a reduced-calorie diet on rhesus monkeys, right, has shown benefits for health and aging. The exact reasons are still under debate. Calorie restriction may reduce free-radical damage in the cells, or stressed cells may muster defenses against environmental damage. The effects in humans are still unknown.

BUILDING BLOCKS

LIFE SPANS OF CELLS

THROUGHOUT THE body, aging's effects can be felt in its organs and systems, but the process of aging begins in its smallest units, the cells. Just how and why cells age is a hot research topic and a subject of much debate. A few mechanisms are becoming more clear, however, as scientists learn more about the molecular life of the cell.

In the early 1960s, biologist Leonard Hayflick cultured human embryonic cells and discovered that the cells would not divide indefinitely. They would multiply until they filled the flask and touched each other; after that, they stopped dividing. When separated into new cultures, they would begin dividing again—but eventually, no matter how much room they had, their divisions slowed and then stopped altogether.

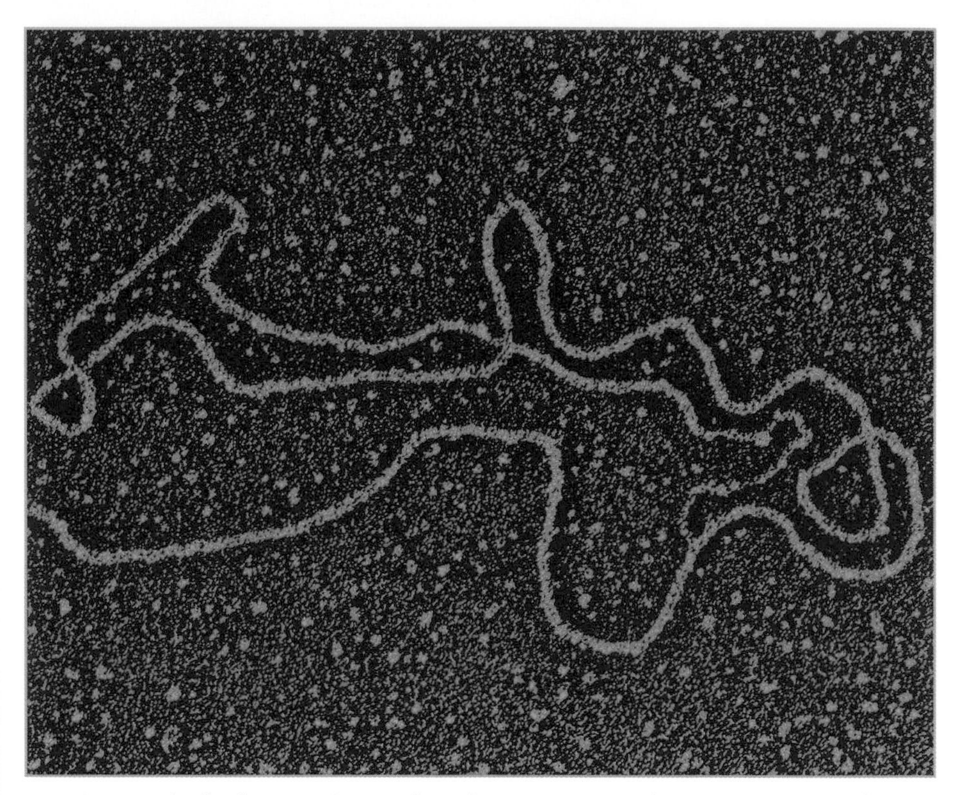

A strand of telomere (in a colored transmission electron micrograph) caps the ends of each chromosome.

The number of times the cells would divide before stopping became known as the Hayflick limit. In the cells of Hayflick's experiment, that limit was reached after an average of 50 divisions. The cells stopped dividing, then they enlarged, and then they lived on for a while before dying.

In studies since then, most—but not all—cells have reached their own Hayflick limit and died. They include cells from many of the human body's tissues, although "immortal" cells such as stem cells, germline cells (which give rise to eggs and sperm), or cancerous cells seem to divide indefinitely. Cells from older people divide fewer times; those cultured from short-lived mice divided only 15 times, while cells from long-lived Galápagos tortoises divided 110 times.

So the question arose: What makes cells slow down and die? Recent research has suggested several causes, generally falling into either the wear-and-tear or the genetic-programming theories. One of the most interesting findings was the discovery of the role of telomeres. Telomeres are stretches of DNA that cap the ends of chromosomes, protecting them from damage and keeping them from fusing with other chromosomes.

TELOMERES

IN THE LATE 1970s, biologist Elizabeth Blackburn studied a simple protozoan with an astonishing 20,000 chromosomes (as opposed to the human's 46). Researchers found they could extend the life of yeast by snipping off the telomeres at the ends of these abundant chromosomes and adding them to the yeast's DNA. Further research led to the discovery of the enzyme telomerase in the 1980s.

SEE ALSO: Chapter One, "The Cell," PAGE 14, Chapter Two, "Changes," PAGE 62

Researchers found that each time a cell divides, about 50 to 100 of the telomere's nucleotides are lopped off; when the telomere reaches a minimum length, cell division stops altogether. This finding was bolstered by the discovery of an enzyme in immortal cells, dubbed telomerase, that repairs telomeres after each division. It seems clear that telomere shortening works as a kind of cellular clock. No telomerase immortality pill is in the works yet, however. The enzyme does not affect nondividing cells, such as those in the brain or heart tissue, and in cells that do divide, it may promote cancer. Telomeres are just one piece of the aging puzzle.

Another piece of the puzzle comes from studies of a gene that suppresses the growth of tumor cells. Ordinarily, this is a good thing, but the gene may also be associated with slowing or stopping the growth of normal cells. Researchers knocked out the gene in some strains of mice and found that their bodies regained some functions seen in younger mice. The mice did not live longer than others, however, because with the tumor-suppressor gene gone, they developed more cancers.

Wear and tear on genes may also be involved in aging. We know that damage and mutations accumulate within the DNA of our cells as we age. DNA is typically redundant, however—the same genes appear on both copies of each chromosome—and it also has the ability to repair itself. Over time, though, the repair genes themselves may pick up damage and become unable to fix various errors in the cells' DNA, causing the myriad changes seen in aging.

This theory of the role of DNA damage in the process of aging is supported by studies of the rare disease called Werner's syndrome. The gene associated with Werner's syndrome codes for an enzyme involved in DNA replication. Children with the disease develop symptoms reminiscent of old age, including baldness, wrinkling skin, cataracts, and atherosclerosis.

A related theory of aging involves toxic by-products of cell metabolism known as free radicals. These unstable elements, made mainly in the cells' energy-producing mitochondria, have unpaired electrons in their outer orbits, which makes them pull electrons from other molecules in an oxidizing reaction, damaging them. Damage to the cells' mitochondrial DNA or to their nuclear DNA, as free radicals accumulate, may be one cause of aging. For this reason, many nutritionists recommend taking moderate amounts of antioxidants, such as vitamins C and E, to combat the damage.

KEEPING HEALTHY

Smoking damages the skin.

IN ADDITION TO wreaking havoc on the lungs, heart, and other organs, cigarette smoking ages the skin of the entire body.

Observers have long noted that smokers tend to have paler, more wrinkled facial skin. Now researchers have also found that smokers' skin is considerably more wrinkled than that of nonsmokers in people of similar age and sun exposure. This characteristic applies even to parts of the body that have been protected from the sun, such as the underarms.

The reasons for this pattern of wrinkled skin among smokers are still under investigation. They probably include the fact that smoking constricts blood vessels, therefore cutting off the oxygen supply to the skin.

There is another reason why being around cigarettes may affect a person's skin. Tobacco smoke boosts an enzyme that breaks down collagen, leading to wrinkles and a dull, inelastic look. ■

BONES AND MUSCLES

Bones and muscles change significantly with age, and those changes affect an older person's daily life perhaps more than any other symptoms

SEE ALSO: Chapter Three, "The Bones," PAGE 76, Chapter Four, "Skeletal Muscle," PAGE 110

JACK LaLANNE

Jack LaLanne is a living tribute to the benefits of exercise.

FOR MORE THAN 70 years, Jack LaLanne has taught Americans that getting fit and staying active are keys to a long, happy life. Before his time, lifting weights was considered radical—even dangerous. In 1936, when LaLanne opened the first health club in the United States, medical experts suggested that weight lifting could turn muscles unsightly, cause heart attacks, and even decrease sex drive. "Time has proven that what I was doing was scientifically correct," says LaLanne today.

Born in 1914 and raised in California, LaLanne first discovered the benefits of a diet and exercise at age 15. After attending a nutrition lecture, he swore off sugar, started a daily exercise regimen, and transformed his body. He opened his first health club six years later in Oakland, California. In 1951, LaLanne introduced fitness to a wider audience when he began hosting a television fitness show, sponsored by a local health food manufacturer. LaLanne's enthusiasm and simple exercises made it a hit. *The Jack LaLanne Show* became nationally syndicated in 1954 and ran for the next 34 years on as many as 200 television stations.

Age didn't slow LaLanne down in his efforts to promote fitness. As he got older, he regularly performed taxing physical stunts. At age 41, he donned handcuffs and swam in San Francisco Bay from Alcatraz to Fisherman's Wharf. At age 60 he did it again, this time wearing handcuffs and shackles and towing a thousand-pound boat. A decade later, he swam, handcuffed and shackled, from Queen's Way Bridge in Long Beach Harbor, California, to the *Queen Mary*. For the duration of the 1.5-mile swim, he towed 70 boats carrying 70 people in honor of his 70th birthday. Today LaLanne is still going strong at 92. "I can't die," he jokes. "It would ruin my image." ■

of aging. Between the ages of 30 and 60, bone density begins to decrease in both men and women.

Bones are constantly being broken down and reformed from within, a process called remodeling. Where bones are under stress, such as during exercise, they become thicker and stronger. In this constant rebuilding process, bones need minerals such as calcium and phosphorous, vitamin D to help absorb the calcium, and hormones such as estrogen, testosterone, and growth hormone.

Beginning in a woman's 30s, more bone is broken down than is reformed. Bones become thinner and more porous from the inside out. Menopause greatly accelerates bone loss, because estrogen levels plummet. In the first years of menopause, bone density can decrease by 3 to 5 percent a year and by 1 to 2 percent thereafter. In most men, the decline will be more gradual, as testosterone slowly decreases. In both sexes, new bone formation slows down with age; lack of weight-bearing exercise, insufficient calcium in the diet, and inadequate exposure to sunlight, a source of vitamin D, contribute to bone loss in older people.

As a result, 55 percent of Americans over 50 have osteoporosis, a condition involving dangerously brittle bones. Of those, 80 percent of them are women. Both men and women grow shorter with age as well, as spinal discs and vertebrae compress. On average, men lose

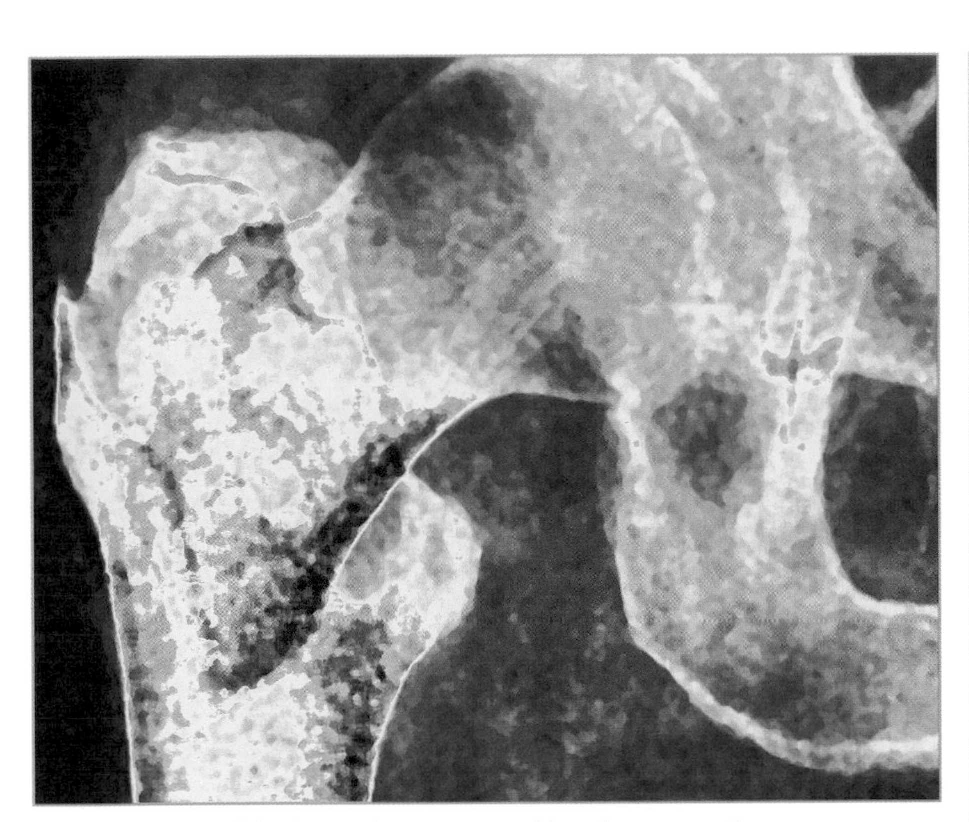

Older bones lose mass and break more easily, as shown in this x-ray of a hip fracture.

1.2 inches and women lose 2 inches between the ages of 30 and 70.

The cartilage that lines the joints shrinks with age as well, and the collagen in it becomes stiffer. These changes mean that joints can't stand up to stress as well as in earlier days. Ligaments and tendons also weaken and stiffen; they rip more easily and don't heal as well in older people. By the age of 70, most people have some degree of osteoarthritis. Wear and tear, old injuries, and obesity make joint cartilage deteriorate and bone ends develop spurs, causing pain and stiffness, particularly in the finger joints, spine, hips, and leg joints.

Muscles change over time as well. After the age of 30 or so, the body begins to lose muscle mass as muscle fibers shrink in number and size. Between the ages of 30 and 75, about half the body's muscle mass disappears while the amount of fat doubles. The greater loss comes in muscle fibers responsible for sudden, powerful contractions. Those people that maintain posture or endurance decline more slowly. In general, these changes don't affect daily life, but an elderly person who becomes immobilized, even briefly, will find the muscles losing conditioning quickly. After just one day of bed rest, an older person may need two weeks of steady exercise to return to strength.

The best ways to counter the loss of bone and muscle are diet and exercise. Adequate levels of calcium and vitamin D can help avoid bone fracture, and low vitamin D has been associated with decreased muscle strength and falls as well. Weight-bearing exercise—such as walking for 30 minutes three times a week—will maintain strength in aging bones and muscles.

EXERCISE FOR OLDER BODIES

	ACTIVITIES	BENEFITS
BALANCE	Standing on one leg, tai chi	Decreases likelihood of falls and fear of falling
CARDIO	Walking, swimming, dancing, racket sports, bicycling	Increases delivery of oxygen to muscles, builds endurance; prevents heart disease and diabetes; alleviates depression
STRENGTH	Biceps curl or shoulder shrugs with hand weights, wall push-ups, no-hands chair squats	Strengthens thigh, hip, and lower back muscles; builds bones and muscles; improves balance; prevents falls
FLEXIBILITY	Tai chi, yoga, stretching	Improves range of motion in joints; prevents injuries; prevents joint and muscle stiffness

STEADY RHYTHM

THE HEART AND LUNGS

THE HEART, BLOOD vessels, and lungs are remarkably durable structures, built to last for a long lifetime. The fact that so many older people develop heart and lung problems has less to do with the aging process than with lifestyle factors, such as lack of exercise, smoking, and obesity. When properly cared for, the body's cardiovascular and respiratory systems can carry most people to the age of 100.

These systems do change with age, however, and their ability to function declines with time. For instance, the valves and walls of the heart (particularly the hard-working left side) become thicker and stiffer. That makes the heart work harder to pump blood. It fills more slowly. The artery walls thicken and stiffen as well, which can contribute to both high blood pressure and atherosclerosis. Even the blood itself changes somewhat with age. Blood volume decreases in the elderly, and anemia becomes more common as the amount of red-blood-cell-producing bone marrow shrinks. These changes will be felt the strongest, however, when the system is stressed, as during intense exercise or illness.

Many heart and blood vessel disorders build up silently over the decades, and in the United States, they have come to represent the leading causes of death and serious illness among the elderly. Heart disease is the number one killer in America. Congestive heart failure, coronary artery disease, cardiac dysrhythmias, and acute myocardial infarction (or heart attack) were four of the five top causes of hospitalization among the elderly in 2004. More than half of those over 65 have hypertension, or high blood pressure, which can damage blood vessels, weaken the heart and kidneys, and lead to stroke.

Atherosclerosis, the buildup of cholesterol and fatty substances in the arteries, can start in youth. It more often becomes a problem in old age, when blood vessels become blocked or when plaques—clusters of cells—break off and travel to the

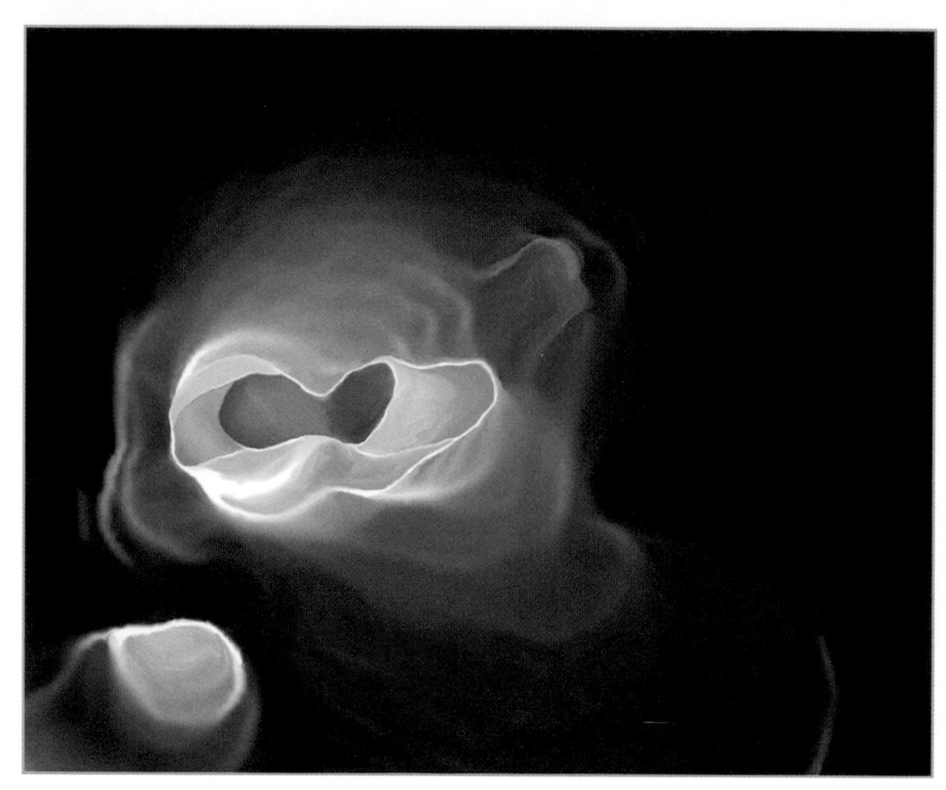

A CT scan of a 59-year-old's carotid artery shows fatty deposits (yellow). Over time this buildup can become atherosclerosis.

CHOLESTEROL

THERE ARE TWO types of cholesterol: low-density lipoprotein (LDL), the "bad cholesterol," and high-density lipoprotein (HDL), the "good" kind. LDL, which builds up in arteries, should measure no more than 100 milligrams per deciliter (mg/dL) and HDL, which may reduce the risk of heart disease, should greater than be 40 mg/dL in men and 50 mg/dL in women. Total cholesterol numbers should ideally be below 200 mg/dL.

SEE ALSO: Chapter Five, "The Heart," PAGE 140, "Blood Vessels," PAGE 150 coronary arteries or brain. High blood pressure and coronary artery disease can also lead to congestive heart failure, when damaged heart muscle can't keep up with the body's demands and blood backs up in the veins, sometimes spilling excess fluid into the tissues, lungs, and lower extremities.

Abnormal heart rhythms also increase with age. In some cases these are transitory and meaningless, while in others they indicate an underlying heart problem. These skipping, irregular heart beats, called arrhythmias, may be fast or slow, brief or prolonged. Some result from problems with the heart's chambers or with the pacing of the electrical current through the heart.

The good news is that most heart problems are strongly linked to lifestyle, which makes them treatable or reversible if caught in time. Stopping smoking, eating a diet low in saturated fats, reducing blood cholesterol, and exercising can prevent or at least control many heart conditions.

AGING LUNGS

Although lung tissue loses its elasticity as the body grows old, these changes are relatively minor and remain unnoticed in a healthy, resting person. Only during exercise or during the stress of illness will the lungs' reduced capacity make itself felt.

Lungs reach their peak fairly early. After age 20 or so, the

MICHAEL DEBAKEY

Heart surgeon Michael DeBakey survived heart surgery at 97.

DR. MICHAEL E. DEBAKEY (1908–), the pioneering American cardiovascular surgeon, became a living example of the resilience of the older body when he successfully underwent one of the procedures he had invented, repair of an aortic aneurysm, at the age of 97. At the time, DeBakey was arguably the most famous surgeon in the world. He had invented techniques and devices that allowed remarkable surgical repairs and procedures to the human heart.

Among the devices he invented were a roller pump that allowed safe blood transfusions without damaging blood cells and a Dacron grafting tool to be used in repairing diseased blood vessels. (The Dacron graft he originally created on his wife's sewing machine.) DeBakey pioneered the carotid endarterectomy, a treatment for blocked arteries in the neck to prevent a stroke, as well as the procedure called coronary bypass surgery, revolutionary at the time but now has become the standard method for treating blocked coronary arteries. As of this writing, his career encompasses more than 60,000 operations, 1,500 articles, and one National Medal of Science.

In December of 2005, DeBakey was working at home when he felt in his shoulder the telltale ripping pain of a dissecting aortic aneurysm—a tear in the inner lining of the aorta. After enduring the condition for several weeks, he finally underwent the risky surgery needed to replace the torn artery with the kind of Dacron graft he invented. Michael E. DeBakey became the oldest patient to undergo and survive the surgical procedure he himself pioneered. After rehabilitation, he returned to work—at the age of 98. ■

CHAPTER GLOSSARY

ANTIOXIDANT. Nutrient or chemical such as beta-carotene or vitamin C. Inhibits oxidation and may protect cells from damage caused by free radicals.

CORONARY BYPASS SURGERY. Surgical procedure whereby healthy vessel tissue is used to bypass, or detour, blocked passages in the coronary arteries, which bring blood to the heart.

DELIRIUM. Temporary mental disturbance marked by hallucinations, disorganized speech, and confusion.

DEMENTIA. A progressive mental condition, such as Alzheimer's disease, characterized by the development of many cognitive defects, such as disorientation and the inability to remember family members or make coherent plans.

ENDERARTERECTOMY. A procedure for treating blocked carotid arteries in the neck, developed as a preventive for stroke.

FREE RADICALS. Atoms or groups of atoms with an odd number of electrons. Free radicals can start a chemical chain reaction within cells that can damage their DNA or other structures.

GERIATRICS. The branch of medicine that is devoted to studying the processes and diseases of aging and old age.

GERONTOLOGY. The study of aging people.

HAYFLICK LIMIT. The natural limit that represents the highest number of times a cell can divide.

HIGH-DENSITY LIPOPROTEIN (HDL). Component of blood cholesterol often called "good cholesterol" because higher amounts are linked to a decreased chance of developing atherosclerosis

LIFE EXPECTANCY. Expected number of years of life based on statistical probability that can be measured from birth or from any other age.

LONGEVITY. The length of life.

LOW-DENSITY LIPOPROTEIN (LDL). Component of blood cholesterol often called "bad cholesterol" because higher amounts are linked to an increased chance of developing atherosclerosis.

MENOPAUSE. The natural ending of a woman's menstrual cycle; more broadly, the time in life when it occurs.

OSTEOPOROSIS. Disorder marked by decreased bone mass and increased susceptibility to fractures.

PRESBYCUSIS. Age-related hearing loss resulting from degenerative changes in the ear.

PRESBYOPIA. Age-related changes in vision that lead to an inability to focus on close objects.

SENESCENCE. The process of growing old and the changes associated with it.

TELOMERASE. An enzyme that restores telomere sequences to the ends of chromosomes.

TELOMERE. The natural end, or cap, of a chromosome, consisting of a repetitive DNA sequence.

number of alveoli (air sacs) and the number of capillaries in the lungs begin to decline. With these changes, the lungs don't collect and distribute as much oxygen as they did in youth. Oxygen levels in the blood decrease slightly, and sensitivity to built-up carbon dioxide also decreases. Many elderly people develop apnea—the tendency to stop breathing briefly—during sleep and are in greater danger of running short of oxygen when ill with a lung disease such as pneumonia or emphysema.

The lungs' immune system also begins to break down with age. Cilia, the small hairlike sensors in the mucous membranes, grow sluggish and let in more foreign particles. Fewer macrophages (immune system cells) scout the airways for invaders. The coughing reflex weakens as well.

Because the lungs pull in all kinds of airborne organisms from outside the body, they are particularly vulnerable to infections. Pneumonia and influenza are special risks, ranking together as the fifth leading cause of death in those over 65. Usually caused by bacteria but also by viruses or fungi, pneumonia can spread in institutions such as nursing homes or in hospitals, where it is particularly dangerous to people already debilitated by other ailments.

A common pathway for infection in the elderly is via aspirated food—food caught in their breathing passages. Bacteria often colonize

SEE ALSO: Chapter Six, "Respiration," PAGE 156, "The Lungs," PAGE 162 the throats of older people and will hitch a ride into the lungs when pieces of food are accidentally inhaled. The infection then inflames the airways, and as fluid begins to fill the air sacs, the lungs struggle to take in enough oxygen to feed the body. With weaker lungs, decreased immune systems, and a feeble coughing reflex, older people can have difficulty fighting off the infection, which then spreads into the bloodstream as sepsis. Influenza, a viral infection, is similarly dangerous to the elderly, in large part because it often leads to pneumonia or bronchitis.

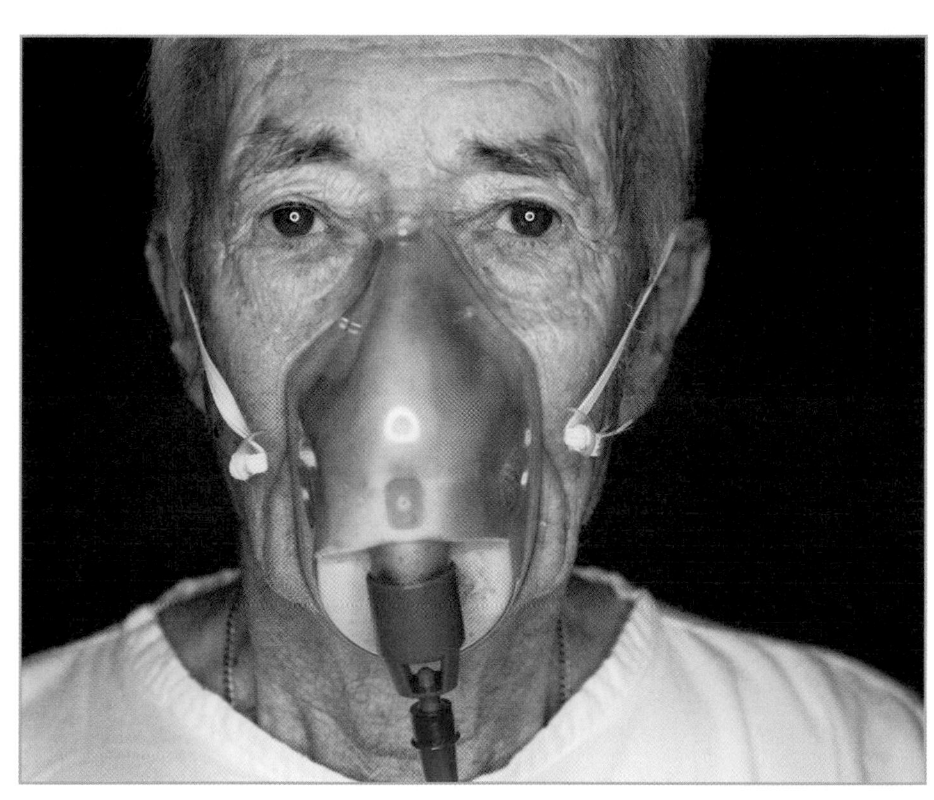

An elderly patient wears an oxygen mask to assist ailing lungs and to draw in enough oxygen.

Many of the deadliest lung diseases in the elderly (including pneumonia) are tied to the damage caused by smoking. This is particularly true of chronic obstructive pulmonary diseases (COPD), typified by obstructive emphysema and chronic bronchitis.

The good news is that the number of smokers in the population has dropped dramatically in recent decades, and fewer than 10 percent of people over 65 currently smoke. The bad news is that many of them are former smokers. The lungs can recover partially when a person stops smoking, but some damage has already been done. COPD, the fourth leading cause of death among Americans of all ages, is particularly prevalent among those over 55. Emphysema is characterized by the destruction of alveoli in the lungs—the sacs where oxygen enters the bloodstream. Bronchitis is characterized by a swelling of the airways and persistent coughing. Both leave the sufferer short of breath and lacking in oxygen.

Aerobic exercise can boost lung function and increase the ability of the cardiovascular system to keep oxygen in the bloodstream. The pneumococcal vaccine can prevent or lessen the effects of common forms of pneumonia, as can the influenza vaccine for that disease.

WHAT CAN GO WRONG

INFLUENZA—the flu—is a highly contagious virus, right, that affects between 5 and 20 percent of Americans each year, typically between the months of December and April: flu season. About 36,000 die annually from the illness and its complications, more than 80 percent of them elderly. The disease hits older people so hard because they often have other complicating diseases such as diabetes or heart disease that are already straining their systems, and because parts of their respiratory systems don't work as well as they used to in replenishing the oxygen in the blood and cleaning the system.

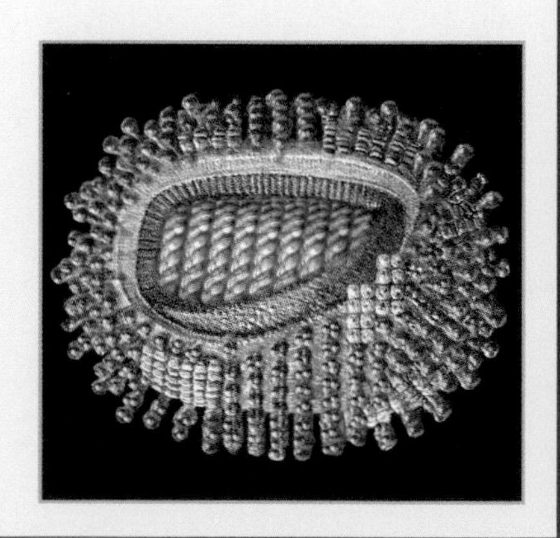

A WISE DIET

FUELING THE BODY AS IT AGES

GOOD NUTRITION, vital at any age, is particularly important as the body ages and its organs work less efficiently. A diet that worked adequately for a 30-year-old may leave the same person at 70 with problems ranging from high blood pressure to dehydration.

With age, the body's metabolism slows. The basal metabolic rate—the energy needed to keep basic body systems going—drops by 20 percent in men and 13 percent in women between the ages of 20 and 90. Body composition changes, too: Bone mass and muscle mass decrease while body fat increases. Some perceptions and sensory experiences are dulled. Many older people are less aware of thirst and therefore become dehydrated; dulling of the senses of smell and taste can make food bland and uninviting, leading to malnutrition. The digestive tract moves food more slowly, which can bring on constipation. And disorders common to old age often dictate a change in lifelong eating habits. People with kidney problems need to be careful about proteins, for instance, while those with diabetes must monitor their carbohydrates.

Living at an around-the-clock care facility, an older couple has a specially crafted diet to meet their nutritional needs.

The older body needs fewer calories but just as much nutrition. (To get a rough estimate of the range of calories the average older person needs each day, multiply body weight in pounds by 12 to 15.) The U.S. Department of Agriculture has modified the food pyramid for older people to include at least eight servings of water or other clear, noncaffeinated liquids at its

SEE ALSO: Chapter Seven, "Digestive System," PAGE 178

WHAT CAN GO WRONG

ADULT ANOREXIA: Eating disorders are not just the problems of adolescents. Seniors also experience problems with taking in enough food. Past the age of 70, most elderly people lose weight, and some lose too much. About 16 percent of older people living outside institutions eat fewer than a thousand calories a day, far too little for anyone to sustain proper function. Any older person who has unintentionally lost ten pounds or more in the last six months is at risk of malnutrition, a condition that is particularly prevalent among those living in nursing homes or living at home alone. The causes of malnutrition in the elderly are many. Appetite naturally decreases in older people because levels of the hormone that creates a feeling of fullness increase, while at the same time both the senses of taste and smell diminish. Food can seem bland and base, with high-fiber whole-grain breads and cereals, bright-colored vegetables, and deep-colored fruits ranking next in importance. Refined carbohydrates, such as sugar, potatoes, and white rice, need to be consumed sparingly; these have little nutrition and are readily converted to fat.

Saturated fats, found in red meat and butter, and partially hydrogenated fats (trans fats), found in many manufactured cookies, crackers, and chips, should make up no more than 10 percent of the diet. These fats are implicated in heart disease. Good fats include monounsaturated fats and omega-3 fats, found in canola oil, peanut butter, flaxseed, salmon, and other fishes. No more than 30 percent of total calories, however, should come from any kind of fat.

At the top of the food pyramid for older people are supplements of calcium, vitamin D, and vitamin B12. Medications often affect nutrition, blocking the absorption of vitamins or of calories in general, and the elderly must keep these matters in balance. Diuretics, for instance, can deprive the body of zinc, magnesium, vitamin B6, potassium, and copper; antacids can block vitamin B12, folate, iron, and total calories. Older people also need to be aware that medicines they have been taking successfully for years may need to be adjusted as they age; amounts that were appropriate for a younger body can be toxic to someone with aging kidneys. Older livers and kidneys also don't tolerate large amounts of alcohol well. Alcoholism, an underdiagnosed problem, affects from 5 to 20 percent of the elderly, and contributes to disabling falls and car accidents in older people. While one alcoholic beverage may decrease the risk of heart disease, excess alcohol blocks the absorption of vitamins and minerals.

+ SUPERFOODS +

THE BEST FOODS for an older body pack the most nutrition into the fewest calories. Water, calcium, and fiber are essential. Good choices include

+ Fruit and vegetable juices
+ Whole-grain bread, brown rice
+ Deep-colored fruits and vegetables, above, like broccoli, tomatoes, strawberries, and blueberries
+ Nuts and legumes (beans, lentils)
+ Moderate protein, especially fish
+ Low-fat or no-fat dairy products

Both obesity and anorexia, or wasting, can affect the elderly, often with more dire consequences than in younger, more resilient individuals. The increase in obesity across the entire U.S. population includes the elderly, with more than 30 percent of those over 60 considered obese. Excess weight is a major factor in many common disorders, including cardiovascular diseases, diabetes, and osteoarthritis. Obesity also makes it more difficult for elderly people to perform even the tasks of daily living, such as bathing and dressing.

WHAT CAN GO WRONG

unappetizing. Chewing and swallowing become more difficult if teeth are lost. Mouths become drier, and muscle strength wanes in the jaws. Other factors, like depression, dementia, and many medications can also contribute to the problem of lack of appetite and resulting malnutrition, as can social issues such as limited mobility, isolation, and low income. It's important to recognize malnutrition before it takes its toll in muscle wasting, anemia, fatigue, and overall loss of function. Techniques to encourage the elderly to eat well include: providing companionship during meals; preparing food with varied textures and colors; using herbs, spices, and strong flavors; avoiding hard-to-open packaging; allowing enough time for meals; catering to food preferences where possible; encouraging exercise to stimulate appetite; and ensuring firm teeth or dentures.

CONTROL

THE NERVOUS SYSTEM AND THE ENDOCRINE SYSTEM

FOR MANY PEOPLE, the most fearsome specters of aging are those of memory loss, dementia, and the erosion of self that comes with the deterioration of the brain. These disorders are not features of normal aging, however. The healthy brain, in fact, works very well in old age. Like the rest of the body, its tissues shrink slightly as cells die off, and it loses about 10 percent of its weight by extreme old age.

This loss matters less in the brain than it would in many other organs, though, because the brain has far more cells than it needs. Its neurons also form new connections as cells die, their dendrites extending to still-living cells.

Some neurotransmitters (the chemicals that transmit messages across synapses) and neuroreceptors (the structures that receive those messages) decrease with age as well, and cerebral blood flow is reduced. These changes, taken together, can produce subtle alterations over time. They may affect short-term memory, verbal fluency, and learning ability, but they need not significantly alter intellectual functioning.

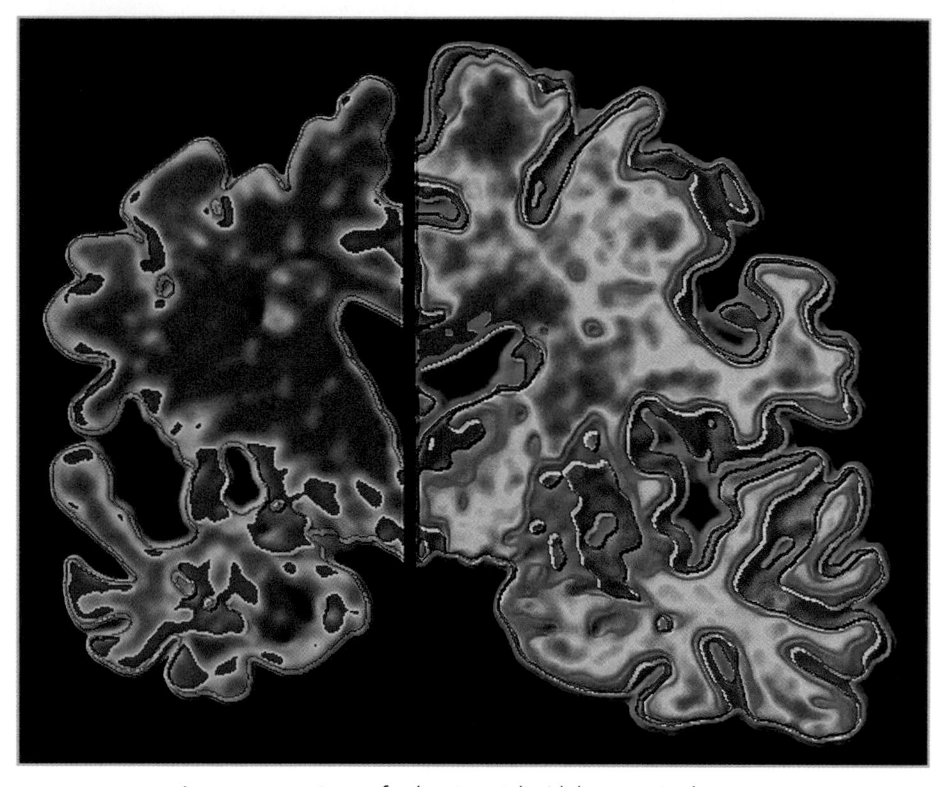

A cross-section of a brain with Alzheimer's disease, left, contrasts with a normal brain, right.

The spinal cord and peripheral nerves, which extend into the rest of the body, decline very little with age. The speed at which signals travel through the nerves slows only slightly with age.

SENSORY LOSS

Perhaps the most noticeable changes to the nervous system over time occur in the senses, particularly vision and hearing. Presbyopia—a

WHAT CAN GO WRONG

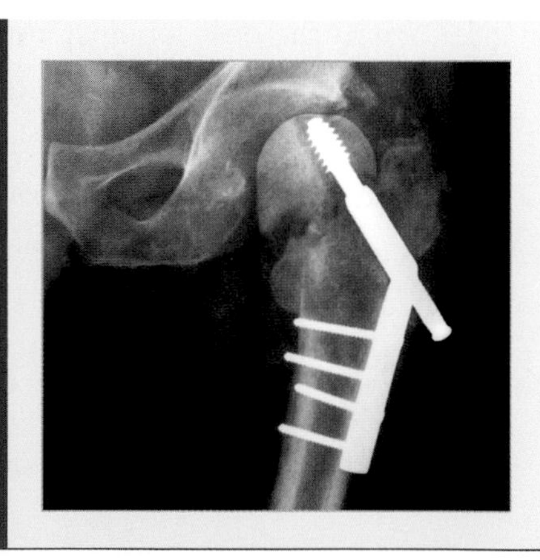

FALLING—an innocuous occurrence in childhood—becomes a fearsome event in old age. Brittle bones, left, may break more easily. Falls are the seventh leading cause of death over 65. Causes for increased falls include changes in sensory acuity, such as deteriorating balance organs in the inner ear and decreased vision and depth perception. Athritis, joint stiffness, or muscular weakness can cause falls. Some medications, like those for hypertension and heart disease, can lead to dizziness. Hazards in the home—loose throw rugs, power cords, and dim lighting—should be fixed. Grab handles and nonskid mats are essential.

SEE ALSO: Chapter Nine, "Nervous System," PAGE 228, "The Brain," PAGE 234 decline in near vision—is one of the few inevitable and virtually universal characteristics of aging. Starting between the ages of 40 and 50, it is the result of a gradual stiffening of the lens of the eye. The lens also becomes denser, blocking more light. The retina loses sensitivity, making it harder for older people to see in dim light. On average, 60-year-olds need three times more reading light than 30-year-olds. With age, the eyes are less able to distinguish fine details; the pupil reacts more slowly as well, so that older people take longer to adjust to sudden brightness or darkness; glare, such as from headlights at night, becomes more bothersome.

Hearing worsens slightly, too, due both to the normal processes of aging and to long-term damage from loud noises. Half of those older than 75 have some form of hearing loss. The cochlea, hair cells, neurons, and blood supply within the ear's hearing apparatus all begin to wear out or decrease with age.

The most common form of hearing loss due to aging, called presbycusis, involves the degeneration of various parts of the cochlea or auditory nerve. This usually comes on gradually. It involves both ears equally and results in the loss of high-frequency sounds. Although it doesn't seem too serious—who needs to hear a bat squeak?—most consonants in speech, such as *k, p,* or *t,* involve quick, high pitches. Without consonants, speech sounds like an inarticulate mumble. Background noise can also become more distracting as the range of hearing diminishes. Fortunately, a variety of hearing aids exists to correct, at least partially, most common forms of hearing loss.

Less noticeable to most people are age-related changes to the senses of taste and smell. Subtle smells become harder to detect, and as taste buds decline in number and sensitivity, so does the experience of different flavors.

KEEPING HEALTHY

Daily Yoga keeps this 84-year-old fit.

EXERCISE PROVIDES a cornucopia of benefits to the elderly, but one of the most surprising and valuable advantages is improved mental performance.

One study of formerly sedentary people over 60 who began walking vigorously for 45 minutes, three times a week, showed major improvements in their frontal-lobe abilities. In 2006, another study examined 460 people at age 79 and found that those in better physical shape were also more mentally sharp.

Other studies have shown that regular physical activity increases memory, mental alertness, problem-solving skills, and attention levels. At the same time, regular physical exercise can combat depression, improve sleep, and boost energy levels—yet more reasons to stay out that chair and keep moving at every age. ■

MENTAL DISORDERS

The most serious neurological problems of old age stem not from natural aging processes but from brain disorders such as dementia or from psychiatric disorders such as depression. Although fewer old people suffer from depression than do younger ones, the disorder can be hard to diagnose in someone who may have other chronic illnesses and be taking several kinds of drugs.

Up to 15 percent of the elderly, and up to 30 percent of those in nursing homes, suffer from some form of depression. Alcoholism, grief from the loss of a spouse, sickness, and drugs can contribute to the illness. The most alarming statistic associated with depression in the elderly is the rate of suicide among older white men. Suicide is 85 percent more common in white men aged 70 to 74 than it is among young men, and the rate increases

PORTRAIT OF ALZHEIMER'S

IN 1906 GERMAN neurologist Alois Alzheimer (1864–1915) reported on some unusual structures he had found in the autopsied brain of a former patient. The 51-year-old woman had been admitted to Alzheimer's clinic several years earlier with symptoms of what was then simply called senility: She was disoriented, her memory was poor, and she had trouble reading and writing. Gradually she had developed hallucinations and began to lose higher mental functions altogether. She died at the age of 55.

Alzheimer found that not only was the patient's cerebral cortex unusually thin, but her brain was also littered with senile plaques (clumps of cells and degenerated nerve fibers) and neurofibrillar tangles (twisted nerve fibers within the neuron cell bodies). His findings, particularly the discovery of the neurofibrillar tangles, revealed that this kind of senility was in fact a distinct, physical brain disorder, the disease now named for him.

DIAGNOSIS

Today more is known about the symptoms and progression of Alzheimer's disease, but its basic causes are still unknown. No cure has yet been found.

Alzheimer's is a progressive, degenerative, and fatal brain disorder, marked by the gradual loss of memory, learning, reasoning, communication, and activities of daily living. In its later stages, it can cause hallucinations, personality changes, and eventually the loss of basic functions such as the ability to walk or swallow. It affects 5 million people in the United States, almost all over the age of 65.

Many older people fear they are developing Alzheimer's when they forget a name or misplace their glasses, but the symptoms of this dementia are more serious and progressive than that. It's the difference, one observer notes, between forgetting where you left your keys and forgetting what a key is.

Artist William Utermohlen painted a self-portrait every year after he was diagnosed with Alzheimer's disease.

SIGNS TO WATCH

Symptoms of Alzheimer's disease do often include memory loss. Performing familiar tasks, such as cooking a meal or making a phone call, can become difficult or confusing. Another sign is forgetting words or substituting unusual words for common ones. Misplacing things in unusual ways, like putting a hat away in a refrigerator, is another sign. Sufferers may also feel disorientated in time and place: Many find themselves lost in a familiar neighborhood, unable to find the way home, or unable to remember the year.

Other signs can include increasingly poor judgment about daily living or finances. Loved ones should also watch for changes in mood, personality, or behavior, including paranoia or anger. Expressing increased passivity and lost interest in daily activities can also be an indicator.

Doctors who suspect that a patient has Alzheimer's disease will typically conduct a series of physical and cognitive tests. Physical exams rule out other problems, such as alcoholism, medication problems, infections, or other brain disorders. Exams to evaluate the patient's mental function will include questions about the date and the person's location, basic calculations, and simple memory tests.

If the patient has Alzheimer's disease, the doctor will usually be able to say if it is in an early, middle, or late stage of the disease. Early stages include mild cognitive decline, such as noticeable memory lapses or a declining ability to plan and organize. In the middle stages, the patient may begin to forget his address or the current date and have difficulty with tasks such as paying bills or choosing clothing. By the late stages of the disease, the patient may lose awareness of his surroundings, develop hallucinations, and become unable to perform basic tasks of daily living. As damage in the brain spreads, he will eventually be deprived of basic physical functions such as the ability to walk or sit without support. This stage is followed by death.

Two years after painting the portrait opposite, Utermohlen could no longer depict a recognizable face.

LOOKING FOR A CURE

Researchers have identified various processes at work in the brain with Alzheimer's, although its ultimate cause—or, more likely, causes—are still frustratingly out of reach. The brains of Alzheimer's patients shrink, particularly in areas involved in memory and cognition. They also develop a shortage of the neurotransmitter acetylcholine, which is involved in memory and reasoning. Researchers have identified genes directly connected with a rare, inherited version of the disorder familial Alzheimer's disease. The more common, sporadic form of the disease may also be linked to mutations on genes that produce a molecule called beta-amyloid (found in plaques). One theory holds that the overproduction of beta-amyloid may inhibit learning and lead somehow to tissue destruction in the brain. Another theory says that free radicals accumulate in the brain tissue and damage neurons.

Scientists have begun to develop medications that can help with symptoms. A class of drugs called cholinesterase inhibitors keep acetylcholine from breaking down in the brain, improving cognitive function somewhat in about half of the patients treated.

Another kind of drug blocks overproduction of glutamate, a neurotransmitter that may kill brain cells if present in excess quantities. Drugs that block the protein's production or improve neuronal regeneration are also under development. However, efforts to come up with a vaccine have not yet yielded a success.

Alzheimer's patients benefit from behavioral and environmental treatments. Caregivers can learn to simplify the patient's surroundings, keep to a routine, label household rooms and objects, and modify living quarters to make them safer.

And the basic measures that keep anyone healthy in old age—exercising, quitting smoking, mental stimulation, social connections—apply to Alzheimer's patients as well. With informed care, they can ease the course of a destructive disease.

even more above age 85. Caregivers and doctors need to be alert to the signs of depression—such as feelings of worthlessness, hopelessness, and emptiness—and not attribute them simply to old age.

Disorders involving mental confusion include delirium and a range of dementias. Delirium, which comes on suddenly, is marked by inattention and sporadic incoherence. It usually stems from a physical problem such as dehydration, infection, or drug withdrawal, and it requires immediate medical attention. Dementia is typically caused by a chronic brain disorder such as Alzheimer's disease or multiple small strokes. Memory is typically affected first, but gradually other mental functions, such as language or the sense of time and place, deteriorate. No cures exist for dementia, but people with the disorder can be helped with supportive treatments.

HORMONAL CHANGES

With age, the mixture of hormones in the bloodstream and the ability of the body's organs to respond to hormone changes. Most noticeable, perhaps, is the drop in estrogen that comes with menopause. Both men and women will experience a number of other hormonal changes as well, however: Alterations in skin, muscles, sexual drive, overall energy, and health are often connected to shifting hormones.

As we age, changing hormones have a visibile affect on the body's skin and shape.

Growth hormone, for instance, declines considerably with age. This decrease may be associated with age-related loss of muscle and bone mass, increased belly fat, and higher risk of heart disease. However, experimental efforts to compensate with growth hormone supplements have led to side effects that include carpal tunnel syndrome and breast enlargement. Aldosterone, produced by the adrenal glands, drops by about 30 percent by the age of 80. This hormone regulates levels of water and salt in the body, and its decrease may play a part in the tendency of older people to become dehydrated. Parathyroid hormone, made by the tiny parathyroid glands in

WHAT CAN GO WRONG

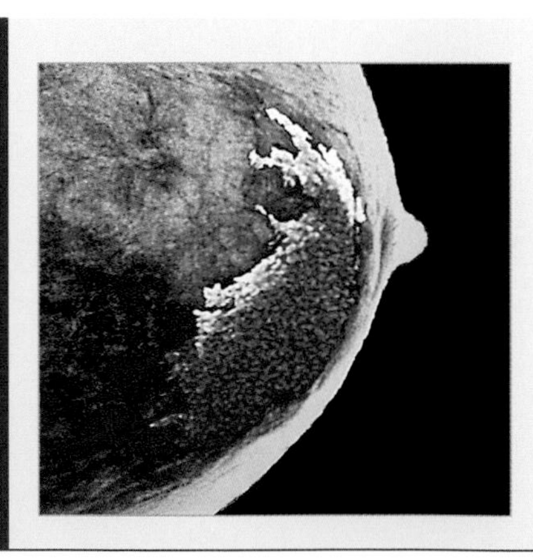

BREAST CANCER, left, is second only to skin cancer in its prevalence among women and second only to lung cancer as the cause of death in women today. It also occurs in a few men.

Most cases of breast cancer occur in women over 60, and the chances of a woman developing the disease at some point in her life are one in eight. Having a close relative with the disease increases the risk two to three times. Hormonal history also plays a significant part; in at least some women, long-term exposure to estrogen increases the risk of cancer. Women who had their first menstrual periods before age 12 or

SEE ALSO: Chapter Ten, "Hormones," PAGE 268

the neck, controls the blood's calcium balance. This hormone typically increases with age, leading to decreased bone mass and, often, osteoporosis. Adding calcium and vitamin D to the diet can help to counteract this bone loss.

Thyroid hormone levels don't change significantly in older people, but thyroid disorders are twice as common. Hypothyroidism, or abnormally low levels of thyroid hormone, can lead to fatigue, depression, weakness, anemia, and other problems. Hyperthyroidism, due to overactive thyroid glands, is not as common in older adults, but can show up as Graves's disease, marked by nervousness, a rapid heartbeat, weight loss, and, sometimes, protruding eyeballs. Confusingly, older people suffering from hyperthyroidism sometimes show the same symptoms as if they were suffering from hypothyroidism—fatigue or depression. This condition is called apathetic thyrotoxicosis. Blood tests can help to determine the condition.

One of the most common hormonal disorders connected to aging is type 2 diabetes mellitus, which affects between 15 and 20 percent of people over 65. This disease has serious complications, including heart disease, blindness, and kidney failure; elderly people with type 2 diabetes are more likely than others to have trouble with activities of daily living.

Reproductive hormones decline in both sexes with aging, but more dramatically in women. Levels of the hormone estrogen, produced mainly by the ovaries, begin to decline in a woman's 30s and drop to very low amounts by the late 40s or early 50s, when women experience menopause, the end of ovulation and menstrual periods.

For many women, the most noticeable aspects of this hormonal change are hot flashes, the still-mysterious dilation of blood vessels in the skin that leaves the woman flushed and sweating. Hot flashes generally fade after a few years, but the effects of decreased estrogen continue into old age and include bone loss, thinning skin, vaginal dryness, and increases in blood cholesterol.

Men do not undergo a dramatic drop in hormones equivalent to menopause, but testosterone does decrease with age, dropping about 1 percent a year from age 30. Sperm production and sex drive decline somewhat over time, though some men can and do father children well into their 70s.

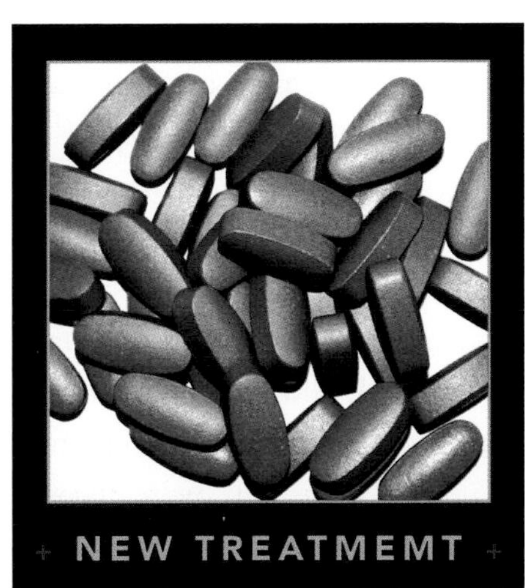

NEW TREATMEMT

PERSISTENT ERECTILE dysfunction, or impotence—the inability to achieve or sustain an erection—affects more than 50 percent of men over 40. Beginning in the late 1990s, its treatment was revolutionized by drugs that block the action of an enzyme called phosphodiesterase-5 (PDE-5) and allow another chemical, cyclic guanosine monophosphate (cGMP), to build up and relax the smooth muscle of the erectile tissue of the penis, allowing an erection.

WHAT CAN GO WRONG

who entered menopause after 55 are at higher risk. Obesity is also a risk factor, possibly because fatty tissue produces small amounts of estrogen.

Breast cancer typically appears as a lump or thickened spot in the breast. Many such lumps are benign; those that are malignant can be either in situ or invasive. Doctors classify each case to indicate its spread and treatment. Stages 0 and I have a 100 percent five-year survival rate, while stage IV, the most invasive, has a 20 percent 5-year survival rate. Overall rates of death by breast cancer are dropping, probably because of our ability to detect and treat the cancer earlier.

Breast cancers are usually found by self-exams or mammograms. In almost all cases, a surgeon will remove the tumor; sometimes radiation therapy or chemotherapy follows, depending on the cancer's stage. Estrogen-blocking drugs can help to prevent a recurrence.

DEFENSES

THE CONTINUING PROTECTION OF THE IMMUNE SYSTEM

As with so many other aspects of the aging human body, the immune system works fairly well in healthy older people but does shows weakness when compromised or attacked. For instance, elderly people have a harder time fighting off infections, and other infirmities (such as tuberculosis or chickenpox) that were acquired during youth may show up again in old age.

Immune cells that combat cancer and invading pathogens decrease, allowing the growth of diseases that might have been quelled in a younger body. Vaccines work less well in the elderly, and autoimmune activity increases, meaning that the body's immune cells are more likely to attack the body itself.

Although scientists have much to learn about how the body's defenses change with age, they suspect that some of these changes may contribute to the overall degeneration of the body's functioning as it grows older.

A family prepares for an afternoon dirt bike ride together. Protective gear is essential to shield an older body that bruises more easily.

DECLINE

Two kinds of immunity work in sequence to defend the body. Natural immunity cells and proteins mobilize the rapid, generalized response to a threat. These forces hold the line until the more adaptable acquired immunity cells can take over. Both kinds of immunity cells—natural and acquired—are compromised during the aging process.

Among the natural immune cells are macrophages: cells that fight tumors and kill and ingest invaders such as bacteria. Though the older body has a full complement of these cells, they don't work as well in clearing out invaders as they do in a younger body, nor are they as effective in fighting cancer cells. This relative inefficiency may account, at least in part, for the increase in instances of cancer among the elderly. The complement system, a set of proteins that combat microorganisms, also responds sluggishly to infections in older people.

The slowdown in an elder's immune system response is even more dramatic in the case of the acquired immunity cells. T cells, for instance, are powerful immune cells that develop in the thymus gland. Cytotoxic T cells attack invaders or damaged cells

BLACK & BLUE

BRUISES SHOW more on an older body. As the body ages, capillary walls weaken and split more easily, leaking the blood that becomes a bruise. Blood vessels also have less protection, because the skin is thinner, with less of a cushion of fat beneath it. Drugs such as aspirin, warfarin, and corticosteroids can also lead to unexpected bruising.

SEE ALSO: Chapter Twelve, "Immune System," PAGE 318

directly, while helper T cells send out chemical signals that call in other immune system defenses. Both play essential roles in combating disease.

The overall population of T cells does stay about the same as people age, but the proportion of cells that function well drops substantially as the body gets older. The chemical signals that prompt T cells to proliferate decrease with age, which mean fewer T cells are actually generated by the body. At the same time, the level of other body chemicals rises in a way that might interfere with a solid immune response.

B cells, responsible for producing antibodies, decrease in number with age as well, and the quality of the antibodies they make declines. These changes mean that they, too, simply don't respond as robustly to foreign antigens as they might have in a younger body.

While functioning T and B cells are on the wane, autoimmune cells begin to increase over time. The body's defenses become less able to distinguish self from other. Although most autoimmune diseases appear in middle age, medical researchers are coming to understand that the proliferation of autoimmune cells in older people may be a contributing factor in many degenerative diseases.

One piece of good news arises amid this overall decline in the immune system: The number of antibodies that trigger allergic reactions also decreases in older people. In general, the elderly experience fewer allergies and less severe allergic symptoms than younger people do.

KEEPING HEALTHY

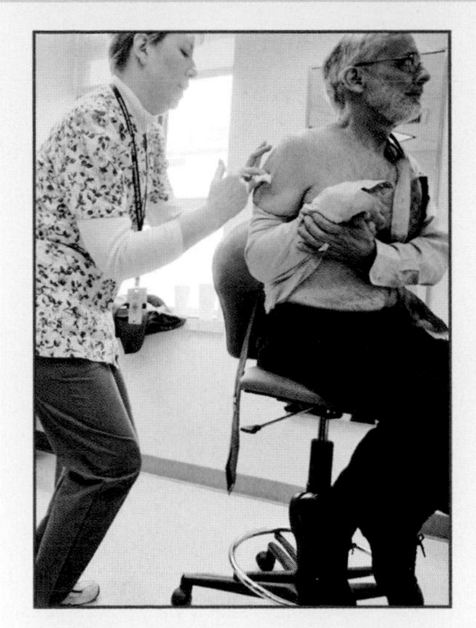

Older people need vaccinations too.

VACCINES ARE NOT just for kids: The elderly can and should be vaccinated as well because they become more prone to infection and serious complications as they age. Among the vaccinations recommended for people over 65 are: influenza (annually in the fall); pneumococcal disease (single dose); tetanus/diphtheria booster (every ten years); chicken pox (two doses for those who have not had the disease); herpes zoster (to prevent shingles); measles/mumps/rubella (for those who have never had these diseases or vaccines); hepatitis B (for those in institutions or those exposed to carriers); and hepatitis A (only for those with certain chronic conditions). Vaccines aren't foolproof: They don't work as well in the elderly as they do in the young. Even those who have been vaccinated need to be careful and take routine precautions against disease. ■

KEEPING WELL

The effects of this immune system decline are felt most strongly when an older person develops illnesses such as pneumonia, endocarditis (an infection that can affect the heart valves), influenza, or urinary tract infections. AIDS and HIV are particularly deadly in older people, attacking an already weakened immune system.

The elderly are also particularly vulnerable to diseases that are transmitted in hospitals. Sixty-five percent of these, often viral or bacterial infections transmitted by air, occur in people over 60. Illnesses can progress at frightening speeds in older people because other chronic conditions and stress may further weaken their diminished immune systems. These conditions leave them more vulnerable to infections, which in turn further taxes the immune cells, and so forth, in a deadly cycle.

Much research remains to be done on the role of the immune system in aging and how to sustain a healthy immune system for longer. Supplements, like vitamin E, may help to restore immunity. As always, a good diet, exercise, and regular checkups can help to prevent the illnesses that push that system past its limits.

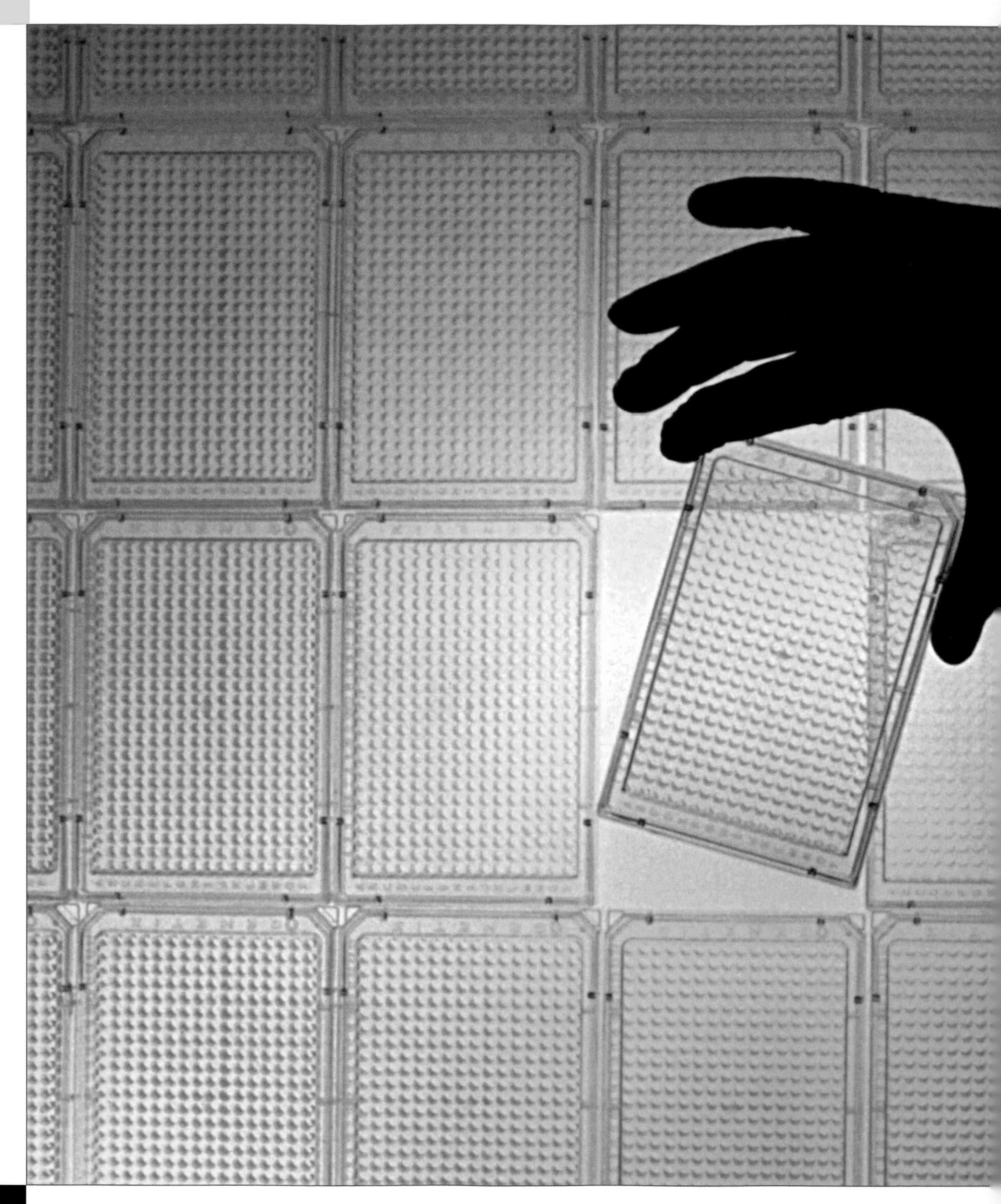

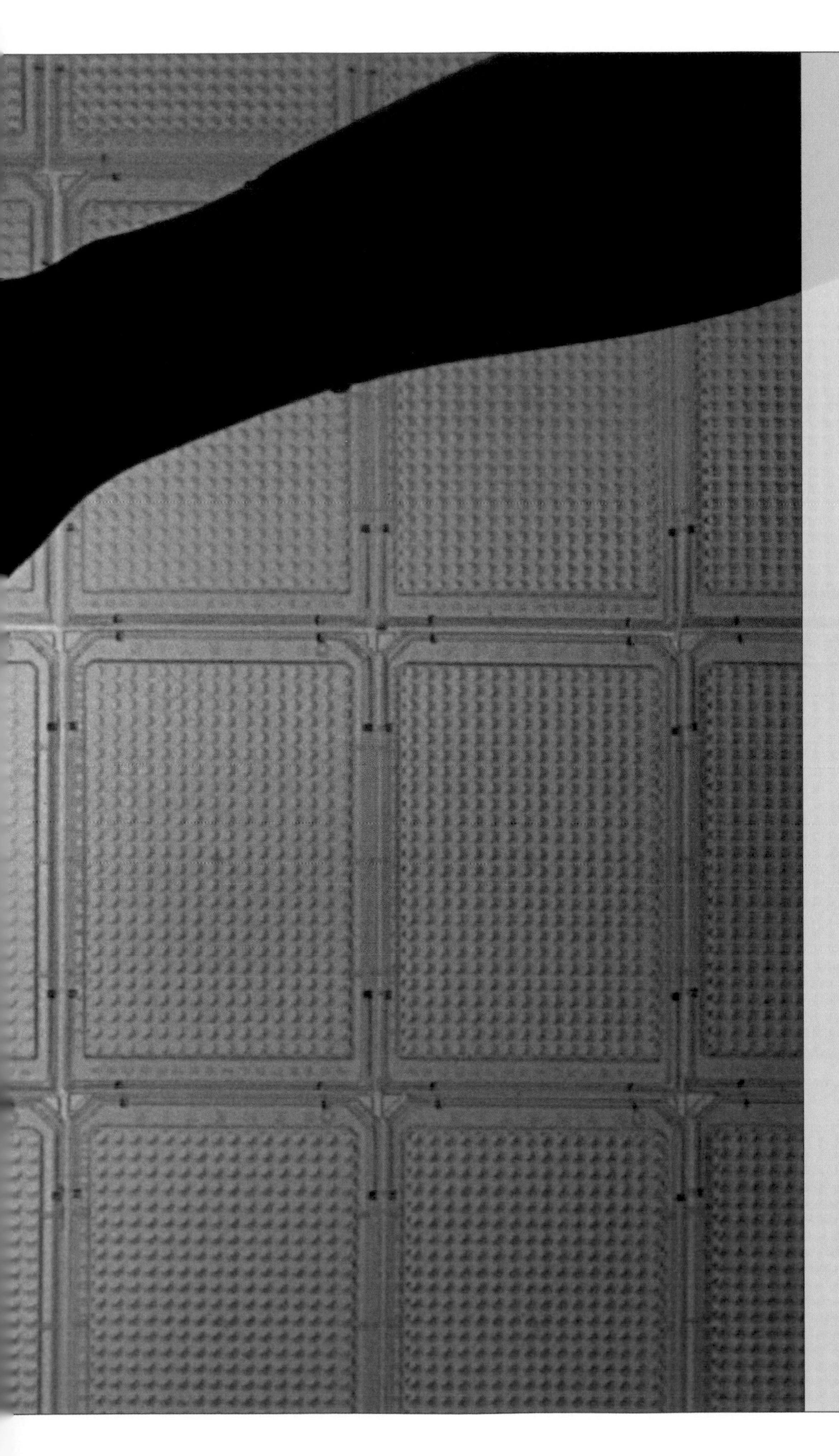

EPILOGUE

THE FUTURE BODY

FROM ACHILLES TO THE *Bionic Woman,* Gilgamesh to Captain Marvel, humans have always loved stories about people who were stronger and smarter, could jump higher, penetrate the impenetrable—and, perhaps, live forever. Superhumans may still be the stuff of myth and comic book, yet scientific and medical revolutions—stem cell research, tissue engineering, genetic engineering, computer-assisted bionics—are bringing the traits of the epic hero closer to reality.

Each tray contains the complete human genome as 23,040 different DNA fragments.

THE FRONTIER

WHERE WE'VE BEEN
WHERE WE'RE GOING

THE 20TH CENTURY overflowed with amazing medical advances that used to exist only in the imagination. Not so long ago, a "cyborg" was a half-human, half-machine creation from science fiction, but today thousands of people walk around with mechanical hips, knees, and hearts. In vitro fertilization (IVF) has become so routine that more than 3 million people worldwide have now been conceived this way. The nearsighted can correct the problem with LASIK surgery, and men with erectile dysfunction can rejuvenate their sex lives with a variety of drugs and treatments. But what does the future have in store for the body? "The next frontier," says Gregory Stock, director of the futuristic Program on Medicine, Technology, and Society at the UCLA medical school, "is our own selves."

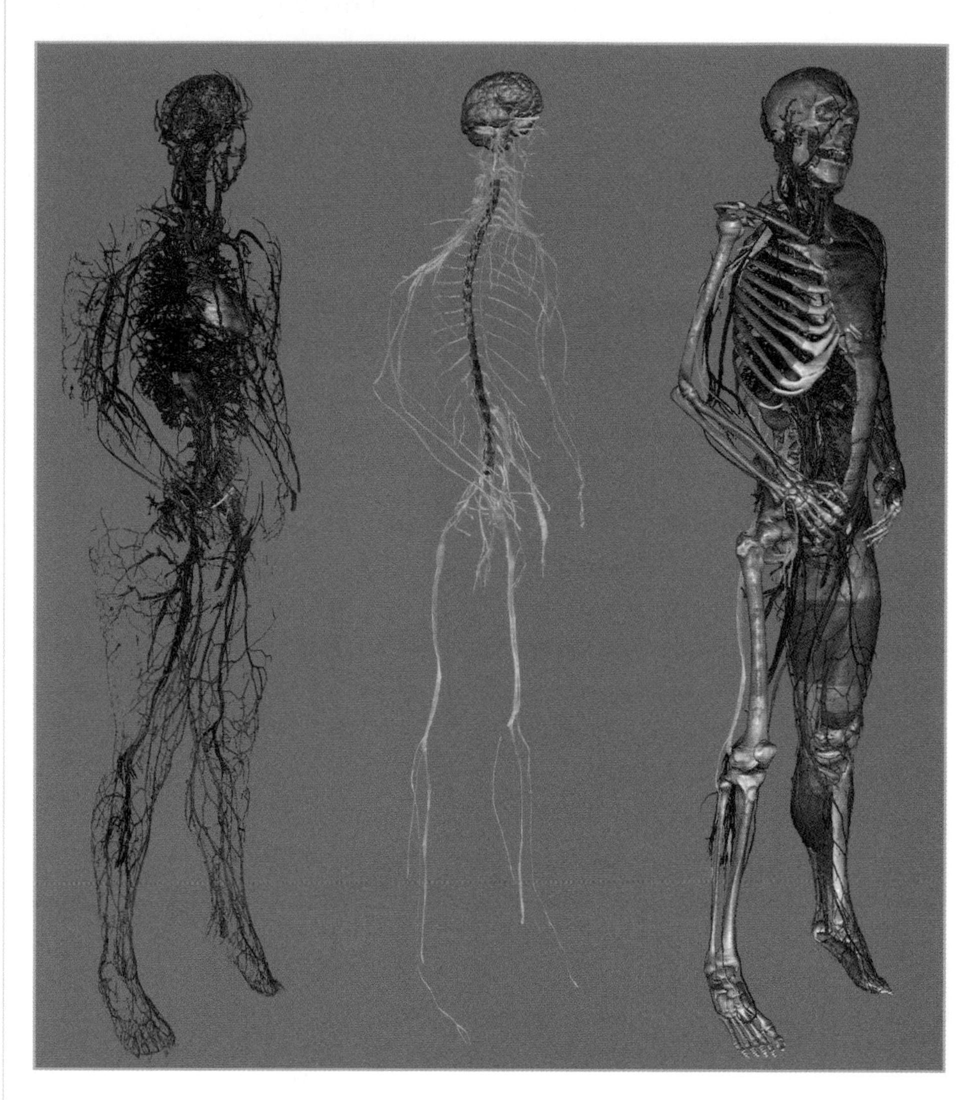

Studying individual systems—such as the circulatory, left, and nervous, center—or multiple ones, right, is possible thanks to computer scanning.

SPEED OF SCIENCE

What is perhaps most astonishing is how fast these scientific accomplishments are taking place. The best-known measure of the speed of technological change is Moore's law, which is based on a 1965 statement by Gordon Moore, a founder of the computer chip company Intel. Moore's Law states that the power of a silicon chip—and by extension, of all information technology—will double every 18 months. Since 1965, that power has doubled 29 times, as of 2005, which means that computing power has increased more than 100 million times in a mere 40 years—an advance unprecedented in human history.

Following the logic of Moore's law means that "you probably have more processing power in your microwave oven than was available to the entire world in 1950," according to *New Scientist* magazine. It means that the cost of sequencing a single letter of DNA (a job requiring massive computing muscle) has halved every 23 months since 1990. It also means that while it took 15 years to sequence the genome of HIV, decoding the genome of the SARS virus took just 31 days.

SEE ALSO: Chapter One, "Inheritance," PAGE 24

Yet the speed and potency of this technological revolution gives many people pause. A great theme of Greek tragedy is the striving for human perfection and the consequences of pursuing it at all costs. Concerns about genetic engineering (changing the DNA of an organism to produce desirable characteristics) echo this theme. The benefits of implants and transplants are incalculable, but taking the next step is more problematic. The possibility of altering the blueprint of life, rather than merely treating symptoms, promises to usher in a new era of science and medicine.

CONCERNS

To some, genetic engineering sounds eerily akin to the eugenics movement, first formulated in 1865 by Sir Francis Galton, who sought to improve hereditary traits through various kinds of intervention. The word *eugenics* derives from Greek roots meaning "well-born," but the eugenics movement led to discrimination against the "feebleminded" and "genetically unfit."

By the 1980s, dramatic new advances in genetics gave rise to new possibilities and perils. Recombinant DNA technology, gene splicing, and other techniques allowed researchers to create the first genetically engineered drugs: human insulin, approved by the Food and Drug Adminstration in 1982, and in 1986, a vaccine for hepatitis B. Researchers moved from elucidating the genes to manipulating them.

Such discoveries raised the possibility that it might someday be possible to alter our genes and fundamentally change the human body. James Watson, co-discoverer of the structure of DNA, framed the issue this way: "If we could make better human beings by knowing how to add genes, why shouldn't we?"

Biotechnology, tissue engineering, stem cell research—it all was moving forward so fast that the President's Council on Bioethics issued a report in 2003 seeking to address some of the moral issues engendered by these marvels. "A certain vague disquiet hovers over the entire enterprise," the report concluded, recommending that certain practices (such as human cloning) be banned. What remains murky are the ethical guidelines for technologies that lie somewhere between therapy and human enhancement. Most of the new treatments (such as gene therapy for cystic fibrosis or SSRI's for depression) were designed to cure disease or disability. But what if others come to be used to make healthy people smarter, stronger, or more attractive?

Is the world that would be created by improving, enhancing, or even reinventing the human body a good thing, a bad thing, or something in between? The ethical complexities of that question will continue to be debated. For now, let's assume that these conundrums have been resolved in the world of the future. Here, then, are four key areas where progress may dramatically transform the human body—not just a hundred years from now, but perhaps next week.

Computers have played a large role in advancing knowledge gleened from these colorful bands of human DNA.

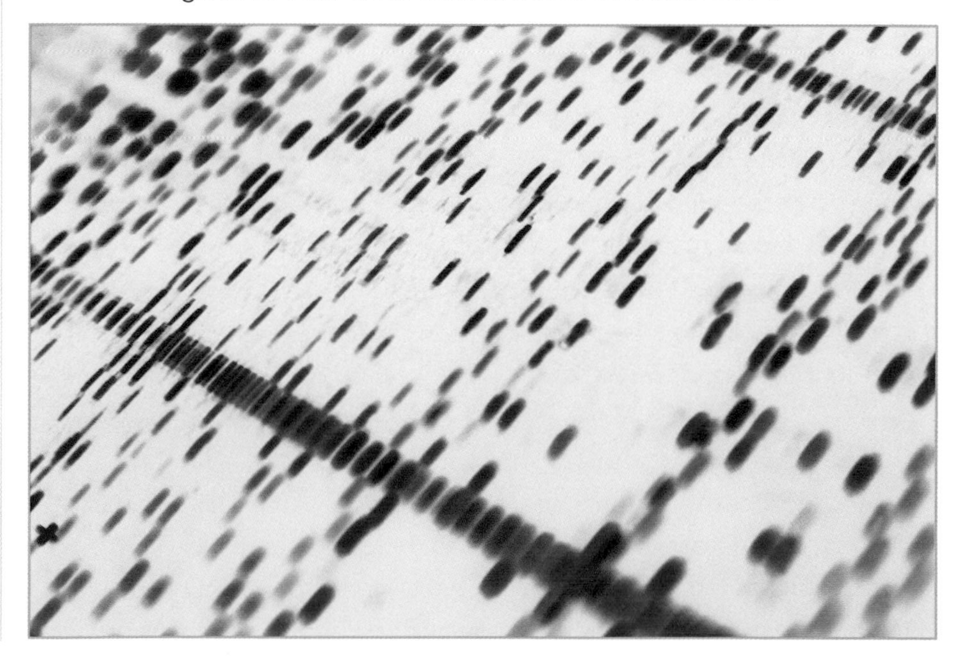

DEEPER PROFILES

UNLOCKING MORE GENETIC SECRETS

FOR GENERATIONS, human reproduction has been a relatively simple process: Find a mate, mate, and then wait. Although people have evolved elaborate behavioral and unconscious methods of assessing the reproductive suitability of a potential partner, the reproductive outcome largely remains a genetic crapshoot. Detecting the detailed genetic make up of a partner isn't in the human's bag of tricks. How does anyone know for sure that a partner isn't carrying a gene for a heritable disorder such as schizophrenia or Down syndrome? In most cases, it's impossible to tell from mere appearances. Consequently, the risks remain high—it's now believed that there are about 6,000 disorders linked to genetic abnormalities of one kind or another. About one in 16 children is born with some kind of gene-based ailment.

GENETIC TESTING

What if, in the future, however, it were possible to significantly reduce the risk of passing on DNA errors? Instead of "submitting to the tyranny of nature's lottery," as Ronald Bailey puts it in his book *Liberation Biology*, what if it were possible to select good genes, to screen out bad ones, or to insert new, improved genes into the cells before a baby is even born?

Although there have been serious problems early on with some pioneering genetic treatments, the promise of new medical interventions called assisted reproductive technologies is truly astonishing. The new techniques seek to intervene at the earliest possible moment in order to positively affect the ultimate outcome of human reproduction. In the case of a procedure called preimplantation genetic diagnosis (PGD), developed in 1989, scientists intervene when the future human being consists of nothing more than four to eight cells—a speck of life that could swim in the period at the end of this sentence.

The first great step toward PGD was in vitro fertilization (IVF), perfected in 1978, which brought together sperm and egg not inside a woman's body, but in a lab dish, under controlled conditions. PGD takes IVF one step further—it's a way of screening out "bad genes" before the embryo is implanted in the mother's womb. The life-affirming possibilities of

With the completion of the Human Genome Project in 2003, genetic research is unlocking new secrets everyday.

SEE ALSO: Chapter One, "Development," PAGE 34, Chapter Eleven, "Conception," PAGE 306

this procedure were described in one specific case in the *Journal of the American Medical Assocation* in February 2002. Dr. Yuri Verlinsky and colleagues at the Reproductive Genetics Institute, in Chicago, treated a 30-year-old woman with a family history of early-onset Alzheimer's disease. Caused by an extremely rare genetic disorder in the amyloid precursor protein (APP) gene, the disease could have resulted in a healthy baby who would have disappeared into dementia as early as age 40. The woman was tested and found to be carrying the mutated gene.

To change her baby's fate, she was given drugs to stimulate hyperovulation, or increased egg production. Then 15 eggs were removed from her body and fertilized with the father's sperm in a dish. The resulting fertilized eggs were grown to the four-cell or eight-cell stage (called blastomeres) and then tested for the presence of the mutated APP gene. Because there is so little DNA in this wee sample, the genetic material was amplified for examination using a technique called polymerase chain reaction. In this way, the doctors were able to find one embryo that was completely free of the malignant gene. This lottery-winning embryo was then implanted in the woman's body and grown to term—a healthy baby free of an unwanted inheritance.

According to Dr. Verlinsky, this was the first known PGD procedure for inherited early-onset Alzheimer's disease that resulted in a healthy, disease-free baby. In the future, other babies will be born free of all kinds of other diseases and disorders. In fact, such procedures are so powerful, Dr. Verlinsky predicts that in the future, "there will be no IVF without PGD." So far, these procedures remain expensive for most people—$10,000 to $20,000 or more for both procedures together. Even so, PGD has made possible the birth of more than 10,000 healthy children worldwide so far.

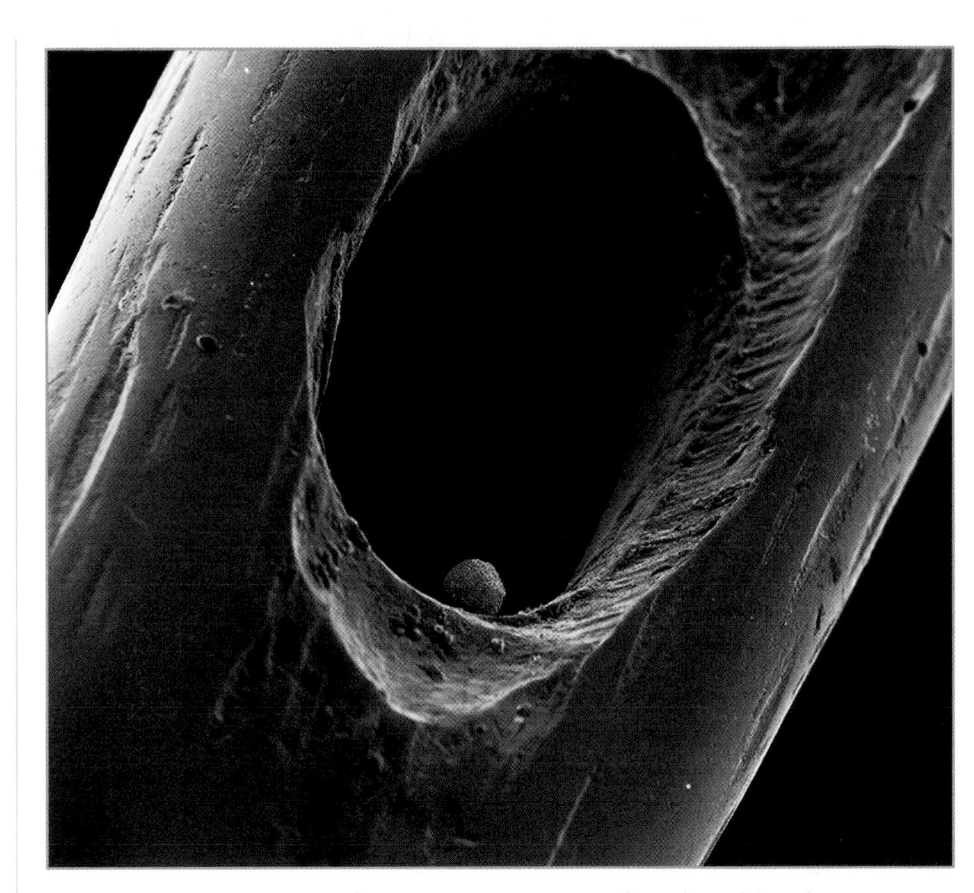

Under high magnification, a microscopic five-day-old embryo can be seen sitting in the eye of a needle.

BIOETHICS COUNCIL

THE PRESIDENT'S COUNCIL on Bioethics was created in 2001 to advise the President on "bioethical issues that may emerge as a consequence of advances in biomedical science and technology." Among its areas of concern are stem cells, cloning, and end-of-life care. The council's 18 members are drawn from diverse fields: science and medicine, law and government, and philosophy and theology.

ETHICS

The President's Council on Bioethics observed that "as genomic knowledge increases and more genes are identified that correlate with diseases, the applications for PGD will likely increase greatly," not only for hemophilia or fragile X syndrome but for subtler traits such as colorblindness or height. "While currently a small practice, PGD is a momentous development. It represents the first fusion of genomics and assisted reproduction—effectively opening the door to the genetic shaping of offspring." This whole new field, still largely unexplored and incompletely regulated, has come to be known as reprogenetics.

Another prenatal screening test was first announced in December 2005 by the Baylor College of Medicine in Houston. It offered testing for more than 150 different genetic abnormalities using a technique called array-CGH, which hunts for chromosomes that have been deleted or duplicated. Genetic material is gathered in much the same way as with an amniocentesis and chorionic villus sampling.

In all, genetic tests for more than 800 conditions are now available, with more hitting the market every year, such as FISH testing (flourescence in situ hydridization), which uses a specific protein, called a probe, that has been designed to "stick" to specific DNA in a cell. The probes glow with a special dye, so that when the sample is observed under a microscope, researchers can clearly see extra chromosomes (characteristic of disorders such as Down syndrome) as well as more subtle genetic disturbances, or duplications or deletions of DNA.

Unfortunately, only about 2 percent of all diseases are caused by mutations of a single gene, such as the defect that causes early-onset Alzheimer's. The rest are the result of multiple genes acting together and in conjunction with various environmental influences. Knowledge about both single-gene

DNA is carefully extracted from a mammalian egg cell through a tiny pipette.

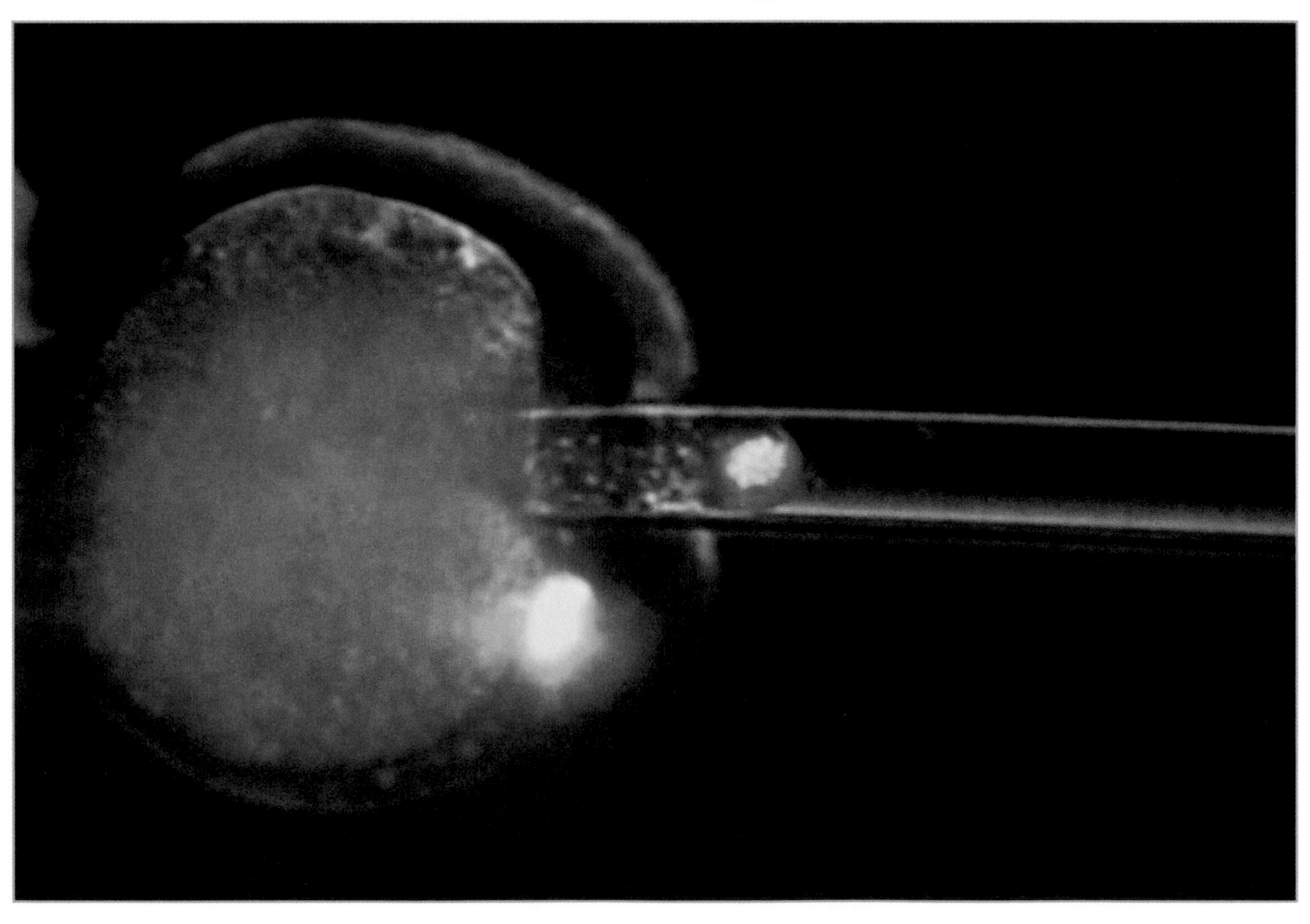

SEE ALSO: Chapter One, "Inheritance," PAGE 24

diseases and multigene diseases, however, is increasing at a rate in keeping with Moore's law—that is to say, at a breathtaking pace. With the completion of the Human Genome Project in 2003, which identified the more than three billion base pairs of human DNA—the rungs on the ladder of life—almost unimaginable possibilities seem to yawn before us.

GENETIC REPAIR

Genetic engineering may not be focused only on the earliest moments of human life. It may be possible to repair defective genes in infants or even adults. Preliminary experimental work in animals, for instance, has shown that it is possible to repair genes that cause conditions such as sickle-cell anemia or muscular dystrophy.

Children with Duchenne muscular dystrophy produce a defective version of the muscle protein dystrophin. This mutation causes their muscles to gradually weaken, and they often die of respiratory failure in early adolescence.

In one pioneering experiment, a short DNA segment of the normal dystrophin gene, with RNA segments attached at both ends, was injected into a golden retriever dying of the disease. The RNA sequences sought out and then bound to the defective DNA on the dog's X chromosome. Then the cell's own DNA repair mechanism spliced in the normal segment to replace the mutation. The dog's cells began cranking out normal dystrophin. (Though this remarkable experiment brought new treatments closer to reality, it did not "cure" the disease in humans. One major remaining obstacle in human use is the diversity of human dystrophin mutations—each patient has a different mutation of the dystrophin gene, so each patient would require a customized gene to correct the mutation.)

Altering the genes of adult organisms is actually more difficult than altering genes in embryos or single fertilized eggs (called "germline interventions"). Gene repairs made at this early stage would require fixing only a few cells, which would then multiply as the organism matured. These permanent genetic changes would not only cure the patient, they would stop him or her from passing the mutation on to descendants (barring some future mutation of the gene). Such a benefit has never before been possible.

PROFILES

What has also never before been possible is truly individualized medical care on a genetic level. In the future, a patient visiting the doctor might carry a record of his or her own complete genome, perhaps imbedded on a microchip (or something like it), making it possible for a doctor to prescribe treatment not just designed for a class of people like the patient in general, but tailored for that individual.

So far, "much of the promise and pitfalls of personalized medicine remain untested," according to one industry report, but two new, rapidly developing fields are bringing this groundbreaking medical practice ever closer. Pharmacogenetics (focusing on single genes) and pharmacogenomics (focusing on combinations of multiple genes) attempt to identify precisely how an individual will respond to specific drugs and medications, given his or her genetic profile. If successful, gene-based, individualized medicine will help physicians to "first, do no harm" by tailoring their treatments in a way that has never been so personal before.

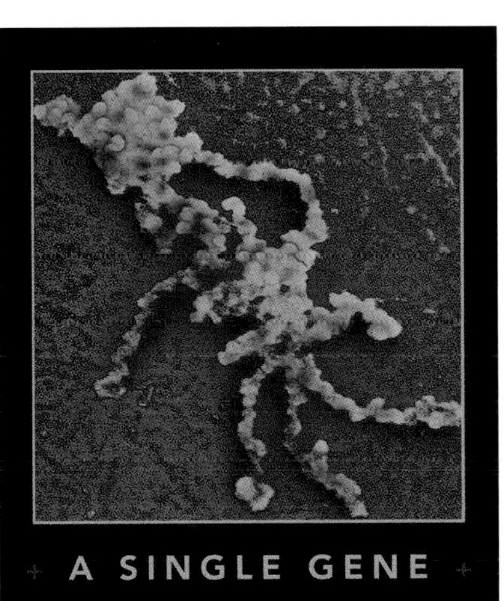

A SINGLE GENE

DEFECTS ON single genes, above, cause roughly 6,000 diseases, such as

+ Cystic fibrosis, one of the most common disorders among Caucasians. About one in 25 people of northern European descent carries the mutation.
+ Tay-Sachs disease, a fatal disorder in children that causes progressive degeneration of the nervous system.

SHARPER MINDS

BUILDING A BETTER BRAIN

HONING THE MIND IN school and play is the first way we build up our brains, but scientists are also investigating other methods of enhancing human mental performance. Whether through substances people ingest or laboratory techniques designed to optimize the pathways of the mind, improved brain function is a part of the future of the body.

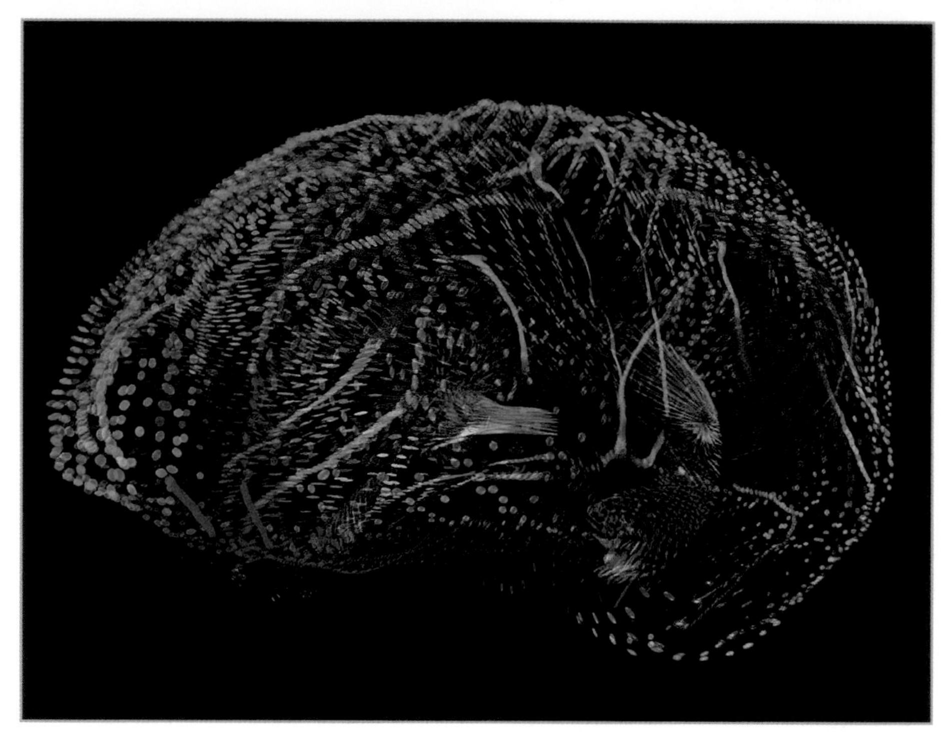

Three-dimensional mapping of the human brain can reveal striking anatomical differences in the gyral patterns of the cerebral cortex.

SMARTER DRUGS

One need not peer into the 23rd century to find performance-enhancing drugs. Perhaps the oldest and most reliable one, caffeine, is in that morning cup of coffee or tea. According to several studies, about 10 percent of college students in the United States also use Ritalin (the commercial name for methylphenidate, an amphetamine-like prescription stimulant) or similar pharmaceuticals to maximize their concentration when studying.

Ritalin is fairly crude in its effect, is quite expensive, and has the potential for abuse. Prescribed primarily to treat attention deficit disorder in children, Ritalin sedates hyperactive kids and increases alertness, improves mood, and boosts energy in other patients. It can also produce physical and mental dependence, and sudden discontinuation can cause alarming withdrawal symptoms such as depression and anxiety. Researchers are working toward smarter, safer drugs that can effectively boost mental function but without the potential to cause addiction or other harmful side effects.

At least 40 different cognition-enhancing drugs, sometimes called neuropharmaceuticals, are currently under development in laboratories around the world. In 2006 one industry newsletter proclaimed that "the brain is poised to become the number-one target for drug development during the next decade." It forecast that the total worldwide market for such brain-boosting drugs would eventually exceed $100 billion.

One class of new drugs, the ampakines, are modified benamide compounds that enhance attention span and altertness by facilitating nerve-signal transmission, but without unpleasant side effects. Other new drugs promise to sharpen memory, heighten wakefulness, and even improve decision-making.

In Britain, an expert panel recently identified 15 molecular pathways that are currently under investigation as targets for cognitive enhancement. Some medical investigations are aimed at finding drugs to help treat cognitive dysfunctions caused by damage from Alzheimer's disease and stroke.

SEE ALSO: Chapter Nine, "The Brain," PAGE 234

Because a main concern with regard to such drugs is that, if they work, healthy people will begin using them to boost their brainpower, columnist William Safire has called these intellect-enhancers "Botox for the brain."

One of these drugs, modafinil, has been called "the first real smart drug." Modafinil was not developed to make people smarter—it is sold by prescription (as Provigil) as a "wakefulness promoter" for people suffering from narcolepsy, a neurological disorder marked by periods of uncontrollable daytime sleepiness. Studies have shown that the drug also has a remarkable effect on mental clarity. It can heighten alertness, brighten mood, and improve both verbal and visual memory.

In one 2003 study at Cambridge University in England, modafinil significantly improved subjects' ability to plan complex problems, recall strings of numbers, and remember abstract patterns. At the same time, the drug does not appear to boost heart rate or blood pressure. Initial testing also indicates that it may be nonaddictive. The testing is ongoing, however, and until further research is available, modafinil is classified as a non-narcotic controlled substance, illegal without a prescription.

MORE CONNECTIONS

Building and maintaining a better brain may also be a low-tech activity. Studies have shown that challenging the intellect—learning a new language, balancing a checkbook without using a calculator—actually builds new neural pathways. A new fad among brain builders is the development of "brain fitness centers" in retirement communities around the country. By paying a monthly fee, just like at a gym, members join these "neurobics clubs" to challenge their minds and thus do battle against intellectual decline with all manner of mental aerobics, from crosswords and acrostics to math games and chess.

Seniors can even hire a personal mental trainer to design a customized workout for the wits. For the do-it-yourselfers, the World Wide Web offers many sites devoted to brain teasers and mental fitness. There are also mind-exercising video games being targeted to an older audience. They feature a wide range of puzzles and games to stimulate the mind.

The scientific community has been exploring the definitive benefits of these activities, both in maintaining and regenerating parts of the brain. The results of studies involving animals are encouraging, but the results from earlier surveys on humans tend to be skewed, as they were inclined to rely on people whose brains were already in fairly good shape. Dozens of new studies are currently underway to assess the effects of these brain health

After stimulation with nerve growth factor, nerve cells sprout neurites (pink and green). Cultures like these are used to investigate neural regeneration.

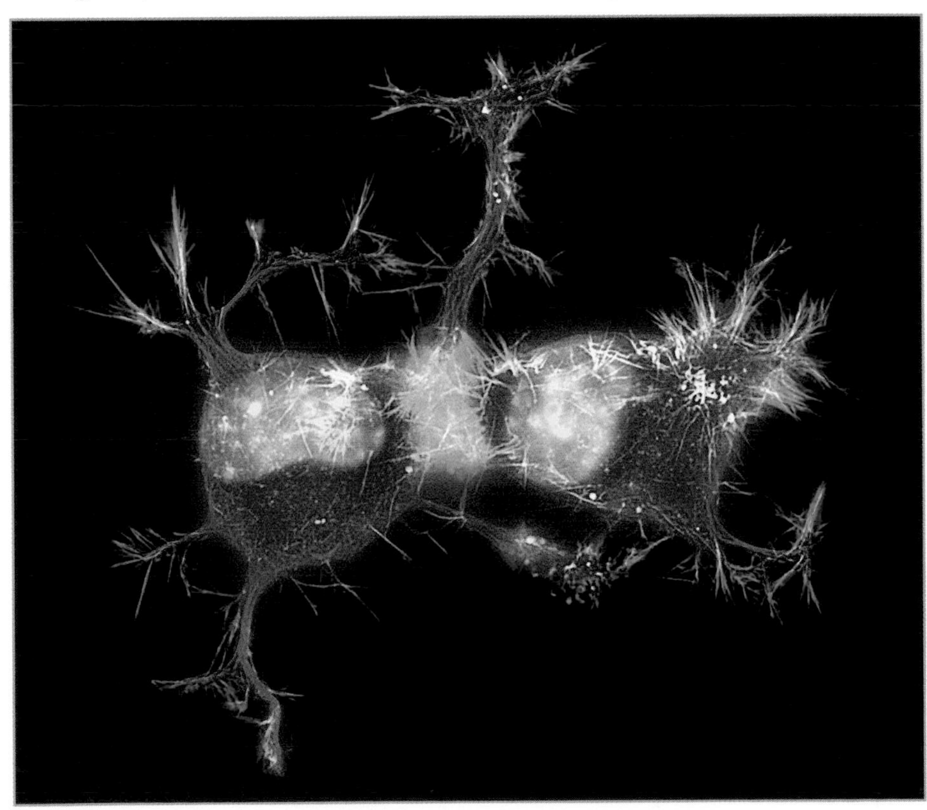

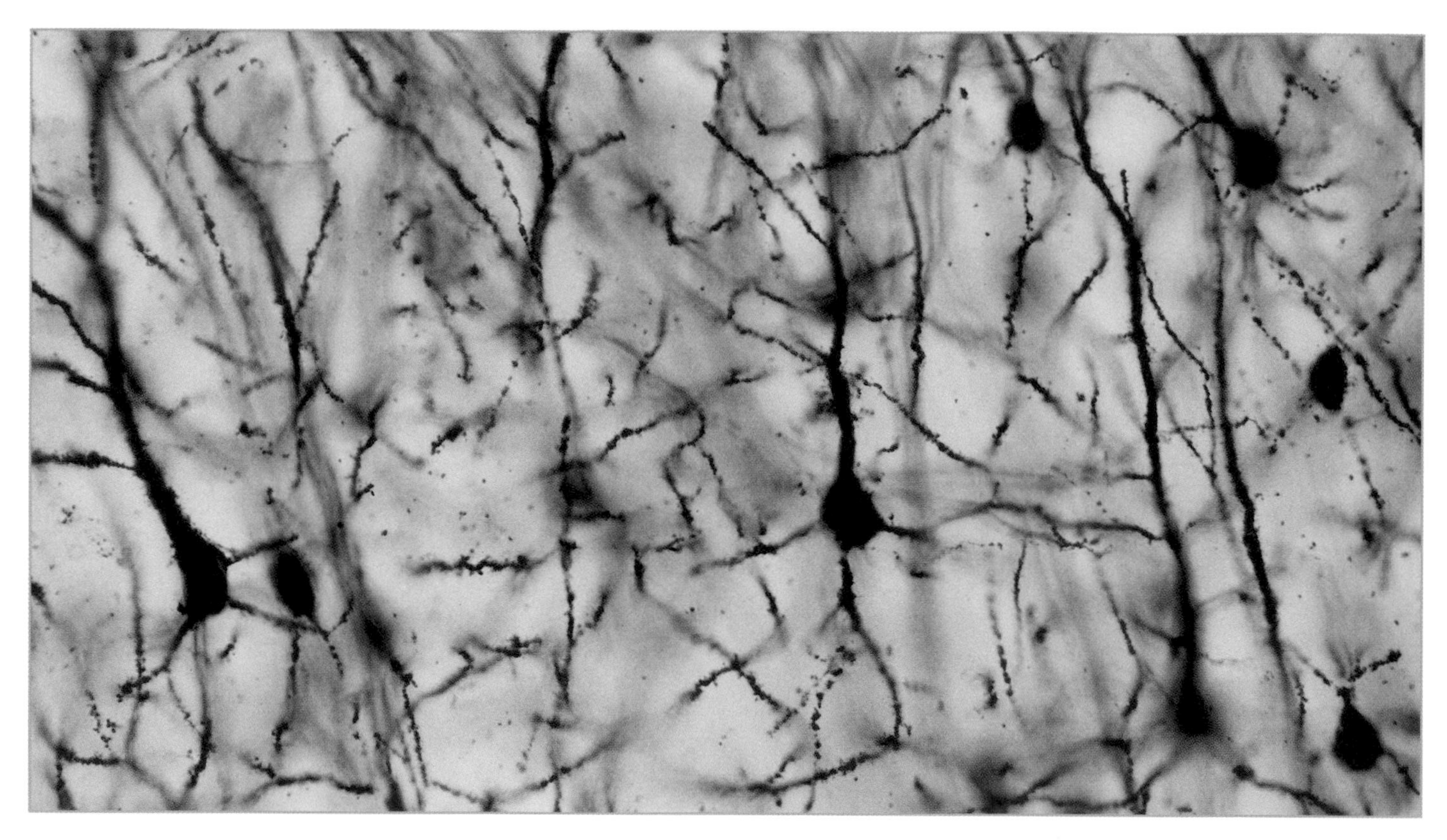

Neurons connect in the cerebral cortex—the brain region that controls conscious thought, memory, and language.

programs and their benefits for the human mind in all its stages.

BRAIN FOOD

In addition to drugs and activities, a healthy diet might lead to a better brain. Certain foods are showing promise in building and maintaining a strong mind. A May 2007 study has shown that natural compounds found in blueberries, tea, grapes, and cocoa can enhance memory. The beneficial effects were increased when regular exercise was added to the mix.

This compound, epicathechin, belongs to a group of chemicals called flavonols. Scientists at the Salk Institute in San Diego, California, working with mice, found that a diet rich in foods containing epicathechin, along with a program of regular exercise, created positive structural changes in the part of the mouse brain involved in learning and memory.

Published in the *Journal of Neuroscience*, their findings suggest that a diet chockful of flavonols may also help reduce the incidence or severity of neurodegenerative diseases related to aging. Earlier research on humans has shown that the epicathecin compound also improves cardiovascular function in people and increases blood flow to the brain, so perhaps epicathechin's brain-stimulating benefits will apply to humans, too.

Research continues into these low-tech solutions to see just how much they can bolster the brain in humans. While waiting for conclusive scientific results, people can safely engage in these activities without potential side effects or risks (as might be incurred with some drugs or herbal supplements). Other studies have shown that eating foods rich in flavonols

+ FLAVONOIDS +

FLAVONOLS are just one of six classes of flavonoid, a group of natural antioxidents that give many foods their natural color. Brightly colored fruits and vegetables, crunchy almonds, and cocoa are rich with flavonoids. Their consumption has been linked to many health benefits, such as lowering the risk of heart disease and certain forms of cancers.

and omega-3 fatty acids and getting regular cardiovascular exercise are beneficial to the maintenance of the bones, muscles, heart, and lungs. Keeping other body parts healthy might be the key to keeping the mind in great shape, too.

SMARTER MICE

Other researchers are taking different measures to alter the brain itself in order to enhance memory. At Princeton University, researchers created "smart mice," which they nicknamed "Doogie" after the title character in a television show, *Doogie Howser, M.D.* (about a child genius who becomes a doctor). The Doogie mice were genetically modified to produce more NMDA (N-methyl-D-asparate) brain receptors, structures that are key to the formation and maintenance of memories. With their enhanced ability to remember, the improved mice became faster learners, quicker and more adept at solving problems than their untreated peers.

In an object-recognition test, two groups of mice were given the opportunity to familiarize themselves with two objects. When the researchers then replaced one of the two objects with a third, unfamiliar one, the Doogie mice recognized the substitution more quickly and spent more time familiarizing themselves with the new object. The unimproved mice spent equal amounts of time exploring both the new object and the old one; they were unable to differentiate between the new object and the one it had replaced. By identifying what appears to be a critical target in the intricate biochemistry of memory, these findings led to a host of other studies of ways to boost the brain's abilities.

Inspired by the work with the Doogie mice, a team of Texas scientists found that mice became smarter after the gene Cdk5, associated with Alzheimer's disease, was turned off. Led by Dr. James Bibb, a research team at the University of Texas Southwestern Medical Center published details of their findings in the journal *Nature Neuroscience* in 2007. Bibb and his colleagues used genetic engineering to breed mice that could be manipulated to switch off the gene. Cdk5 controls a brain enzyme that has been linked to diseases such as Alzheimer's that are signaled by the death of neurons in the brain. The altered mice were better at learning tasks—navigating a water maze, for instance—than their unaltered peers. Their associative memories were also stronger. The scientists attributed the mice's increased brain power to an increased sensitivity to their surroundings, which the familiar objects more recognizable

The team is now exploring how these mechanisms can have practical applications for human beings without requring genetic alterations or interventions. Bibb and his team are hopeful that these discoveries will lead to the development of drugs to improve the performance of people with cognitive deficits.

Researchers engineered the brain of a "Doogie" mouse to have more brain receptors and better memory.

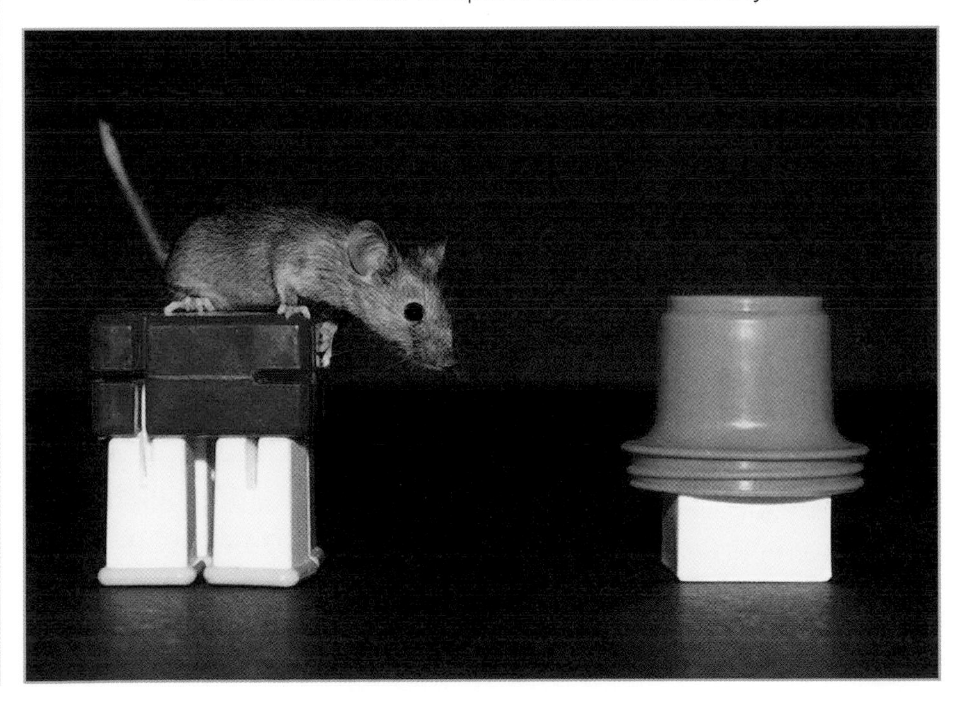

BIONIC BODIES

ROBOTIC REPLACEMENTS

IN THE 1970s TV series *The Six Million Dollar Man*, an injured test pilot is "rebuilt" using bionic devices that transform him into a sort of souped-up human—better, faster, and stronger than his off-the-shelf peers. His eye is replaced with a zoom lens with night vision capability. His damaged legs are exchanged for bionic limbs that enable him to run faster than a speeding car and to make fantastic leaps. His battered left arm is replaced by a mechanized one with the strength of a bulldozer. Although the show captured the public's imagination, everyone at the time understood it was only science fiction. Now, thirty years later, the line is not so clear when it comes to bionics.

Having lost a hand at age 18, Jay Schiller demonstrates the bionic capabilities of his prosthetic hand by playing the piano.

GOING BIONIC

In the spring 2006, leading researchers in the rapidly expanding bionics field explained at a conference (tellingly titled, "The Six Billion Dollar Man") just how much of what was once fiction is now science. Daniel Palanker, of Stanford University, described his "optoelectronic retinal prosthesis," a device that enables people blinded by macular degeneration to see.

Biomedical engineer William Craelius's Dextra bionic arm features a hand and fingers that can be independently controlled by commands sent from the brain to nerves that communicate with a computer imbedded in the artifical limb. The result: Amputees can type or play the piano. Mechanical engineer H. Kazerooni described his work for the Department of Defense developing an "exoskeleton orthotic system." A kind of robotic suit, it enables the wearer to tote 170 pounds, but feel as though it were only five.

SEE ALSO: Chapter Four, "Skeletal Muscle," PAGE 110

The biggest difficulty in bionics is communicating between the brain and the mechanical limb, Dr. Craelius says, but his bionic hand (the first to allow a person to use existing nerve pathways to control mechanical fingers) shows that the challenge is not insurmountable. The computer-loaded Dextra hand moves in response to electrical signals generated by the user's remaining muscles and tendons.

MIND & MACHINES

Other researchers are trying to cross the brain-machine divide in different ways. A team at UCLA has imbedded a wireless implant the size of a grain of rice beneath a subject's skin. This remote control can communicate directly between nerves and a bionic device.

Italian researchers are developing the bionic Cyberhand, a fully sensitized, five-fingered hand that is hard-wired directly into the nervous system, allowing instructions from the brain to control the hand (at least partially) and sensory feedback from the hand to reach the brain. Meanwhile, Moore's law has led to the rapid miniaturization of components, so that "the number of transistors we can fit onto an integrated circuit doubles about every 18 months," Craelius says. "At this pace, within the decade, the processing for complex bionic activity will be implantable in the brain or elsewhere in the body."

At the U.S. military's Defense Advanced Research Projects Agency (DARPA), research is investigating ways to seamlessly merge mind and machine. In one experiment, an owl monkey named Belle has had tiny probes implanted directly into the motor cortex of her brain. She is able to move a mechanical arm just by thinking about moving it. (As of this writing, Belle can only move—not manipulate—objects with her robotic arm).

Robotic suits have been developed to aid recovering stroke victims rehabilitate paralyzed arms.

This study may lead to the development of another way to cross the brain-machine interface—perhaps with an implanted computer chip—so that someday quadraplegics might be able to manipulate a thought-operated exoskeleton, allowing them to move and manipulate objects.

DARPA researchers are also trying to design what the military calls the Metabolically Dominant Soldier. The goal is to create a soldier who is essentially unstoppable: impervious to pain, fatigue, wounds, and bleeding. So DARPA researchers are working on drugs that will block intense pain in ten seconds, stop bleeding within minutes, and enable soldiers to function without sleep for days on end. Other researchers are trying to enhance the body in more fundamental ways—for instance, by enabling soldiers to complete not just 50 pull-ups but 300. How? By increasing the number of mitochondria (the tiny power plants) in muscle cells. In this way, an

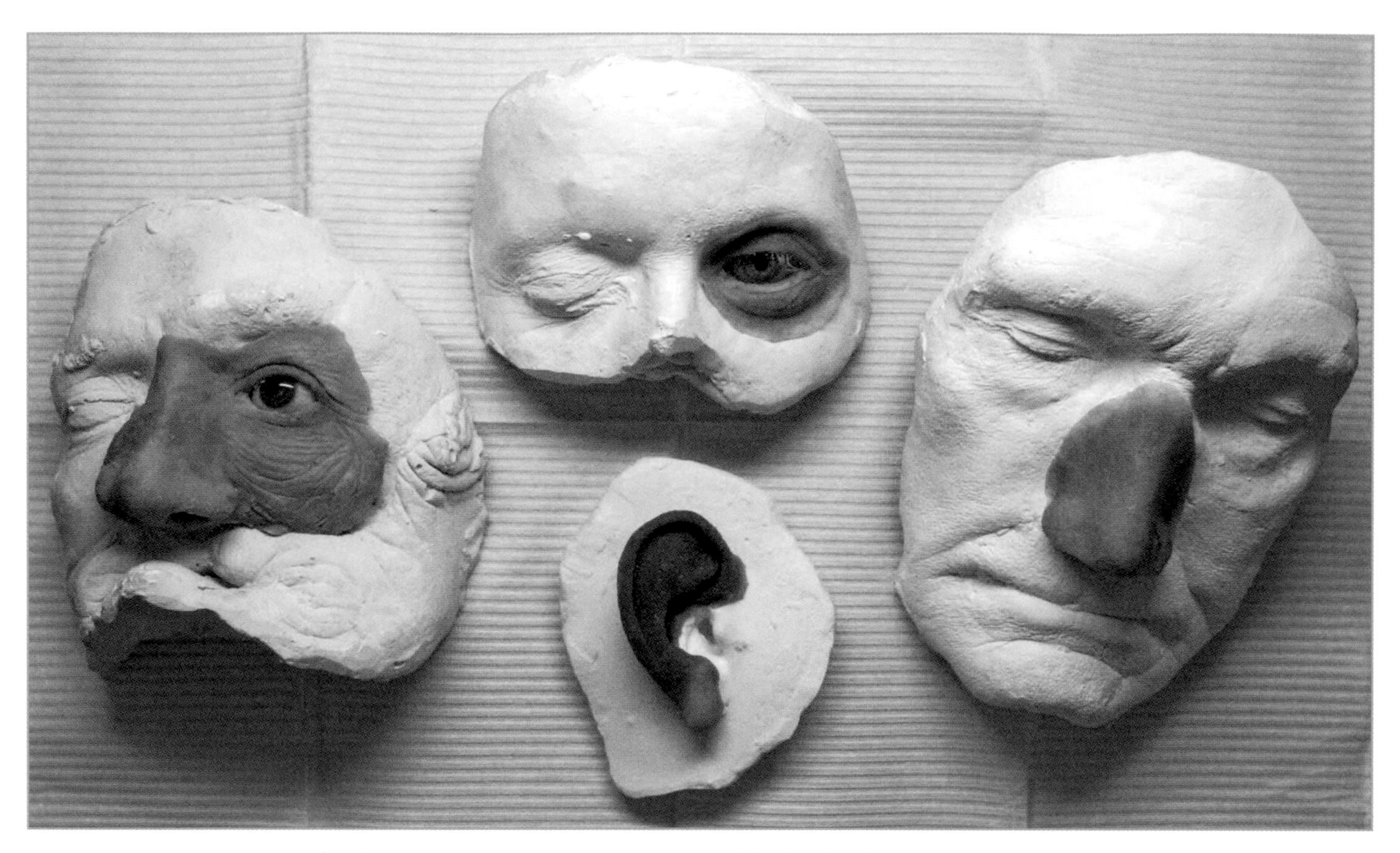

Facial prothestics have become even more lifelike thanks to plastics and silicone.

ordinary soldier could become as metabolically fit as an Olympian.

REGENERATION

All these bionic breakthroughs may be overshadowed by the possibilities of regenerating human tissue through embryonic stem cells. Embryonic stem cells were first isolated in 1998, and their discovery has resulted in an explosion of research all over the world. At the very earliest stages of development, an embryo is a mass of "blank," undifferentiated stem cells, which have the capacity to grow into any of the more than 200 kinds of cells found in the human body. Stem cells, in other words, are pure potential.

The hunt is under way to deduce methods by which these cells could be used to grow made to order tissue to repair organs that have been damaged—resulting in a rejuvenated heart, liver, or pancreas, for instance—or even whole organs to replace those that have been destroyed. If a new organ were grown from a person's own stem cells, it would be genetically identical to that person, so there would be no problem with tissue rejection (a serious problem in organ transplants).

Because these organs would be brand-new—free from the damages of time and abuse—they hold out the promise of dramatically extending the human life span by replacing damaged organs. Just imagine: A 60-year-old man with a brand-new heart grown from his very own cells. The promise of this research could mean that the fountain of youth has been found living inside our own bodies all along.

Much of this research is no longer theoretical—it's already in progress. Stem cells have been grown into beating heart muscle at the Technion-Israel Institute of Technology. In animals, heart tissue grown from stem cells can be integrated into a damaged heart, where it helps regulate activity, in effect becoming a biological pacemaker. There are more than 70 diseases and injuries that could be helped by stem cell research,

SEE ALSO: Chapter One, "Inheritance," PAGE 24, "Development," PAGE 34 including Parkinson's disease, Alzheimer's disease, spinal cord injuries, stroke, heart disease, arthritis, and severe burns.

CONTROVERSY

Despite the potential of embryonic stem cells, research remains controversial. Harvesting these cells requires the destruction of a fertilized egg, which some people find morally objectionable. But researchers may find a way around this ethical dilemma—for instance, a different variety of stem cells can be found in amniotic fluid and blood from fetal umbilical cords. These stem cells are less plastic (that is, less able to differentiate into different kinds of cells) than the stem cells found in embryos. Harvesting them from discarded afterbirth material may one day provide a conflict-free solution.

It is already possible to "bank" umbilical cord blood shortly after birth, in case the child gets sick later and the stem cells are needed for a bone marrow transplant, which can can treat certain blood diseases like sickle cell anemia and leukemia. Imagine, then, a not-so-distant world where even more remarkable things are possible.

In the future, will every maternity ward come equipped with a walk-in freezer, so that newborns' parents can make a deposit of the baby's own cells into a cryogenically frozen "stem cell bank," to regrow tissue or whole organs if they're needed later in life?

HOPE FROM SKIN

The debate over the use of embryonic stem cells took a fascinating turn in June 2007 when three teams of scientists announced that they had found a way to create what are essentially embryonic stem cells from the skin cells of mice. The advance, which does not require the creation or destruction of embryos, could completely change the way stem cell research is performed and eventually resolve the ethical concerns that currently surround it. All adult cells contain instructions for converting back into embryonic stem cells. The breakthrough discovered how to activate those instructions by turning "on" or "off" certain genes.

In 2006, Shinya Yamanaka of Kyoto University identified four genes that can turn other genes on and off to return a skin cell to an embryonic stem cell. Yamanaka and two U.S. teams (one at the Whitehead Institute for Biomedical Research in Cambridge, Massachusetts, and the other at Massachusetts General Hospital and the Harvard Stem Cell Institute) published their findings in the journals *Nature* and *Cell Stem Cell* that they were able to use genetically engineered viruses to activate the four key genes. About one in 10,000 of the cells infected with the virus became an IPS cell, or "induced pluripotent stem" cell.

Scientists caution that many hurdles remain before the technology can be applied to humans. For one thing, the viruses used in the mice can cause cancer, so other viruses and small molecules, are being tried to manipulate the four key genes. It may also be that in humans, the reprogramming process uses a different set of genes. It took 20 years after mouse embryonic stem cells were discovered before their human equivalent was found, so while this breakthrough is potentially revolutionary, scientists caution that research using human embryonic stem cells should continue. One day, however, it just might be possible to make stem cells, capable of generating all the parts that make up a human, directly from a person's own skin.

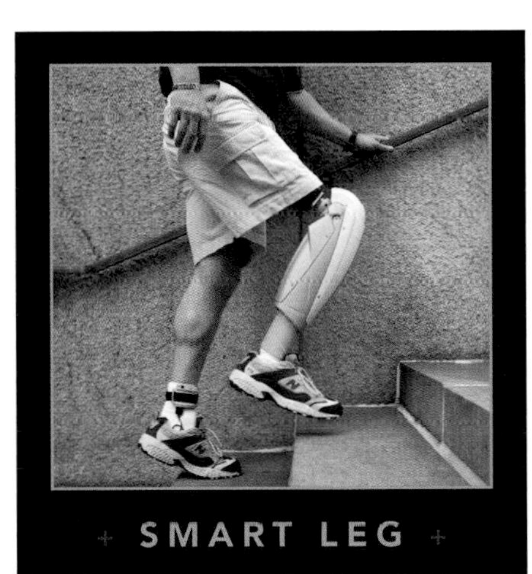

SMART LEG

ANOTHER STEP FORWARD: Prosthetic legs can make use of artificial intelligence. Former Army Ranger Bill Dunham lost his right leg in 1989. In 2006 he was outfitted with a "Power Knee," above, a smart prosthesis that works in unison with his sound leg. A sensor in his left shoe transmits information to the artificial leg, enabling it to mimic movements.

LONGER LIVES

ADDING MORE YEARS

THE SLOW DECLINE of cells and tissues with age seems as inexorable as gravity. Imagine: What if there were some future breakthrough—perhaps the discovery of a "Methuselah gene"—that could dramatically increase human life span? A position statement issued in 2002 by 51 prominent aging researchers proclaimed that "scientific knowledge holds the promise that means may eventually be discovered to slow the rate of aging. If successful, these interventions are likely to postpone age-related diseases and disorders and extend the period of healthy life."

Unfortunately, there do seem to be limits to the capacity to control the aging process—specifically Hayflick's limit, the biological law that states that human cells can't divide indefinitely. They divide only about 50 times and then stop, which means that, on the cellular level, the body is deeply programmed to wear out. (Programmed cell death is at least partly protective because out-of-control cell growth is one of the key characteristics of cancer.) Even so, human life span has dramatically increased in recent times, so who's to say it won't continue to lengthen? The past several generations have been living through the first "longevity revolution" in human history.

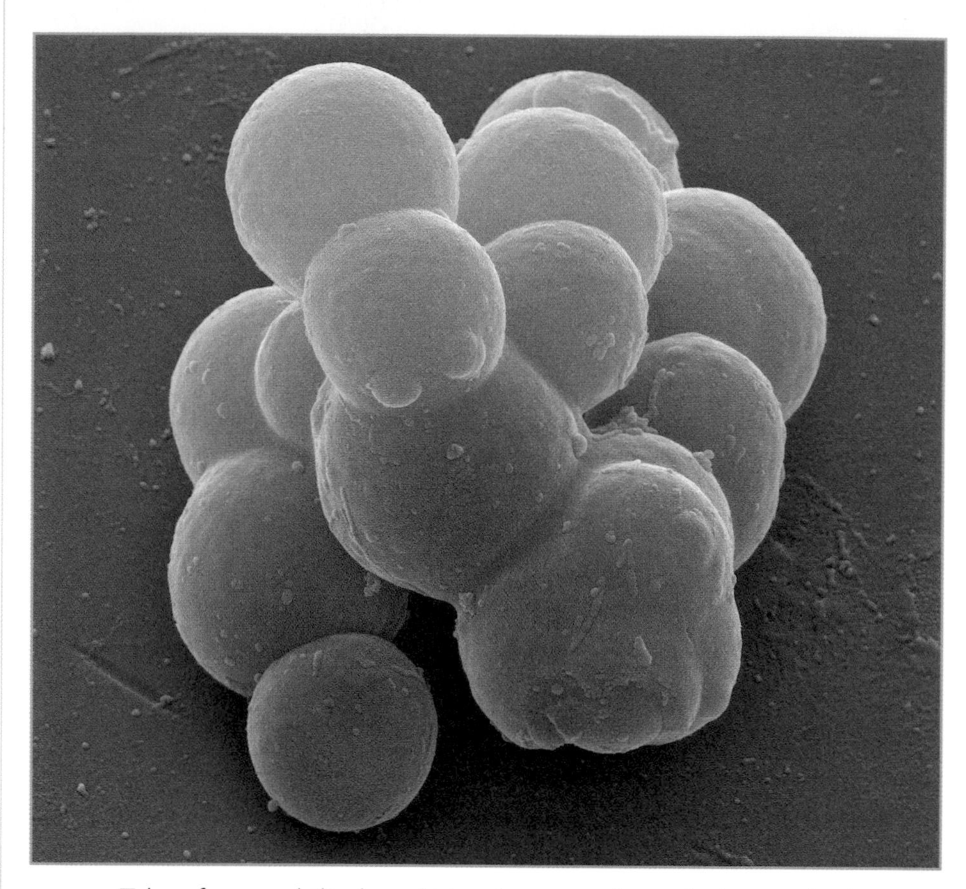

Taken from umbilical cord blood, stem cells are being studied to see whether they hold the key to the secrets of life extension.

In the Stone Age, life expectancy wasn't much longer than 20 years. Even millennia later, during the era of classical Greece and Rome, life expectancy had still increased to only about 28 years. By the beginning of the 20th century, life expectancy for an American male had grown to 49 years. Yet by the end of the century, it soared to 77, an increase of almost 60 percent in a mere hundred years.

Researchers have reported that human life expectancy has been increasing at about two and a half years per decade for the past 160 years. Better nutrition, clean water, vaccines and medications, and a decline in smoking (at least in the U.S.) all contributed to a longer life span. The invention of sewers and modern sanitation greatly reduced infant mortality and the prevalance of communicable diseases. The human lifespan record, set by cigarette-smoking Frenchwoman Jeanne Calment, who died in 1997, stands at 122 years and 164 days. Is the upper limit of human life span being approached? Or will the future hold even more dramatic breakthroughs in aging

SEE ALSO: Chapter One, "The Cell," PAGE 14, Chapter Thirteen, "Maturity," PAGE 348

Depicted here in Canterbury Cathedral, Methuselah is the oldest person in the Bible, making his name synonymous with long life spans.

research, enabling us to reach what one gerontologist calls "actuarial escape velocity"?

In recent years there has been considerable excitement about the possibility that the breakthrough secret to longevity may lie in the genes themselves. In experiments done on relatively humble life-forms, such as nematode worms, fruit flies, and mice, alterations of a single gene have been shown to dramatically extend life. The most remarkable changes occurred in the humblest species, the nematode, in which alteration of a single gene was shown to double life span and an alteration in two genes nearly tripled it. Single-gene alterations in mice have been shown to increase life span by 25 to 50 percent. "It now seems possible that the rate of aging may be governed by highly conserved general mechanisms across many species," one team of researchers concluded, "and that single-gene alterations that extend life may ultimately be discovered in humans."

CALORIE CONTROL

Yet despite the amazing promise of genetic manipulation in extending human life, so far the only thing that has been consistently shown to extend life span (at least in animals) is something far more mundane: eating less. This phenomenon was discovered in lab rats back in the 1930s, when animals on severely restricted diets were shown to live 30 to 50 percent longer than those allowed to eat whatever they wanted. In experiments reported in 2002, when gene alterations were combined with caloric restriction in mice, the animals' life span increased by nearly 75 percent, the greatest life-span extension so far reported in mammals.

"With nearly seven decades of laboratory research, this is by far the most-studied and best-described avenue of age-retardation, though scientists still lack a clear understanding of how it works," one panel of experts reported. Extending life span by reducing food intake has been demonstrated not just in rats, but also in monkeys, dogs, and other mammals, which maintain a lively, "youthful" level of activity at an age well beyond that of their normally fed peers. A restricted diet has also

METHUSELAH

IN THE BIBLE, Methuselah has the distinction of having the longest life span: 969 years. He is introduced as the son of Enoch and the father of Lamech (the father of Noah), who was born when Methuselah was still comparatively young: 187. Because of Methuselah's long life, his name in modern times has become associated with life-extension research.

been shown to slow the age-related decline of neurological functions, muscle function, immune function, and "nearly every other measurable marker of aging."

But will this practice work in human beings? Studies continue to examine the effects in humans, but have yet to reach any hard conclusions. Researchers at the National Institute on Aging are now trying to develop drugs that mimic the effects of caloric restriction. The best-studied candidate for a so-called "caloric restriction mimetic" is something called 2-deoxy-D-glucose, or 2DG. It works by interfering with the way cells process the sugar glucose. "2DG tricks the cell into a metabolic state similar to that seen during caloric restriction, even though the body is taking in normal amounts of food," according to a report in *Scientific American*.

Why would going on a diet, even an extreme one, cause the body to extend its own life? It could have to do with the "disposable soma" theory of aging, proposed by aging expert Thomas Kirkwood of the University of Newcastle in England. Organisms balance the need to procreate against the need to maintain the body, or soma, Kirkwood argues. When resources are plentiful, the body can do both. When food is limited, the body invokes processes that inhibit growth and reproduction in order to preserve the soma, which then reduces body size and fertility but also leads to longer life.

DNA is bound to the columns on this purification chip before being transfered to a biochip for use in genetic screening and disease diagnosis.

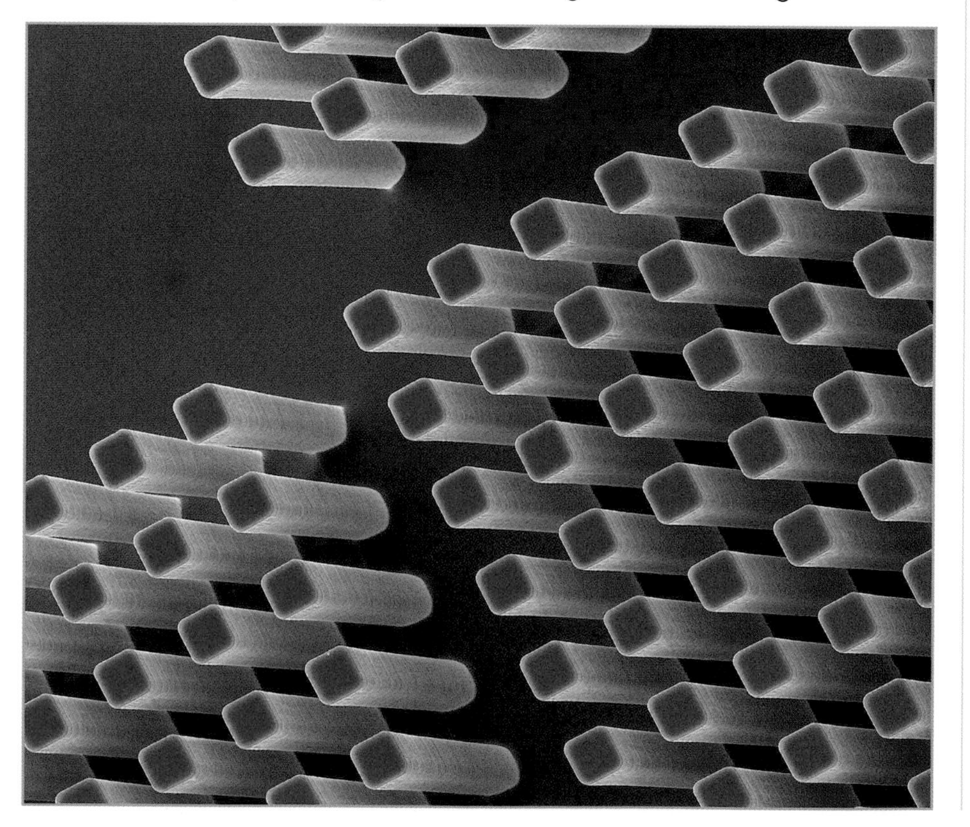

CELLS

There is now fairly wide scientific agreement that one of the prime causes of aging, if not the cause, is microscopic damage done to the cells by free radicals. Oxygen free radicals—oxygen molecules that have one unpaired electron, and that are therefore chemically very active—cause gradual deterioration of many of the body's cells and tissues, including breakdowns in protein synthesis and even minor errors in DNA replication. A biologist at the University of California at Berkeley, Bruce Ames, has estimated that free radicals damage the DNA inside one of our cells some 10,000 times per day.

Antioxidants (substances that protect cells against oxidative damage by free radicals) shield the body against this unceasing microscopic onslaught and thus appear to slow down the cellular aging process. The body can create its own antioxidants and take them in through diet. While it may make sense, however, to fight aging by adding antioxidants (such as vitamin E or C) to the diet, "remember that your cells already know about the dangers of free radicals, and have done so through much of their evolutionary history," observes Dr. Kirkwood. "Compared to the potent antioxidants your cells manufacture already, the antioxidants you take on a spoon

SEE ALSO: Chapter 1, "The Cell," PAGE 14

or in a tablet may make little difference, even if they reach your cells in a biologically active form."

Might there not be a more direct way to rejuvenate cells battered by free radicals? In recent years scientists have begun to focus their search on the tiny cellular power plants called mitochondria. These membrane-wrapped organelles provide energy to the cells by breaking down sugars, but in the process they also produce plenty of cellular waste, including free radicals. One odd thing about mitochondria is that they have their own tiny set of DNA, much smaller than the full-blown DNA in the cell's nucleus (mitochondrial DNA only produce 13 proteins, for instance). But the cell is much better able to repair damage done to nuclear DNA than to mitochondrial DNA. Mitochondrial DNA damage accumulates over time, steadily degrading the mitochondria's ability to produce energy, which appears to be part of the reason cells age and die.

Extreme cold preserves stem cell lines at the U.K. Stem Cell Bank, the first of its kind in the world.

The whole idea that mitochondria might play a central role in the mystery of human aging got a boost in 2004 in an experiment designed by scientists at the Karolinska Institute in Stockholm. The researchers created mice with a defect in their ability to replicate mitochondrial DNA. As a consequence, the animals produced three to five times more errors in their mitochondrial DNA than normal mice. The results were that the aging process was greatly accelerated in the test subjects. Though normal lab mice live two to three years, the defective mice started showing signs of premature aging after only 25 weeks, including weight loss, hair loss, curvature of the spine, and osteoporosis. Most only lived less than a year.

Could rejuvenating damaged mitochondria in the cells be the key to a longer life? In 2003, two researchers at the University of Virginia devised a vector loaded with healthy, youthful genes, which disperses throughout the body and targets only aging, damaged mitochondria. One writer described their work as perhaps "one of the great biomedical breakthroughs of the twenty-first century: a technique to replace old, worn-out mitochrondria with fresh, new, revved-up ones." As such research continues, it may one day become possible that "mitochondrial flushes" and treatments as yet unknown will offer humans that long, sweet drink from the wellspring of life so coveted by humans since the days of Gilgamesh and Achilles.

GLOSSARY

AT THE END OF EACH DEFINITION IS A REFERENCE TO THE CHAPTER IN WHICH THE TERM APPEARS.

ABSORPTION. The passage of nutrients from the GI tract into the blood and lymph systems. (Chapter Seven)

ADENOSINE TRIPHOSPHATE (ATP). Molecule that provides energy to living cells, essential to transforming chemical energy from food into mechanical energy and heat. (Chapter Four)

ADRENAL CORTEX. Outer portion of an adrenal gland; produces cortisol. (Chapter Ten)

ADRENAL GLANDS. Pair of glands that controls the stress response mechanism through the secretion of cortisol, epinephrine (also known as adrenaline), and norepinephrine (also known as noradrenaline). (Chapter Ten)

ADRENAL MEDULLA. The inner portion of an adrenal gland. Synthesizes and secretes the hormones epinephrine and norepinephrine. (Chapter Ten)

ADRENOCORTICOTROPIC HORMONE (ACTH). Released by the pituitary gland in response to CRF. Targets adrenal glands to boost synthesis of corticosteroids. (Chapter Ten)

AGONIST. A muscle that drives a particular movement. (Chapter Four)

ALBINISM. Abnormal partial or total absence of pigment in skin, hair, and eyes. (Chapter Two)

ALLELE. A variant form of the same gene on a matching chromosome; different alleles lead to variations in inherited characteristics, such as eye color. (Chapter One)

ALLERGY. A hypersensitivity or overly aggressive immune response to a perceived invader, or antigen. (Chapter Twelve)

ALOPECIA. Hair loss. (Chapter Two)

ALVEOLUS (pl. alveoli). One of the tiny, thin-walled air cavities of the lungs; alveoli cluster in alveolar sacs around a common air passageway. (Chapter Six)

AMYGDALA. A structure in the brain's limbic system that plays an important role in emotional learning. (Chapter Nine)

ANASTOMOSIS. An end-to-end joining of blood vessels (or lymphatic vessels or nerves). (Chapter Five)

ANGIOPLASTY. A technique for opening space within a clogged artery usually by temporarily inflating a balloon inside the artery. (Chapter Five)

ANTAGONIST. A muscle that reverses or opposes the action of another muscle. (Chapter Four)

ANTIBODY. A protein molecule produced by a plasma cell (progeny of B lymphocytes) in response to a specific antigen; also called an immunoglobulin. (Chapter Twelve)

ANTIGEN. A substance (chemical constituent of a microbe, other chemical material) recognized as a threat and thus provoking an immune response. (Chapter Twelve)

ANTIGEN-PRESENTING CELL. A fixed or migratory cell that presents antigens to T cells in an immune response.

ANTIOXIDANT. A nutrient or chemical such as beta-carotene or vitamin C that inhibits oxidation and may protect cells from damage by free radicals. (Chapter Thirteen)

APNEA. Brief cessation of breathing. Can be caused by airway obstructions or problems in the part of the brain that controls breathing. (Chapter Six)

APOCRINE GLANDS. Large sweat glands found in armpits and genital areas. (Chapter Two)

APOPTOSIS. Planned cell death. (Chapter One)

APPENDICULAR SKELETON. Group of bones in the skeleton that includes the limbs and their girdles. (Chapter Three)

ARACHNOID MATER. The weblike middle layer of the membranes that protect the brain and spinal cord. (Chapter Nine)

ARRECTOR PILI. Smooth muscles attached to hairs that contract when a person is cold or frightened, causing "goose bumps" and hair to stand on end. (Chapter Two)

ARTERIOLE. A tiny, muscular artery that carries blood to capillaries. (Chapter Five)

ARTERIOSCLEROSIS. A group of diseases characterized by stiff, thickened arterial walls. (Chapter Five, Chapter Thirteen)

ARTERY. Blood vessel that carries blood away from the heart. (Chapter Five)

ARTICULAR CARTILAGE. Occurs in joints where bone meets another bone. Cushions and reduces friction. (Chapter Three)

ARTICULATION. Joint or junction of two or more bones. (Chapter Three)

ATHEROSCLEROSIS. A disease characterized by the buildup of plaque on arterial walls. (Chapter Five, Chapter Thirteen)

ATRIOVENTRICULAR (AV) NODE. A specialized mass of electrically conducting cells located between the heart's atria and ventricles. (Chapter Five)

ATRIOVENTRICULAR VALVES. Valves between the atria and the ventricles. (Chapter Five)

ATRIUM (pl. atria). Either one of the heart's two upper chambers. (Chapter Five)

ATROPHY. Decrease in muscle bulk because of lack of exercise or loss of its motor nerve. (Chapter Four)

AUTOIMMUNITY. An immune response that attacks healthy body tissues. (Chapter Twelve)

AUTONOMIC NERVOUS SYSTEM. The division of the peripheral nervous system that controls cardiac muscles, smooth muscles, and glands. (Chapter Nine)

AUTORHYTHMIC FIBERS. Muscle cells in the heart that produce electrical signals without outside stimulus. (Chapter Four, Chapter Five)

AXIAL SKELETON. Group of bones in the skeleton that includes the skull, spine (or vertebral column), and rib cage (or bony thorax). (Chapter Three)

AXON. The hairlike extension of a neuron (or nerve cell) that sends out impulses. (Chapter One, Chapter Nine)

B CELL. A lymphocyte that can develop into an antibody-producing plasma cell or a memory cell. (Chapter Twelve)

BASOPHIL. The smallest and least common type of white blood cell. Releases histamines and other inflammation mediators. (Chapter Twelve)

BLASTOCYST. A hollow ball of cells that will develop into an embryo. It consists of a trophoblast (outer cells), blastocele (inner cavity), and inner cell mass. (Chapter One, Chapter Twelve)

BOLUS. Food mass that travels from the mouth to the stomach. (Chapter Seven)

BRAIN STEM. The portion of the brain just above the spinal cord, consisting of the midbrain, pons, and medulla. (Chapter Nine)

BRONCHIOLES. The small, final branches of the bronchi within the lungs that lead to the lobules and alveolar sacs. (Chapter Six)

BRONCHUS (pl. bronchi). Branching air passageway; primary bronchi lead from the trachea into each lung and then split repeatedly into smaller bronchi. (Chapter Six)

BRUSH BORDER. In the small intestine, the fuzzy line created by microvilli between the lumen (interior) and its mucosa. (Chapter Seven)

BURSA. Sac filled with synovial fluid found around joints between tendon and bone. Decreases friction. (Chapter Three)

CAPILLARY. A microscopic blood vessel that connects arterioles to venules; the site for interchange between blood and tissues. (Chapter Five)

CARDIAC CYCLE. The events of one complete heartbeat. (Chapter Five)

CARDIAC MUSCLE. Cells that form the wall of the heart and that contract and release rhythmically, controlled by the autonomic nervous system and often indwelling pacemaker. (Chapter Four, Chapter Five)

CARDIAC OUTPUT. Blood pumped by a ventricle in one minute. (Chapter Five)

CARRIER-PROTEINS. Substances that ferry insoluble hormones to targets. (Chapter Ten)

CARTILAGE. Smooth, rubbery connective tissue. Found in the skeleton, as well as other body structures. (Chapter Three, Chapter Four)

CELL-MEDIATED IMMUNITY. Wing of the adaptive immune system in which activated T cells and other effector cells destroy microbes and other antigens. (Chapter Twelve)

CENTRAL NERVOUS SYSTEM. The brain and spinal cord. (Chapter Nine)

CENTRIOLE. Two tubular structures, arranged at right angles, within a cell's centrosome. (Chapter One)

CENTROSOME. A region of the cell containing centrioles that forms the spindle during cell division. (Chapter One)

CEREBELLUM. Part of the brain behind the medulla and pons, governs coordinated muscle activity. (Chapter Nine)

CEREBRAL CORTEX. In the brain, outer layer of the cerebral hemispheres, responsible for conscious experience, thought, and planning. (Chapter Nine)

CEREBRUM. The two cerebral hemispheres that make up most of the brain. (Chapter Nine)

CERVIX. The lower neck of the uterus, which extends into the vagina. (Chapter Eleven)

CHROMATID. One of a pair of replicated chromosomes. (Chapter One)

CHROMATIN. The threadlike genetic material, made of DNA and histone proteins, that is present in the nuclei of nondividing cells. (Chapter One)

CHROMOSOME. The structure found in the nucleus of a cell, made of DNA and contains genes In humans, chromosomes come in 23 pairs, making 46 chromosomes altogether in all cells but gametes. (Chapter One)

CILIA. Tiny, hairlike projections from a cell that can move the cell or move substances along its surface. (Chapter Six)

CIRCULAR FOLDS. The transverse ridges in the mucosa of the small intestine that increase the surface area for absorption. (Chapter Seven)

CLITORIS. A small, sensitive structure, richly innervated and composed of erectile tissue; located on the exterior female genitalia. (Chapter Eleven)

COLLAGEN. A strong, flexible fiber, found mainly in the skin, connective tissue, and bones. The most abundant of the three types of fibers found in the connective tissue. (Chapter Two)

COLLECTING DUCT. In the kidney, a duct that carries urine from the nephron into a duct that leads to the renal pelvis. (Chapter Eight)

COMPACT BONE. Hard, dense tissue that forms outer bone layer. (Chapter Three)

COMPLIANCE. The ability to stretch or yield. (Chapter Six)

CONDUCTING ZONE. All the air passageways of the respiratory system that do not play a role in gas exchange, such as the pharynx, the trachea, and the bronchi. (Chapter Six)

CONTRACTION. Muscle movement where filaments inside muscle cells slide over one another, making muscles denser and shorter (Chapter Four)

CORONARY BYPASS SURGERY. A surgical procedure whereby healthy vessel tissue is used to bypass, or detour, blocked passages in the coronary arteries, which bring blood to the heart. (Chapter Thirteen)

CORPUS LUTEUM. A ruptured ovarian follicle that secretes progesterone and estrogen after ovulation has occurred. (Chapter Eleven)

CORTICOTROPIN-RELEASING HORMONE (CRH). Stimulatory hormone produced by the hypothalamus. (Chapter Ten)

CORTISOL. Anti-inflammatory and stress hormone produced by the adrenal glands, circulates to the hypothalamus in order to shut down production of CRH and ACTH. Also called hydrocortisone. (Chapter Ten)

CREATINE PHOSPHATE. A molecule found only in muscle tissue that helps in ATP production. (Chapter Four)

CYTOKINE. Protein produced by the lymph system that affects the activity of other cells and is important in controlling inflammatory responses. Interleukins and interferons are cytokines. (Chapter Ten, Chapter Twelve)

CYTOPLASM. The contents of a cell, except for the nucleus. (Chapter One)

CYTOSKELETON. A network of fibers in the cytoplasm that helps maintain a cell's structure. (Chapter One)

CYTOSOL. In a cell, the fluid portion of cytoplasm holding the organelles. (Chapter One)

DELIRIUM. A temporary mental disturbance marked by disorganized speech, hallucinations, and confusion. (Chapter Thirteen)

DEMENTIA. A progressive mental condition, such as Alzheimer's disease, characterized by the development of many cognitive defects, such as the inability to remember family members or make coherent plans. (Chapter Thirteen)

DENDRITE. A branching extension of the neuron cell body that receives electrical signals. (Chapter Nine)

DEOXYRIBONUCLEIC ACID (DNA). A nucleic acid molecule that makes up the 46 human chromosomes. (Chapter One)

DERMAL PAPILLA. Sensory receptor. A nipple-like projection of the dermis that may contain blood capillaries or corpuscles of touch (also called Meissner corpuscles). (Chapter Two)

DERMIS. A skin layer of dense irregular connective tissue beneath the epidermis. (Chapter Two)

DIAPHYSIS. Shaft of a long bone. (Chapter Three)

DIASTOLE. The relaxation phase of the cardiac cycle. (Chapter Five)

DIENCEPHALON. Part of the forebrain that includes the thalamus, hypothalamus, and epithalamus. (Chapter Nine)

DISTAL CONVOLUTED TUBULE. In the kidney, the final section of the renal tubule. (Chapter Eight)

DUODENUM. The first 10 inches of the small intestine. (Chapter Seven)

DURA MATER. Outer membrane covering the brain and spinal cord. (Chapter Nine)

ECCRINE GLANDS. Sweat glands found on much of the body; abundant in the palms of the hand, soles of the feet, and forehead. (Chapter Two)

ECHOCARDIOGRAM. An image of the heart's structure and function produced by ultrasound. (Chapter Five)

EJACULATION. A smooth-muscle contraction that expels semen from the male reproductive tract through the urethra during male orgasm. (Chapter Eleven)

ELECTROCARDIOGRAM. A recording of the heart's electrical activity. (Chapter Five)

EMBOLISM. A blood clot that forms in a blood vessel and travels to another part of the body. (Chapter Five, Chapter Nine)

EMBRYO. The human organism from early stages of development to the end of the eighth week after fertilization occurs. (Chapter One, Chapter Eleven)

ENDOCARDIUM. The smooth, innermost layer of the heart wall. (Chapter Five)

ENDOPLASMIC RETICULUM. A network of channels and flattened sacs running through a cell's cytoplasm; it serves to store, transport, and package molecules. (Chapter One)

ENDOTHELIUM. The inner lining of many body structures, including the heart and blood vessels. (Chapter Five)

ENZYME. A protein that acts as a biological catalyst to speed up a chemical reaction. (Chapter Two)

EPICARDIUM. The thin membrane covering the heart. (Chapter Five)

EPIDERMIS. The thin outermost layer of the skin. (Chapter Two)

EPIPHYSIS. Bone end containing spongy bone, red bone marrow. (Chapter Three)

EPIPHYSEAL PLATE. Cartilage separating diaphysis and epiphysis; area of bone growth. (Chapter Three)

EPITHELIUM. A primary tissue that covers the body's surface, lines its internal cavities, and forms glands. Also referred to as epithelial tissue. (Chapter Two)

ERYTHROCYTE. Red blood cell. (Chapter Three, Chapter Five)

ENDERARTERECTOMY. A procedure for treating blocked carotid arteries in the neck, developed as a preventive for stroke. (Chapter Thirteen)

ENDOCRINE GLANDS. Ductless glands that secrete hormones directly into the bloodstream or lymph system for a specific physiological purpose, process, or response. (Chapter Ten)

ENDOMETRIUM. The inner uterine membrane where a fertilized egg will embed; shed monthly and regenerated if no pregnancy occurs. (Chapter Eleven)

EPIDIDYMIS. A curled portion of the male duct system in which sperm mature. Emerges from the testes and connects to the vas deferens. (Chapter Eleven)

ESTROGEN. Female sex hormone produced by the ovaries. (Chapter Eleven)

FALLOPIAN TUBES. The tubes through which the egg travels from the ovary to the uterus. (Chapter Eleven)

FASCICLE. Bundle of muscle fibers. (Chapter Four)

FAST-TWITCH MUSCLE FIBERS. The muscle cells responsible for quick bursts of energy. (Chapter Four)

FAUCES. The narrow opening from the mouth into the pharynx. (Chapter Six)

FERTILIZATION. The joining of the sperm and egg nuclei. (Chapter One, Chapter Eleven)

FETUS. Term for the developing human organism from nine weeks until birth. (Chapter One, Chapter Eleven)

FIBRILLATION. Rapid, uncoordinated contractions of heart fibers. (Chapter Five)

FIBRIN. An insoluble protein formed during blood clotting. (Chapter Five)

FIBRINOGEN. A clotting factor in plasma; converted to fibrin. (Chapter Five)

FOLLICLE. An ovarian structure that houses a developing egg and contains layers of follicle cells. (Chapter Eleven)

FOLLICLE-STIMULATING HORMONE (FSH). Secreted by the anterior pituitary gland, FSH stimulates egg development and estrogen secretion in females; initiates sperm production in males. (Chapter Ten, Chapter Eleven)

FONTANELLES. "Soft spots" found on an infant's skull where sutures have not yet closed. (Chapter Three)

FORAMEN. Central channel in bones through which blood vessels, nerves, and ligaments pass. (Chapter Three)

FOREBRAIN. The forward portion of the brain that includes the cerebrum and diencephalon. (Chapter Nine)

FREE RADICALS. Atoms or groups of atoms with an odd number of electrons. Free radicals can start a chemical chain reaction within cells that can damage their DNA or other structures. (Chapter Thirteen)

GAMETE. A male or female reproductive cell; the sperm or egg. (Chapter One, Chapter Eleven)

GANGLIA (sing. ganglion). Groups of nerve cell bodies outside of the central nervous system. (Chapter Nine)

GENE. The basic unit of heredity; a piece of DNA that codes for a specific protein or segment of protein. Found in chromosomes. (Chapter One)

GENITALIA. The internal and external reproductive organs. (Chapter Eleven)

GENOME. All of the genes of an entire organism. (Chapter One)

GERIATRICS. The branch of medicine that is devoted to studying the processes and diseases of aging and old age. (Chapter Thirteen)

GERONTOLOGY. The study of aging people. (Chapter Thirteen)

GLAND. An organ that regulates the secretion or excretion of substances within the body for further use or for elimination. (Chapter Ten)

GLOMERULAR (BOWMAN'S) CAPSULE. In the kidney, a double-layered helmet enclosing the glomerulus. (Chapter Eight)

GLOMERULUS (pl. glomeruli). Ball of capillaries in the nephron where blood enters. (Chapter Eight)

GLOTTIS. The vocal cords and the space between them. (Chapter Six)

GOLGI COMPLEX. A cell organelle that modifies and delivers proteins and lipids within the cell and to the plasma membrane. (Chapter One)

GONADOTROPIN. Secreted by the pituitary gland and prompts gonadal activity, including the onset of sexual maturity. (Chapter Ten)

GONADOTROPIN-RELEASING HORMONE. Released by the hypothalamus to prompt the pituitary gland to release FSH and LH. (Chapter Ten, Chapter Eleven)

GROWTH HORMONE. Hormone that stimulates muscle and bone cells to grow. (Chapter Ten)

HAIR FOLLICLE. Structure surrounding the root of a hair, from which hair develops. (Chapter Two)

HAYFLICK LIMIT. The natural limit that represents the highest number of times a cell can divide. (Chapter Thirteen)

HEMANGIOMA. Common type of harmless, red, vascular birthmark. (Chapter Two)

HEMOGLOBIN. A part of red blood cells, made of protein and pigment, that transports oxygen and some carbon dioxide. (Chapter Five, Chapter Six)

HETEROZYGOUS. Having different alleles that control the same trait on matching chromosomes. (Chapter One)

HIGH-DENSITY LIPOPROTEIN (HDL). A component of blood cholesterol often called "good cholesterol" because higher amounts are linked to a decreased chance of developing atherosclerosis. (Chapter Thirteen)

HIPPOCAMPUS. A structure, seahorse-shaped, in the brain's limbic system that is involved in learning, memory, and emotion. (Chapter Nine)

HISTAMINE. A substance released by the immune system that dilates blood vessels. (Chapter Twelve)

HOMOZYGOUS. Having the same allele that controls the same trait on matching chromosomes. (Chapter One)

HORMONE. Chemical messengers, generally amino acid or steroidal molecules, that are released into the bloodstream to maintain and regulate individual bodily functions. (Chapter Ten)

HUMORAL IMMUNITY. Wing of the adaptive immune system in which immunity is provided by antibodies. (Chapter Twelve)

HYPERTENSION. High blood pressure. (Chapter Five)

HYPERTROPHY. Increase in muscle bulk by means of exercise. (Chapter Four)

HYPOTHALAMUS. A structure in the brain's diencephalon that monitors the autonomic nervous system. Also functions as a neuroendocrine organ that manages all endocrine functions, working to regulate processes and maintain homeostasis within the body. (Chapter Nine, Chapter Ten)

ILIUM. The terminal section of the small intestine, which connects to the colon. (Chapter Seven)

IMMUNIZATION. The conferring of immunity, whether actively (through administration of a vaccine), or passively (via the placenta, breast milk, or through the administration of of an "antiserum"). (Chapter Twelve)

IMMUNODEFICIENCY. Deficient function of the immune system. (Chapter Twelve)

INFLAMMATION. A nonspecific defensive response to injury. Indicated by redness, increased temperature, swelling, and pain. (Chapter Twelve)

INTESTINAL GLAND. A gland in the intestinal mucosa that secretes digestive enzymes (also called a crypt of Lieberkühn). (Chapter Seven)

INVOLUNTARY. Type of action regulated by the nervous system, but without deliberate control. (Chapter Four)

JEJUNUM. The middle section of the small intestine. (Chapter Seven)

KERATIN. An insoluble protein found in the skin, hair, and nails, that makes structures hard and water- repellent. Found in epidermal cells. (Chapter Two)

LACTEAL. The lipid-absorbing lymphatic vessel in a villus. (Chapter Seven)

LARYNX. The respiratory passageway between the pharynx and the trachea, containing the vocal cords; also known as the voice box. (Chapter Six)

LEUKOCYTE. White blood cell. (Chapter Three, Chapter Five, Chapter Twelve)

LIFE EXPECTANCY. An expected number of years of life based on statistical probability that can be measured from birth or from any other age. (Chapter Thirteen)

LIGAMENT. Band of fibrous tissue that connects bone to bone. (Chapter Three)

LIMBIC SYSTEM. A part of the forebrain containing various structures involved in emotions and behavior. (Chapter Nine)

LIPASE. A lipid-digesting enzyme, it splits fatty acids from triglycerides and phospholipids. (Chapter Seven)

LOW-DENSITY LIPOPROTEIN (LDL). A component of blood cholesterol often called "bad cholesterol" because higher amounts are linked to an increased chance of developing atherosclerosis. (Chapter Thirteen)

LUNULA. The crescent-shaped white area at the base of a nail. Part of the nail-generating nail matrix. (Chapter Two)

LUTEINIZING HORMONE (LH). Secreted by the anterior pituitary, LH stimulates ovulation in females, and the testosterone production in males. (Chapter Ten, Chapter Eleven)

LYMPH NODE. One of many bean-shaped structures situated along the lymph system. (Chapter Twelve)

LYMPHOCYTE. A white blood cell in the blood, lymph, and immune organs involved in both cell- and antibody-mediated immune responses. (Chapter Twelve)

LYSOSOME. A cell organelle containing powerful digestive enzymes. (Chapter One)

MACROPHAGE. A large phagocyte derived from a monocyte; it phagocytizes bacteria and other debris, and it acts as an antigen-presenter for T and B cells. (Chapter Five, Chapter Twelve)

MAJOR HISTOCOMPATIBILITY COMPLEX (MHC) PROTEINS. Proteins that occur on all cells in the body in a configuration unique to each individual person. (Chapter Twelve)

MAST CELLS. Immune cells that detect foreign substances in the body and activate local inflammatory responses to them. (Chapter Twelve)

MEDIASTINUM. The portion of the chest cavity between the lungs, containg the heart, trachea, bronchi, esophagus, and other structures. (Chapter Six)

MEDULLA OBLONGATA. The lowest part of the brain stem. (Chapter Nine)

MEDULLARY CAVITY. The core of the bone shaft (diaphysis), contains yellow marrow. (Chapter Three)

MEIOSIS. A type of cell division that occurs only in egg and sperm cells; it involves two nuclear divisions and produces four daughter cells with half the number of chromosomes as the body's other cells. (Chapter One, Chapter Eleven)

MELANOCYTE. Epidermal cells that produce melanin, or pigment. (Chapter Two)

MELANOCYTE-STIMULATING HORMONE (MSH). Stimulates production of the pigment melanin in skin and hair and also acts as a neurotransmitter in connection with appetite and sexual arousal. (Chapter Two, Chapter Ten)

MENINGES. Protective coverings of the brain and spinal cord. (Chapter Nine)

MENISCUS. Curved fibrous cartilage in some joints such as the knee. (Chapter Three)

MENOPAUSE. A period of life when, due to changes in hormonal secretions and cycles, female ovulation and menstruation cease. (Chapter Ten, Chapter Eleven, Chapter Thirteen)

MENSTRUAL CYCLE. A series of uterine changes, including ovulation and bleeding, which occur cyclically in response to the fluctuating blood levels of ovarian hormones. (Chapter Eleven)

MICROVILLI. Miniscule projections atop epithelial cells (for example, on villi) that increase area for absorption. (Chapter Seven)

MIDBRAIN. The brain stem between the pons and diencephalon. (Chapter Nine)

MIGRATING MOBILITY COMPLEX (MMC). A peristaltic wave sequence in the small intestine, in which a wave travels a short distance before dying out, then followed by another wave. (Chapter Seven)

MITOCHONDRION (pl. mitochondria). A double-membraned cell organelle that produces energy for the cell. (Chapter One)

MITOSIS. A type of cell division in which each daughter cell receives the same amount of DNA as the parent cell and is genetically identical to the parent. (Chapter One)

MITRAL VALVE. The valve between the left atrium and left ventricle. (Chapter Five)

MOLE. A growth on the skin that can be pink, tan, dark brown, or black. Occurs when cells in the skin grow in a cluster. Some moles, called congenital nevi, are present at birth. (Chapter Two)

MOTOR NEURON. A nerve that carries impulses from the brain and spinal cord to effectors, either muscles or glands. (Chapter Four, Chapter Nine)

MUSCLE TONE. Minute, involuntary contractions in a muscle group, experienced as muscle firmness and essential to optimal health. (Chapter Four)

MUTATION. Any change in a gene that alters its sequence of bases. (Chapter One)

MYELIN SHEATH. A multilayered fatty covering that insulates most nerve fibers. (Chapter Nine)

MYOCARDIAL INFARCTION. A heart attack consisting of damage to or death of an area of heart muscle due to inadequate blood supply. (Chapter Five)

MYOCARDIUM. Muscular wall of the heart. (Chapter Five)

MYOFIBRIL. The organelle inside a muscle cell where filaments slide over each other to create a muscle contraction. (Chapter Four)

NATURAL KILLER (NK) CELL. Defensive cells of the body's innate immune system; destroy pathogens, cancer cells, and transplanted foreign cells before activation of the adaptive defenses. (Chapter Twelve)

NEBULIZATION. The reduction of medicine to a fine spray; often used for lung disorders such as asthma. (Chapter Six)

NECROSIS. Cell death resulting from disease or injury. (Chapter One)

NUCLEOLUS (PL. nucleoli). A spherical body within a cell nucleus that is the site for production of ribosomal units. (Chapter One)

NUCLEOTIDE. A compound consisting of a nitrogenous base, a phosphate group, and a sugar; DNA and RNA are chains of nucleotides. (Chapter One)

NEPHRON. Functional unit of the kidney, which filters blood plasma. (Chapter Eight)

NEUROGLIA. Cells of the nervous system that support and protect neurons; also called glial cells. (Chapter Nine)

NEUROMUSCULAR JUNCTION. In muscle cells, the connection point between a neuron and muscle fiber. (Chapter Four)

NEURON. A nerve cell. (Chapter One, Chapter Nine)

NEUROTRANSMITTER. Chemical released by a neuron at a synapse. (Chapter Nine)

NEUTROPHIL. Most common kind of white blood cell, the neutrophil phagocytizes and destroys bacteria and other pathogens. (Chapter Twelve)

NONSPECIFIC IMMUNE SYSTEM. A generalized immune response to a pathogen, requiring no previous encounter. Also called "innate defense system." (Chapter Twelve)

OOCYTE. Immature female sex cell, also known as ovum or egg. Contains a woman's genetic material. (Chapter One, Chapter Eleven)

OOGENESIS. Process of egg production in the female. (Chapter Eleven)

ORGANELLE. A structure within a cell that performs a specific job. (Chapter One)

OSTEOARTHRITIS. Form of arthritis generally associated with aging in which joints wear out. (Chapter Three)

OSTEOBLAST. Bone-forming cells. (Chapter Three)

OSTEOCLAST. Bone-destroying cells. (Chapter Three)

OSTEOPOROSIS. A disorder marked by decreased bone mass and increased susceptibility to fractures. (Chapter Three, Chapter Thirteen)

OVARY. The chamber for egg production in the female reproductive system; the female gonad. (Chapter Ten, Chapter Eleven)

OVULATION. The monthly release of an immature egg from the ovary. (Chapter Eleven)

OXYTOCIN. Multipurpose hormone released by the hypothalamus. Stimulates uterine contractions during childbirth; milk production during breastfeeding; orgasm after sex. (Chapter Ten, Chapter Eleven)

PANCREAS. Organ that produces endocrine secretions such as insulin, somatostatin, and glucagon, in order to regulate blood sugar levels. Also produces digestive enzymes. (Chapter Seven, Chapter Ten)

PANCREATIC AMYLASE. A starch-cleaving enzyme secreted by the pancreas. (Chapter Seven)

PARASYMPATHETIC DIVISION. Subdivision of the autonomic nervous system responsible for overseeing the conservation and restoration of the body's energy. (Chapter Nine)

PARATHYROID GLAND. Regulates increases in calcium levels through the secretion of parathyroid hormone (PTH). (Chapter Ten)

PENIS. A male organ used in reproduction and urination. Consists of a connecting root, a shaft, and an enlarged tip. (Chapter Eight, Chapter Eleven)

PEPSIN. A protein-digesting enzyme secreted in the stomach and activated by hydrochloric acid. (Chapter Seven)

PERICARDIUM. Double-layered membrane that encloses the heart. (Chapter Five)

PERIPHERAL NERVOUS SYSTEM. Portion of the nervous system, consisting of nerves and ganglia, that lies outside the brain. (Chapter Nine)

PERISTALSIS. Rhythmic and successive muscular contractions in hollow muscular structures such as blood vessels or intestines. (Chapter Four)

PERIOSTEUM. Tough, fibrous connective tissue that covers and nourishes bones. (Chapter Three)

PEROXISOME. A cell organelle containing enzymes that utilize molecular oxygen to oxidize organic compounds. (Chapter One)

PHAGOCYTE. White blood cell which engulfs foreign matter and cellular debris. (Chapter Five, Chapter Twelve)

PHARYNX. The respiratory and digestive passageway that starts behind the nasal cavity and extends to the larynx and esophagus. (Chapter Six)

PIA MATER. The innermost of the brain's three protective coverings. (Chapter Nine)

PINEAL GLAND. Secretes melatonin, a hormone considered causally linked to sleep cycles. (Chapter Nine, Chapter Ten)

PITUITARY GLAND. A neuroendocrine gland with a variety of functions, including regulating the gonads, thyroid, lactation, and water balance. (Chapter Ten)

PLACENTA. A temporary organ formed from maternal tissues during pregnancy to nourish the fetus. Delivers nutrients, removes wastes, and produces pregnancy hormones. (Chapter One, Chapter Eleven)

PLAQUE. A deposit of accumulated cholesterol, fats, and other substances on the inner lining of the artery wall. (Chapter Five)

PLASMA. Fluid component of blood. (Chapter Five)

PLASMA MEMBRANE. An outer membrane that separates a cell's contents from the outside environment. (Chapter One)

PLASMIN. Enzyme in blood plasma that dissolves the fibrin in blood clots. (Chapter Five)

PLATELETS. Fragments of cells, found in blood, that aid in clotting. (Chapter Five)

PLEURA. The double-layered membrane that covers the lungs and lines the chest cavity; pleural fluid fills the narrow space between the two layers. (Chapter Six)

PNEUMOTHORAX. A condition in which gas enters the pleural cavity through a rupture or puncture. (Chapter Six)

PLEXUS. A network of nerves. (Chapter Nine)

PODOCYTE. Cell with footlike projections through which filtrate passes into the glomerular capsule. (Chapter Eight)

PONS. Part of the brain stem between the medulla and midbrain. (Chapter Nine)

PREGNANCY. The 280-day time period during which a woman carries a developing baby in the womb. (Chapter One, Chapter Eleven)

PRESBYCUSIS. Age-related hearing loss resulting from degenerative changes in the ear. (Chapter Thirteen)

PRESBYOPIA. Age-related changes in vision that lead to an inability to focus on close objects. (Chapter Thirteen)

PROGESTERONE. A female hormone that prepares the uterus for a fertilized egg. (Chapter Ten, Chapter Eleven)

PROLACTIN (PRL). Protein hormone released by the pituitary gland that triggers milk production in the mother's breasts after birth. (Chapter One, Chapter Ten, Chapter Eleven)

PROSTATE. Chestnut-size gland that encircles the male urethra and is responsible for a part of the process of semen production. (Chapter Eight, Chapter Eleven)

PROTEASOME. A tiny cell organelle that destroys unwanted proteins. (Chapter One)

PROTHROMBIN. An inactive blood-clotting factor produced by the liver that can be changed into thrombin. (Chapter Five)

PROXIMAL CONVOLUTED TUBULE. First section of the renal tubule. (Chapter Eight)

PRURITUS. Itching; has many possible causes such as dry skin, allergies, or underlying skin diseases. (Chapter Two)

PUBERTY. The period of life in which the levels of gonadal hormones, estrogen and testosterone, rise. This causes the reproductive organs to fully develop and gain function and any secondary sex characteristics to appear. (Chapter Ten, Chapter Eleven)

PULMONARY VENTILATION. Breathing; the rhythm of inhalation and exhalation. (Chapter Six)

RECEPTOR. Specialized cell or nerve-cell part that responds to sensory input and converts it to an electrical signal. (Chapter Nine)

RED BONE MARROW. Spongy tissue in bones where blood cells are formed. (Chapter Three)

RELAXATION. Movement where filaments inside muscle cells slide away from each other, making muscles longer and looser. (Chapter Four)

RENAL CORPUSCLE. Combination of glomerulus and enclosing glomerular capsule. (Chapter Eight)

RENAL CORTEX. Smooth outer husk of the kidney, where most nephrons are located. (Chapter Eight)

RENAL MEDULLA. Dark red inner portion of the kidney. (Chapter Eight)

RENAL PELVIS. The cavity in the middle of the kidney where urine drains into the ureter. (Chapter Eight)

RESPIRATION. The process of supplying the body with oxygen and ridding it of carbon dioxide. External respiration involves bringing oxygen into the lungs and then into the bloodstream; internal respiration is the exchange of gases between the body's cells and the blood. (Chapter Six)

RESPIRATORY CENTERS. Brain regions in the medulla oblongata and pons of the brain stem that control the rate and depth of breathing. (Chapter Six)

RESPIRATORY MEMBRANE. The tissue joining an alveolus and a capillary, through which gases pass. (Chapter Six)

RESPIRATORY ZONE. Microscopic structures within the lung—primarily the alveoli and their bronchioles—that are the site of gas exchange. (Chapter Six)

RIBONUCLEIC ACID (RNA). Transmits instructions from DNA to the cytoplasm, where proteins are made, and does other metabolic tasks. (Chapter One)

RIBOSOME. A cell organelle that is the site of protein synthesis. (Chapter One)

RIGOR MORTIS. Muscle stiffness that occurs in tissue 48 to 60 hours after death, caused when muscle cells deprived of ATP remain in a contracted position. (Chapter Four)

SALIVARY AMYLASE. A starch-cleaving enzyme secreted by the salivary glands; its action ends when the bolus reaches the stomach. (Chapter Seven)

SARCOLEMMA. The outer membrane of a muscle fiber. (Chapter Four)

SCROTUM. The external sac found in the male, hanging below the abdomen. It houses and prtoects the testes while helping regulate their temperature. (Chapter Eleven)

SEMEN. Milky, slightly acidic fluid containing sperm and various other secretions that sustain the sperm cells in the female reproductive tract. (Chapter Eleven)

SEMILUNAR VALVES. The heart valves between the ventricles and the aorta or pulmonary trunk. (Chapter Five)

SENESCENCE. The process of growing old and the changes associated with it. (Chapter Thirteen)

SENSORY NEURON. A nerve cell that carries sensory information into the brain and spinal cord. (Chapter Nine)

SEPTUM. The wall dividing the left and right chambers of the heart. (Chapter Five)

SINOATRIAL (SA) NODE. A specialized mass of cardiac muscle cells in the right atrium; generates the electrical current that causes the heart to contract. (Chapter Five)

SKELETAL MUSCLE. Cells composed of striated fibers, designed to contract and release, moves voluntarily, connected primarily to bones by tendons and served by motor neurons. (Chapter Four)

SLOW-TWITCH MUSCLE FIBERS. The muscle cells responsible for steady endurance energy. (Chapter Four)

SMOOTH MUSCLE. Found primarily in vessels and hollow organs and served by the autonomic nervous system. Made of nonstriated fibers. (Chapter Four, Chapter Five, Chapter Seven, Chapter Eight)

SOMATIC NERVOUS SYSTEM. The division of the peripheral nervous system that activiates skeletal muscles. (Chapter Nine)

SPECIFIC IMMUNE SYSTEM. The specialized wing of the immune system, which targets specific microorganisms and other antigens. Also called "adaptive defense system," or "acquired defense system." (Chapter Twelve)

SPLEEN. Largest of the lymphoid organs, it houses B and T cells, phagocytizes aging and damaged blood cells, and filters out and destroys certain pathogens. (Chapter Twelve)

SPERM. The male gamete, which holds a father's genetic material. (Chapter One, Chapter Eleven)

SPERMATOGENESIS. The process of sperm formation. (Chapter Eleven)

SPINAL CORD. The bundle of nervous tissue that runs down the center of the vertebral column, carrying messages to and from the brain. (Chapter Nine)

SPONGY BONE. Lattice-like porous bone, gives strength with minimal weight; found in ends and inner portions of long bones; contains red bone marrow. Also called cancellous bone. (Chapter Three)

STENOSIS. The narrowing of a blood vessel or heart valve. (Chapter Five)

STRIATION. Dark bands visible in some muscle cells. (Chapter Four)

SUBCUTANEOUS TISSUE. A continuous sheet of fatty connective tissue beneath the dermis, it attaches the skin to the rest of the body, provides cushioning for the skin and organs, and stores energy reserves. (Chapter Two)

SURFACTANT. Slippery fluid produced by cells in the alveoli. Reduces the surface tension of water and prevents the alveoli from collapsing. (Chapter Six)

SUTURE. A fused, fibrous joint that does not move. (Chapter Three)

SYMPATHETIC DIVISION. The subdivision of the autonomic nervous system responsible for overseeing activation of body systems in response to stress. (Chapter Nine)

SYMPHYSIS. The place where two bones join and are held firmly together so they function as one bone. Also known as a cartilaginous joint. (Chapter Three)

SYNAPSE. The junction between two neurons or between a neuron and an effector, such as a gland or muscle. (Chapter Nine)

SYNOVIAL FLUID. Lubricating fluid present in the joint cavity. (Chapter Three)

SYNOVIAL JOINTS. Freely movable joints in the body. (Chapter Three)

SYSTOLE. The contraction phase of the cardiac cycle. (Chapter Five)

T CELL. A lymphocyte that controls major functions of the adaptive immune system and mediates cell-mediated immune responses as helper, cytotoxic, suppressor, or memory T cells. (Chapter Twelve)

TELOMERASE. An enzyme that restores telomere sequences to the ends of chromosomes. (Chapter Thirteen)

TELOMERE. The natural end, or cap, of a chromosome, consisting of a repetitive DNA sequence. (Chapter Thirteen)

TENDON. White fibrous cord attaching muscle to bone. (Chapter Three, Chapter Four)

TESTES. The male gonad, which produces and cultivates spermatozoa, immature sperm cells. (Chapter Ten, Chapter Eleven)

TESTOSTERONE. The male sex hormone, produced by the testes. (Chapter Ten, Chapter Eleven)

THALAMUS. A structure made of two egg-shaped masses of gray matter in the brain; a relay station for sensory information flowing into the brain. (Chapter Nine)

THROMBIN. Enzyme that induces clotting by converting fibrinogen to fibrin. (Chapter Five)

THROMBUS. Blood clot. (Chapter Two, Chapter Five, Chapter Twelve)

THYMUS GLAND. Organ in the chest at the base of the neck, part of the immune system where T lymphocytes develop an ability to recognize self antigens. Also an organ in the endocrine system (Chapter Ten, Chapter Twelve)

THYROID GLAND. Regulates metabolism through the secretion of thyroid hormone (TH). (ChapterTen)

TRACHEA. Air passageway from the larynx to the bronchi. (Chapter Six)

TRICUSPID VALVE. Valve between the right atrium and right ventricle. (Chapter Five)

URETER. Duct that transports urine from the kidney to the bladder. (Chapter Eight)

URETHRA. The tube through which urine moves from the bladder to outside the body. (Chapter Eight)

URINARY BLADDER. A reservoir for the temporary storage of urine. (Chapter Eight)

UTERUS. Organ within the female reproductive system in which the fertilized egg is received and in which the embryo will develop after conception. (Chapter Eleven)

VACCINE. A preparation of killed microorganisms, living attenuated organisms, or living fully virulent organisms that is administered to produce of artificially increase immuinity to a specific disease. (Chapter Twelve)

VAGINA. The canal in the female system extends from the cervix to the exterior of the body; also referred to as the birth canal. (Chapter One, Chapter Eleven)

VAS DEFERENS. In the male system, the duct that propels sperm from the epididymis to the urethra. (Chapter Eleven)

VASOPRESSIN. Produced by the hypothalamus, maintains water balance. Also referred to as antidiuretic hormone (ADH). (Chapter Eight, Chapter Ten)

VEIN. Blood vessel that carries blood from the body to the heart. (Chapter Five)

VENTRICLES. (a) The two lower chambers of the heart. (Chapter Five); (b) Large interior spaces in the forebrain and brainstem filled with cerebrospinal fluid. (Chapter Nine)

VENULE. Small vein that leads from the capillaries to the larger veins. (Chapter Five)

VILLI. Hairlike projections of the intestinal mucosa through which nutrients are absorbed. (Chapter Seven)

VULVA. The external female genitalia that surrounds and protects the vaginal opening. (Chapter Eleven)

ZYGOTE. The fertilized egg that will develop into an embryo. (Chapter One, Chapter Eleven)

MEDICAL MILESTONES

Going back thousands of years, the history of medicine overflows with amazing discoveries and inventions, all of which broke new ground at the time and then became part of everyday life. Here is just a sample.

ca 5000 B.C.
Skulls from Europe and South America bear signs that Stone Age man trephinated, or drilled holes, in the head, one of the earliest surgical forms.

ca 1760 B.C.
In Babylonia, the Code of Hammurabi lays out the first known code of medical ethics. Fees for services and penalty structures are also included in the Code.

ca 1550 B.C.
An Egyptian, 110-page scroll (now known as the Eber's Papyrus) details remedies for fever, pain, swelling, and burns.

ca 500 B.C.
The Indian physician Sushruta describes the earliest form of rhinoplasty, or nose job, to rebuild amputated noses.

ca 424 B.C.
Roman historian Herodotus describes an early use of prosthetics, a Persian man who wore a wooden peg in place of his amputated foot.

ca 375 B.C.
Ancient Greek physician Hippocrates, known as the father of medicine, dies. The Hippocratic Oath, an ethical code attributed to him, has served as a foundation for many medical codes of conduct.

ca A.D. 30
Celsus, one of Rome's greatest medical writers, published *De Re Medica,* which describes four signs of inflammation.

ca A.D.129
Greek physician Galen is born. His writings will form much of the foundation of Western medicine until the mid-17th century.

ca A.D.150
The ancient Egyptians, Babylonians, and Romans use catgut to bind up wounds.

ca A.D. 850
Hunayn ibn Ishaq, an Arab physician in the court at Baghdad, translates Galen's works from Greek to Arabic.

ca 865
Persian physician, ar-Razi (Rhazes) is born. Author of many medical treatises, he is the first to distinguish smallpox as a separated ailment from measles.

1284
Salvino D'Armate invents the first wearable eye glasses.

1543
Andreas Vesalius publishes *On the Fabric of the Human Body*, the first accurately illustrated book of anatomy based on human dissection.

1545
French physician Ambroise Paré, considered by many to be the father of modern surgery, publishes a treatise on how to treat gunshot wounds. Because his work is written in French and not Latin, it does not gain wide acceptance until after his death.

1628
British physician William Harvey discovers how blood circulates, disproving Galen's theories of circulation.

1664
Thomas Willis publishes "Anatomy of the Brain, with a Description of the Nerves and Their Function," the most accurately complete descriptions of the nervous system of his time.

1665
English scientist Robert Hooke publishes his book *Micrographia,* which contains the first description of cells.

1677
Dutch scientist Anton van Leeuwenhoek observes semen through a microscope and identifies sperm cells.

1761
Josef Leopold Auenbrugger publishes his description of chest percussion. His theory gains popular acceptance in 1808 when Jean Nicolas Corvissart des Marets publishes it in French.

1791
Nicholas Dubois de Chemant patents the first true dentures.

1819
French physician René Théophile Hyacinthe Laënnec develops the first stethoscope.

1822
After observing the action of the stomach through a bullet wound, American physician William Beaumont proves that digestion in the stomach is a chemical process involving hydrochloric acid.

1827
German scientist Karl Ernst von Baer announces that he has discovered the female sex cell, the egg.

1830
Sir Charles Bell publishes *The Nervous System of the Human Body*, in which he describes the various functions of different sensory nerves, distinguishing between the different types of impulses.

1839
German physiologist Theodor Schwann concludes that all living organisms are made from cells.

1849
Arnold Berthold conducts experiments on roosters and capons that reveal the effects of the hormone testosterone on male behavior.

1861
Ignaz Philip Semmelweis publishes his findings on the virtues of handwashing for reducing mortality rates in births.

1869
Gustav Simon, through demonstrations on living patients, proves for the first time, that the human body can function with only one kidney.

1880
Swedish anatomist Ivar Sandstrom makes the last discovery of a major organ in the human body, the parathyroid gland.

1880
Henry Faulds and William James Herschel publish the first public description of the uniqueness and permanence of fingerprints.

1892
Juan Vucetich, captures the first criminal based on a fingerprint identification.

1896
Rontgën discovers the X-ray. He will win the Nobel Prize in 1901 for his discovery.

1898–1930
Russian scientist Ivan Pavlov's experiments with dogs reveal the existance of the conditioned reflex.

1901
Karl Landsteiner, an Austrian America, discovers different blood groups and develops a system for blood typing, making blood transfusions a routine medical procedure.

1902
Ernest Starling and William Bayliss discover the first hormone, secretin. Two years later, Starling will officially coin the word *hormone*, from the Greek word meaning "to excite."

1903
Willem Einthoven invents the earliest version of the electrocardiogram, the "string galvanometer," for which he will win the 1924 Nobel Prize.

1906
Alzheimer's disease is identified by Dr. Alois Alzheimer, a German physician.

1912
Alexis Carrel awarded the Nobel Prize in physiology for his pioneering work with heart surgery.

1921
American Earle Dickson invents the Band-Aid bandage.

1921–1926
Otto Loewi first demonstrates the role of chemicals (now called neurotransmitters) in nerve cell transmission.

1928
Scottish scientist Alexander Fleming discovers penicillin, an antibiotic that will eventually be widely adopted for use in medicine.

1929
The electroencephalograph (EEG), which records the electrical activity of the brain, is invented by Hans Berger.

1943
Dutch physician Willem Kolff develops his first version of a dialysis machine. He worked for many years to perfect the device, which extended the lives of those suffering from renal failure.

1948–1953
The Kinsey Reports on human sexuality are published.

1952
Virgina Apgar develops her scoring system for evaluating the health of newborn babies.

1952
Paul Zoll restores a heartbeat using electrodes, the first primitive instance of a defibrillator.

1953
Based on x-ray diffraction images created by Rosalind Franklin and Maurice Wilkins, James Watson and Francis Crick postulate their theory of DNA's double helix structure. Watson, Crick, and Williams will win the Nobel Prize in 1962.

1953
Gibbon incorporates the first heart-lung machine into heart surgery.

1954
Joseph Murray performs the first organ transplant with long-term success.

1955
U.S. physician Jonas Salk's polio vaccine is proven to be effective and safe. By 1995, the disease would be eradicated from the Western Hemisphere.

1960
The FDA approves the use of the first oral contraceptive.

1961
Cell biologist Leonard Hayflick discovers what is now known as the "Hayflick Limit," which refers to the maximum number of times a cell can divide before it dies.

1963
The first liver transplant is performed by Dr. Thomas Starzyl. Survival rates after the procedure will remain low until the development of powerful immunosuppressant drugs.

1964
The United States Government officially recognized smoking as having harmful effects and begins placing warnings on cigarette packages.

1967
Vaccination for mumps becomes routine, drastically reducing the number of cases.

1972
Computerized tomography, better known as CT-scanning, is introduced as a non-invasive medical imaging technique.

1974
Physician Henry J. Heimlich invents the Heimlich maneuver, a technique to stop someone from choking that has saved many lives.

1978
Louise Brown, the world's first "test tube baby," is born. Brown's parents conceived her through in vitro fertilization.

1982
The first artificial heart is invented by Robert Jarvik.

1983
The cause of AIDS, the human immunodeficiency virus (HIV), is first isolated and identified.

1990
The Human Genome Project is initiated. Scientists strive to map the units that make up human genes.

1990
Ashanthi De Silva becomes the first person to receive gene therapy for a rare genetic disorder, an ADA deficiency.

2000
Initial sequencing of the human genome is completed.

2006
The FDA approves the first HPV vaccine that will protect women from one common form of cervical cancer.

FURTHER READING

ANATOMY AND PHYSIOLOGY

Goldberg, Stephen. *Clinical Anatomy Made Ridiculously Simple.* Miami, Florida: MedMaster, Inc., 2007.

Marieb, Elaine N. *Human Anatomy & Physiology.* 6th ed. San Francisco, California: Pearson Benjamin Cummings, 2004.

Totora, Gerard J., and Sandra Reynolds Grabowski. *Principles of Anatomy and Physiology.* 10th ed. Hoboken, New Jersey: John Wiley & Sons, Inc., 2003.

BIOTECHNOLOGY

Bailey, Ronald. *Liberation Biology: The Scientific and Moral Case for the Biotech Revolution.* Amherst, New York: Prometheus Books, 2005.

Garreau, Joel. *Radical Evolution: The Promise and Peril of Enhancing Our Minds, Our Bodies—and What It Means to be Human.* New York: Doubleday, 2005.

Kurzwell, Ray, and Terry Grossman. *Fantastic Voyage: Live Long Enough to Live Forever.* Emmaus, Pennsylvania: Rodale, 2004.

BRAIN AND NERVOUS SYSTEM

Damasio, Antonio. *The Feeling of What Happens: Body and Emotion in the Making of Consciousness.* San Diego, California: Harcourt, 1999.

Damasio, Antionio. *Looking for Spinoza: Joy, Sorrow, and the Feeling Brain.* San Diego, California: Harcourt, 2003.

Restak, Richard. *The New Brain: How the Modern Age is Rewiring Your Mind.* New York, New York: Rodale, 2003.

Society for Neuroscience. *Brain Facts: A Primer on the Brain and the Nervous System.* 5th ed. Washington, D.C.: Society for Neuroscience, 2006.

Squire, Larry, Floyd Bloom, Susan McConnell, James Roberts, Nicolas Spitzer, and Michael Zigmond, eds. *Fundamental Neuroscience.* 2nd ed. San Diego, California: Academic Press, 2002

CARDIOLOGY

Fuster, Valentin, et al, eds. *Hurst's The Heart: Manual of Cardiology.* 11th ed. New York: The McGraw-Hill Companies, 2004.

Willis, Fredrick A., M.D., and Thomas E. Keys, M.A., eds. *Classics of Cardiology.* New York, New York: Dover Publications, Inc., 1961.

DERMATOLOGY

Bolognia, Jean L., Joseph L. Jorizzo, and Ronald P. Rapini, eds. *Dermatology.* 2-vol. Philadelphia, Pennsylvania: Mosby, 2003.

Johnson, Richard Allen, Klaus Wolff, and Dick Suurmond. *Fitzpatrick's Color Atlas and Synopsis of Clinical Dermatology: Clinical and Serious Diseases.* 5th ed. New York: McGraw-Hill Professional, 2005.

DIGESTIVE SYSTEM

Feldman, Mark, Lawrence S. Friedman, and Lawrence J. Brandt, eds. *Sleisenger and Fordtran's Gastrointestinal Liver Disease.* 8th ed. Philadelphia, Pennsylvania: Saunders, 2006.

King, John E., ed. *Mayo Clinic on Digestive Health.* 2nd ed. Rochester, Minnesota: Mayo Clinic Health Information, 2004.

ENDOCRINOLOGY

Larsen, P. Reed, Henry Kronenburg, Shlomo Melmed, and Kenneth Polonsky. *Williams Textbook of Endocrinology.* 10th ed. Philadelphia, Pennsylvania: Saunders, 2002.

Becker, Kenneth, et al. *Principles and Practices in Endocrinology and Metabolism*, 3rd ed. New York, New York: Lippincott, Williams & Wilkins, 2001.

GENERAL HEALTHCARE AND FIRST AID

American Medical Association. *American Medical Association Family Medical Guide.* 4th ed. Hoboken, New Jersey: John Wiley & Sons, 2004.

Litin, Scott, ed. *Mayo Clinic Family Health Book.* 3rd ed. Rochester, Minnesota: Mayo Clinic Health Information, 2003.

The Merck Manual of Medical Information: Second Home Edition. New York, New York: Pocket Books, 2003.

GENETICS AND HUMAN DEVELOPMENT

Gilbert, Scott F. *Developmental Biology.* 4th ed. Sunderland, Massachusetts: Sinauer Associates, Inc., 1994.

Klug, William S., Michael R. Cummings, and Charlotte Spencer, eds. *Concepts of Genetics.* 8th ed. San Francisco, California: Benjamin Cummings, 2005.

McLaren, Anne. *Germ Cells and Soma: A New Look at an Old Problem.* New Haven, Connnecticut: Yale University of Press, 1981.

Moore, Keith and T.V.N. Persaud. *Developing Human: Clinically Oriented Embryology.* 7th ed. Philadelphia, Pennsylvania: Saunders, 1993.

O'Rahilly, Ronan and Fabiola Müller. *Human Embryology and Teratology.* 3rd ed. New York: Wiley-Liss, 1992.

Wells, Spencer. *Deep Ancestry: Inside the Genographic Project.* Washington, D.C.: National Geographic, 2006.

HISTORY OF MEDICINE

Adler, Robert. *Medical Firsts: From Hippocrates to the Human Genome.* Hoboken, New Jersey: John Wiley & Sons, 2004.

Friedman, Meyer, and Gerald Friedland. *Medicine's 10 Greatest Discoveries.* New Haven, Connecticut: Yale University Press, 2000.

Rifkin, Benjamin, and Michael Ackerman. *Human Anatomy: From the Renaissance to the Digital Age.* New York, New York: Harry N. Abrams, 2006.

Straus, Eugene W., and Alex Straus. *Medical Marvels: The 100 Greatest Advances in Medicine.* Amherst, New York: Prometheus Books, 2006.

IMMUNOLOGY

Austen, Frank, et al. *Samter's Immunologic Diseases.* 6th ed. New York, New York: Lippincott, Williams & Wilkins, 2001.

Kumar, Vinay, Abul Abbas, and Nelson Fausto, eds. *Robbins and Cotran Pathologic Basis of Disease.* 7th ed. Philadelphia, Pennsylvania: Saunders, 2005.

MEN'S AND WOMEN'S HEALTH

American Medical Women's Association. *Women's Complete Health Book.* New York, New York: Dell Press, 1995.

Centola, G.M., and K. A. Ginsburg, eds. *Evaluation and Treatment of the Infertile Male.* Cambridge, United Kingdom: Cambridge University Press, 1996.

Mayo Clinic Guide to a Healthy Pregnancy. New York, New York: Collins, 2004.

Scialli, Anthony, ed. *The National Women's Health Resource Center Book of Women's Health: Your Comprehensive Guide to Health and Well-Being.* New York, New York: William Morrow, 1999.

Simon, Harvey B.. *The Harvard Medical School's Guide to Men's Health.* New York, New York: Free Press, 2004.

RESPIRATORY SYSTEM

Mason, Robert J., V. Courtney Broaddus, John F. Murray, and Jay A. Nadal. *Murray and Nadel's Textbook of Respiratory Medicine.* 4th ed. Philadelphia, Pennsylvania: Saunders, 2005.

Fishman, Alfred P., et al., eds. *Fishman's Pulmonary Disease and Disorders.* 3rd ed. 2-vol. New York, New York: McGraw-Hill, 1998

THE WEBSITE

Throughout *Body: The Complete Human,* you'll find weblinks to the official National Geographic Human Body website—http://nationalgeographic.com/humanbody. A visit to the website delivers even more information on the body, its systems, and its organs through its interactive format. Scientifically accurate, animated models and graphics show organs and systems at work. Watch as a piece of food is converted into energy through digestion. A click on the heart can make it beat faster and pump more blood. See how diseases can affect the inner workings of the brain. A perfect complement to *Body: The Complete Human,* the Human Body website brings a whole new dimension to this fascinating subject.

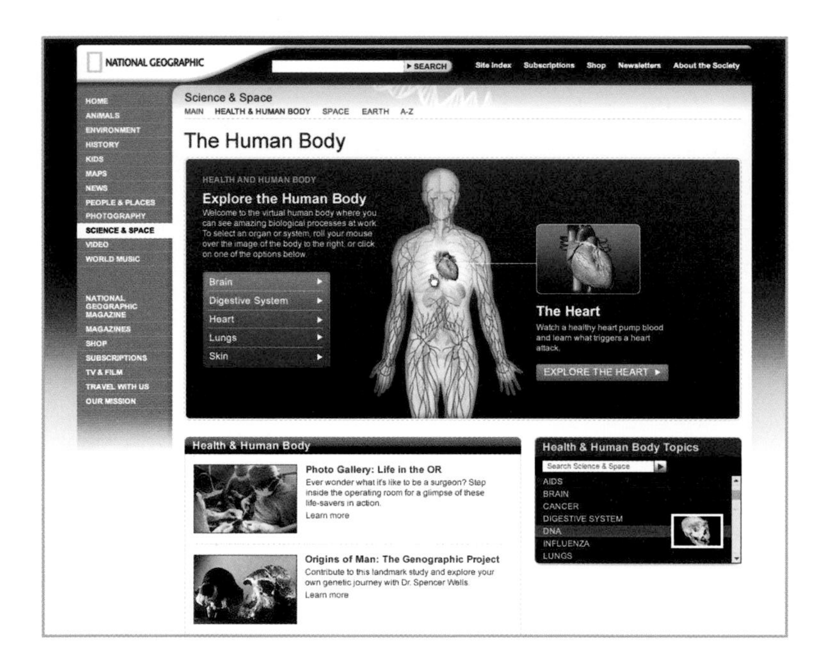

Visit the National Geographic Human Body website at

HTTP://NATIONALGEOGRAPHIC.COM/HUMANBODY

BIOGRAPHIES

BOARD OF ADVISERS

WILLIAM AYERS, M.D. is Emeritus Professor of Medicine and consultant in educational affairs to the Academic Dean at Georgetown University School of Medicine in Washington, D.C. He is a Fellow of the American College of Physicians and was a Robert Wood Johnson Fellow in Health Policy.

ROBERT M. BRIGGS, M.D. graduated from Yale University School of Medicine and completed his residency at Stanford University. Before retiring, Dr. Briggs was chair of the Department of Plastic Surgery at the St. Barnabus Medical Center in Livingston, New Jersey. He served as Clinical Assistant Professor of Surgery at Robert Wood Johnson Medical School.

ELIZABETH BROWNELL, M.D. is a graduate of St. George's University School of Medicine, Grenada, West Indies, and did her residency at the Medical College of Wisconsin in Milwaukee, where she later became Assistant Professor of Clinical Medicine. She is a practicing family physician at Waukesha Memorial Hospital in Wisconsin.

TED M. BURNS, M.D. is Associate Professor of Neurology at the University of Virginia. He graduated from the Kansas University School of Medicine and did a residency in Neurology and a fellowship in Electrodiagnostic and Neuromuscular Medicine at the University of Virginia, followed by a fellowship in Peripheral Nerve Disease at the Mayo Clinic in Rochester, Minnesota.

ROBERT A. CHASE, M.D. is the Emile Holman Professor of Surgery and Professor of Anatomy, Emeritus at the Stanford University School of Medicine. After joining the faculty at Yale in 1947, Dr. Chase established the Section of Plastic Surgery there. He later joined the faculty at Stanford as Professor and Chairman of Surgery and was named the first Holman professor.

D. ROBERT DUFOUR, M.D. is Emeritus Professor of Pathology at George Washington University Medical Center School of Medicine and Health Sciences. He is also Consultant Pathologist at the Veterans Affairs Medical Center in Washington, D.C., where he served as Chief of Pathology and Laboratory Medicine Service.

DAVID E. FLEISCHER, M.D. is a gastroenterologist at the Mayo Clinic, Arizona and a Professor of Medicine at the Mayo Clinic College of Medicine. He is past President of the American Society for Gastrointestinal Endoscopy and a previous Robert Wood Johnson Health Policy Fellow.

JEAN L. FOURCROY, M.D., PH.D, M.P.H., is a retired Navy captain who now serves as staff urologist at Walter Reed Army Medical Center in Washington, D.C. She is a former Medical Officer with the Food and Drug Administration. She currently serves on the Board of the U.S. Anti-Doping Agency.

DONALD S. KARCHER, M.D. is Acting Chair of the Department of Pathology at George Washington University Medical Center School of Medicine and Health Sciences, where he also serves as Chief of the Clinical Chemistry and Flow Cytometry Services and Director of Laboratories.

JOHN E. MORLEY, M.B., B.CH. is Dammert Professor of Gerontology and Director, Division of Geriatric Medicine and Endocrinology at Saint Louis University Medical Center. He is also Director of the Geriatric Research, Education, and Clinical Center at the St. Louis Veterans Affairs Medical Center.

POLLY E. PARSONS, M.D. is Professor of Medicine, Chief of Pulmonary and Critical Care Medicine, and Chair of the Department of Medicine at the University of Vermont College of Medicine. She graduated from the University of Arizona College of Medicine and completed a residency in Internal Medicine and a fellowship in Pulmonary Medicine at the University of Colorado.

ROBERT S. ZOHLMAN, M.D. retired as Clinical Professor of Medicine at the University of California, San Francisco. He graduated from the Albert Einstein College of Medicine and completed his residency at Montefiore Hospital in New York. He completed a fellowship in Nephrology at the University of California, San Francisco.

AUTHORS

STEFAN BECHTEL is the author or coauthor of seven books, which have sold two million copies and been translated into ten languages. His latest book, *Roar Of The Heavens*, is about Hurricane Camille.

PATRICIA DANIELS has written extensively on science and health. Among her publications are *Human Body, Childhood Medical Guide, Medical Advisor, Caring for Your Parents: The Complete AARP Guide,* and the *National Geographic Encyclopedia of Space*.

TRISHA GURA is a molecular biologist and former journalist for the *Chicago Tribune*. She earned a Knight Science Journalism Fellowship at MIT and Harvard. Today, she is a Resident Scholar at Brandeis University, a freelance writer, and author of *Lying in Weight: The Hidden Epidemic of Eating Disorders*.

SUSAN TYLER HITCHCOCK has written 13 books, including National Geographic's *Geography of Religion*. Her most recent book is *Frankenstein: A Cultural History*. Today, she currently works as an editorial project director for the National Geographic Society.

LISA STEIN is an online news editor with *Scientific American*. She was former deputy national editor at *U.S. News and World* Report. In addition, Ms. Stein has written and edited for many media outlets, including *TV Guide*, (as Washington Bureau Chief) and *National Journal*.

JOHN THOMPSON has a background in English and biology. He is author of nine National Geographic books. He was editor for *The Journals of Captain John Smith: A Jamestown Biography,* and is now working on a book about the western Dakotas.

RICHARD RESTAK, M.D., is a neurologist and bestselling author of 18 books, including *The New Brain.* He is Clinical Professor of Neurology at the George Washington University Medical Center School of Medicine and Health Sciences, has served on many national advisory councils for brain research, and has been a consultant to NBC's *Today* show.

INDEX

Boldface indicates illustrations.

B

C

D

G

H

I

J

K

L

M

N

O

P

T

ILLUSTRATIONS CREDITS

1 (LE), Photographer's Choice/Getty Images; 1 (CTR), Veronica Burnmeister/Visuals Unlimited/ Getty Images; 1 (RT), Stockbyte/Punchstock; 2-3, George Doyle/Stockbyte/Getty Images; 8, Anatomical Travelogue/Photo Researchers, Inc.; 12-13, Alfred Pasieka/Photo Researchers, Inc.; 14 (LE), Photo Researchers, Inc.; 14 (RT), Omikron/Photo Researchers, Inc.; 15 (UP), Jennifer C. Waters/Photo Researchers, Inc.; 15 (CTR), N. Kedersha/Photo Researchers, Inc.; 15 (LO), Quest/Photo Researchers, Inc.; 16 (UP), Carl Rohrig; 16 (LO), Science Photo Library/ Photo Researchers, Inc.; 17, Don W. Fawcett/ Photo Researchers, Inc.; 18, Lennart Nilsson; 19 (LE), A. Barrington Brown/Photo Researchers, Inc.; 19 (RT), Museum of London; 20 (UP), Jennifer C. Waters/Photo Researchers, Inc.; 20 (CTR), Jennifer C. Waters/Photo Researchers, Inc.; 20 (LO), Jennifer C. Waters/Photo Researchers, Inc.; 21, Moredun Scientific/Photo Researchers, Inc.; 22, Jodi Hilton/CORBIS; 24, Kirk Moldoff; 25, Richard Nowitz/NG Image Collection; 26, L. Willatt, East Anglican Regional Genetics Service/Photo Researchers, Inc.; 27, Emma Rian/zefa/CORBIS; 28, NG Maps; 29, David Edwards/NG Image Collection; 30, Eye of Science/Photo Researchers, Inc.; 31 (UP), Mika/ zefa/CORBIS; 31 (LO), Photo Insolite Realite/ Photo Researchers, Inc.; 32, Karen Kasmauski; 33, Karen Kasmauski; 34, Veronique Estiot/ Photo Researchers, Inc.; 35, Hulton-Deutsch/ CORBIS; 36, Alfred Pasieka/Photo Researchers, Inc.; 37, Dr. Yorgas Nikas/Photo Researchers, Inc.; 38, Max Aguilera-Hellweg; 39, Science Faction/Getty Images; 40 (UP), Albert Forlag Bonniers; 40 (CTR), Lennart Nilsson; 40 (LO), Lennart Nilsson; 41 (UP), Albert Forlag Bonniers; 41 (CTR), Lennart Nilsson; 41 (LO), Lennart Nilsson; 42, Tim Flach/Stone/Getty Images; 43, Library of Congress/Photo Researchers, Inc.; 44-45, Jorg Steffens/CORBIS; 47, Eye of Science/ Photo Researchers, Inc.; 49, Jim Zuckerman/ CORBIS; 50-51 (A), Sarah Leen; 50-51 (B), Sarah Leen; 50-51 (C), Sarah Leen; 51 (UP), Jaimie D. Travis/Shutterstock; 52, SPL/Photo Researchers, Inc.; 53, Martin Dohrn/Photo Researchers, Inc.; 55, Steve Gschmeissner/Photo Researchers, Inc.; 56 (UP LE), Vienna General Hospital; 56 (UP RT), Psoriasis diffusa. Farblithographie von Anton Elfinger (vgl. Kat-Nr. 5); 57, Michael P. Gadomski/Photo Researchers, Inc.; 58, Ed Kashi/ CORBIS; 60, Sarah Leen; 61, Biophoto Associates/Photo Researchers, Inc.; 62, Coneyl Jay/Photo Researchers, Inc.; 63, Image Source/ Punchstock; 64, Steve Gschmeissner/Photo Researchers, Inc.; 65, LADA/Photo Researchers, Inc.; 67, Neil Bromhall/Photo Researchers, Inc.; 68-69, Jeffry W. Myers/CORBIS; 72 (UP), Visuals Unlimited/Getty Images; 72 (LO), Anatomical Travelogue/Photo Researchers, Inc.; 73, Anatomical Travelogue/Photo Researchers, Inc.; 74, Vesalius, Andreas. De corporis humani fabrica libri septem; 75, Photo Researchers, Inc.; 77, CNRI/Photo Researchers, Inc.; 78 (UP), Michael Ross/Photo Researchers, Inc.; 78-79 (LO), ESRF-CREATIS/Photo Researchers, Inc.; 80, British Museum/Heritage-Images; 81, Imagesource/Punchstock; 83, AP/Wide World Photos; 84, CNRI/Photo Researchers, Inc.; 85, CNRI/Photo Researchers, Inc.; 86 (UP), CORBIS; 86 (CTR), CORBIS; 86 (LO), CORBIS; 87 (UP), CORBIS; 87 (CTR), CORBIS; 87 (LO), CORBIS; 89 (UP), Steve Prezant/CORBIS; 89 (LO), Chris Bjornberg/Photo Researchers, Inc.; 91, Dave Roberts/Photo Researchers, Inc.; 92, Bettmann/CORBIS; 93, Illustration Copyright © 2007 Nucleus Medical Art, All rights reserved, www.nucleusinc.com; 95, Library of Congress/ Photo Researchers, Inc.; 98 (UP), Rubberball/ Punchstock; 101, Michael Nichols/NG Image Collection; 102, Living Art Enterprises/Photo Researchers, Inc.; 104-105, Blaise Hayward/ Photodisc Red/Punchstock; 106, CORBIS; 107 (UP), Innerspace Imaging/Photo Researchers, Inc.; 107 (CTR), CNRI/Photo Researchers, Inc.; 107 (LO), SPL/Photo Researchers, Inc.; 108, Anatomical Travelogue/Photo Researchers, Inc.; 110 (LO), Hulton-Deutsch/CORBIS; 111 (UP), Anatomical Travelogue/Photo Researchers, Inc.; 111 (LO), Kent Wood/Photo Researchers, Inc.; 112, Zsolt Nyulaszi/Shutterstock; 113, Oxford Science Archive/Heritage-Images; 114, Orla/ Shutterstock; 115, 3D4Medical/Getty Images; 116, Darren Robb/The Image Bank/Getty Images; 118, Wellcome Photo Library; 119, Eliane/zefa/CORBIS; 120, Sheila Terry/Photo Researchers, Inc.; 121, John Martin Rare Book Room, Hardin Library for the Health Sciences, University of Iowa; 123 (UP), Image 100/ Punchstock; 123 (CTR), Image 100/Punchstock; 123 (LO), Anne Domdey/CORBIS; 124-125, Kurt Kormann/zefa/CORBIS; 126, Martin Dohm/Royal College of Surgeons/Photo Researchers, Inc.; 128, SPL/Photo Researchers, Inc.; 129, Photo Researchers, Inc.; 130 (UP), Samuel Ashfield/Photo Researchers, Inc.; 130 (LO), CNRI/Photo Researchers, Inc.; 132, Corbis/Punchstock; 133 (UP), Photo Insolite Realite/Photo Researchers, Inc.; 133 (CTR), SPL/ Photo Researchers, Inc.; 133 (LO), NIBSC/Photo Researchers, Inc.; 134, Abeles/Photo Researchers, Inc.; 135, Tek Image/Photo Researchers, Inc.; 136, The Granger Collection, NY; 137, Visuals Unlimited/CORBIS; 138, Toby Melville/Reuters/ CORBIS; 139, Mauro Fermariello/Photo Researchers, Inc.; 140 (LO), CORBIS; 141 (UP), Stock Montage/Getty Images; 141 (LO), James Cavallini/Photo Researchers, Inc.; 142, Alfred Pasieka/Photo Researchers, Inc.; 143, The Alan Mason Chesney Medical Archives of the Johns Hopkins Medical Institutions; 144, Al Fenn/ Time Life Pictures/Getty Images; 145, SPL/Photo Researchers, Inc.; 146, Toshiba American Medical Systems, Inc.; 147, Hybrid Medical Animation/ Photo Researchers, Inc.; 148, CORBIS; 149, AJPhoto/Photo Researchers, Inc.; 150, Scott Camazine/Photo Researchers, Inc.; 151 (UP), SPL/Photo Researchers, Inc.; 151 (LO), Susumu Nishinaga/Photo Researchers, Inc.; 152 (UP), CNRI/Photo Researchers, Inc.; 152 (LO), Jean-Luc Kokel/Photo Researchers, Inc.; 153, Adrianna Williams/zefa/CORBIS; 154-155, Jason Hosking/zefa/CORBIS; 157, Yoav Levy/ Phototake; 158, Michael DeYoung/CORBIS; 159, Hulton Archive/Getty Images; 160, Bettmann/CORBIS; 161 (UP), Spauln/ Shutterstock; 162, Alfred Pasieka/Photo Researchers, Inc.; 163, CORBIS; 164 (LO), James Cavallini/Photo Researchers, Inc.; 165 (UP), Hulton Archive/Getty Images; 165 (LO), Du Cane Medical Imaging Ltd./Photo Researchers, Inc.; 166 (UP), Dr. Kessel & Dr. Kardon/Tissues & Organs/Visuals Unlimited/Getty Images; 166 (LO), James Leynse/CORBIS; 167, Ralph Hutchings/Getty Images; 168 (LE), Anatomical Travelogue/Photo Researchers, Inc.; 168 (RT), Bettmann/CORBIS; 169, Liette Parent/ Shutterstock; 170, George Contorakes/CORBIS; 171, Bettmann/CORBIS; 172, Imagesource/ Punchstock; 173, David Scharf/Photo Researchers, Inc.; 174, Pixland/Punchstock; 175, Steve Taylor/ Stone/Getty Images; 176-177, Chris Johns, NGS; 179, Sam Abell; 180, Godelon/Photo Researchers, Inc.; 181, Phanie/Photo Researchers, Inc.; 183, Digital Vision/Punchstock; 184, Anatomical Travelogue/Photo Researchers, Inc.; 185, Wellcome Library, London; 186 (UP), Alfred Pasieka/Photo Researchers, Inc.; 187 (UP), Stockbyte/Punchstock; 188, Alfred Pasieka/ Photo Researchers, Inc.; 189, Bettmann/CORBIS; 190 (UP), Susumu Nishinaga/Photo Researchers, Inc.; 190 (LO), J. James/Photo Researchers, Inc.; 191, Professor Pietro M. Motta/Photo Researchers, Inc.; 192, Karen Kasmauski; 193, Karen Kasmauski; 194, Lennart Nilsson; 197, Visuals Unlimited/Getty Images; 198 (UP), Richard J. Green/Photo Researchers, Inc.; 198 (CTR), SPL/ Photo Researchers, Inc.; 198 (LO), James Cavallini/Photo Researchers, Inc.; 199, Anatomical Travelogue/Photo Researchers, Inc.; 200, Bettmann/CORBIS; 201 (UP), Karen

Kasmauski; 201 (LO), Professors P. M. Motta and T. Fujita/SPLPhoto Researchers, Inc.; 202, CNRI/Photo Researchers, Inc.; 203, Visuals Unlimited/Getty Images; 204, Susumu Nishinaga/Photo Researchers, Inc.; 205, Zephyr/Photo Researchers, Inc.; 206-207, Science Faction/Getty Images; 209, Mc Pherson Colin/CORBIS; 210 (RT), Naile Goelbasi/Photographer's Choice/Getty Images; 210 (LE), CNRI/Photo Researchers, Inc.; 211, Saturn Stills/Photo Researchers, Inc.; 212 (RT), Lester V. Bergman/CORBIS; 213, Ghislain & Marie David De Lossy/The Image Bank/Getty Images; 214, Howard Sochurek/CORBIS; 215, Lester V. Bergman/CORBIS; 216, Bettmann/CORBIS; 217, AJPhoto/Photo Researchers, Inc.; 219 (UP RT), Alexander Tsiaras/Photo Researchers, Inc.; 219 (LO), Lester V. Bergman/CORBIS; 220, Bettmann/CORBIS; 221, Owen Franken/CORBIS; 222 (UP), Anatomical Travelogue/Photo Researchers, Inc.; 222 (LO), Nancy Kedersha/Photo Researchers, Inc.; 223, SPL/Photo Researchers, Inc.; 224, Visuals Limited/CORBIS; 225 (UP), Volker Steger/Photo Researchers, Inc.; 225 (LO), TH Foto-Werbung/Photo Researchers, Inc.; 226-227, Jens Nieth/zefa/CORBIS; 229, 3D4Medical/Getty Images; 230, Science Source/Photo Researchers, Inc.; 231 (LO), Steve McCurry; 232, Michael W. Davidson/Photo Researchers, Inc.; 233 (LO), Zephyr/Photo Researchers, Inc.; 234, Jerome Prevost/CORBIS; 236 (UP), Joe McNally; 236 (LO), GJLP/Photo Researchers, Inc.; 237, Science Source/Photo Researchers, Inc.; 238, Science Source/Photo Researchers, Inc.; 239, Helen King/CORBIS; 240, Cary Sol Wolinsky; 241, US National Library of Medicine/Science Photo Library/Photo Researchers, Inc.; 242, Sinclair Stammers/Science Photo Library/Photo Researchers, Inc.; 243, Allan Hobson/Science Photo Library/Photo Researchers, Inc.; 244, Louie Psihoyos/CORBIS; 245, Joe McNally; 246 (LO), National Library of Medicine/Photo Researchers, Inc.; 247, Reuters/CORBIS; 249, Anatomical Travelogue/Photo Researchers, Inc.; 250, Astrid & Hanns-Frieder Michler/Photo Researchers, Inc.; 251, Bettmann/CORBIS; 252, Wally McNamee/CORBIS; 253, Michael Donne/Photo Researchers, Inc.; 255, John Dominis/Time & Life Pictures/Getty Images; 257, Omikron/Photo Researchers, Inc.; 258, Stefan Schuetz/zefa/CORBIS; 259 (UP), Suzanne Tucker/Shutterstock; 259 (LO), Prof. P.M. Motta/Univ. "La Sapienza", Rome/Photo Researchers, Inc.; 260, Neal Preston/CORBIS; 262-263, Alfred Pasieka/Photo Researchers, Inc.; 265, Anatomical Travelogue/Photo Researchers, Inc.; 267 (UP), Biophoto Associates/Photo Researchers, Inc.; 267 (LO), Wellcome Library, London; 268, Alfred Pasieka/Photo Researchers, Inc.; 269, Caro/Alamy Ltd; 270 (UP), Scott Camazine/Photo Researchers, Inc.; 271, Phil Date/Shutterstock; 272, Asia Images/Getty Images; 274, Bart's Medical Library/Phototake; 275, Lennart Nilsson; 276, Zephyr/Photo Researchers, Inc.; 277, Bettmann/CORBIS; 278 (LO), Tom Grill/CORBIS; 278 (UP), Anatomical Travelogue/Photo Researchers, Inc.; 279, CNRI/Photo Researchers, Inc.; 280, Anatomical Travelogue/Photo Researchers, Inc.; 281, Science Source/Photo Researchers, Inc.; 282, CORBIS; 283, Joel Kiesel/Getty Images; 284, Dr. Mark J. Winter/Photo Researchers, Inc.; 285, CNRI/Photo Researchers, Inc.; 286, Bettmann/CORBIS; 287, Richard T. Nowitz/CORBIS; 288, Anatomical Travelogue/Photo Researchers, Inc.; 289, Rich Pilling/Getty Images; 290, Penni Gladstone/San Francisco Chronicle/CORBIS; 291, Richard T. Nowitz/CORBIS; 292-293, Stockbyte/Punchstock; 295, Reunion des Musees/Art Resource, NY; 297 (UP), James King-Holmes/Photo Researchers, Inc.; 297 (LO), James King-Holmes/Photo Researchers, Inc.; 298, Science Source/Photo Researchers, Inc.; 300, 3D4Medical/Getty Images; 301, SPL/Photo Researchers, Inc.; 302 (UP), Claude Edelmann/Photo Researchers, Inc.; 302 (CTR), Petit Format/Photo Researchers, Inc.; 302 (LO), Claude Edelmann/Photo Researchers, Inc.; 303 (UP), Claude Edelmann/Photo Researchers, Inc.; 303 (CTR), Claude Edelmann/Photo Researchers, Inc.; 303 (LO), Claude Edelmann/Photo Researchers, Inc.; 305, Time Life Pictures/Getty Images; 306, Anatomical Travelgoue/Photo Researchers, Inc.; 307, James Cavallini/Photo Researchers, Inc.; 308, Bettmann/CORBIS; 309 (UP), Lennart Nilsson; 309 (CTR), Visuals Unlimited/CORBIS; 309 (LO), Lennart Nilsson; 310, Hank Morgan/Photo Researchers, Inc.; 311, Bettmann/CORBIS; 313 (UP), Norbert Schaefer/CORBIS; 313 (LO), Zephyr/Photo Researchers, Inc.; 314, Wellcome Library, London; 315, Taxi Japan/Getty Images; 316-317, SPL/Photo Researchers, Inc.; 318, Jean-Loup Charmet/Photo Researchers, Inc.; 319 (UP), Steve Gschmeissner/Photo Researchers, Inc.; 319 (CTR), Steve Gschmeissner/Photo Researchers, Inc.; 319 (LO), J. King-Holmes/Photo Researchers, Inc.; 320 (UP), Reuters/CORBIS; 320 (LO), Ian Boddy/Photo Researchers, Inc.; 322, Steve Gschmeissner/Photo Researchers, Inc.; 323, Tim Flach/Stone//Getty Images; 324, Dr. Dennis Kunkel/Visuals Unlimited; 325, Corbis/Punchstock; 326 (UP LE), Erich Schrempp/Science Photo Library/Photo Researchers, Inc.; 326 (UP RT), Erich Schrempp/Science Photo Library/Photo Researchers, Inc.; 326 (LO LE), Erich Schrempp/Science Photo Library/Photo Researchers, Inc.; 326 (LO RT), Erich Schrempp/Science Photo Library/Photo Researchers, Inc.; 327, Patrik Giardino/Iconica/Getty Images; 328, Dr. Klaus Boller/Photo Researchers, Inc.; 329, The Granger Collection, NY; 331, AP/Wide World Photos; 332, Zephyr/Science Library/Photo Researchers, Inc.; 334, Lisa Spindler Photography Inc/Getty Images; 335, Steve Gschmeissner/Photo Researchers, Inc.; 336, Phototake Inc/Alamy Ltd; 337, AP/Wide World Photos; 338, Owen Franken/CORBIS; 339, The Granger Collection, NY; 340 (LO), Scott Camazine/Photo Researchers, Inc.; 340 (UP), SuperStock; 341 (LO), The Granger Collection, NY; 341 (UP), SPL/Photo Researchers, Inc.; 343, Phanie/Photo Researchers, Inc.; 344 (UP), Max Aguilera-Hellweg; 344 (LO), Du Cane Medical Imaging Ltd./Photo Researchers, Inc.; 346-347, Dennis M. Sabangan/CORBIS; 348, Blend/Punchstock; 349, Jean Pierre Fizet/Sygma/CORBIS; 350, Scala/Art Resource, NY; 351 (LE), Joel Sartore/NG Image Collection; 351 (RT), Jeff Miller/University of Wisconsin-Madison; 352, Dr. Gopal Murti/Science Photo Library/Photo Researchers, Inc.; 353, Gary Gladstone/CORBIS; 354, 1986 George Rose/Getty Images; 355, James Cavallini/Photo Researchers, Inc.; 356, Zephyr/Photo Researchers, Inc.; 357, F. Carter Smith/Sygma/CORBIS; 359 (UP), Corbis/Punchstock; 359 (LO), RussellKightley/Photo Researchers, Inc.; 360, Tom Gannam/AP/Wide World Photos; 361, Elena Elisseeva/Shutterstock; 362 (UP), A. Pakieka/Photo Researchers, Inc.; 362 (LO), Dario Sablijak/Shutterstock; 363, David McLain; 364, Estate of William Utermohlen (CK); 365, Estate of William Utermohlen (CK); 366 (UP), Christopher Gramann/zefa/CORBIS; 366 (LO), Zephyr/Photo Researchers, Inc.; 367, Bartosz Wardzinski/Shutterstock; 368, David McLain; 369, Lynn Johnson; 370-371, James King-Holmes/Photo Researchers, Inc.; 372 (LE), Touch of Life Technologies; 372 (CTR), Touch of Life Technologies; 372 (RT), Touch of Life Technologies; 373, Andrew Brookes/CORBIS; 374, Andrew Brookes/CORBIS; 375, Yorgos Nikas, M.D.; 376, AP/Wide World Photos/HO, Advanced Cell Technology; 377, Biophoto Associates/Photo Researchers, Inc.; 378, Arthur Toga/UCLA/Visuals Unlimited; 379, Dr. Torsten Wittmann/Photo Researchers, Inc.; 380, CNRI/Photo Researchers, Inc.; 381, AP/Wide World Photos/Princeton University; 382, AP/Wide World Photos/Mike Derer; 383, AP/Wide World Photos/Katsumi Kasahara; 384, AP/Wide World Photos/LexingtonHerald-Leader, David Stephenson; 385, AP/Wide World Photos/Dima Gavrysh; 386, Juergen Berger, Max-Planck Institute/Photo Researchers, Inc.; 387, The prophets Methuselah and Jareth from the south transept window, 12th century,/Canterbury Cathedral, Kent, UK/The Bridgeman Art Library; 388, Dennis Kunkel/Phototake; 389, Max Aguilera-Hellweg.

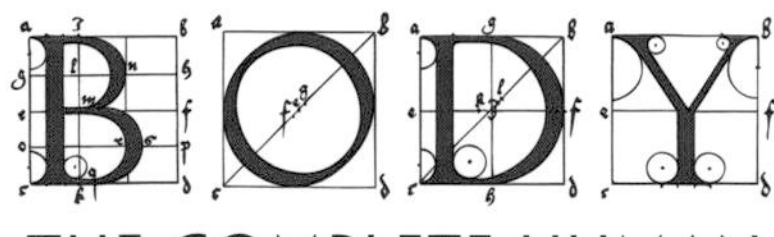

THE COMPLETE HUMAN

by Patricia Daniels, Trisha Gura, Susan Tyler Hitchcock,
Lisa Stein, John Thompson, and Stefan Bechtel
Foreword by Richard Restak, M.D.

Published by the National Geographic Society
John M. Fahey, Jr., *President and Chief Executive Officer*
Gilbert M. Grosvenor, *Chairman of the Board*
Nina D. Hoffman, *Executive Vice President;*
President, Book Publishing Group

Prepared by the Book Division
Kevin Mulroy, *Senior Vice President and Publisher*
Leah Bendavid-Val, *Director of Photography Publishing and Illustrations*
Marianne R. Koszorus, *Director of Design*

Barbara Brownell Grogan, *Executive Editor*
Elizabeth Newhouse, *Director of Travel Publishing*
Carl Mehler, *Director of Maps*

Staff for This Book
Amy Briggs, *Editor*
Suzanne Crawford, Karin Kinney, *Text Editors*
Kris Hanneman, *Illustrations Editor*
Melissa Farris, *Art Director*
Cameron Zotter, *Designer*
Dan O'Toole, Suzanne Poole, *Researchers*
Richard S. Wain, *Production Project Manager*
Marshall Kiker, *Illustrations Specialist*
Connie D. Binder, *Indexer*
Stephanie Hanlon, *Editorial Intern*
Nicole DiPatrizio, *Design Intern*

Jennifer A. Thornton, *Managing Editor*
Gary Colbert, *Production Director*

Manufacturing and Quality Management
Christopher A. Liedel, *Chief Financial Officer*
Phillip L. Schlosser, *Vice President*
John T. Dunn, *Technical Director*
Chris Brown, *Director*
Maryclare Tracy, *Manager*
Nicole Elliott, *Manager*